# Invasive Alien Plants

An Ecological Appraisal for the Indian Subcontinent

FSC
www.fsc.org
MIX
Paper from
responsible sources
FSC® C013604

## CABI Invasive Species Series

Invasive species are plants, animals or microorganisms not native to an ecosystem, whose introduction has threatened biodiversity, food security, health or economic development. Many ecosystems are affected by invasive species and they pose one of the biggest threats to biodiversity worldwide. Globalization through increased trade, transport, travel and tourism will inevitably increase the intentional or accidental introduction of organisms to new environments, and it is widely predicted that climate change will further increase the threat posed by invasive species. To help control and mitigate the effects of invasive species, scientists need access to information that not only provides an overview of and background to the field, but also keeps them up to date with the latest research findings.

This series addresses all topics relating to invasive species, including biosecurity surveillance, mapping and modelling, economics of invasive species and species interactions in plant invasions. Aimed at researchers, upper-level students and policy makers, titles in the series provide international coverage of topics related to invasive species, including both a synthesis of facts and discussions of future research perspectives and possible solutions.

### Titles Available

1. Invasive Alien Plants : An Ecological Appraisal for the Indian Subcontinent
   Edited by J.R. Bhatt, J.S. Singh, R.S. Tripathi, S.P. Singh and R.K. Kohli

### Titles in Preparation

2. Invasive Plant Ecology and Management: Linking Processes to Practice
   Edited by T.A. Monaco and R.L. Sheley

# Invasive Alien Plants

## An Ecological Appraisal for the Indian Subcontinent

*Edited by*

**J.R. Bhatt**

*Ministry of Environment and Forests, India*

**J.S. Singh**

*Banaras Hindu University, India*

**S.P. Singh**

*Uttarakhand State Institute of Technology and Science, India*

**R.S. Tripathi**

*National Botanical Research Institute, India*

and

**R.K. Kohli**

*Panjab University, India*

www.cabi.org

**CABI is a trading name of CAB International**

CABI
Nosworthy Way
Wallingford
Oxfordshire OX10 8DE
UK

Tel: +44 (0)1491 832111
Fax: +44 (0)1491 833508
E-mail: cabi@cabi.org
Website: www.cabi.org

CABI
875 Massachusetts Avenue
7th Floor
Cambridge, MA 02139
USA

Tel: +1 617 395 4056
Fax: +1 617 354 6875
E-mail: cabi-nao@cabi.org



A catalogue record for this book is available from the British Library, London, UK.

**Library of Congress Cataloging-in-Publication Data**

Invasive alien plants : an ecological appraisal for the Indian subcontinent / edited by J.R. Bhatt ... [et al.].
p. cm. -- (CABI invasive species series ; 1)
Includes bibliographical references and index.
ISBN 978-1-84593-907-6 (alk. paper)
1. Invasive plants--India. 2. India--Environmental conditions. 3. Plant ecology--India. 4. Plant diversity--India. 5. Invasive plants--Control--India. 6. Invasive plants--Government policy--India. 7. Environmental policy--India. I. Bhatt, J.R. II. Title. III. Series.

SB613.I4I57 2012
333.95'330954--dc23

2011026515

ISBN-13: 978 1 84593 907 6

Commissioning editor: David Hemming
Editorial assistant: Alexandra Lainsbury and Gwenan Spearing
Production editor: Fiona Chippendale

Typeset by Columns Design Ltd, Reading.
Printed and bound in the UK by CPI Group (UK) Ltd, Croydon, CR0 4YY.

# Contents

# Contributors

**Adhikari, D.**, Centre for Advanced Studies in Botany, North-Eastern Hill University, Shillong-793022, India; E-mail: dibyenduadhikari@gmail.com

**Arul Pragasan, L.**, Department of Ecology and Environmental Sciences, Pondicherry University, Puducherry-605014, India; E-mail: arulpragasan@yahoo.co.in

**Barik, S.K.**, Centre for Advanced Studies in Botany, North-Eastern Hill University, Shillong-793022, India; E-mail: sarojkbarik@yahoo.com

**Batish, D.R.**, Department of Botany, Panjab University, Chandigarh-160014, India; E-mail: daizybatish@yahoo.com

**Bhatt, J.R.**, Director, Ministry of Environment and Forests, CGO Complex, Lodi Road, New Delhi, India; E-mail: jrbhatt@nic.in

**Dar, G.H.**, Department of Botany, University of Kashmir, Srinagar-190006, Jammu and Kashmir, India; E-mail: ghdar99@yahoo.com

**Hamal, Irshad A.**, Vice-chancellor, Baba Ghulam Shah Badshah University, Rajouri, Jammu and Kashmir, India; E-mail: vc_bgsbu@rediffmail.com

**Hassan Ganie, A.**, Department of Botany, University of Kashmir, Srinagar-190006, Jammu and Kashmir, India; E-mail: aijazku@gmail.com

**Kathuria, Vinish**, Associate Professor, SJM School of Management, IIT Bombay, Powai, Mumbai-400076, India; E-mail: vinish67@yahoo.com

**Kaur, Gurpreet**, Department of Environment Studies, Panjab University, Chandigarh-160014, India; E-mail: gurpreet_env@rediffmail.com

**Kaur, Shalinder**, Department of Environment Studies, Panjab University, Chandigarh-160014, India; E-mail: kaur_shalu@yahoo.com

**Khan, M.L.**, Department of Forestry, North-Eastern Regional Research Institute of Science & Technology (Deemed University), Nirjuli-791109 (Itanagar), Arunachal Pradesh, India; E-mail: khanml@yahoo.com

**Khuroo, A.K.**, Department of Botany, University of Kashmir, Srinagar-190006, Jammu and Kashmir, India; E-mail: anzarak@gmail.com

**Kohli, R.K.**, Chairman, Department of Botany, Panjab University, Chandigarh-160014, India; E-mail: rkkohli45@yahoo.com

**Kumar, P.**, G.B. Pant Institute of Himalayan Environment & Development, Kosi-Katarmal, Almora-263643 (Uttarakhand), India; E-mail: pramodgbpihed@gmail.com

**Kushwaha, S.P.S.**, Forestry and Ecology Division, Indian Institute of Remote Sensing, Indian Space Research Organisation, 4 Kalidas Marg, Dehradun-248001, India; E-mail: spskushwaha@gmail.com

**Muthumperumal, C.**, Department of Ecology and Environmental Sciences, Pondicherry University, Puducherry-605014, India; E-mail: cmperumal@gmail.com

**Negi, G.C.S.**, G.B. Pant Institute of Himalayan Environment & Development, Kosi-Katarmal, Almora-263643 (Uttarakhand), India; E-mail: gcsnegi@yahoo.co.in

**Pandey, C.N.**, Additional Principal Chief Conservator of Forests, Aranya Bhawan, Sector 10A, Gandhi Nagar, Gujarat, India; E-mail: cnpandey@gmail.com

**Pandey, R.**, Research Coordinator, Mangrove Research, Gujarat Ecological Education and Research (GEER) Foundation, Indroda Nature Park, P.O. Sec. 7, Gandhi Nagar, Gujarat, India; E-mail: richa81181@yahoo.co.in

**Pant, M.**, G.B. Pant Institute of Himalayan Environment & Development, Kosi-Katarmal, Almora-263643 (Uttarakhand), India; E-mail: mpant922@yahoo.co.in

**Parthasarathy, N.**, Department of Ecology and Environmental Sciences, Pondicherry University, Puducherry-605014, India; E-mail: parthapu@yahoo.com

**Patterson Edward, J.K.**, Suganthi Devadason Marine Research Institute, 44 Beach Road, Tuticorin-628001, Tamil Nadu, India; E-mail: edwardjkpatterson@sdmri.in

**Raghubanshi, A.S.**, Institute of Environment and Sustainable Development, Banaras Hindu University, Varanasi-221005 UP, India; E-mail: raghubansh@gmail.com

**Rai, J.P.N.**, Department of Environmental Sciences, G.B. Pant University of Agriculture and Technology, Pantnagar-263145, Uttarakhand, India; E-mail: jamesraionline@gmail.com

**Rao, R.R.**, INSA Honorary Scientist, 328, B-4, Kendriya Vihar, Yelahanka, Bangalore-560064, India; E-mail: raocimap@gmail.com

**Rashid, Irfan**, Department of Botany, University of Kashmir, Srinagar-190006, Jammu and Kashmir, India; E-mail: ecoirfan@yahoo.co.in

**Rasool, Nazima**, Department of Botany, University of Kashmir, Srinagar-190006, Jammu and Kashmir, India; E-mail: rasoolnazima@gmail.com

**Reshi, Zafar A.**, Department of Botany, University of Kashmir, Srinagar-190006, Jammu and Kashmir, India; E-mail: zreshi@yahoo.com

**Sagar, Kavitha**, 328, B-4, Kendriya Vihar, Yelahanka, Bangalore-560064, India; E-mail: kavcsa@gmail.com

**Sathyanarayana, N.**, Sir M. Visvesvaraya Institute of Technology, Yelahanka, Bangalore, India; E-mail: sathya_dixit@yahoo.com

**Shah, Manzoor A.**, Department of Botany, University of Kashmir, Srinagar-190006, Jammu and Kashmir, India; E-mail: mashah75@yahoo.com

**Shankar, U.**, Department of Botany, North-Eastern Hill University, Shillong-793 022, India; E-mail: arshuma@yahoo.com

**Sharma, G.P.**, Department of Environmental Biology, University of Delhi, Delhi-110007, India; E-mail: gyanprakashsharma@gmail.com

**Singh, H.P.**, Department of Environment Studies, Panjab University, Chandigarh-160014, India; E-mail: hpsingh_01@yahoo.com

**Singh, Ishwar**, Conservator of Forests (Central), Northern Regional Office of the Ministry of Environment and Forests, Chandigarh, India; E-mail: singh.ishwar0@gmail.com

**Singh, J.S.**, Department of Botany, Banaras Hindu University, Varanasi, India; E-mail: jssingh@bhu.ac.in; singh.js1@gmail.com

**Singh, S.P.**, Advisor, Planning Commission, Uttarakhand State Institute of Technology and Science, SGRR Education Mission, Patel Nagar, Dehradun-248001, India; E-mail: surps@yahoo.com

**Tripathi, R.S.**, National Botanical Research Institute, Lucknow-226 001, India; E-mail: tripathirs@yahoo.co.uk

**Wafai, B.A.**, Department of Botany, University of Kashmir, Srinagar-190006, Jammu and Kashmir, India; E-mail: bawafai@yahoo.co.in

**Yadav, A.S.**, Department of Botany, R.R. Autonomous College, Alwar-301001, Rajasthan, India; E-mail: atarsingh_1010@rediffmail.com

# 1 Plant Invasion in India: an Overview

**R.K. Kohli[1], D.R. Batish[1], J.S. Singh[2], H.P. Singh[3] and J.R. Bhatt[4]**

[1]*Department of Botany, Panjab University, Chandigarh, India;* [2]*Department of Botany, Banaras Hindu University, Varanasi, India;* [3]*Department of Environment Studies, Panjab University, Chandigarh, India;* [4]*Director, Ministry of Environment and Forests, CGO Complex, Lodi Road, New Delhi, India*

## Introduction

The Earth's flora is dynamic and has been constantly changing over a period of time. Changes may be natural or human-aided, although in the recent past the latter has played a vital role. In fact, the movement of plants from one part of the earth to the other has become very common and frequent owing to better trade and transport facilities. Plant species that move from one geographical region to the other (either accidentally or intentionally), establish and proliferate there and threaten native ecosystems, habitats and species are known as invasive alien plants (hereafter referred to as invasive plants) (Richardson *et al.*, 2000).

The problem of invasive plants has become global and is largely human-aided (Vitousek *et al.*, 1997). Invasive plants are responsible for global environmental changes, biodiversity crisis, species endangerment and disruption of ecosystem processes essential for human welfare (Drake *et al.*, 1989; Parker *et al.*, 1999; Mack *et al.*, 2000; Mooney, 2005; Charles and Dukes, 2007; Herron *et al.*, 2007, Pejchar and Mooney, 2009). The impact of invasive plants on global biodiversity is second only to habitat fragmentation and is a major global issue. Furthermore, the economic costs due to invasive species are also enormous, though not widely studied (Pimentel *et al.*, 2005). Plant movement or introduction to an alien environment is not a new phenomenon but has been an important part of our history. A number of economically important plants such as crops and ornamentals are introduced plants that have proved especially beneficial and non-harmful to the human race. However, many introduced plant species initially considered to be valuable have become a nuisance and difficult to manage. Example include *Lantana camara*, a South American shrub now known to be a very troublesome invader but now introduced to several parts of the world as an ornamental. There are also examples of accidental entry of some invasive plants as contaminants of imported food grains or agricultural/horticultural material or in ships' ballast water. An invasive plant such as *Parthenium hysterophorus* – one of the most serious invasive plants – is an example of accidental entry.

Environmental problems such as climate change, disturbances and changing landscape patterns have further escalated the process of biotic invasion. The harms caused by biotic invasions are enormous as they interfere with the socio-economic system, human and animal health and food security of the region. Various international and national organizations, such as World Conservation Union (WCU, formerly IUCN – International Union for Conservation of Nature and Natural Resources), Convention on Biological Diversity (CBD) and Global

Invasive Species Programme (GISP), are concerned with the prevention and control of this global problem.

## Status and Vulnerability of the Indian Region to Plant Invasion

The Indian region, because of its diverse climatic and environmental conditions, is highly vulnerable to biotic invasion. Moreover, a burgeoning population, high rate of trade and transport, coupled with greater movement of people favour the accidental and intentional entry of plant species in this region. The recent fast rate of economic growth of the country is also expected to leave its mark on loss of plant diversity and increased invasion of alien species. Three hot spots of biodiversity, although supporting rich floral and faunal diversity including endemic species, also reflect a high rate of habitat degaradation where opportunist invasive species can easily establish themselves. Consequently, a numer of invasive species have made their abode in the region. The three main reasons for the greater invasibility of this region could be listed as follows:

- excessive human populations that migrate frequently and carry seeds or propagules of invasive plants from one place to another;
- availability of fragmented/disturbed habitats or species-poor regions due to habitat fragmentation and degradation that provide habitat for alien species; and
- favourable environmental and climatic conditions owing to the diversity of seasons and climatic variations that give invasive species an opportunity to establish.

Invasive plants such as *Ageratum conyzoides*, *Eupatorium adenophorum*, *Eupatorium odoratum*, *Lantana camara*, *Mikania micrantha* and *Parthenium hysterophorus* have caused havoc in terrestrial ecosystems, and *Eichhornia crassipes*, *Ipomoea* spp. and *Salvinia molesta* in aquatic ecosystems (Raghubanshi *et al.*, 2005). Despite several known invasive plants in India, there is no complete listing of the status of invasive plants. According to one source, around 18% of Indian flora are adventive aliens, of which 55% are from America, 30% from Asia and 15% from Europe and Central Asia (Nayar, 1977). More recently, Reddy (2008) has identified 173 invasive plants belonging to 117 genera and 44 families, based on his comprehensive work. Tropical America (74%) and tropical Africa (11%) have contributed most to the invasive flora of the country.

Besides, there are several other reports available that provide information on the invasive flora of a particular region/area. For example, Kohli *et al.* (2004, 2006) have provided information on the status of invasive plants in the north-western Himalayan region and have stated that three invasive weeds – *Ageratum conyzoides*, *Parthenium hysterophorus* and *Lantana camara* – found in this region have caused considerable harm to the fragile biodiversity. Several reasons, such as fragile soil, anthropogenic activity, tourism, pollution, rapid industrialization/urbanization and the livestock-dependent lifestyle of the aboriginals such as the Gaddi or Gujjar communities, are responsible for the establishment of invasives and declining biodiversity (Kohli *et al.*, 2009). Khuroo *et al.* (2007) reported 571 alien species belonging to 352 genera and 104 families from the Kashmir Himalayas, their origins traceable from Europe, Asia and Africa. Negi and Hajra (2007) reported 308 woody and 128 herbaceous exotic species from the Doon Valley of the north-western Himalayas of which many are harmul invasives that have created several environmental, socio-economic and health problems.

## Major Invasive Plants and their Attributes

Based on the available databases and some regional reports, nearly 60 invasive plants have been identified from the Indian region (Table 1.1), the majority from South and tropical America and Australia, Africa,

Europe and even the Asian region. Table 1.1 includes *Arundo donax* – a grass of Indian origin that can pose a major threat to biodiversity owing to its capacity to invade huge areas around water bodies. Most of the invasive plants, irrespective of their origin, belong to the family Asteraceae, while families such as Poaceae, Solanaceae and Fabaceae also predominate. Furthermore, invasive plants belong to a variety of life forms such as herbs, shrubs, trees, climbers/vines, grasses and aquatic plants (Table 1.1). Important invasive plants that have created havoc in a number of habitats include terrestrial herbaceous weeds (e.g. *Ageratum conyzoides* and *Parthenium hysterophorus*), shrubs (e.g. *Lantana camara* and *Chromolaena odorata*), trees (e.g. *Prosopis juliflora* and *Leucaena leucocephala*), vines (e.g. *Mikania micarantha*) and aquatic plants (e.g. *Eichhornia crassipes*). These have entered the alien environment by one of two main pathways – either intentionally to serve some human purpose or accidentally (through import of agricultural/horticultural material, human beings, ballast water, etc.).

*Lantana camara* is perhaps the best-known example of a serious weed having been intentionally introduced for ornamental value, in this case from tropical America to other parts of the world. *Lantana camara* is now rated as one of the worst invasive species identified by the Global Invasive Species Database and is also included in the top 100 invasive species of the world (GISD, 2010). In India, it was introduced at the beginning of 19th century as an ornamental plant, and is now found widespread in almost the whole of the subcontinent and has encroached even at high altitudes (> 1700 m) in the Himalayas (Kohli *et al.*, 2006). *Lantana* is a serious invader of forests, grasslands, agricultural land and vacant urban areas, and also in protected areas, and this has implications for the vegetation structure and dynamics (Kohli *et al.*, 2004; Sharma and Raghubanshi, 2007; Sahu and Singh, 2008; Love *et al.*, 2009).

Another weed introduced in India as an ornamental plant is *Chromolaena odorata* (National Focal Point for APFISN, 2005), and this is also included in the list of the top 100 worst invaders (GISD, 2010). It is also one of the most obnoxious weeds in the Western Ghats, north-eastern parts of the country and impacts on coconut, rubber, coffee and teak plantations (Singh, 1998). *Mikania micrantha* (Asterceae), commonly known as mile-a-minute, was introduced for camouflaging airfields during the Second World War (Randerson, 2003). It is now a very noxious weed in plantations and forests, especially in the southern and north-western parts of the country (Muniappan and Viraktamath, 1993). Trees such as *Leucaena leucocephala* and *Prosopis juliflora* introduced under various forestry programmes are also now of invasive proportions.

Of species introduced accidentally to the region, *Parthenium hysterophorus* is one of the most harmful weeds and the best-known example (Kohli and Rani, 1994). It grows rapidly in vacant areas, agricultural land, pastures, urban areas and natural and man-made forests, where it forms its own monoculture stands. It is found in every state and union territory of India except at higher altitudes and has now covered nearly 2,025,000 ha of land (Aneja *et al.*, 1991). It spreads mainly through its minute seeds, which are blown easily by wind and water. Likewise, *Ageratum conyzoides* is a fast-spreading weed, is the most serious invader of agricultural land and is a problem in hilly tracts. The weed also spreads through its seeds, which are minute and are carried by wind and water, as is *Parthenium hysterophorus*. *Eichhornia crassipes*, *Salvinia molesta*, *Alternanthera philoxeroides* and *Ipomoea* sp. are invaders of aquatic ecosystems and wetlands; these have done much harm to the biodiversity of aquatic ecosystems (Reddy, 2008).

There are several attributes of invasive plant species that enable them to spread in alien environments (see Tables 1.2 and 1.3, respectively, for established invasive plants and those in the process of establishment). Some of the key features of invasive plants are discussed below.

**Table 1.1.** Prominent invasive alien plants of the Indian subcontinent.

| Family name | Botanical name (common name) | Nativity | Life form |
|---|---|---|---|
| *Amaranthaceae* | *Chenopodium album* (lamb's-quarters) | Europe | Herb |
| | *Alternanthera philoxeroides* (alligator weed) | South America | Aquatic herb |
| *Amaryllidaceae* | *Zephyranthes citrina* (yellow rain lily) | Central, S. America | Herb |
| *Apocynaceae* | *Cryptostegia grandiflora* (rubber vine) | Madagascar | Vine-climber |
| *Araceae* | *Pistia stratiotes* (tropical duckweed) | South America | Aquatic plant |
| *Asparagaceae* | *Asparagus densiflorus* (asparagus fern) | South Africa | Herb |
| *Asteraceae* | *Ageratum conyzoides* (billy goat weed) | Tropical America | Herb |
| | *Ambrosia artemisiifolia* (small ragweed) | USA, Canada, Mexico | Herb |
| | *Anthemis cotula* (stinking mayweed) | Europe | Herb |
| | *Chromolaena odoratum* (Siam weed) | Central, S. America | Shrub |
| | *Cirsium arvense* (creeping thistle) | Europe | Herb |
| | *Eupatorium adenophorum* (Crofton weed) | Central America | Shrub |
| | *Eupatorium cannabinum* (hemp-agrimony) | British Isles | Herb |
| | *Gymnocoronis spilanthoides* (Senegal tea plant) | South America | Aquatic plant |
| | *Leucanthemum vulgare* (ox-eye daisy) | Europe | Herb |
| | *Mikania micrantha* (mile-a-minute weed) | Central, S. America | Vine/climber |
| | *Parthenium hysterophorus* (ragweed parthenium) | Tropical America | Herb |
| | *Sphagneticola trilobata* (Singapore daisy) | Central America | Herb |
| | *Synedrella vialis* (straggler daisy) | South America | Herb |
| | *Tagetes minuta* (Mexican marigold) | South America | Herb |
| *Azollaceae* | *Azolla pinnata* (mosquito fern) | Non-specific | Aquatic plant |
| *Bignoniaceae* | *Macfadyena unguis-cati* (cat's claw vine) | Central America | Climber |
| *Cabombaceae* | *Cabomba caroliniana* (green cabomba) | South America | Aquatic plant |
| *Convolvulaceae* | *Ipomoea aquatica* (water spinach) | China | Climber |
| | *Merremia peltata* (merremia) | Africa | Climber |
| *Elaeagnaceae* | *Elaeagnus umbellata* (Japanese silverberry) | China, Korea, Japan | Tree/shrub |
| *Euphorbiaceae* | *Ricinus communis* (castor bean) | North-eastern Africa | Tree/shrub |
| | *Sapium sebiferum* (Chinese tallow) | China | Tree |
| *Fabaceae* | *Acacia farnesiana* (sweet acacia) | Trop. America | Tree/shrub |
| | *Acacia mearnsii* (black wattle) | Australia | Tree |
| | *Acacia melanoxylon* (blackwood acacia) | Australia | Tree |
| | *Leucaena leucocephala* (wild tamarind) | Trop. America | Tree |
| | *Mimosa diplotricha* (giant sensitive plant) | South America | Climber/shrub |
| | *Mimosa pudica* (touch-me-not) | South America | Herb |
| | *Prosopis juliflora* (mesquite) | Central, S. America | Tree |
| | *Ulex europeus* (gorse) | Europe | Tree |
| *Hydrocharitaceae* | *Hydrilla verticillata* (water thyme) | Asia, North Australia | Aquatic plant |
| *Limnocharitaceae* | *Limnocharis flava* (yellow velvetleaf) | South America | Aquatic plant |
| *Melastomataceae* | *Clidemia hirta* (Koster's curse) | South America | Shrub |
| | *Miconia calvescens* (velvet tree) | Tropical America | Tree |
| *Moraceae* | *Broussonetia papyrifera* (paper mulberry) | China | Tree |
| *Myrtaceae* | *Eugenia uniflora* (Surinam cherry) | South America | Tree |
| | *Psidium guajava* (apple guava) | Central, S. America | Tree |
| *Onogranaceae* | *Ludwigia peruviana* (Peruvian primrose willow) | South America | Aquatic plant |
| *Poaceae* | *Arundo donax* (giant cane) | Indian subcontinent | Grass |
| | *Imperata cylindrica* (Cogon grass) | Asia/Africa? | Grass |
| | *Paspalum vaginatum* (seashore paspalaum) | North America | Grass |
| | *Pennisetum clandestinum* (Kikuyu grass) | Tropical Africa | Grass |
| | *Phalaris arundinacea* (reed canary grass) | Europe | Grass |
| | *Spartina alterniflora* (smooth cord grass) | South America | Grass |
| *Pontederiaceae* | *Eichhornia crassipes* (water hyacinth) | South America | Aquatic plant |
| | *Coffea arabica* (Arabic coffee) | Africa | Shrub |
| *Rubiaceae* | *Coffea canephora* (robusta coffee) | Africa | Shrub |
| *Salviniaceae* | *Salvinia molesta* (water fern) | South America | Aquatic plant |
| *Solanaceae* | *Physalis peruviana* (cape gooseberry) | South America | Shrub |
| | *Solanum mauritianum* (wild tobacco tree) | South America | Tree |
| | *Solanum sisymbriifolium* (sticky nightshade) | South America | Herb |
| | *Solanum viarum* (tropical soda apple) | South America | Shrub |

**Table 1.2.** Attributes of selected established invasive plants in India.

| Plant | Country of origin | Pathway of invasion | Invasive potential |
|---|---|---|---|
| *Ageratum conyzoides* | Tropical America | Unknown, probably as ornamental | Fast-growing herb; aggressive stolon formation; high regenerative and reproductive potential through large number (av. 94,772/plant) of minute, lightweight seeds, bearing pappi that help in dispersion through wind; wide ecological amplitude; strongly allelopathic |
| *Parthenium hysterophorus* | Tropical America | Food grains | Fast-growing herb; wide ecological amplitude; high regenerative and reproductive potential through large number of minute seeds (15,000/plant) dispersed through wind, water, biotic factors and transport; strongly allelopathic |
| *Lantana camara* | Tropical America | Introduced as ornamental | Shrubby habit; flowers exhibit genetic variations; forms suckers; high reproductive and regenerative potential; produces numerous seeds disseminated by birds, animals and even humans; wide ecological amplitude; strongly allelopathic |
| *Chromolaena odorata* | Tropical America | Introduced as ornamental | Highly vigorous; high regenerative and reproductive potential; prolific seed producer (260,000/m$^2$) with prolonged viability |
| *Mikania macrantha* | South America | Brought for camouflaging army during WWII | Vigorous, rampant growth; produces 40,000 seeds/year; spreads vegetatively by rooting at nodes |
| *Eichhornia crassipes* | Tropical America | Ornamental aquatic plant | Spreads fast due to quick vegetative (stolons) and sexual reproduction in water bodies |
| *Alternanthera philoxeroides* | South America | Accidently, as contaminant | Aquatic herb, spreads through vegetative reproduction by formation of fragmentation and axillary buds |
| *Salvinia molesta* | South America, Tropical America | Accidental/ ornamental | Reproduces very fast, mostly vegetatively by fragmentation |

- Fast growth and reproduction: invasive alien plants often have better growth and reproductive traits, and thus they spread very fast in the invaded region. They often have either high regenerative potential (*Parthenium hysterophorus*) or special organs for vegetative reproduction, such as stolons (*Ageratum conyzoides*), root suckers (*Lantana camara*), rhizomes, bulbs or turions.
- Ability to adapt to diverse environmental conditions: invasive plants have high ecological amplitude and thus have enormous adaptability to a wide range of environmental conditions. These can survive under stressed conditions and they also have the ability to modify growth patterns in response to changing environments such as soil condition, moisture status or limited space availability.

**Table 1.3.** Attributes of selected potential invasive plants in India in the early stages of establishment.

| Plant | Country of origin | Pathway of invasion | Invasive potential |
|---|---|---|---|
| *Leucaena leucocephala* | South America | Agroforestry/social forestry | Self-fertile and produces large number of seeds growing in vicinity of parent plant forming monoculture; ability to re-sprout after cutting; seed production through year |
| *Broussonetia papyrifera* | China, Japan | Avenue tree | Prolific growth, multiplying rapidly vegetatively by suckering activity of root system |
| *Tagetes minuta* | South America | Essential oils | Spreads very fast by seeds and forms huge populations showing vigorous growth |
| *Prosopis juliflora* | Tropical America | Timber | Spread rapidly due to aggressive growth |
| *Sapium sebiferum* | China, Japan | Avenue tree | Spreads very fast by suckers, re-sprouts very rapidly and produces large number of seeds over a year with effective dispersal |
| *Synedrella nodiflora* | Tropical America | – | Spreads through seed dispersal |
| *Anthemis cotula* | Europe, S. Africa | – | Spreads through seed dispersal |
| *Potamogeton crispus* | Eurasia, Africa, Australia, N. America | Aquatic weed | Tolerance to low light and low water temperature allows it to outcompete native plants; spread by plant fragments attached to water craft |
| *Imperata cylindrica* | SE Asia | Unknown | Aggressive and invasive nature is attributed to its rhizome, which produces numerous buds that sprout into new shoots |
| *Galinsoga ciliata* | Tropical America | Unknown | Rapid establishment of weed is due to: (i) production of large number of seeds in a wide range of environmental circumstances; and (ii) vegetative reproduction |
| *Mimosa invisa* | Tropical America | By tea gardeners for soil enrichment | Spreads very fast through seeds that remain dormant for up to 50 years |
| *Tithonia diversifolia* | Tropical America | Ornamental | Spreads through seed dispersal and vegetatively from basal stem |
| *Eupatorium riparium* | Tropical America | – | Spreads through seed dispersal |

- Absence of natural enemies or predators: invasive alien plants in new geographical areas are devoid of their natural predators or pests that co-evolved with them in their native environment. In the absence of these enemies their populations grow unchecked. Consequently, these form huge monocultures. In fact, absence of natural enemies forms the basis of the Natural Enemies hypothesis, as proposed by Heirro and Callaway (2003).
- Competitive advantage on account of allelopathy: many invasive alien plants are allelopathic. In other words, they release toxic chemicals to the environment that in turn hamper the growth and establishment of native flora. *Parthenium hysterophorus* is a potent allelopathic plant (Kohli and Rani, 1994). *Lantana camara* and *Ageratum conyzoides* are likewise reported as being strongly allelopathic (Ambika *et al.*, 2003; Kohli *et al.*, 2006). In fact, the allelopathic nature of many invasive alien plants forms the basis for the Novel Weapon hypothesis (Heirro and Callaway, 2003).

- Efficient seed dispersal mechanism: invasive alien plants have a better and more efficient mechanism of dispersal of their seeds or propagules, by virtue of which these spread very rapidly from one place to another.

## Potential Invasive Plants in the Process of Establishment

There are several plants that have not proved especially harmful to date but which possess the potential to become serious invaders. Early warnings regarding these species may help in managing them well before they damage natural ecosystems. *Tagetes minuta*, a native of South America, was introduced to various parts of the world, including India, for essential oils and medicinal value. However, it has been reported to be a noxious and invasive weed at many sites (Soule, 1993; Holm *et al.*, 1997). In India it seems to have spread from cultivated areas, as it can be seen growing luxuriantly in disturbed sites in the north-western parts of the country (personal observation of the authors). *Anthemis cotula*, an ornamental plant from Europe, is now spreading fast in the Kashmir valley – a Himalayan biodiversity hot spot (Shah and Zafar, 2007). *Leucanthemum vulgare*, another ornamental plant from Europe, is also spreading rapidly in this valley and inhibiting the regeneration of the native forest floor vegetation (Khuroo *et al.*, 2010). Jaryan *et al.* (2007) reported that *Sapium sebiferum*, a woody perennial, has the tendency to become invasive, particularly in the Himalayan region. *Broussonetia papyrifera*, commonly known as paper mulberry – a shrubby deciduous tree of Asian origin, possesses an invasive character and is spreading rapidly in the city of Chandigarh, India (personal observation of the authors).

## Management Tools

When looking at the alarming rate at which invasive plants are spreading in India and also worldwide, effective management measures are required to control them. There are separate ways of dealing with those already established and those possessing the potential to be invasive but not of immediate risk. Above all, preventive measures are of utmost importance and require great attention at every level to prevent the entry of invasive species.

### Management of potential invasive species

- Environment Impact Assessment of introduced plants could be very useful in managing invasive species in the initial stages of their establishment.
- Gathering information regarding the pathways of invasion of potential invasive species; appropriate steps can be taken to prevent the entry of potential invasive plants based on this information.
- Following the international standards and quarantine measures based on the International Plant Protection Convention (IPCC) or the Convention on International Trade on Endangered Species of Flora and Fauna (CITES) to prevent the entry of potential plant invaders. Further information regarding these can be obtained from international organizations on invasive species such as The Global Invasive Species Programme (www.gisp.org), IUCN's Invasive Species Specialist Group (www.issg.org) and its Global Invasive Species Database (GISD).
- Creating adequate awareness of the public about potential invasive plants and their management. To this end, online websites, scientific documents and books containing information on invasive plants could be quite useful. Success stories on control of weeds like *Parthenium hysterophorus* can also be used for dissemination of information (Batish *et al.*, 2004).

### Management of established species

Management of established invasive plants is very difficult. Based on the local situation, geographic region and climatic conditions, various methods such as physical, biological, chemical or cultural methods have been suggested, but these are of limited use. It is thus very important to follow an integrated approach and common interactive protocols for the management of invasive plants across the world. Some suggestions that may help in the management of invasive plants are listed below.

- understanding the life history of invasive plant species and their phenology, to help in devising control measures for their management;
- studying the ecology and biology of invasive plants, with special reference to their reproductive biology, dispersal ability and efficiency;
- understanding various biotic and abiotic factors limiting the growth of such plants;
- pests/pathogens of such plants can be explored from their native countries for their biological control;
- mass awareness programmes through booklets/pamphlets containing information on established local invasive species to discourage their further spread; and
- management, in a coordinated manner, through involvement of the general public and government officials or policy makers.

## Conclusions and the Way Forward

From the above discussion, it is clear that a number of invasive plants have become established in India and several others are in the process. These cause major environmental, socio-economic and health-related harms to the native people, and it is thus important to take the following steps.

- Prepare a list of invasive plant species and gather comprehensive information on their status, pathways of entry and spread.
- Determine means of spread of invasive plants in order to prevent their further spread.
- Determine the environmental and socio-economic impact of the invasive plants and devise management tools for their control.

## References


Ambika, S.R., Poornima, S., Palaniraj, R., Sati, S.C. and Narwal, S.S. (2003) Allelopathic plants. 10. *Lantana camara* L. *Allelopathy Journal* 12, 147–162.

Aneja, K.R., Dhawan, S.R. and Sharma, A.B. (1991) Deadly weed *Parthenium hysterophorus* L. and its distribution. *Indian Journal of Weed Science* 23, 14–18.

Batish, D.R., Singh, H.P., Kohli, R.K., Johar, V. and Yadav, S. (2004) Management of invasive exotic weeds requires community participation. *Weed Technology* 18, 1445–1448.

Charles, H. and Dukes, J.S. (2007) Impacts of invasive species on ecosystem services. In: Nentwig, W. (ed.) *Biological Invasions* (Ecological Studies, Vol. 193). Springer-Verlag, Berlin, pp. 217–237.

Drake, J.A., Mooney, H.A., di-Castri, F., Groves, R.H., Kruger, F.J. and Rejmanek, M. (1989) *Biological Invasions: A Global Perspective.* John Wiley and Sons, Chichester, UK.

GISD (2010) *Global Invasive Species Database.* Available online at http://www.issg.org/database

Heirro, J.L. and Callaway, R.M. (2003) Allelopathy and exotic plant invasion. *Plant and Soil* 256, 29–39.

Herron, P.M., Martine, C.T., Latimer, A.M. and Leicht-Young, S.A. (2007) Invasive plants and their ecological strategies: prediction and explanation of woody plant invasion in New England. *Diversity and Distributions* 13, 633–644.

Holm, L., Doll, J., Holm, E., Pancho, J. and Herberger, J. (1997) *World Weeds. Natural Histories and Distribution.* John Wiley and Sons, Inc., New York.

Jaryan, V., Chopra, S., Uniyal, S.K. and Singh, R.D. (2007) Spreading fast yet unnoticed: Are we in for another invasion? *Current Science* 93, 1483–1484.

Khuroo, A.A., Rashid, I., Reshi, Z., Dar, G.H. and Wafai, B.A. (2007) The alien flora of Kashmir Himalaya. *Biological Invasions* 9, 269–292.

Khuroo, A.A., Malik, A.H., Reshi, Z.A. and Dar, G.H. (2010) From ornamental to detrimental: plant invasion of *Leucanthemum vulgare* Lam. (Ox-eye Daisy) in Kashmir valley, India. *Current Science* 98, 600–602.

Kohli, R.K. and Rani, D. (1994) *Parthenium hysterophorus* – a review. *Research Bulletin (Science) Punjab University* 44, 105–149.

Kohli, R.K., Dogra, K.S., Batish, D.R. and Singh, H.P. (2004) Impact of invasive plants on the structure and composition of natural vegetation of north western Indian Himalayas. *Weed Technology* 18, 1296–1300.

Kohli, R.K., Batish, D.R., Singh, H.P. and Dogra, K.S. (2006) Status, invasiveness and environmental threats of three tropical American invasive weeds (*Parthenium hysterophorus* L., *Ageratum conyzoides* L., *Lantana camara* L.) in India. *Biological Invasions* 8, 1501–1510.

Kohli, R.K., Jose, S., Singh, H.P. and Batish, D.R. (2009) *Invasive Plants and Forest Ecosystems.* CRC Press, Boca Raton, Florida, pp. 437.

Love, A. Babu, S. and Babu, C.R. (2009) Management of *Lantana*, an invasive alien weed, in forest ecosystems of India. *Current Science* 97, 1421–1429.

Mack, R.N., Simberloff, D., Lonsdale, W.M., Evans, H., Clout, M. and Bazzaz, F.A. (2000) Biotic invasions: Causes, epidemiology, global consequences and control. *Ecological Applications* 10, 689–710.

Mooney, H.A. (2005) Invasive alien species: the nature of the problem. In: Mooney, H.A., Mack, R.N., McNeely, J.A., Neville, L.E., Schei, P.J. and Waage, J.K. (eds) *Invasive Alien Species: A New Synthesis.* Island Press, California, pp. 1–15.

Muniappan, R. and Viraktamath, C.A. (1993) Invasive alien weeds in the Western Ghats. *Current Science* 64, 555–557.

National Focal Point for APFISN, India (2005) *Stocktaking of National Forest Invasive Species Activities, India (India Country Report 101005).* Ministry of Environment and Forests, New Delhi, India.

Nayar, M.P. (1977) Changing patterns of the Indian flora. *Bulletin of the Botanical Survey of India* 19, 145–154.

Negi, P.S. and Hajra, P.K. (2007) Alien flora of Doon Valley, Northwest Himalaya. *Current Science* 92, 968–978.

Parker, I.M., Simberloff, D., Londale, W.M., Goodell, K., Wonhem, M., Kareiva, P.M. *et al.* (1999) Impact: towards a framework for understanding the ecological effects of invaders. *Biological Invasions* 1, 3–19.

Pejchar, L. and Mooney, H.A. (2009) Invasive species, ecosystem services and human well-being. *Trends in Ecology and Evolution* 24, 487–504.

Pimentel, D., Zuniga, R. and Morrison, D. (2005) Update on the environmental and economic costs associated with alien invasive species in the United States. *Ecological Economics* 52, 273–288.

Raghubanshi, A.S., Rai, L.C., Gaur, J.P. and Singh, J.S. (2005) Invasive alien species and biodiversity in India. *Current Science* 88, 539–540.

Randerson, J. (2003) Fungus in your tea, sir? *New Scientist* 178, 10.

Reddy, C.S. (2008) Catalogue of invasive alien flora of India. *Life Science Journal* 5, 8–89.

Richardson, D.M., Pysek, P., Rejmánek, M., Barbour, M.G., Panetta, F.D. and West, C.J. (2000) Naturalization and invasion of alien plants: Concepts and definitions. *Diversity and Distributions* 6, 93–107.

Sahu, P.K. and Singh, J.S. (2008) Structural attributes of lantana-invaded forest plots in Achanakmar-Amarkantak Biosphere Reserve, Central India. *Current Science* 94, 494–500.

Shah, M.A. and Zafar, R. (2007) Invasion by alien *Anthemis cotula* L. in a biodiversity hotspot: Release from native foes or relief from alien friends? *Current Science* 92, 21–22.

Sharma, G.P. and Raghubanshi, A.S. (2007) Effect of *Lantana camara* L. cover on local depletion of tree population in the *vindhyan* tropical dry deciduous forest in India. *Applied Ecology and Environmental Research* 5, 109–121.

Singh, S.P. (1998) A review of biological suppression of *Chromolaena odorata* K & R in India. In: Ferrar, P., Muniaapan, R. and Jayanth, K.P. (eds) *Proceedings of the 4th International Workshop on Biological Control and Management of Chromolaena odorata.* University of Guam, Guam, pp. 86–92.

Soule, J.A. (1993) *Tagetes minuta*: A Potential New Herb from South America. In Janick, J. and Simon, J.E. (eds) *New Crops.* John Wiley, New York, pp. 649–654.

Vitousek, P.M., D'Antonio, C., Loope, L.L., Rejmanek, M. and Westbrooks, R. (1997) Introduced species: a significant component of human-caused global change. *New Zealand Journal of Ecology* 21, 1–16.

# 2 Biology, Ecology and Spread of the Invasive Weed *Parthenium hysterophorus* in India

**D. R. Batish[1], R.K. Kohli[1], H.P. Singh[2] and Gurpreet Kaur[2]**

[1]*Department of Botany, Panjab University, Chandigarh, India;*
[2]*Department of Environment Studies, Panjab University, Chandigarh, India*

## Introduction

*Parthenium hysterophorus* L. (family *Asteraceae*, and hereafter referred to as *Parthenium*) is an obnoxious invasive weed from tropical America that has spread to various tropical and subtropical parts of the world. It is rated as one of the most serious weeds of the 20th century and also as a biological pollutant because of its adverse effects on human health. *Parthenium* is a weed of global importance (Adkins *et al.*, 2005). Owing to its invasive nature, the weed is included in the Global Invasive Species database of IUCN. It is known by several common and vernacular names such as parthenium weed, ragweed parthenium, starweed, bastard feverfew, Santa Maria feverfew, gajar ghas, safed topi, chatak chandni, white top weed and congress grass. It has created many problems in India and Australia and is spreading in Ethiopia. The weed spreads very rapidly over large areas and quickly forms its own monoculture. It occurs widely in different habitats varying from hot and arid, semi-arid to humid and from low- to middle- to high-altitude regions.

## Geographical Distribution

The exact area of origin of *Parthenium* is difficult to pinpoint because of its very wide distribution. However, it is presumed to be a native of central Argentina, the West Indies and the Gulf of Mexico. The weed has been reported from China, Taiwan, Nepal, Pakistan, Sri Lanka, Bangladesh, Vietnam, Pacific islands and African countries including Ethiopia, Kenya, Madagascar, South Africa, Somalia, Mozambique and Zimbabwe and also USA and several countries of South and Central America (Kohli and Rani, 1994; Dhileepan and Strathie, 2009). In Australia, *Parthenium* is referred to as a Weed of National Significance (Commonwealth of Australia and the National Weeds Strategy Executive Committee, 2000), while in India it is known to be one of the most serious weeds in terms of its potential to replace other plant species.

## Description

Parthenium weed is an annual herb with a tendency to be perennial, normally attaining a height of 1 m but sometimes reaching up to 2 m under favourable conditions (rainy season). It is an erect plant with a deeply penetrating root system and with an angular, grooved and profusely branched stem bearing dissected pale green leaves that resemble carrot leaves during initial growth. The leaves are pinnately and irregularly dissected and pubescent on both sides. The stem is also pubescent and covered by hairy structures known as trichomes. Four different kinds of trichomes observed on the upper and lower leaf surfaces carry great taxonomic significance for the recognition of different species of the genus *Parthenium* (Sahu, 1982). Trichomes are considered as storehouses for toxic chemicals found in the weed such as parthenin (a sesquiterpene lactone). The flowers are creamy-whitish and are arranged in capitula. Pollen grains are produced in clusters and are anemophilous. Typically three types of pollen grains, i.e. colporate, oblate-spheroidal and spinulose, have been reported (Agashe and Vinay, 1975). The weed prolifically produces wedge-shaped seeds that are small, dark brown to black, bearing two thin, white scales.

A single plant can produce around 15,000 seeds (even up to 100,000) that are disseminated by wind, water and also through transport (Haseler, 1976). The seed bank of the plant is also huge, seeds can remain viable for long periods and about 50% of the seed bank remains viable for 6 years (Navie *et al.*, 1998). The weed completes its life cycle quickly under favourable conditions (Fig. 2.1). Its active phase begins in the summer season and reaches its peak during the monsoon season, when huge stands of the weed can be seen in flowering and fruit setting. Although the weed grows throughout the year, however, its incidence is less during the winter (Fig. 2.1).

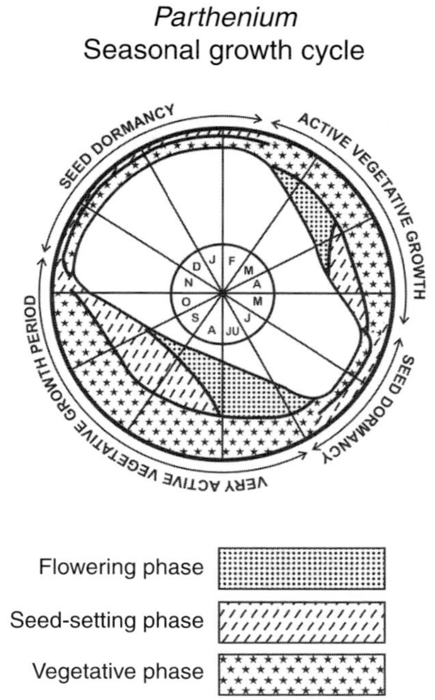


**Fig. 2.1.** Life cycle pattern of *Parthenium hysterophorus*.

## Ecological and Biological Features

The weed can grow under a wide range of environmental conditions. Soil moisture, however, seems to be the only limiting factor for its germination and growth. The ideal conditions for growth are high moisture content, high humidity and temperature around 25°C. The weed can survive under both a very low temperature of –2°C and at the high temperatures of arid zones (Nunez and Mata, 1976; Williams and Groves, 1980). It can grow under a wide range of soil pH, varying from 2.5 to 10.0, though the optimum pH for growth is 5.5–7.0 (Challa, 1987). The plant also has the ability to tolerate high salt-stress conditions (Hegde and Patil, 1982). Under water-stressed conditions, the plant remains in a rosette state and bolting occurs with the onset of the rains/availability of moisture. Furthermore, it has an amazing ability to regenerate quickly from the root stumps/parts (even petiole or midrib) remaining in the soil. The weed can grow and produce seeds throughout the year if moisture is sufficient, as it is a day-neutral plant. Being a $C_3$ type of plant, *Parthenium* has a tendency to become a $C_4$ (Hegde and Patil, 1982; Patil and Hegde, 1983).

## Pathways of Invasion and Invasiveness

*Parthenium* seems to have entered other countries via contaminated cereal grains (PL-480) or pasture seeds imported from the USA (GISD, 2010). Its entry into the Indian subcontinent is controversial. In India it was first observed in 1951 as a stray plant on the outskirts of Puna, and was identified by R.S. Rao in 1956 (Rao, 1956). Accordingly, it was presumed to have entered India accidentally along with imported wheat grains. However, there are evidences suggesting its presence in India much earlier than that of its first report. Catalogues from East India Gardens, Calcutta show its accidental introduction in 1810. Likewise, a herbarium list in Forest Research Institute, Dehradun collected in 1888 confirms its presence well before 1951. According to a recent report, the earliest record of *Parthenium* dates back to 1814, when William Roxburgh described this plant in his book *Hortus Bengalensis*. This report indicates that *Parthenium* entered India accidentally in 1810 (Paul, 2010). In all likelihood, the weed remained dormant for a long time until described by Rao in 1956. After this the weed spread to various parts of India and colonized there. By 1963 the weed had spread to the entire state of Jammu and Kashmir (Hakoo, 1963), and had completely naturalized in the union territory of Delhi by 1969 (Vaid and Naithani, 1970). It is now found almost throughout India up to an altitude of about 1600 m (Kohli *et al.*, 2004). It can be seen growing in open and vacant areas, degraded and disturbed land, roadsides, railway tracks, pastures, along water channels and near office premises and educational institutions. Some areas infested with *Parthenium* weed are shown in Fig. 2.2. It can also be seen growing luxuriantly in agricultural fields, forest nurseries and reserve forests.

It is an aggressive exotic weed because of its potential for spread, fast growth rate, high reproductive potential and adaptive nature. Being a rapid invader it invades almost all accessible land and reduces biodiversity, affects landscapes and reduces soil quality (Kohli and Rani, 1994). In agricultural fields the weed depletes soil nutrients and releases toxic substances into the soil that affect the emergence, early growth and development of crops and associated species. However, the longer viability of the seeds allows these to thrive

A B C D

**Fig. 2.2.** *Parthenium*-infested areas in different ecosystems. A, wasteland; B, forest area; C, agricultural field; D, along a road divider.

well even under adverse environmental conditions. Although the weed is known to colonize wastelands, especially those disturbed by human interference, it also encroaches upon agricultural areas, grasslands, urban areas, forestry and plantation sites. Upon invading any terrain it outcompetes the local vegetation and creates its own territory, forming huge, impenetrable monoculture stands affecting not only biodiversity but also crop production, animal husbandry and human health.

The survival strategies of this weed include:

- vigorous growth (both vegetative and reproductive);
- adaptability to diverse environmental conditions;
- quick regrowth upon removal of weed parts;
- unpalatable nature because of toxic principles; and
- allelopathic impact that enable it to outcompete other vegetation.

## Allelopathic Potential

*Parthenium* is known to be a potential allelopathic plant, which is evident from its ability to dominate vegetation. A number of studies have established the allelopathic potential of this weed (Kanchan and Jyachandra 1979a, b, 1980a, b; Mersie and Singh, 1987; Kohli and Batish, 1994, Batish *et al.*, 2002b, Singh *et al.*, 2005a). Various water-soluble allelochemicals such as phenolics and sequiterpene lactones from the plant have been identified as putative allelochemicals (Kanchan, 1975); even the pollen is allelopathic in nature. Artificially pollen clusters dusted on the stigmatic surface of *Crotalaria pellida* and *Desmodium heteracarpon* reduced their fruit setting and, on the leaves, reduced chlorophyll content (Kanchan and Jayachandra, 1980c). Residues of *Parthenium* in soil not only reduce crop growth but also affect soil chemistry (Batish *et al.*, 2002b). Singh *et al.* (2003) have reported that not only decaying – and even burnt – residues of *Parthenium* interfere with crop growth. Singh *et al.* (2005a) have demonstrated that allelochemicals released from its residues incorporated in soil adversely affected the growth of three *Brassica* species, *B. rapa*, *B. oleracea* and *B. campestris*. The allelopathic nature of weed gives it an advantage in invading other areas to gain rapid dominance.

*Parthenium* releases various allelochemicals such as phenolic acids (caffeic, p-coumaric, ferulic, anisic and p-hydroxy benzoic acids) and sesquiterpenes (parthenin, ambrosin, dihydro-isoparthenin and hymenin) from fresh plant parts and residues. Among the sesquiterpenes, parthenin is the major bioactive constituent of the weed that produces a number of its characteristics such as allergenicity, allelopathy and livestock toxicity (Kohli and Rani, 1994). Parthenin is present in all plant parts except the roots.

## Hazards

### Environmental

Owing to the invasive capacity of *Parthenium*, a number of native plant species are affected leading to imbalance in the ecosystem, which in turn affects various ecological functions. The weed poses a serious threat to the diversity of other plants (Kohli *et al.*, 2004; Sridhara *et al.*, 2005). The competitive and allelopathic ability of the weed allows it to replace dominant flora in a wide range of habitats (Kohli *et al.*, 2006), including indigenous grasses and weeds, due to its ability to dominate the area upon invasion. This reduces fodder and forage production leading to problems for livestock (Nath, 1981). This, in turn, disturbs the food chain and alters the trophic structure resulting in ecological imbalance in the area. Even soil fertility is affected as the weed is known to interfere with soil nutrients due to its allelopathic nature (Batish *et al.*, 2002b). The huge monoculture strands of the weed prevent the growth of native vegetation leading to loss of species diversity, as indicated by various ecological indices such as Shannon's, dominance and richness (Arora, 1999).

### Agricultural

Though a weed of unattended land, *Parthenium* poses operational hazards in agricultural fields leading to loss of yield in rice, maize, sugarcane, pea, castor, sunflower, brinjal, coconut and orchards of mango, grapes and guava (Krishnamurthy *et al.*, 1975). Besides, the weed releases toxic substances into the soil of agricultural fields causing food and fodder scarcity. Various studies have shown that different parts of the weed reduce both the growth and yield of agricultural crops by releasing into the soil water-soluble phenolics and sesquiterpene lactones such as parthenin (Kanchan and Jayachandra, 1979a, b, 1980a, b). Germination and yield of pulse crops such as black gram (*Phaseolus mungo*), green gram (*Phaseolus aureus*) and guar (*Cyamopsis tetragonoloba*) were considerably reduced in soil previously infested with *Parthenium* (Kohli and Batish, 1994). Channappagoudar *et al.* (1990) estimated a significant loss of grain yield of sorghum, from nearly 6.5 to 4.3 t/ha, and grain weight was reduced by nearly 30%. The weed was declared as noxious by the Karnataka government on 23 October 1975 under the Karnataka Agricultural Pests and Disease Act 1968.

### Health

*Parthenium* is known to cause a number of health problems in both humans and livestock. In humans it causes diseases such as allergenic eczematous contact dermatitis (AECD), rhinitis and asthma (Lonkar *et al.*, 1974; Rao *et al.*, 1985). The disease AECD is caused by prolonged skin contact with *Parthenium*, and moving away from this weed is the only solution. Rhinitis and asthma on the other hand are caused by inhalation of pollens or even trichomes during breathing, as these remain suspended in the air (Towers and Subba Rao, 1992). Parthenin found in trichomes has also been implicated as a health-hazardous compound owing to the presence of a cyclopentenene group that causes chromosomal damage, uncouples phosphorylation and inhibits cellular enzyme activity (Evans, 1997). A study conducted in Punjab (North India) shows that most bronchial asthmatic patients were sensitive to *Parthenium* (Suresh *et al.*, 1994). Animals feeding on *Parthenium* produce tainted milk due to the presence of parthenin, which causes Indian childhood cirrhosis (ICC) (Evans, 1997).

*Parthenium* is also very toxic to livestock and other animals, causing acute to chronic toxicity; symptoms may appear as ulceration, nausea, loss of appetite or restlessness and can even prove fatal (Narasimhan *et al.*, 1980, 1985; Kadhane *et al.*, 1992). Normally the weed is unpalatable to livestock, but in the absence of fodder they have no option but to consume it and suffer later from its toxicity.

## Economic Importance

*Parthenium*, when exploited by suitable means and technology, can prove to be a useful asset. In contrast to the various health hazards enlisted above, the plant has been utilized in the past as a cure for several diseases. The people of the West Indies have been using the weed as a remedy against various afflictions such as ulcerated sores, certain skin diseases, facial neuralgia, fever and anaemia (Arny, 1897). The chemical constituents of the weed have also been used as an insecticide and for curing skin diseases such as psoriasis (Kohli and Rani, 1994).

Furthermore, the properties of the weed in stimulating menstrual function and reducing fever and neuralgic pain have also been described. Zutshi *et al.* (1975) explored the antimicrobial activity of essential oils from *P. hysterophorus* against *Escherichia coli*, *Vibrio cholerae* and *Klebsiella aerogenes*. Parthenin is biologically very active; its anti-amoebic activity against axenic and polyxenic cultures of *Entamoeba histolytica* is claimed to be comparable to that of metronidazole, a standard drug prescribed for treatment of intestinal amoebic parasites (Sharma and Bhutani, 1988). Its anti-malarial activity has been found effective against *Plasmodium falciparum* (Hooper *et al.*, 1990), and it also possesses herbicidal and pesticidal properties (Batish *et al.*, 1997; Datta and Saxena, 2001; Batish *et al.*, 2002a).

Aqueous extracts of pollen grains exhibited antifungal activity by inhibiting sporangial germination and zoospore motility in *Sclerospora graminicola* infesting *Pennisetum typhoides* (Char and Bhat, 1975). *Parthenium* served as an indicator of aerial lead pollution in Caracas (Venezuela), along with *Eleusine indica* and *Tillamdsia ecurvata* (Tuges and Lois, 1976). Presumably the weed prefers a lead-rich environment and aids in removing lead from the air. The weed exhibits larvicidal properties against the root-knot nematode *Neloidogyne incognita* (Rao *et al.*, 1986). The weed not only had a negative influence on egg hatching but also caused significant destruction of the larvae.

## Control Measures

Management of *Parthenium* is a challenge for scientists. A number of physical, biological, chemical and ecological methods have been tried in the past, but none has succeeded to date. The problems related to the management of this weed are numerous:

- The weed occupies a huge area that is difficult to control.
- At any given time the weed occurs in several life forms, such as young seedlings, mature plants with seeds, plants in vegetative stage, etc.
- As already mentioned, it is a prolific seed producer and the seeds are normally wind and water disseminated, so it is difficult to check their spread to distant areas.
- It is adaptable to diverse environmental conditions.
- It causes allergic reactions in humans and hence no one is willing either to uproot it or to apply any other means of control.

The various methods employed to manage *Parthenium*, such as physical, chemical, biological and cultural, will now be described.

### Physical methods

Techniques such as uprooting by hand, the use of harvesting or shrub-masters and physical cutting have been employed. These are applicable to small areas and should be carried out prior to the flowering stage. Manual uprooting is generally avoided or recommended only with precautions, due to the plants allergenic properties. Burning is another physical control method used in many places, but the burnt residues left in the fields compromise soil quality and therefore this technique is not recommended (Singh *et al.*, 2003).

### Chemical methods

Control by this option is both effective and economical compared with other methods, but there are potential environmental risks including ill-effects on human health, and this option should therefore be used cautiously. The use of natural plant products is another option as these are biodegradable, environmentally safe, efficacious and economical. Singh *et al.* (2005b) reported that volatile essential oil from lemon-scented eucalyptus (*Eucalyptus citriodora*) possesses good potential for the control of *Parthenium*.

### Biological methods

In several countries, such as India, Australia and Ethiopia, this approach has been extensively tried for the control of *Parthenium* (Dhileepan, 2009). Among various biological agents used, the Mexican beetle (*Zygogramma bicolorata*) and selected fungal-based mycoherbicides have been trialled for this purpose (Evans, 1997), although with limited success. Another strategy used for the control of *Parthenium* is to replace this species with other, competitively superior, species such as *Cassia* spp. (Joshi, 1991). However, the success rate is very limited in this strategy, and in fact there is no single method currently available for the control of *Parthenium*. An integrated approach involving different management options could, however, be successful, although it should be based on ecological principles. Batish *et al.* (2004) indicated that in order to control *Parthenium* on a long-term basis, community participation is key in addition to the integrated use of other methods.

## References

Adkins, S.W., Navie, S.C. and Dhileepan, K. (2005) *Parthenium* weed in Australia: Research progress and prospects. In: Ramachandra Prasad, T.V. *et al.* (eds) *Proceedings of the Second International Conference on* Parthenium *Management*, Bangalore, India. University of Agricultural Sciences, Bangalore, India, pp. 11–27.

Agashe, S.N. and Vinay, P. (1975) Aeropalynological studies on Bangalore city. I. Pollen morphology of *Parthenium hysterophorus* Linn. *Current Science* 44, 216–217.

Arny, H.V. (1897) *Parthenium hysterophorus. American Journal of Pharmacy* 69, 168.

Arora, V. (1999) Comparative assessment of ecophysiological function between *Ageratum conyzoides* L. and *Parthenium hysterophorus* L. PhD thesis, Panjab University, Chandigarh, India.

Batish, D.R., Kohli, R.K., Singh, H.P. and Sexena, D.B. (1997) Studies on herbicidal activity of parthenin – a constituent of *Parthenium hysterophorus* toward billy-goat weed. *Current Science* 73, 369–371.

Batish, D.R., Singh, H.P., Kohli, R.K., Saxena, D.B. and Kaur, S. (2002a) Allelopathic effects of parthenin against two weedy species, *Avena fatua* and *Bidens pilosa. Environmental and Experimental Botany* 47, 149–155.

Batish, D.R., Singh, H.P., Pandher, J.K., Arora, V. and Kohli, R.K. (2002b) Phytotoxic effect of *Parthenium* residues on the selected soil properties and growth of chickpea and radish. *Weed Biology and M*anagement 2, 73–78.

Batish, D.R., Singh, H.P., Kohli, R.K., Johar, V. and Yadav, S. (2004) Management of invasive exotic weeds requires community participation. *Weed Technology* 18, 1445–1448.

Challa, C. (1987) Chemical control of weeds in grape (*Vitis vinifera* L.) nurseries. *Pesticides* 21, 27–29.

Channappagoudar, B.B., Panchal, Y.C., Manjunath, S. and Koti, R.V. (1990) Studies on influence of parthenium on sorghum growth under irrigated conditions. *Farming Systems* 6, 102–104.

Char, M.B.S. and Bhat, S.S. (1975) Antifungal activity of pollen. *Die Naturwissenschaften* 62, 536.

Commonwealth of Australia and the National Weeds Strategy Executive Committee (2000) Agriculture & Resource Management Council of Australia & New Zealand, Australian & New Zealand Environment & Conservation Council and Forestry Ministers. *Weeds of National Significance, Parthenium weed (Parthenium hysterophorus) Strategic Plan.* National Weeds Strategy Executive Committee, Launceston, Australia.

Datta, S. and Saxena, D.B. (2001) Pesticidal properties of parthenin (from *Parthenium hysterophorus*) and related compounds. *Pest Management Science* 57, 95–101.

Dhileepan, K. (2009) Managing parthenium weed across diverse landscapes: Prospects and Limitations. In: Inderjit, R. (ed.) *Management of Invasive Weeds.* Springer Science and Business Media B.V., Berlin, pp. 227–259.

Dhileepan, K. and Strathie, L. (2009) *Parthenium hysterophorus* L. (Asteraceae). In: Muniappan, R., Reddy, G.V.P. and Raman, A. (eds) *Biological Control of Tropical Weeds Using Arthropods.* Cambridge University Press, Cambridge, UK.

Evans, H.C. (1997) *Parthenium hysterophorus*: A review of its weed status and the possibilities for biological control. *Biocontrol News and Information* 18, 89–98.

GISD (2010) Global Invasive Species Database, available at http://www.issg.org/database

Hakoo, M.L. (1963) A diploid *Parthenium* in Jammu. *Current Science* 32, 273.

Haseler, W.H. (1976) *Parthenium hysterophorus* L. in Australia. *Pest Articles and News Summaries* 22, 515–517.

Hegde, B.A. and Patil, T.M. (1982) Effect of salt stress on the structure and the carbon flow mechanism in a noxious weed *Parthenium hysterophorus* L. *Weed Research* 22, 51–56.

Hooper, M., Kirby, G.C., Kulkarni, M.M., Kulkarni, S.N., Nagasampagi, B.A., O'Neill, M.J. *et al.* (1990) Antimalarial activity of *Parthenium* and its derivatives. *European Journal of Medicinal Chemistry* 25, 717–723.

Joshi, S. (1991) Biological control of *Parthenium hysterophorus* L. (Asteraceae) by *Cassia uniflora* Mill (Leguminosae) in Bangalore, India. *Tropical Pest Management* 37, 182–184.

Kadhane, D.L., Jangde, C.R., Sudekar, R.D. and Joshirao, M.K. (1992) *Parthenium* toxicity in buffalo calves. *Journal of Soils and Crops* 2, 69–71.

Kanchan, S.D. (1975) Growth inhibitors from

*Parthenium hysterophorus* Linn. *Current Science* 44, 358–359.

Kanchan, S.D. and Jayachandra (1979a) Allelopathic effects of *Parthenium hysterophorus* L. I. Exudation of inhibitors through roots. *Plant and Soil* 53, 27–35.

Kanchan, S.D. and Jayachandra (1979b) Allelopathic effects of *Parthenium hysterophorus* L. III. Inhibitory effects of weed residue. *Plant and Soil* 53, 37–47.

Kanchan, S.D. and Jayachandra (1980a) Allelopathic effects of *Parthenium hysterophorus* L. II. Leaching of inhibitors from aerial vegetative parts. *Plant and Soil* 55, 61–66.

Kanchan, S.D. and Jayachandra (1980b) Allelopathic effects of *Parthenium hysterophorus* L. IV. Identification of inhibitors. *Plant and Soil* 55, 67–75.

Kanchan, S. and Jayachandra (1980c) Pollen allelopathy – a new phenomenon. *New Phytologist* 84, 739–746.

Kohli, R.K. and Batish, D.R. (1994) Exhibition of allelopathy by *Parthenium hysterophorus* L. in agroecosystems. *Journal of Tropical Ecology* 35, 295–307.

Kohli, R.K. and Rani, D. (1994) *Parthenium hysterophorus* – a review. *Research Bulletin*, Panjab University, Chandigarh, India.

Kohli, R.K., Dogra, K.S., Batish, D.R. and Singh, H.P. (2004) Impact of invasive plants on the structure and composition of natural vegetation of Northwestern Indian Himalayas. *Weed Technology* 18, 1296–1300.

Kohli, R.K., Batish, D.R., Singh, H.P. and Dogra, K.S. (2006) Status, invasiveness and environmental threats of three tropical American invasive weeds (*Parthenium hysterophorus* L., *Ageratum conyzoides* L., *Lantana camara* L.) in India. *Biological Invasions* 8, 1501–1510.

Krishnamurthy, K., Ramachandra Prasad, T.V. and Muniyappa, T. (1975) Agricultural and health hazards of *Parthenium. Current Research* 4, 169–171.

Lonkar, A., Mitchell, J.C. and Calnan, C.B. (1974) Contact dermatitis from *Parthenium hysterophorus. Transactions of the St. John's Dermatological Society* 60, 43–53.

Mersie, W. and Singh, M. (1987) Allelopathic effect of parthenium (*Parthenium hysterophorus* L.) extract and residue on some agronomic crops and weeds. *Journal of Chemical Ecology* 13, 1739–1747.

Narasimhan, T.R. *et al.* (1980) Toxicity of *Parthenium hysterophorus* L. in cattle and buffaloes. *Indian Journal of Animal Sciences* 50, 173–178.

Narasimhan, T.R., Harindernath, N., Premlatha, S., Keshva Murthy, B.S. and Subba Rao, P.V. (1985) Effect of parthenin on mitochondrial oxidative phosphorylation. *Biochemistry International* 11, 239–244.

Nath, R. (1981) Note on the effect of *Parthenium* extract on seed germination and seedling growth in crops. *Indian Journal of Agricultural Sciences* 51, 601–603.

Navie, S.C., Panetta, F.D., McFadyen, R.E. and Adkins, S.W. (1998) Behaviour of buried and surface-lying seeds of parthenium weed (*Parthenium hysterophorus* L.). *Weed Research* 38, 338–341.

Nunez, S.A. and Mata, O.A. (1976) Times of emergence of main weeds of Northern Tamsulings region. *Agricultura técnica en México* 3, 437.

Patil, T.M. and Hegde, B.A. (1983) Pattern of starch distribution, $CO_2$ composition, concentration and photochemical reduction of tetranitro tetrazolium blue in *Parthenium hysterophorus. Photosynthetica* 17, 64–66.

Paul, T.K. (2010) The earliest record of *Parthenium hysterophorus* L. (Asteraceae) in India. *Current Science* 98, 1272.

Rao, K.C., Mani, A. and Rao, P. (1986) Effect of plant products on egg hatch and larval mortality of *Meloidogyne incognita*. In: *Proceedings of Indian Academy of Sciences (Animal Sciences)* 95, 397–402.

Rao, M., Parkash, O. and Subba Rao, P.V. (1985) Reaginic allergy to *Parthenium* pollen evaluation by skin test and radio allergosorbent test. *Clinical Allergy* 15, 449–454.

Rao, R.S. (1956) *Parthenium hysterophorus* Linn. – a new record for India. *Journal of the Bombay Natural History Society* 54, 218–220.

Sahu, T.R. (1982) Trichome studies in *Parthenium hysterophorus* and their taxonomic importance. *Fedder Report* 93, 437–441.

Sharma, G.L. and Bhutani, K.K. (1988) Plant-based antiamoebic drugs. Part II. Amoebicidal activity of parthenin isolated from *Parthenium hysterophorus. Plant Medica* 54, 120–122.

Singh, H.P., Batish, D.R., Pandher, J.K. and Kohli, R.K. (2003) Assessment of allelopathic properties of *Parthenium hysterophorus* residues. *Agriculture, Ecosystem and Environment* 95, 537–541.

Singh, H.P., Batish, D.R., Pandher, J.K. and Kohli, R.K. (2005a) Phytotoxic effects of *Parthenium hysterophorus* residues on three *Brassica* species. *Weed Biology and Management* 5, 105–109.

Singh. H.P., Batish, D.R., Setia, N. and Kohli, R.K. (2005b) Herbicidal activity of volatile oils from *Eucalyptus citriodora* against *Parthenium hysterophorus. Annals of Applied Biology* 146, 89–94.

Sridhara, S., Basavaraja, B.K. and Ganeshaiah, K.N. (2005) Temporal variation in relative dominance of *Parthenium hysterophorus* and its effect on native biodiversity. In: Ramachandra Prasad, T.V. *et al.* (eds) *Proceedings of the Second International Conference on Parthenium Management*, Bangalore, India. University of Agricultural Sciences, Bangalore, India, pp. 240–242.

Suresh, P.V., Gupta, D., Behera, D. and Jindal, S.K. (1994) Bronchial provocation with *Parthenium* pollen extract in bronchial asthma. *Indian Journal of Chest Diseases and Allied Sciences* 36, 104.

Towers, G.H.N. and Subba Rao, P.V. (1992) Impact of the pan-tropical weed, *Parthenium hysterophorus* L. on human affairs. In: Richardson, R.G. (ed.) *Proceedings of the 1st International Weed Control Congress*, Melbourne, Australia. Weed Science Society of Victoria, Australia, pp. 134–138.

Tuges, B. and Lois, J. (1976) Plants as indicator of lead in the atmosphere. *Acta Botánica Venezuelica* 11, 105–204.

Vaid, K.M. and Naithani, H.B. (1970) *Parthenium hysterophorus* Linn. – a new record for the north western Himalayas. *Indian Forester* 96, 791–972.

Williams, J.D. and Groves, R.H. (1980) The influence of temperature and photoperiod on growth and development of *Parthenium hysterophorus* L. *Weed Research* 20, 47–52.

Zutshi, S.K., Joshi, S.K. and Bokadia, M.M. (1975) Antimicrobial activity of some Indian essential oils. *Indian Journal of Pharmacology* 37, 129–130.

# 3 Invasive Species: Ecology and Impact of *Lantana camara* Invasions

**Gyan P. Sharma[1] and Akhilesh S. Raghubanshi[2]**

[1]*Department of Environmental Biology, University of Delhi, Delhi, India;* [2]*Institute of Environment and Sustainable Development, Banaras Hindu University, Varanasi, India*

## Introduction

Invasion of exotic species is among the most important global problems experienced by natural ecosystems. Although biological invasion is a natural process, the recent enhanced rate of invasions is clearly a human-instigated phenomenon and constitutes one of the most important effects that humans have exerted on the planet. Disturbance, whether regular or episodic, is a natural feature of dynamic ecosystems (Sousa, 1984; Gurevitch and Padilla, 2004), but it also facilitates the invasion process.

As a result of the rapid modification of natural habitats, the pace of invasion has accelerated particularly during the past century (Schei, 1996). Extinction of species related to invasion is an outcome of human activities. Invasive species are the second largest threat to global biodiversity after habitat destruction, and the number one cause of species extinction in most island states (Schei, 1996). In the past, many of the irretrievable losses of native biodiversity due to biological invasion have gone unrecorded but, today, there is an increasing realization of the ecological costs of this process. Over 40% of the plants listed as threatened and endangered species in the USA are at risk from invasive species (Wilcove *et al.*, 1998).

Alien plants are widespread throughout the world, in agriculture, forest and natural areas. Alien species usually have much greater potential for invasion than the indigenous plants, as they may be more effective competitors and experience relatively little pressure from natural predators compared with those that have evolved in their native land. The majority of invaders pose threats to the invading ecosystem by virtue of their aggressive qualities, which can include superior growth by effectively competing for resources, efficient dispersal and rapid establishment. These invaders come in many shapes and sizes. They may be trees, shrubs, small herbaceous plants, or aquatic species, but they have in common the ability to spread and reproduce rapidly, thus overcoming biological, physical and environmental thresholds. Some of the invaders arrive by accident, but the majority of introductions have been intentional, through a number of sources.

The objective of this chapter is to focus on the definition of invasive species and how these spread, their biological characteristics and deleterious effects on ecosystem attributes and, in light of the aforementioned characteristics, how *Lantana camara* L. (hereafter known as *Lantana*) fits within the scenario and what management options are available to deal with the problems of plant invasion.

Batianoff and Butler (2003) have rated *Lantana* among the most invasive species in Queensland, Australia. Also, it is among the world's 100 worst invasive alien species, as recognized by the Invasive Species Specialist Group (IUCN, 2001). Although *Lantana* is regarded as one of the principal invasive species, limited comprehensive studies on its biology and ecology are available. This chapter also reviews some of the key features of *Lantana* that make this a particularly serious invasive species, with some quantitative case studies.

These case studies discuss the following: (i) the effect of canopy opening on invasive cover; (ii) the effect of invasion on herb species composition and soil properties; (iii) the effect of invasion intensity on the decline of tree species populations; and (iv) the effect of plant invasion on soil process, especially nitrogen (N)-mineralization.

## What are invasive species and why are they important?

Organisms immigrating to new localities and their descendants have been referred to as alien, adventive, exotic, introduced and non-indigenous (Mack *et al.*, 2000). Species whose native status and origin are not clear are called cryptogenic species (Carlton, 1996). A taxon can be considered successfully naturalized after overcoming geographical, environmental and reproduction barriers, while an invasive species requires, in addition, to overcome dispersal barriers within the new region (Richardson *et al.*, 2000). According to Rejmanek (2000), invasive taxa represent a subset of naturalized taxa. Invasion is usually discovered once the plant has already naturalized. The biotic invaders tend to establish a new range in which they proliferate, spread and persist, to the detriment of the environment. Although defined variously by different authors, we consider the definition of alien invasive species given by GISP (2003) as most pertinent to the present discussion: 'Invasive alien species are nonnative organisms that cause, or have the potential to cause, harm to the environment, economies, or human health'. Thus establishment and spread of these species threatens ecosystems, habitats or species with economic/environmental harm (GISP, 2000). Invasion of exotic species is among the most important global-scale problems experienced by natural ecosystems. The growing human population and improved transcontinental transport have increased the scales of movement of non-indigenous organisms, and the current enhanced rate of invasion constitutes one of the most important effects that humans have had on the earth's ecosystem. In the past, many of the irretrievable losses of native biodiversity have gone unrecorded, but today there is an increasing realization of the ecological costs of biological invasion.

Over 40% of the species on the list of threatened and endangered species are there due to invasive species (Wilcove, 1998). Rejmanek and Randall (1994) estimated that 20% or more of plant species are non-indigenous in many continental areas, and 50% or more on many islands. As many as 10% of the 260,000 vascular plant species are estimated to be potential invaders (Rapoport, 1991). Around 18% of the Indian flora constitutes adventive aliens, of which 55% are American, 10% Asian, 20% Asian and Malaysian and 15% European and Central Asian species (Nayar, 1977). Although large numbers of exotics have become naturalized in India and have affected the distribution of native flora to some extent, only a few have conspicuously altered the vegetation patterns of the country. *Ageratum* sp., *Calatropis procera*, *Cytisus scoparius*, *Chromolaena odorata*, *Eupatorium adenophorum*, *Hyptis suaveolens*, *Lantana camara*, *Mikania micrantha*, *Mimosa invisa*, *Parthenium hysterophorus* and *Prosopis juliflora* among terrestrial exotics, and *Eichhornia crassipes* and *Pistia stratiotes* among aquatics, have posed serious threats to the native flora.

Not all non-native species are, however, harmful. For example, over 70% of the world's food comes from just nine crops (wheat, maize, rice, potato, barley, cassava, soyabean, sugarcane and oats), each of which is cultivated far beyond its place of origin (Prescott-Allen and Prescott-Allen, 1990). In New Zealand, 95% of export earning derives from alien species (NSDS, 1996). Despite all the benefits provided by non-native organisms, invasive species are the second greatest threat to biodiversity globally after habitat destruction, and constitute the number one cause of species extinctions in most island states (Schei, 1996; Fritts and Rodda, 1998; Gurevitch and Padilla, 2004). The problem apears alarming because many biological invasions are, effectively, irreversible.

## The invasion process

The essential first step in invasion by an alien plant is its introduction to an area beyond its previous geographical range. Introduction of non-native species may occur through: (i) accidental introduction; (ii) import for a limited purpose and subsequent escape; or (iii) deliberate introduction on a large scale (Levine, 1989). Once introduced, the invader colonizes the new habitat, produces new self-perpetuating populations and is naturalized by becoming incorporated into the resident flora. This is followed by spread to new locations (Fig. 3.1).

Recently, Wilson *et al.* (2009) categorized extra-range dispersal into six types of dispersal pathway: leading edge, corridor, jump, extreme long distance, mass and cultivation, to improve our understanding

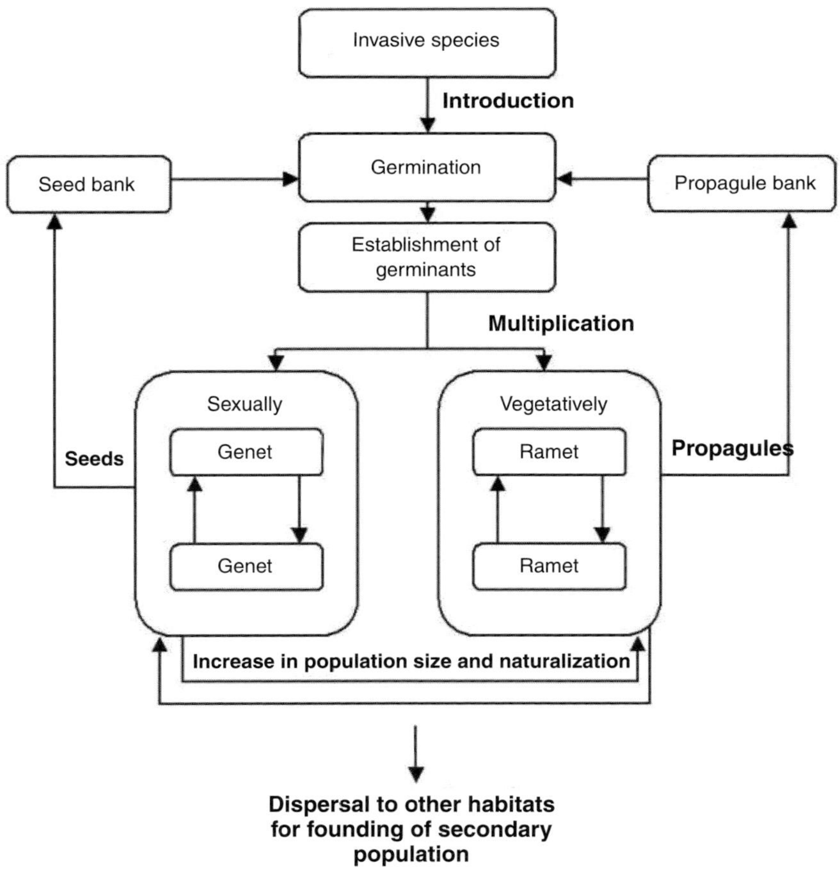


**Fig. 3.1.** Recruitment model for invasive plant species. The invasive species has to overcome geographical, environmental and reproductive barriers, face competition from indigenous species and then proceed to colonize new locations (adapted from Sharma *et al.*, 2005a).

of the properties and mechanisms of extra-range dispersal of species and provide a possible management scenario. The invader is likely to exist for some time as a single or small localized population. The concept of dormant invader has been applied to species that are present in an area for an extended period of time before becoming a significant invader. At some point in time, however, the invader will enter a period of rapid expansion, both in terms of total population size and the number and size of individual infestations. Finally, an invader will reach a stage at which it will become a major problem (Fig. 3.2). The invasion is not necessarily a smooth process – major episodes of population expansion may be punctuated by uneventful periods. A statistical rule, known as the 10th rule, holds that 1 in 10 of imported species become introduced, 1 in 10 of those introduced become established and 1 in 10 of those established become pests (Williamson, 1993). Box 3.1 includes examples of the rapid spread of invasive species and their impact on native flora.

### Species and community invasibility

In a classical 'naturalization–invasion continuum' model, Richardson and Pyšek (2006) visualized various barriers that a plant must overcome to become alien, casual, naturalized or invasive (see Fig. 3.3 for explanation) in a new environment, and the model highlights the dynamics of plant invasions, provides insights into plant traits (elements of species invasiveness) and features of the environment (components of community invasibility) and emphasizes that none of components can be fully assessed without reference to the others (Fig. 3.3). Therefore, to understand plant invasion in totality one must consider both species invasibility and community invasibility.

The focus of the present chapter is specifically on *L. camara*, as this species was recognized unanimously as one of the chief invaders in India, during several meetings on plant invasion (Raghubanshi *et al.* 2005; Corbett, 2009). We review and highlight *L. camara*'s invasiveness in light of both species and community invasibility.

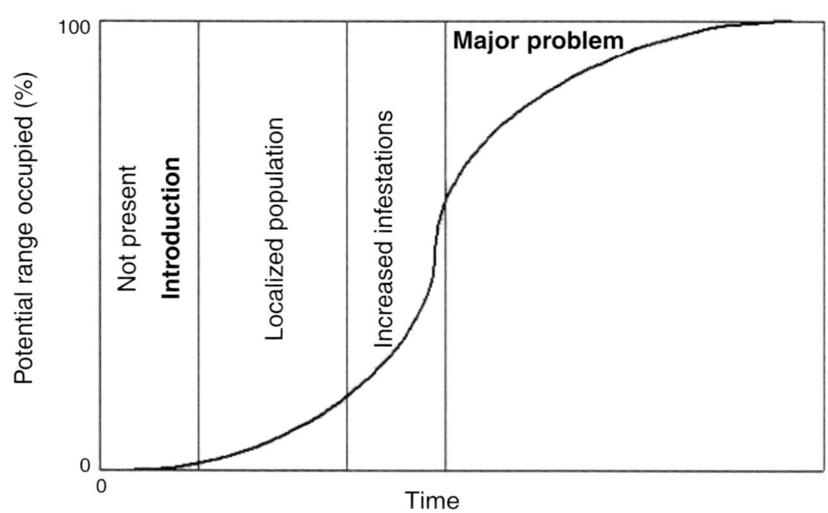


**Fig. 3.2.** Diagrammatic representation of generalized phases of invasion (adapted from Sharma *et al.*, 2005a).

**Box 3.1.** Examples of the spread of invasive species displacing native flora (compiled from Sharma *et al.*, 2005a).

- As much as 425 $km^2$ of multilayered forest in the Amazon basin in Brazil is at risk from African grasses (*Melinis minutiflora*, *Hyparrhenia rufa*, *Panicum* sp. and *Rhynchelytrum repens*) (Eiten and Goodland, 1979).
- *Mimosa pigra* has transformed 80,000 ha of tropical wetland in North Australia into monotonous tall shrubland (Braithwaite *et al.*, 1989).
- *Myrica faya* is native to the Azores and Canary islands. It was originally brought to Hawaii by immigrants from Portugal in the late 19th century and was first observed on volcanoes in the National Park in 1961; by 1977 it had covered 600 ha in the Park despite intensive control efforts, and by 1985 it had covered 12,200 ha in the Park and 34,365 ha throughout the Hawaiin islands (Whiteaker and Gardner, 1985).
- *Lantana* infests 4 million ha in Australia and it has also infested millions of hectares of natural grazing lands in 47 other countries (Weeds of National Significance, 2001).
- Of the 463 grasses introduced to improve pasture in North Australia, only 5% increased pasture productivity, over 60% of the remaining species naturalized and around 13% of the introduced species survived in the wild to become weeds (Lonsdale, 1994).
- The Australian paperbark tree (*Melaleuca quinquenervia*), which increased its range in south Florida by >20 ha per day and replaced other native species, now covers about 160,000 ha (Schmitz *et al.*, 1997).
- *Miconia calvescens* covers 75% of Tahiti, where it has the nickname 'the green cancer' (Miconia, 2001).
- *Schinus terebinthifolius* is displacing the native vegetation of both uplands and wetlands in south Florida (Myers and Evel, 1990) and now covers 243,810 ha (Austin, 1978).
- *Casuarina equisetifolia* interferes with the nesting activities of turtles and American crocodiles in coastal communities of southern Florida, and now infests nearly 151,065 ha (Myers and Evel, 1990).
- *Imperata cylindrica* was imported into Florida in the 1940s for erosion control and as a source of forage. It proved inadequate for either purpose and now displaces native plants (Coile and Shilling, 1993).
- An important invader in the Indian subcontinent, *Parthenium hysterophorus*, has spread to virtually every state in India and the area currently infested is estimated at 2,025,000 ha. It not only replaced native plant species but is also a health hazard (Aneja *et al.*, 1991; see also Chapter 2, this volume).

## Geographical Distribution, Biology and Habitat Characteristics

### Habitat

*Lantana* is a member of the family *Verbenaceae* and is a pantropical weed affecting pastures and native forests in more than 60 countries worldwide (Parsons and Cuthbertson, 2001). It occurs in diverse habitats and on a variety of soil types. The plant generally grows best in open, unshaded situations, such as degraded land, pasture, edges of tropical and subtropical forests, warm temperate forests, beach fronts and forests recovering from fire or logging. It also invades forest plantations and riparian zones. In its native range in tropical America, *Lantana* generally occurs in small clumps ≤1 m in diameter (Palmer and Pullen, 1995). In its naturalized range, it often forms dense, monospecific thickets, 1–4 m high and ~1–4 m in diameter (Swarbrick *et al.*, 1998), although some varieties climb trees and reach heights of 8–15 m (Swarbrick *et al.*, 1998). In India, *Lantana* was first introduced in the early nineteenth century as an ornamental plant but is now growing densely throughout the country (Thakur *et al.*, 1992). In Australia, the plant was first reported in 1841 and, by 1897, it was recognized as one of the most troublesome weeds (Van Oosterhout *et al.*, 2004). It is now spreading to form impenetrable

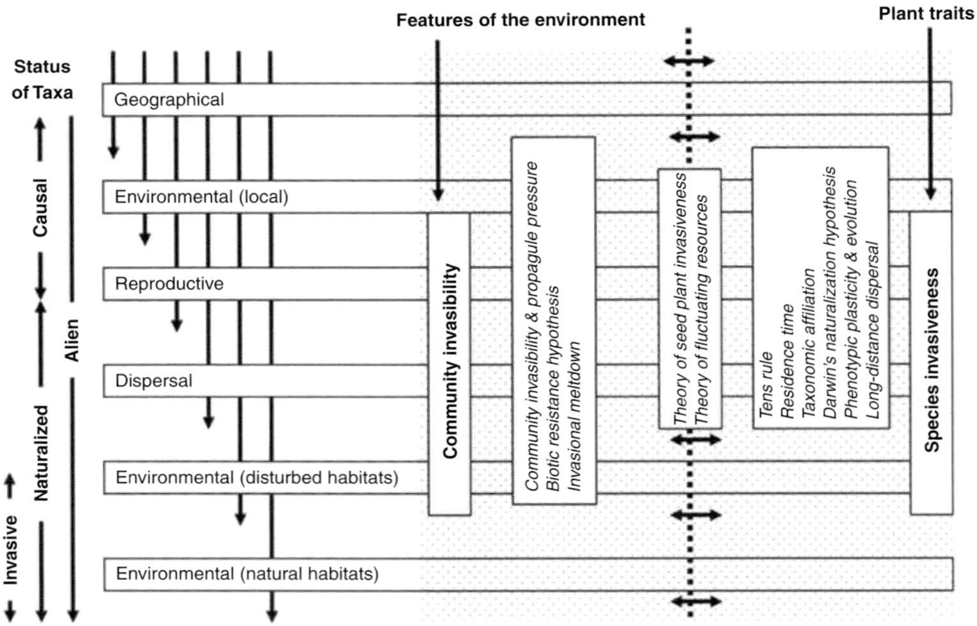


**Fig. 3.3.** The 'Naturalization–invasion continuum' model, modified from Richardson and Pyšek (2006). The definitions given in this figure are adapted from Richardson and Pyšek (2006). 'Alien plants are those whose presence in an area is the result of human-mediated transport. Casual alien plants are alien plants that may flourish and even reproduce occasionally outside cultivation but that eventually die out because they do not form self-replacing populations; they rely on repeated introductions for their persistence. Naturalized plants are those aliens that form self-replacing populations for at least 10 years without direct intervention by people (or despite human intervention) by recruitment from seeds or ramets capable of independent growth. Invasive plants are a subset of naturalized plants that produce reproductive offspring, often in large numbers, at considerable distances from parent plants, and thus have the potential to spread over a large area'.

thickets on the edges of forest and covers ~$4 \times 10^6$ ha across Australia (Van Oosterhout *et al.*, 2004). Globally, it infests millions of hectares of grazing land and is of serious concern in 14 major crops including coffee, tea, rice, cotton and sugarcane. Disturbed areas, such as roadsides, railway tracks and canals, are also favourable for the species (Munir, 1996). It does not appear to have an upper temperature or rainfall limit and is often found in tropical areas receiving 3000 mm of rainfall per year. *Lantana* seldom occurs where temperatures frequently fall to <5°C (Cilliers, 1983).

*Lantana* cannot survive under dense and intact canopies of taller native forest species and it is susceptible to frosts, low temperature and saline soils. This plant tends to rot in boggy or hydromorphic soils and is sensitive to aridity (Van Oosterhout *et al.*, 2004).

## Reproductive biology

The plant usually flowers as early as the first growing season after its establishment and, in most places, flowers all year round if adequate moisture and light are available (Gujral and Vasudevan, 1983; Duggin and Gentle, 1998). In cooler or drier regions, flowering occurs only in the warmer or wetter months (Winder, 1980; Swarbrick *et al.*, 1998). Inflorescences are produced in pairs in leaf axils. Wide variations in flower colour, size and shape have been been reported (Van Oosterhout *et al.*, 2004), with flowers opening as yellow and changing to pink, white or red depending on variety; in some forms, a yellow ring is present around the opening of the corollary tube (Sinha and Sharma, 1984).

A visual clue to pollinating insects is provided by the yellow coloration of the flowers, and colour change in the flower may be stimulated after pollination (Mohan Ram and Mathur, 1984). *Lantana* is a major source of nectar for many species of butterflies and moths, which enhance the pollination success of the species. Differences in corolla length, inflorescence diameter and number of flowers at inflorescence attract some butterfly species to visit *Lantana* more frequently than others. According to this view, different varieties of *Lantana* may attract different species of pollinators and, thus, little cross-pollination would occur between species or varieties of *Lantana* in both naturalized (Dronamraju, 1958) and native ranges (Schemske, 1976). Thrips play a more important role in the pollination of *Lantana* than Lepidoptera or butterflies, as the former are present year-round (Sinha and Sharma, 1984; Mathur and Mohan Ram, 1986). Honeybees, sunbirds (India) and humming birds (Brazil) are believed to play a minor role in pollination (Winder, 1980). Goulson and Derwent (2004) advocated that fruit-set is limited by pollinator abundance. They found that substantial populations of *Lantana* in Australia appear to be pollinated by honeybees, with a significant positive correlation between *Lantana* seed-set and honeybee presence. Individual *Lantana* flowers are capable of self-pollination, but need insects for rapid and higher pollination.

Pollination in *Lantana* results in 85% fruit-set, with an Australian study showing that each inflorescence bears around eight fruits (Barrows, 1976); whereas, in India, ~25–28 fruits are produced (Sharma G., 2003, personal observation), showing intraspecific variation (Thorp, 2000). The fruit is a round, fleshy, drupe, ~0.5 cm wide, initially green but turning purple then blue–black. Each drupe contains 1–2 seeds that are ~0.1–0.2 cm long. *Lantana* seeds germinate at any time of the year with sufficient soil moisture, light and warmth (Gentle and Duggin, 1997; Duggin and Gentle, 1998; Parsons and Cuthbertson, 2001). The germination rate of *Lantana* is low (4–45%) due to seed dormancy, low seed viability and/or meiotic instability (Graaff, 1986; Duggin and Gentle, 1998; Sahu and Panda, 1998). These low germination rates are offset by the extremely low rates of seedling mortality experienced in the field (Sahu and Panda, 1998).

## Biological attributes: key factors for invasion

*Lantana* possesses a number of attributes in its life cycle that characterize it as an invader. There is still uncertainty as to which attributes make some species more invasive or what makes some ecosystems more vulnerable to invasion than others (Drake *et al.*, 1989; Lodge, 1993). However, it is always desirable to develop the capability of predicting the invasiveness of an alien plant on the basis of a small number of biological characteristics. Based on the available information, the biological attributes conferring invasiveness for *Lantana* can be summarized in seven ways.

### *Fitness homeostasis and phenotypic plasticity*

The ability of an individual or population to maintain relatively constant fitness over a range of environments is known as homeostatic fitness. Phenotypic plasticity is the ability of a genotype to modify its growth and development in response to changes in the environment (Dorken and Barrett, 2004). Accordingly, plastic responses in vegetative structure are thought to promote survival and propagation of an exotic species in the new heterogeneous environment (Santamaria, 2002). Values such as mean 'relative physiological performance' or mean 'relative ecological performance' across environmental gradients can be used as an alternative for population fitness homoeostasis (Austin *et al.*, 1985). Artificial defoliation of *Lantana* during spring compared with that in autumn produced more stems and resulted in a greater allocation of biomass to reproductive structures. This suggests that the species compensates for defoliation, exhibiting its invasive potential (Broughton, 2003).

### *Interaction with animals*

The success of *Lantana* may be attributed to the presence of a range of pollinators, accounting for the high percentage of fruit-set. Once formed in high numbers, the seeds of *Lantana* are dispersed efficiently through the participation of a variety of animal dispersal agents that feed on its fruit. The process of invasion is further improved by nutrient additions, with animal droppings, canopy removal and soil disturbance creating a good seedbed. Gentle and Duggin (1997) and Duggin and Gentle (1998) demonstrated that physical soil disturbance associated with cattle grazing can also increase resource availability due to the removal of competitive biomass. *Lantana* itself benefits from the destructive foraging activities of vertebrates such as pigs, cattle, goats, horses, sheep and deer, through enhanced vegetative propagation (Thaman, 1974; Fensham *et al.*, 1994).

### *Geographical range*

*Lantana* has a widespread distribution (35°N–35°S) beyond its native range, having becoming naturalized in ~60 countries (Day *et al.*, 2003). The distribution of *Lantana* spp., using the CLIMAX model (Day *et al.*, 2003) and Myers' biodiversity hot spots (Myers *et al.*, 2000), shows substantial overlap, which could indicate a severe threat for ecosystems in hot-spot areas. *Lantana* covers an altitudinal range of up to 2000 m in the Pulnis hills in southern India, but Mathews (1972) could not determine the genotypic and the phenotypic differences in the individuals occurring at different altitudes. The distribution of *Lantana* is still expanding, with many countries and islands that were listed in 1974 as not having *Lantana* (e.g. Galapagos Islands, Solomon Islands, Palau, Saipan, Tinian, Yap and Futuna Island; Thaman, 1974) being infested with *Lantana* more recently (Waterhouse and Norris, 1987). The density of *Lantana* infestations within its native range is also increasing, which has been recognized as an additional threat to ecosystems.

### *Vegetative reproduction*

Once established, the rapid vegetative growth of *Lantana* facilitates the formation of large, impenetrable clumps (Van Oosterhout *et al.*, 2004) and high seed production. The more common means of vegetative spread is through layering, where horizontal stems produce roots when they come in contact with soil (Swarbrick *et al.*, 1998); in addition, suckering can also occur. Prostrate stems can root at the nodes if covered by moist soil, fallen leaves or other debris. In Australia, it is commonly well spread by landholders dumping vegetative material in bushland (Day *et al.*, 2003). *Lantana* stems or leaves could develop roots and grow into plants and eventually flower.

### *Fire tolerance*

Although *Lantana* burns readily during hot, dry conditions, even when green (Gujral and Vasudevan, 1983), moderate and low-intensity fires can promote the persistence and spread of *Lantana* thickets, rather than reducing them. Moreover, the removal of competing neighbourhood plant species and increases in soil nutrients following burning can increase its germination (Gentle and Duggin, 1997; Duggin and Gentle, 1998). Under conditions of increased soil fertility (Duggin and Gentle, 1998), its re-establishment is encouraged following mechanical or chemical control of mature plants.

### *Competitive ability*

Under conditions of high light, soil moisture and soil nutrients, the mortality rate of mature Lantana plants in its naturalized range is very low (Sahu and Panda, 1998).

Lantana infestations are very persistent and, in forest communities, have the potential to block succession and displace native species, resulting in a reduction in biodiversity (Loyn and French, 1991; Duggin and Gentle, 1998). At some sites, Lantana infestations have been so persistent that they impede the regeneration of rainforest (Lamb, 1991). Lantana is a very effective competitor with native colonizers (Duggin and Gentle, 1998) and is capable of interrupting the regeneration processes of other indigenous species by decreasing germination, reducing early growth rates and increasing mortality. This results in marked changes in the structural and floristic composition of natural communities. Therefore, as the density of Lantana in forests increases, species richness decreases (Fensham et al., 1994). Lantana does not invade intact forests, but is found at its margins (Humphries and Stanton, 1992).

#### *Allelopathy*

The allelopathic effect of *Lantana* is well documented and results in severe reductions in seedling recruitment of nearly all species under its cover (Anonymous, 1962). No growth, or only stunted growth, has been observed for other species growing close to *Lantana* due to allelopathic effects, as shown by the fern *Cyclosorus dentatus* Forsk. (Pteridophyta), milkweed vine (*Morrenia odorata* Lindl.; *Asclepiadaceae*), rye (*Lolium multiflorum* Lam.; *Poaceae*) and many crops such as wheat, maize and soybean (Achhireddy and Singh, 1984; Achhireddy *et al*., 1985). *Lantana* contains as many as 14 phenolic compounds that can reduce seed germination and growth in young plants (Jain *et al*., 1989). A series of aromatic alkaloids and phenolics can be extracted from the various plant parts of *Lantana* (Khan *et al*., 2003). Allelochemicals promote or inhibit crop growth based on their concentration (Ambika *et al*., 2003), and the concentration increases from roots, to the stem, to the leaves (Chaudhary and Bhansali, 2002), making the leaf toxic to grazing animals (Ambika *et al*., 2003).

## Factors Influencing the Invasive Success of *L. camara*

Here we provide a qualitative and quantitative overview of the effect of habitat fragmentation on the invasive success of *Lantana* in the dry deciduous forests of India.

### Disturbance

Dry tropical forests are one of the most threatened and least protected ecosystems of the globe (Wright, 2005) and are highly disturbed. These forests supply approximately 90% of the fuel wood and fodder needs of the local population, and thus fuel wood collection may be a major contributor to forest degradation.

Frequent small-scale disturbances, such as sporadic tree felling and lopping of trees, can create localized patches or canopy gaps since such lopped wood piles can easily be removed from the edge of the forest by local inhabitants, this prevailing situation leading to increased canopy gaps at the perimeter. Previous studies opine that disturbance may include structural changes in the community (e.g. amount of biomass removed from or destroyed in disturbed patches) or changes in soil characteristics, or both (Armesto and Pickett, 1985; Gentle and Duggin, 1997; Duggin and Gentle, 1998). Consequent to such disturbances the habitat is then fragmented – the Vindhyan highlands are now a mosaic of disturbed open-canopy and relatively undisturbed closed-canopy patches at several spatial scales (Jha and Singh, 1990). Microclimatic changes near the perimeter, such as increased light, provide greater solar flux (Zuidema *et al*., 1996), which favours conditions for edge species (Laurance, 1991). The soil moisture regime is also altered at the perimeter. These changed microclimatic conditions in fragmented forests favour the establishment of opportunist and exotic species (Kapos, 1989; Laurance, 1991; Chen *et al*., 1992; Malcom, 1994).

The presence of disturbance in the form of canopy openings increases resource availability and also modifies the microclimate, which is consistent with the disturbance patch invasion model (Gentle and Duggin, 1997). This model shows that the removal of competitive biomass and disruption of interspecific competitive interactions creates patches of increased resources. Many exotic weeds benefit from perimeter environments and exert substantial pressures, including competition, on the range of native species (Ries *et al.*, 2004; Harper *et al.*, 2005; Fischer *et al.*, 2006). The success of an invader depends on the disturbance (Rejmanek, 1989) and the community involved (Forcella and Harvey, 1983; Orians, 1986; Hobbs, 1989). Studies in the Vindhyan highlands confirm that canopy openings, resulting from local disturbance, create patches of greater light availability; these canopy openings act as windows for invasive species (Johnstone, 1986), particularly *Lantana* in our case, because it provides a pertinent light regime and suitable moisture level that is lacking in both open and closed forest patches. Tree canopy cover varied from 23 to 65% and *Lantana* cover varied from 1 to 52% from the perimeter to the interior of the forest fragment (Sharma, 2006; see also Raizada *et al.*, 2008). The effect of distance from edge to interior was significant in regard to cover for both tree canopy and *Lantana*; there was a significant negative relation between the two types. At the large local spatial scale (values of all the studied fragments integrated) with increasing tree canopy cover, *Lantana* cover decreased and, in locations having tree canopy cover >63%, no *Lantana* was reported (for data see Sharma, 2006).

## Light and moisture

Dry deciduous forests are heterogenous in terms of disturbed open-canopy and relatively undisturbed closed-canopy patches at several spatial scales (Jha and Singh, 1990). Heterogeneous canopy opening results in a marked heterogeneity in terms of irradiance and temperature. Light availability at the forest floor has been recognized as a key factor influencing the intrinsic traits of inhabiting species (Bazzaz, 1979; Augspurger, 1984; Jones *et al.*, 1994; Kitajima, 1994; Walters and Reich, 1996), including invasive *Lantana* (Sharma, 2006; Raizada *et al.*, 2008). Canopy opening resulted from local disturbance, creating patches of greater light availability (Rejmanek, 1989) that act as an increased resource (Davis *et al.*, 2000). One study in the Vindhyan highlands revealed that the perimeter had 70–85% more light compared with intact forest. Chandrasekaran and Swamy (2002) also reported that light availability under relatively less canopy enhanced the growth of individual *Lantana* species. Maximum and minimum soil moisture content recorded for this field study was 30 and 7.5%, respectively and *Lantana* existed between a critical moisture level of 10 and 25%); in forest areas having soil moisture <10%, no *Lantana* was reported.

Another study suggests that *Lantana* exists only above a critical moisture level (>10%) and that maintenance of moisture above this critical level under field conditions is possible only when stands have a critical level of tree canopy cover (Sharma, 2006). Any decrease in moisture below this critical level may be attributed to large, open patches in the forest that receive considerable light irradiance and also have lower soil organic matter, which may influence the water-holding capacity of the soil. Both these factors could potentially restrict the establishment and colonization of *Lantana* in the intact forest (Sharma, 2006). *Lantana* performs best at intermediate level of shade and this enhances its overall performance, particularly growth rate (Duggin and Gentle, 1998; Islam *et al.*, 2001).

## Ecosystem structure

### *Vegetation*

Subsequent to plant species invasion marked changes in community composition, diversity and functioning occur (D'Antonio and Vitousek, 1992). In Vindhyan dry deciduous forest, diversity and species richness showed a decreasing trend with increasing *Lantana* cover. All species diversity and richness significantly decreased with increasing *Lantana* cover, and a similar trend was followed for herb species diversity and richness. Although tree species diversity was not significantly affected, tree species richness showed a significant decrease with increasing *Lantana* cover. The study in Vindhyan dry deciduous forest showed that Shannon–Weaver diversity ranged from 1.35 to 1.45, all-species richness from 51.6 to 57.0 in size classes <1 ha, 1–10 ha and >10 ha, whereas *Lantana* cover ranged from 29.7 to 37.6% (Sharma, 2006). Maximum *Lantana* cover was reported at medium fragment-size class, indicating that a critical size of fragment is required for the existence of maximum *Lantana* cover and that below the critical fragment size *Lantana* cover begins to decrease (Sharma, 2006).

The results of the above study concur with the experimental model of most studies (Stohlgren, 2002), that with increasing invasive cover diversity decreases. One study of tropical semi-evergreen forest in Bangladesh revealed that invasion by *Lantana* in forest gaps after human perturbations caused an abrupt decline in species richness (Islam *et al.*, 2001). Thus it is possible that the deleterious effects of *Lantana* on the diversity of understorey vegetation (herbaceous and tree species seedlings) are amplified in areas of relatively less tree canopy cover, as *Lantana* cover increases with decreasing tree canopy cover and thus alters the overall micro-niche under its canopy (e.g. light availability, temperature, soil moisture, etc.). We came to such a conclusion – that *Lantana* thrives in less diverse areas, because diversity following invasion by *Lantana* decreased relative to that in control areas where *Lantana* was absent.

Dense cover created by horizontal stratification of Lantana reduces the intensity and duration of light, preventing the establishment of tree species seedlings (Sharma and Raghubanshi, 2006), resulting in marked changes in structural and floristic composition.

*Lantana* canopy certainly lowers the level of sunlight reaching ground level, which may affect soil temperature. An increase in temperature triggers germination by changing the internal enzymatic kinetics and thus the biochemistry of seed cells (Vazquez-Yanes and Orozeo-Segovia, 1982; Khurana and Singh, 2001). However, a decrease in temperature due to *Lantana* canopy could perturb seedling recruitment, and this may possibly be one of the reasons that seedling establishment of the tree species is inhibited.

*Lantana* bushes can burn readily, altering fire regimes in favour ot its persistence (Hiremath and Sundaram, 2005). *Lantana* bushes when burned create considerable heat, causing seed and seedling mortality in the area (Moore and Wein, 1977). Thus, interaction of these factors with biotic pressure might inhibit both seed germination and seedling establishment, which may result in population loss.

There are several ways in which *Lantana* could selectively suppress tree recruitment. Seeds of some tree species become entrapped in the *Lantana* mat, and when few seeds reach the ground there is the problem of scarcity of light. At ground level there is an accumulation of *Lantana* litter and other dry, broken dry *Lantana* debris, which could lead to allelopathic suppression in germination of seeds of tree species (Gentle and Duggin, 1997).

The results of one study in Vindhyan dry deciduous forest suggest not only that many species have a small local population, but also that several exhibit declining or even severely declining population at different levels of *Lantana* invasion. Such invasive species' cover may create demographic instability among the tree species and reduce tree diversity, and could even change the structure of the forest in the short term. Almost all species with a declining population (*Acacia auriculiformis*, *Adina cordifolia*, *Boswellia serrata*, *Briedelia retusa*, *Buchanania lanzan*, *Cassia fistula*, *Elaeodendron glaucum*, *Emblica officinalis*, *Eriolena quinquelaris*, *Hardwickia binata*, *Miliusa tomentosa*, *Schrebera swietenioides*) require high to moderate light for proper growth (Troup, 1921). The growth pattern of *Lantana* restricts the establishment of tree seedlings, as most of the tree species in the tropical dry deciduous forest require high light levels (Troup, 1921). Light has long been recognized as an important plant resource (Maximov, 1929; Blankenship, 2002) that may interact with other plant resources to affect plant performance (Cole, 2003). Below certain thresholds, however, light limitation alone can prevent seedling survival regardless of other resource levels (Tilman, 1982). This is due to the creation of a photosynthetically inactive light regime at ground level (Fetcher *et al.*, 1983; Turton and Duff, 1992). It is likely that herbs and tree species seedlings are influenced by the level of light that reaches the forest floor, and this may be one of the mechanisms responsible for the decline of herbaceous and tree species seedlings in patches invaded by *Lantana*.

Seedlings of most tree species of the tropical dry deciduous forest are adapted to grow in relatively open conditions, because of the poor canopy cover and deciduous nature, but *Lantana*, requiring the same light regime and possessing allelochemicals, out-competes native species. The presence of *Lantana* shrubs as a dense understorey perturbs the seedling recruitment of native tree species in the forest, leading to differential depletion of native trees. A decline of 42 and 60% in species number was observed during the first (2002) and second (2003) census, respectively, in the Vindhyan highlands (Sharma, 2006; Sharma and Raghubanshi, 2007).

The growth architectural pattern of *Lantana* alters the light regime and microhabitat in its vicinity by the addition of wood debris and litter to the ground, which is an important factor controlling herbaceous layer species composition (Roberts, 2004). Excessive levels of leaf litter deposition make the site unfavourable for herbaceous layer species (Beatty and Sholes, 1988; Roberts and Gilliam, 1995). These types of perturbation lead to damage in pre-existing plants (Halpern and Spies, 1995), alter propagule availability (Stickney, 1990) and also limit recruitment due to lack of suitable habitats. *Lantana*, due to its strong allelopathic properties, has the potential to interrupt regeneration process of other species by decreasing germination, reducing early growth rates and selectively increasing mortality of other plant species (Sharma *et al.*, 2005a, b). These result in a reduction of species diversity and cover (Loyn and French, 1991; Duggin and Gentle, 1998).

Species such as *Anagallis arvensis*, *Barleria cristata*, *Dichanthium annulatum*, *Physalis minima*, *Setaria pumila* and *Sporobolus diander*, with limited distribution as *Lantana* cover increases, can be more susceptible to loss from physical damage or altered habitat conditions (Meier *et al.*, 1995; Sharma, 2006; Sharma and Raghubanshi, 2007). Species such as *Dichanthium annulatum*, *Evolvulus alsinoides*, *Leucas aspera*, *Malvastrum tricuspidatum*, *Rungia pectinata*, *Sida acuta*, *Sida cordifolia*, *Sida rhombifolia* and *Tephrosia purpurea* that decrease and ultimately go locally extinct with increasing levels of *Lantana* cover may not recover quickly because of the altered physical environmental conditions (Sharma, 2006; Sharma and Raghubanshi, 2007). Species such as *Alysicarpus vaginalis*, *Andrographis echioides*, *Begonia picta*, *Ceropegia bulbosa*, *Cyanotis axillaris*, *Justicia simplex* and *Lindernia ciliata*, which are specific to sites devoid of *Lantana* cover, are native species and are highly sensitive to environmental perturbations, and they become completely eliminated after *Lantana* invasion (Sharma, 2006; Sharma and Raghubanshi, 2007).

## Ecosystem function

Alteration of natural species composition creates opportunities for plant invasion, which lead further to altered community structure. Changing structure due to invasive species will contribute to changed ecosystem processes; invasive species may contribute to modification of ecosystem processes by providing a positive feedback that enhances their spread (D'Antonio and Vitousek, 1992). Explanations in regard to the possibility that the complex interaction between plants and soil that cause soil properties to change in response to changes in plant species may influence invasivness (Ehrenfeld *et al.*, 2001). Changes/shifts in plant communities composition resulting from exotic invasion are likely to be associated with changes in soil properties, as these are associated with natural succession processes (Ehrenfeld *et al.*, 2001).

The changing plant community will probably alter the quantity and quality of litter inputs and other soil properties (Bohlen, 2006), which can affect the system by altering rates of litter decomposition and the accumulation and storage of organic matter, as well as nutrient release rates from the decomposing litter, possibly facilitating a feedback system for continued invasion (Kourtev *et al.*, 2003; Wardle and Bardgett, 2004).

In Vindhyan forest it was observed that accumulation of litter beneath *Lantana* canopies builds up soil organic matter. Accumulation of soil N closely follows that of soil organic matter because, on average, 99% of N in terrestrial ecosystems is organically bound (Rosswall, 1976). Raghubanshi (1992) reported a strong positive relation between total N and organic C contents of soil in the dry deciduous forest ecosystem. Several studies have shown that soil nutrient levels play an important role in determining community invasibility (Shea and Chesson, 2002; Reinhart and Callaway, 2006). This self-perpetuating altered microhabitat could probably provide *Lantana* with increased resources, leading to its successful proliferation. Thus *Lantana* changes microhabitat conditions, which could probably help towards its successful proliferation. This dynamic alteration of stand structure and microhabitat leads to changed ecosystem processes.

Despite the ubiquity of the plant-mediated changes in soil properties, there has been little research documenting such effects following exotic invasions. Vitousek (1990) first pointed out that exotic species could alter soil processes. However, changes in soil biogeochemistry following shifts in species composition could be another pathway of change. This alternative mechanism would act indirectly via the effects of the introduced species on the soil biota and/or on soil physical conditions, rather than directly through the traits of the invading species. Either mechanism could allow introduced species to create a feedback system (Wilson and Agnew, 1992).

If changes in soil ecology either accelerate its own growth or promote its competitive superiority to native species, a positive feedback system could develop that promoted the spread of the exotic species. According to this theory, the establishment of an initial population of the exotic species would begin a process of changing structure and function of the soil biota, which would promote the spread of the exotic and/or the competitive reduction of native species population. Either mechanism would enhance the spread of the exotic species. Thus, changes in soil ecology induced by the interaction of the exotic species and the soil could provide an explanation for the rapid increase in the cover/density of invasive *Lantana*.

In Vindhyan forests the N-mineralization rate significantly altered with increasing *Lantana* cover. Similar difference in soil processes and invasion have also been reported (Vinton and Burke, 1995; Asner and Beatty, 1996; Kourtev *et al.*, 1998). Pooled data sets from different locations in the fragment also revealed that ammonification, nitrification and N-mineralization were enhanced with increasing *Lantana* cover (Sharma and Raghubanshi, 2009a). In a study carried out in short-grass steppe, higher N-mineralization rates were found in soil beneath the invasive species *Kochia scoparia* and *Melinis minutiflora* (Vinton and Burke, 1995; Asner and Beatty, 1996). Kourtev *et al.* (1998) also reported increased soil nitrification rates associated with the exotic species *Berberis thunbergii* and *Microstegium vimineum* in the deciduous forest of New Jersey, USA.

The results of one study in Vindhyan dry deciduous forest suggest that *Lantana* cover alters the soil nitrogen dynamics successively with increasing cover. Higher turnover rates of *Lantana* litter, which reflect the rate of nutrient cycling, were reported for *Lantana* in oak forests (Rawat and Singh, 1988). Rawat *et al.* (1994) also showed that high turnover rates, high rates of litter decomposition and efficient translocation of nutrients result in high production values for *Lantana*. The changes in decomposition and nutrient cycling associated with *Lantana* could have positive impacts on nutrient availability that can, in turn, lead further to greater decomposition in a system invaded by an exotic species (Hobbie and Vitousek, 2000).

Rawat *et al.* (1994) reported that the nitrogen content of *Lantana* is higher than for other native species within its habitat. Invasive species often maintain higher concentration of leaf nitrogen (Vituousek *et al.*, 1987; Nagel and Griffin, 2001; Ashton *et al.*, 2005) and, consequently, are expected to decompose more rapidly and release more nitrogen to the soil than native species.

Thus, with increasing *Lantana* cover the levels of litter increase, escalating nutrient availability and rate. Although, when the return of nitrogen from litter decay from invasive species to the soil pool is greater than that from indigenous plants, nitrogen availability at the soil surface may increase and increase the rates of nutrient cycling in invaded areas (Vitousek and Walker, 1989; Witkowski, 1991). Furthermore, this self-perpetuating nutrient availability due to fast N-mineralization results in greater availability of nutrients, subsequently leading to the dense proliferation of *Lantana*. Levine *et al.* (2006) also advocated this concept and proposed the 'push and pull' theory of invasion on the basis of such spread.

As *Lantana* is a fast-growing invader (Rawat *et al.*, 1994), it will capture more resources than slow-growing species, resulting in a positive feedback system of nutrient availability, which will be enhanced with increasing *Lantana* cover. It has been suggested that the combined influence of increased nitrogen and changes in light regimes associated with invasive species in grassland can facilitate the further invasion and encroachment of such invasive species (Siemann and Rogers, 2003).

In conclusion, there is: (i) a general increase in *Lantana* cover with increasing fragmentation; and (ii) an inverse relationship between tree canopy cover and *Lantana* cover. Maximum relative growth rate for *Lantana* was found under low shade, corresponding to conditions at the forest perimeter. *Lantana* invasion in fragmented forest may create demographic instability among tree species, resulting in reduced tree diversity. The presence of *Lantana* in dry deciduous forests alters the spatial pattern of herbaceous vegetation, as *Lantana* alters the microhabitat. *Lantana* alters light and nutrient availability, which may favour its self-perpetuation over the regeneration of other species. Thus, *Lantana* invasion could change the structure of the forest in the short term (Sharma and Raghubanshi, 2009ab). More extensive *Lantana* cover also increases nutrient availability due to higher rates of soil N-mineralization, which leads to a denser proliferation of *Lantana* as other vegetation cannot compete with its rate of growth (Fig. 3.4; Sharma and Raghubanshi, 2009b).

In conclusion, invasion of *Lantana* in the dry deciduous forest is changing forest structure, resulting in a feedback system that accelerates the spread of *Lantana* by promoting its competitive superiority over native species. This, in turn, is leading to species diversity loss and the creation of a homogeneous, monospecific *Lantana*-invaded understory in the forest (Sharma and Raghubanshi, 2009a,b).

## Management Options

Attempts to halt or reduce invasions can be divided into three stages: (i) keeping them out; (ii) if they get in, finding and trying to eradicate them quickly; and (iii) if they can not be eradicated, managing them at low levels (Simberloff *et al.*, 2005). Higgins *et al.* (2000) scaled up a spatially explicit individually based model that predicted rates and pattern of spread of alien plants to a landscape simulation model, which permitted integration of the spatial demography of an alien plant population

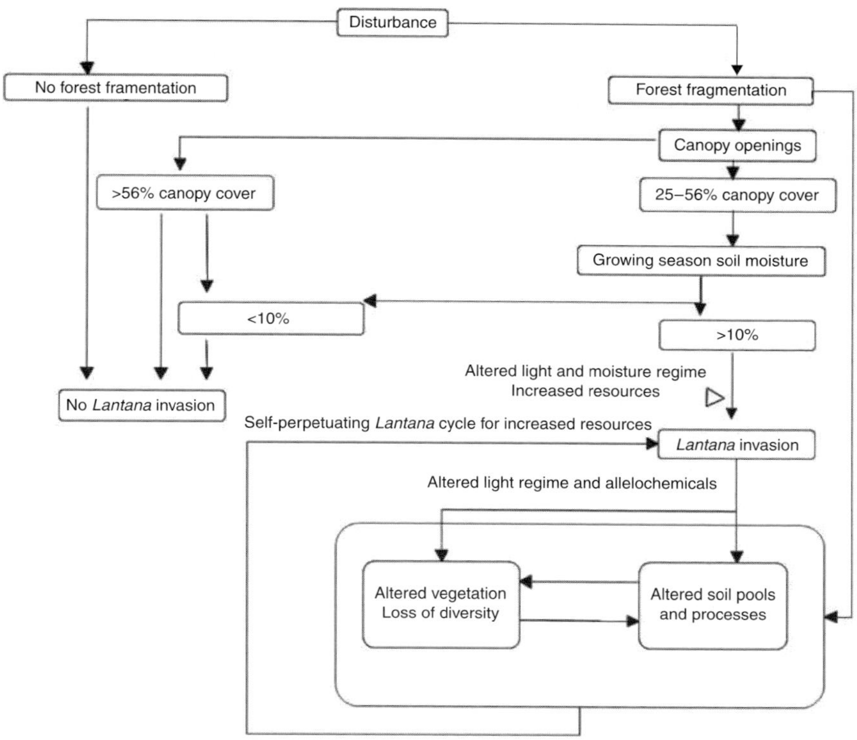


**Fig. 3.4.** Schematic depiction of the proposed effect of *Lantana camara* invasion on ecosystem structure and function in the dry deciduous forests of India (Sharma, 2006).

with information on the structure and dynamics of the ecosystems being invaded. The results demonstrated that the most rapid and cost-effective management strategy for controlling invasive plants would be first to clear low-density stands of juvenile plants, followed by higher-density stands of juvenile plants and then high-density stands of adult plants.

The distribution of *Lantana* is limited by its following environmental attributes: (i) inability to survive under dense, intact canopies of taller native forest species; (ii) low tolerance to saline soils; (iii) tendency to rot in boggy or hydromorphic soils; and (iv) high sensitivity to aridity. Shading by intact canopies is an effective barrier against *Lantana* invasion and is the most appropriate strategy for managing it (Duggin and Gentle, 1998). This strategy will also play an important role in maintaining conservation values and will also provide further opportunities for the expansion and development of forest areas. Apart from managing and manipulating ecosystems, manual and mechanical removal, along with chemical and biological control options, have been much explored for the control of *Lantana*. All these control strategies, however, have associated drawbacks. Manual removal is labour intensive and a low-efficiency technique, while mechanical control is inefficient in dealing with very extensive invasions and is also difficult in undulating, rocky terrain. Chemical control involves the use of inorganic/organic herbicides, and a serious disadvantage is the high cost of most chemical control programmes. Safety for other plant species and the environment is of vital importance when using herbicides to control invasive species.

Biological control of *Lantana* began in 1902 and, since then, 41 agents have been released in some 50 countries (Julien and Griffiths, 1998) but still no biocontrol agent has completely stopped its infestation. The success of biological control programme is often not clear-cut, because complete control is achieved only in certain years and/or at some locations (Syrett *et al.*, 2000). In India the biocontrol agent *Teleonemia scrupulosa* Stal. released for *Lantana* control failed, since it could not cope with the vigorous regrowth of *Lantana* at the onset of the monsoon rains or the control agent itself suffered was largely destroyed during the winter months (Sharma, 1988). According to Mathur and Mohan Ram (1986), the introduction of biocontrol agents to reduce thrips populations is an attempt to decrease pollination and the quantity of seed produced by *Lantana*. Furthermore, Goulson and Derwent (2004) argued that cost-effective control of *Lantana* could be achieved through the judicious employment of honeybees.

One of the best ways to minimize the impact of *Lantana* on the ecosystem is to exploit its biomass so that its presence can be reduced significantly while providing economic benefits. Research shows that *Lantana* extracts have broad-spectrum biocidal potential. Whole- or part-plant extracts have shown antibacterial (Sukul and Chaudhuri, 2001), nematicidal (Shaukat *et al.*, 2003), insecticidal (Ogendo *et al.*, 2003), anti-feedant, growth-inhibitory (Sharma *et al.*, 2003), molluscicidal (Pal *et al.*, 2002), mosquito-repellent (Ogendo *et al.*, 2003), fungitoxic (Gehlot, 2002) and larvicidal (Fatope *et al.*, 2002) properties. Essential oils from *Lantana* can be exploited for insect control in stored products (Bouda *et al.*, 2001). Phyto-extracts from *Lantana* completely inhibited the sporulation of *Alternaria alternata* (Fr.) Kiessl, which causes early potato blight (Chaudhary *et al.*, 2001). *Lantana* also suppressed the growth of another invasive weed, water hyacinth (Saxena, 2000). *Lantana* biomass can also be exploited to increase the nutrient status of the soil (Sharma *et al.*, 2003), and a substantial amount of *Lantana* biomass could be used in biogas production (Saini *et al.*, 2003). Apart from the use of plant extracts for biological control, *Lantana* stems can also be used in the making of baskets, waste receptacles, flowerpots, fruit plates (ATREE, 2003), laying roofs in village huts, and also hedging and furniture. *Lantana* products could attract the urban market with an eco-friendly label that could be utilized by corporate, public and welfare organizations (ATREE, 2003); they could be utilized as an excellent alternative to plastic and bamboo, as they have the potential to alleviate harvesting pressure on bamboo. Proper documentation and extension work is needed to enhance the use of *Lantana* plant parts by establishing a centre for the development of a sustainable market for *Lantana* products, so that the species may be controlled while providing economic benefits to those involved in its eradication.

## Concluding Remarks

Valuable progress has been made within the last two decades in reducing the risks from invasive species. International efforts, conducted under the Scientific Committee on the Problems of the Environment (SCOPE), have provided an international perspective and local initiatives to cope with the problems of invasion. The ecology of biological invasion is a challenging area of study, identified by conservation agencies around the world, that needs to be addressed in regard to biodiversity conservation. *Lantana* must be considered as a weed of national significance since it has widespread distribution, and substantial impact on agriculture and biodiversity; however, there is a scarcity of knowledge on the extent of *Lantana* invasion and its impacts. Ecosystem-level consequences of *Lantana* invasion are little understood and there is an urgent need for studies on biological invasion on this and other invasive species in India. A national strategy to identify and catalogue invasive species such as *Lantana* must be developed in consultation with representatives from local and state conservation agencies and community groups (non-governmental organizations engaged in environmental conservation strategies). A national strategy must incorporate the following components.

**1.** Impact minimization.

- Develop best practice guidelines.
- Develop effective biological control agents.
- Investigate alternative control methods.
- Identify strategic management control areas.
- Areas not yet invaded should be seen as priority sites for good management to prevent invasions.

**2.** Increase in community awareness.

- Gain support from locals for eradication.
- Foster a culture of information sharing among states at local and regional levels.
- Take a holistic view to target the whole community that could take into account ongoing and long-term development of strategies.

**3.** Prevention of spread.

- Determine the level and extent of infestation in different regions.
- Check for contamination in imported goods.
- Quickly report new infestations.
- Investigate further the ecology and biology of *Lantana*.
- Investigate the dynamics of weed spread.

**4.** Management coordination.

- Develop management strategies incorporating cooperation at all levels.
- Implement strategies at the right time and in the right place.
- Urge cooperation to maximize use of resources.

Since species invasion is an international problem, cooperation from other countries/ agencies is necessary for the successful implementation of management and action plans (see also Rashid *et al.*, 2009). In particular, the involvement of neighbouring countries is essential in order to check trans-boundary movement of invasive species, and also to develop a common strategy to control invasive species such as *Lantana*.

## Acknowledgements

The data discussed and collated here are from research carried by G.P. Sharma for his PhD thesis; he gratefully acknowledges the Council of Scientific and Industrial Research (CSIR), New Delhi, India for their research fellowship and the University of Delhi for funding support. We thank Prof. J.S. Singh for his critical comments and suggestions over the entire course of this study.

## References

Achhireddy, N.R. and Singh, M. (1984) Allelopathic effects of Lantana (*Lantana camara*) on milkweed vine (*Morrenia odorata*). *Weed Science* 32, 757–761.

Achhireddy, N.R., Singh, M., Achhireddy, L.L., Nigg, H.N. and Nagy, S. (1985) Isolation and partial characterisation of phytotoxic compounds from Lantana (*Lantana camara* L.). *Journal of Chemical Ecology* 11, 979–988.

Ambika, S.R., Poornima, S., Palaniraj, R., Sati, S.C. and Narwal, S.S. (2003) Allelopathic plants. 10. *Lantana camara* L. *Allelopathy Journal* 12, 147–161.

Aneja, K.R., Dhawan, S.R. and Sharma, A.B. (1991) Deadly weed, *Parthenium hysterophorus* L. and its distribution. *Indian Journal of Weed Science* 23, 14–18.

Anonymous (1962) *Lantana* Linn. (Verbenaceae). In: Sastri, B.N. (ed.) *The Wealth of India: A Dictionary of Raw Materials and Industrial Products*, vol. VI. Council of Scientific and Industrial Research, New Delhi, India, pp. 31–34.

Armesto, J.J. and Pickett, S.T.A (1985) Experiments on distribution in old-field plant communities: Impact on species richness and abundance. *Ecology* 66, 230–240.

Ashton, I.W., Hyatt, L.A., Howe, K.M., Gurevitch, J. and Lerdau, M. (2005) Invasive species accelerate decomposition and litter nitrogen loss in a mixed deciduous forest. *Ecological Applications* 15, 1263–1272.

Asner, G.P. and Beatty, S.W. (1996) Effects of an African grass invasion on Hawaiian shrubland nitrogen biogeochemistry. *Plant and Soil* 186, 205–211.

ATREE (2003) Ashoka Trust for Research in Ecology and Environment. Retrieved from www.atree.org, 28 November 2004.

Augspurger, C.K. (1984) Light requirements of neotropical tree seedlings: a comparative study of growth and survival. *Journal of Ecology* 72, 777–795.

Austin, D.F. (1978) Exotic plants and their effects in southeastern Florida. *Environmental Conservation* 5, 25–34.

Austin, M.P., Groves, R.H., Fresco, M.F. and Kaye, P.E. (1985) Relative growth of six thistle species along a nutrient gradient with multispecies competition. *Journal of Ecology* 73, 667–684.

Barrows, E.M. (1976) Nectar robbing and pollination of *Lantana camara* (Verbenaceae). *Biotropica* 8, 132–135.

Batianoff, G.N. and Butler, D.W. (2003) Impact assessment and analysis of sixty-six priority invasive weeds in southeast Queensland. *Plant Protection Quarantine* 18, 11–17.

Bazzaz, F.A. (1979) The physiological ecology of plant succession. *Annual Review of Ecology and Systematics* 10, 351–371.

Beatty, S.W. and Sholes, O.D.V. (1998) Leaf litter effects on plant species composition of deciduous forest treefall pits. *Canadian Journal of Forest Research* 18, 553–559.

Blankenship, R.E. (2002) *Molecular Mechanisms of Photosynthesis*. Blackwell Science, Oxford, UK.

Bohlen, P.J. (2006) Biological invasions: linking the above ground and the below ground processes. *Applied Soil Ecology* 32, 1–5.

Bouda, H., Tapondjou, L.A., Fontem, D.A. and Gumedoz, M.Y.D. (2001) Effects of essential oils from leaves of *Ageratum conyzoides*, *Lantana camara* and *Chromolena odorata* on the mortality of *Sitopphilus zeamais* (Coleoptera: Curculionidae). *Journal of Stored Product Research* 37, 103–109.

Braithwaite, R.W., Lonsdale, W.M. and Estbergs, J.A. (1989) Alien vegetation and native biota in tropical Australia: the impact of *Mimosa pigra*. *Biological Conservation* 48, 189–210.

Broughton, S. (2003) Effect of artificial defoliation on the growth and biomass of *Lantana camara* L. (Verbenaceae). *Plant Protection Quarantine* 18, 110–115.

Carlton, J.T. (1996) Biological invasions and cryptogenic species. *Ecology* 77, 1653–1655.

Chandrasekaran, S. and Swamy, P.S. (2002) Biomass, litterfall and above ground net primary productivity of herbaceous communities in varied ecosystems at Kodayar in the western ghats of Tamilnadu. *Agriculture, Ecosystems and Environment* 88, 61–71.

Chaudhary, B.L. and Bhansali, E. (2002) Effect of different concentrations of *Lantana camara* Linn. extract on spore germination of *Physcomitrium japonicum* Hedw. in half Knop's liquid medium and double distilled water. *Research Bulletin Punjab University Science* 52, 161–165.

Chaudhary, R.F., Patel, R.L. and Chaudhary, S.M. (2001) In vitro evaluation of different plant extracts against *Alternaria alternata* causing early blight of potato. *Journal of the Indian Potato Association* 30, 141–142.

Chen, J., Franklin, J.E. and Spies, T.A. (1992) Vegetation responses to edge environments in old-growth Douglas-fir forest. *Ecological Applications* 2, 387–396.

Cilliers, C.J. (1983) The weed, *Lantana camara* L., and the insect natural enemies imported for its biological control into South Africa. *Journal of the Entomological Society of South Africa* 46, 131–138.

Coile, N.C. and Shilling, D.G. (1993) Cogongrass, *Imperata cylindrica* (L.) Beauv.: A good grass gone bad. Botanical Circular 28, Florida Department of Agricultural and Consumer Services, Division of Plant Industry, Tallahassee, Florida, p. 3.

Cole, P.G. (2003) Environmental constraints on the distribution of the non-native invasive grass, *Microstegium vimineum*. PhD thesis, Department of Ecology and Evolutionary Biology, University of Tennessee, Knoxville, Tennessee.

Corbett, J. (2009) Workshop on invasive species, *International Conference on Invasive Management*, Punjab University, India, 2009.

D'Antonio, C.M. and Vitousek, P.M. (1992) Biological invasion by exotic grasses, the grass/fire cycle and global change. *Annual Review of Ecology and Systematics* 23, 63–87.

Davis, M.A., Grime, J.P. and Thompson, K. (2000) Fluctuating resources in plant communities: a general theory of invasibility. *Journal of Ecology* 88, 528–534.

Day, M., Wiley, C.J., Playford, J. and Zalucki, M.P. (2003) *Lantana: Current Management Status and Future Prospects*. ACIAR, Canberra, ACT, Australia.

Dorken, M.E. and Barrett, S.C.H. (2004) Phenotypic plasticity of vegetative and reproductive traits in monoecious and dioecious populations of *Sagittaria latifolia* (Alismataceae): a clonal aquatic plant. *Journal of Ecology* 92, 32–44.

Drake, J.A., di Castri, F., Grooves, R.H., Druger, F.J., Mooney, H.A., Rejmanek M. *et al.* (1989) *Biological Invasion: a Global Perspective*. John Wiley and Sons, Chichester, UK.

Dronamraju, K.R. (1958) The visits of insects to different coloured flowers of *Lantana camara* L. *Current Science* 11, 452–453.

Duggin, J.A. and Gentle, C.B. (1998) Experimental evidence on the importance of disturbance intensity for invasion of *Lantana camara* L. in dry rainforest–open forest ecotones in north-eastern NSW, Australia. *Forest Ecology and Management* 109, 279–292.

Ehrenfeld, J.G., Kourtev, P. and Huang, W. (2001) Changes in soil functions following invasions of exotic understorey plants in deciduous forests. *Ecological Applications* 11, 1287–1300.

Eiten, G. and Goodland, R. (1979) Ecology and management of semi arid ecosystem in Brazil. In: Walker, B.H. (ed.) *Management of Semi Arid Ecosystems*. Elsevier, Amsterdam, pp. 277–300.

Fatope, M.O., Salihu, L., Asante, S.K. and Takeda, Y. (2002) Larvicidal activity of extracts and triterpenoids from *Lantana camara*. *Pharmaceutical Biology* 40, 564–567.

Fensham, R.J., Fairfax, R.J. and Cannell, R.J. (1994) The invasion of *Lantana camara* L. in Forty Mile Scrub National Park, North Queensland. *Australian Journal of Ecology* 19, 297–305.

Fetcher, N., Strain, B.R. and Oberbauer, S.F. (1983) Effects of light regime on the growth, leaf morphology, and water relations of seedlings of two species of tropical trees. *Oecologia* 58, 314–319.

Fischer, J., Lindenmayer, D.B. and Manning, A.D. (2006) Biodiversity, ecosystem functioning and resilience: ten guiding principles for commodity production landscapes. *Frontiers in Ecology and Environment* 4, 80–86.

Forcella, F. and Harvey, S.J. (1983) Eurasian weed infestation in Western Montana in relation to vegetation and disturbance. *Madrono* 30, 102–109.

Fritts, T.H. and Rodda, G.H. (1998) The role of introduced species in the degradation of island ecosystems: a case study of Guam. *Annual Review of Ecology and Systematics* 29, 113–140.

Gehlot, P. (2002) Fungitoxic properties of *Lantana camara* L. plant extract against some phytopathogenic fungi. *Geobios* 29, 187–188.

Gentle, C.B. and Duggin, J.A. (1997) *Lantana camara* L. invasions in dry rainforest–open forest ecotones: the role of disturbances associated with fire and cattle grazing. *Australian Journal of Ecology* 22, 298–306.

GISP (2001) Global strategy on invasive alien species. In: *Global Invasive Species Programme Workshop*, September 2000, Cape Town, South Africa.

GISP (2003) The IAS problem. In: *The Global Invasive Species Programme*, http://www.gisp.org/about/IAS.asp (accessed 2 August 2005).

Goulson, D. and Derwent, L.C. (2004) Synergistic interactions between exotic honeybee and exotic weed: pollination of *Lantana camara* in Australia. *Weed Research* 44, 195–202.

Graaff, J.L. (1986) *Lantana camara*, the plant and some methods for its control. *South African Forest Journal* 13, 26–30.

Gujral, G.S. and Vasudevan, P. (1983) *Lantana camara* L., a problem weed. *Journal of Scientific and Industrial Research* 42, 281–286.

Gurevitch, J. and Padilla, D.K. (2004) Are invasive species a major cause of extinctions? *Trends in Ecology and Evolution* 19, 470–474.

Halpern, C.B. and Spies, T.A. (1995) Plant species diversity in natural and managed forests of the Pacific Northwest. *Ecological Applications* 5, 913–934.

Harper, K.A., Macdonald, S.E., Burton, P.J. *et al.* (2005) Edge influence on forest structure and composition in fragmented landscapes. *Conservation Biology* 19, 768–782.

Higgins, S.I., Richardson, D.M. and Cowling, R.M. (2000) Using a dynamic landscape model for planning the management of alien plant invasions. *Ecological Applications* 10, 1833–1848.

Hiremath, A.J. and Sundaram, B. (2005) The fire–Lantana cycle hypothesis in Indian forests. *Conservation and Society* 3, 26–42.

Hobbie, S.E. and Vitousek, P.M. (2000) Nutrient limitation of decomposition in Hawaiian forests. *Ecology* 81, 1867–1877.

Hobbs, R.J. (1989) The nature and effects of disturbance relative to invasion. In: Drake, J.A. *et al.* (eds) *Biological Invasions: Global Perspective*. John Wiley, Chichester, UK, pp. 389–405.

Humphries, S.E. and Stanton, J.P. (1992) Weed assessment in the wet tropics world heritage area of North Queensland. Report. Available from URL: http://www.hear.org/pier/references/pierref000096.htm (accessed 23 November 2004).

Islam, K.R., Ahmed, M.R., Bhuiyan, M.K. and Badruddin, A. (2001) Deforestation effects on vegetative regeneration and soil quality in tropical semi-evergreen degraded and protected forests of Bangladesh. *Land Degradation and Development* 12, 45–56.

Jain, R., Singh, M. and Dezman, D.J. (1989) Qualitative and quantitative characterization of phenolic compounds from Lantana (*Lantana camara*) leaves. *Weed Science* 37, 302–307.

Jha, C.S. and Singh, J.S. (1990) Composition and dynamics of dry tropical forest in relation to soil texture. *Journal of Vegetation Science* 1, 609–614.

Johnstone, I.M. (1986) Plant invasion windows: a time based classification of invasion potential. *Biological Review* 61, 369–394.

Jones, R.H., Sharitz, R.R., Dixon, P.M., Segal, D.S. and Schneider, R.L. (1994) Woody plant regeneration in four floodplain forests. *Ecological Monographs* 64, 345–367.

Julien, M.H. and Griffiths, M.W. (1998) *Biological Control of Weeds. A World Catalogue of Agents and their Target Weeds*, 4th edn. CABI Publishing, Wallingford, UK.

Kapos, V. (1989) Effects of isolation on water status of forest patches in Brazilian Amazon. *Journal of Tropical Ecology* 5, 173–185.

Khan, M., Srivastava, S.K., Jain, N., Syamasundar, K.V. and Yadav, A.K. (2003) A chemical composition of fruit and stem essential oil of *Lantana camara* from northern India. *Flavour and Fragrance Journal* 18, 376–379.

Khurana, E. and Singh, J.S. (2001) Ecology of tree seed and seedlings: implication for tropical forest conservation and restoration. *Current Science* 80, 748–757.

Kitajima, K. (1994) Relative importance of photosynthetic traits and allocation patterns as correlates of seedling shade tolerance of 13 tropical trees. *Oecologia* 98, 419–428.

Kourtev, P.S., Ehrenfeld, J.G. and Huang, W.Z. (1998) Effects of exotic plant species on soil properties in hardwood forests of New Jersey. *Water, Air and Soil Pollution* 105, 493–501.

Kourtev, P.S., Ehrenfeld, J.G. and Haggblom, M. (2003) Experimental analysis of the effect of exotic and native plant species on the structure and function of soil microbial communities. *Soil Biology and Biochemistry* 35, 895–905.

Lamb, D. (1991) Forest regeneration research for reserve management: some questions deserving answers. In: Goudberg, N., Bonell, M. and Benzaken, D. (eds) *Tropical Rainforest Research in Australia: Present Status and Future Directions for the Institute for Tropical Rainforest Studies.* Institute for Tropical Rainforest Studies, Townsville, QLD, Australia, pp. 177–181.

Laurance, W.F. (1991) Edge effects in tropical forest fragments: application of a model for design of nature reserves. *Biological Conservation* 57, 205–219.

Levine, J.M., Pachepsky, E., Kendall, B.E., Yelenik, S.G. and Lambers, J.H.R. (2006) Plant–soil feedbacks and invasive spread. *Ecology Letters* 9, 1005–1014.

Levine, S.A. (1989) Analysis of risk for invasions and control program. In: Drake, J.A. *et al.* (eds) *Biological Invasion: a Global Perspective*, SCOPE 37. John Wiley, Chichester, UK, pp. 425–435.

Lodge, D.M. (1993) Biological invasions: lessons for ecology. *Trends in Ecology and Evolution* 8, 133–137.

Lonsdale, W.M. (1994) Inviting trouble: Introduced pasture species in Northern Australia. *Australian Journal of Ecology* 19, 345–354.

Loyn, R.H. and French, K. (1991) Birds and environmental weeds in southeastern Australia. *Plant Protection Quarantine* 6, 137–149.

Mack, R.N., Simberloff, D., Lonsdale, W.M., Evans, H., Clout, M. and Bazzaz, F.A. (2000) Biotic invasions: Causes, epidemiology, global consequences and control. *Ecological Applications* 10, 689–710.

Malcom, J.R. (1994) Edge effects in Central Amazonian Forest fragments. *Ecology* 75, 2438–2445.

Mathews, K.M. (1972) The high altitude ecology of *Lantana*. *Indian Forester* 97, 170–171.

Mathur, G. and Mohan Ram, H.Y. (1986) Floral biology and pollination of *Lantana camara*. *Phytomorphology* 36, 79–100.

Maximov, N.A. (1929) *The Plant in Relation to Water*. George Allen and Unwin, London.

Meier, A.J., Bratton, S.P. and Duffy, D.C. (1995) Possible ecological mechanisms for loss of vernal-herb diversity in logged eastern deciduous forests. *Ecological Applications* 5, 935–946.

Miconia (2001) The green cancer spreads. *Environment Hawaii, 2001*, www.environment-hawaii.org/301emma.htm

Mohan Ram, H.Y. and Mathur, G. (1984) Flower colour changes in *Lantana camara*. *Journal of Experimental Botany* 35, 1656–1662.

Moore, J.M. and Wein, R.W. (1977) Viable seed

population by soil depth and potential site recolonization after disturbance. *Canadian Journal of Botany* 55, 2408–2412.

Munir, A.A. (1996) A taxonomic review of *Lantana camara* L. and *L. montevidensis* (Spreng.) Briq. (Verbenaceae) in Australia. *Journal of Adelaide Botanical Garden* 17, 1–27.

Myers, N., Mittermeier, R.A., Mittermeier, C.G., Da Fonseca, G.A.B. and Kent, J. (2000) Biodiversity hotspots for conservation priorities. *Nature* 403, 853–858.

Myers, R.L. and Evel, J.J. (1990) Problems, prospects, and strategies for conservation. In: Myers, R.L. and Ewel, J.J. (eds) *Ecosystems of Florida*. University Press of Florida, Gainesville, Florida, pp. 619–632.

Nagel, J.M. and Griffin, K.L. (2001) Construction cost and invasive potential: comparing *Lythrum salicaria* (Lathyraceae) with co-occuring native species along pond banks. *American Journal of Botany* 88, 2252–2258.

Nayar, M.P. (1977) Changing patterns of the Indian flora. *Bulletin of Botanical Survey of India* 19, 145–154.

New Zealand Department of Statistics (1996) *New Zealand Official Yearbook*. Wellington, New Zealand, 99th edn.

Ogendo, J.O., Belmain, S.R., Deng, A.L. and Walker, D.J. (2003) Comparison of toxic effects of *Lantana camara* L. with *Tephrosia vogelii* Hook and synthetic pesticide against *Sitophilus zeamais* Motschulsky (Coleoptera: Curculionidae) in stored maize grain. *Insect Science Application* 23, 127–135.

Orians, G.H. (1986) Site characteristics favoring invasion. In: Mooney, H.A. and Drake, J.A. (eds) *Ecology of Biological Invasion of North America and Hawaii. Ecological Studies* 58, 133–145.

Pal, S., Koley, K.M. and Pathak, A.K. (2002) Molluscidal activity of *Azadirachta indica* leaves and *Lantana camara* leaves powder against *Indoplanorbis exustus. Indian Journal of Animal Science* 72, 561–562.

Palmer, W.A. and Pullen, K.R. (1995) The phytophagous arthropods associated with *Lantana camara, L. hirsuta, L. urticifolia* and *L. urticoides* (Verbenaceae) in North America. *Biological Control* 5, 54–72.

Parsons, W.T. and Cuthbertson, E.G. (2001) Common Lantana. In: *Noxious Weeds of Australia*. CSIRO, Melbourne, Australia, pp. 627–632.

Prescott-Allen, R. and Prescott-Allen, E. (1990) How many plants feed the world? *Conservation Biology*, 4, 365–374.

Raghubanshi, A.S. (1992) Effect of topography on selected soil properties and nitrogen mineralization in a dry tropical forest. *Soil Biology and Biochemistry* 24, 145–150.

Raghubanshi, A.S., Rai, L.C., Gaur, J.P. and Singh, J.S. (2005) Invasive alien species and biodiversity in India. *Current Science* 88, 539–540.

Raizada, P., Sharma, G.P. and Raghubanshi, A.S. (2008) Ingress of lantana in dry tropical forest fragments: edge and shade effects. *Current Science* 94, 180–182.

Rapoport, E.H. (1991) Tropical versus temperate weeds: a glance into the present and future. In: Ramakrishnan, P.S. (ed.) *Ecology of Biological Invasion in the Tropics*. International Scientific Publications, New Delhi, India, pp. 441–451.

Rashid, I., Sharma, G.P., Reshi, Z.A., Esler, K.J. and Simpson, A. (2009) A standardized response to biological invasions. *Science* 325, 146–147.

Rawat, Y.S. and Singh, J.S. (1988) Structure and function of oak forests in Central Himalaya. I. Dry matter dynamics. *Annals of Botany* 62, 397–411.

Rawat, Y.S., Bhatt, Y.D., Pande, P. and Singh, S.P. (1994) Production and nutrient cycling in *Arundinaria falcata* and *Lantana camara*: the two converted ecosystems in central Himalaya. *Tropical Ecology* 35, 53–67.

Reinhart, K.O. and Callaway, R.M. (2006) Soil biota and invasive plants. *New Phytologist* 170, 445–457.

Rejmanek, M. (1989) Invasibility of plant communities. In: Drake, J.A. *et al.* (eds) *Biological Invasion: a Global Perspective*, SCOPE 37. John Wiley and Sons, Chichester, UK, pp. 281–300.

Rejmanek, M. (2000) Invasive plants: approaches and predictions. *Australian Journal of Ecology* 25, 497–506.

Rejmanek, M. and Randall, J.M. (1994) Invasive alien plants in California: 1993 Summary and comparison with other areas in North America. *Madrono* 41, 161–177.

Richardson, D.M. and Pyšek, P. (2006) Plant invasions: merging the concepts of species invasiveness and community invasibility. *Progress in Physical Geography* 30, 409–431.

Richardson, D.M., Pyšek, P., Rejmanek, M., Barbour, M.G., Panetta, F.D. and West, C.J. (2000) Naturalization and invasion of alien plants: concept and definitions. *Diversity and Distribution* 6, 93–107.

Ries, L., Fletcher, R.J., Battin, J. and Sisk, T.D.

(2004) Ecological responses to habitat edges: mechanisms, models, and variability explained. *Annual Review of Ecology and Evolution* 35, 491–522.

Roberts, M.R. (2004) Response of herbaceous layer to natural disturbance in North American forests. *Canadian Journal of Botany* 82, 1273–1283.

Roberts, M.R. and Gilliam, F.S. (1995) Patterns and mechanisms of plant diversity in forested ecosystems: implications for forest management. *Ecological Applications* 5, 969–977.

Rosswall, T. (1976) The internal nitrogen cycle between microorganisms, vegetation and soil. In: Svensson, B.H. and Soderlund, R. (eds) *Nitrogen, Phosphorus and Sulfur – Global Cycle. Ecological Bulletin*, Stockholm, pp. 157–167.

Sahu, A.K. and Panda, S. (1998) Population dynamics of a few dominant plant species around industrial complexes, in West Bengal, India. *Bombay Natural History Society Journal* 95, 15–18.

Saini, R., Kanwar, S.S., Sharma, O.P. and Gupta, M.K. (2003) Biomethanation of lantana weed and biotransformation of its toxins. *World Journal of Microbiology and Biotechnology* 19, 209–213.

Santamaria, L. (2002) Why are most aquatic plants widely distributed? Dispersal, clonal growth and small scale heterogeneity in a stressful environment. *Acta Oecologia* 23, 137–154.

Saxena, M.K. (2000) Aqueous leachate of *Lantana camara* kills water hyacinth. *Journal of Chemical Ecology* 26, 2435–2447.

Schei, P.J. (1996) Conclusions and recommendations from the UN/Norway Conference on Alien species. *Science International* 63, 32.

Schemske, D.W. (1976) Pollinator specificity in *Lantana camara* and *L. trifolia* (Verbenaceae). *Biotropica* 8, 260–264.

Schmitz, D.C., Simberloff, D., Hoffstetter, R.H., Haller, W. and Sutton, D. (1997) The ecological impact of non-indigenous plants. In: Schmitz, D.C., Simberloff, D. and Brown, T.C. (eds) *Strangers in Paradise*. Island Press, Washington, DC, pp. 39–61.

Sharma, G.P. (2006) Habitat fragmentation induced plant species invasion in Vindhyan highlands. PhD thesis, Department of Botany, Banaras Hindu University, Varanasi, India.

Sharma, G.P. and Raghubanshi, A.S. (2006) Tree population structure, regeneration and expected future composition at different levels of *Lantana camara* L. invasion in the Vindhyan tropical dry deciduous forest of India. *Lyonia* 11, 25–37.

Sharma, G.P. and Raghubanshi, A.S. (2007) Effect of *Lantana camara* L. cover on plant species depletion in the Vindhyan tropical dry deciduous forest of India. *Applied Ecology and Environmental Research* 5, 109–121.

Sharma, G.P. and Raghubanshi, A.S. (2009a) Lantana invasion alters the soil pools and processes: a case study with special reference to nitrogen dynamics in the tropical dry deciduous forest of India. *Applied Soil Ecology* 42, 134–140.

Sharma, G.P. and Raghubanshi, A.S. (2009b) How lantana invades a dry deciduous forest in India: a case study from Vindhyan highlands. *Tropical Ecology* 51, 305–316.

Sharma, G.P., Singh, J.S. and Raghubanshi, A.S. (2005a) Plant invasions: Emerging trends and future implications. *Current Science* 88, 726–734.

Sharma, G.P., Raghubanshi, A.S. and Singh, J.S. (2005b) Lantana invasion: an overview. *Weed Biology and Management* 5, 157–167.

Sharma, O.P. (1988) How to combat lantana (*Lantana camara* L.) menace? – a current perspective. *Journal of Scientific and Industrial Research* 47, 611–616.

Sharma, P.K., Ladha, J.K., Verma, T.S., Bhagat, R.M. and Padre, A.T. (2003) Rice wheat productivity and nutrient status in a Lantana (*Lantana* sp.) amended soil. *Biology and Fertility of Soils* 37, 108–114.

Shaukat, S.S., Siddiqui, I.A., Ali, N.I. and Khan, G.H. (2003) Nematicidal and allelopathic responces of *Lantana camara* root extract. *Phytopathologia Mediterranea* 42, 71–78.

Shea, K. and Chesson, P. (2002) Community ecology theory as a framework for biological invasions. *Trends in Ecology and Evolution* 17, 170–176.

Siemann, E. and Rogers, W.E. (2003) Changes in light and nitrogen availability under pioneer trees may indirectly facilitate tree invasions of grasslands. *Journal of Ecology* 91, 923–931.

Simberloff, D., Parker, I.M. and Windle, P.M. (2005) Introduced species policy, management and future research needs. *Frontiers in Ecology and Environment* 3, 12–20.

Sinha, S. and Sharma, A. (1984) *Lantana camara* L. – a review. *Feddes Report* 95, 621–633.

Sousa, W.P. (1984) The role of disturbance in natural communities. *Annual Review of Ecology and Systematics* 15, 353–391.

Stickney, P.F. (1990) Early development of vegetation following holocaustic fire in northern Rocky Mountain forests. *Northwest Science* 64, 243–246.

Stohlgren, T.J. (2002) Beyond theories of plant invasions: lessons from natural landscapes. *Comments in Theoretical Biology* 7, 355–379.

Sukul, S. and Chaudhuri, S. (2001) Antibacterial natural products from leaves of *Lantana camara* L. with activity comparable to some therapeutically used antibiotics. *Indian Journal of Pharmaceutical Science* 63, 20–23.

Swarbrick, J.T., Willson, B.W. and Hannan-Jones, M.A. (1998) *Lantana camara* L. In: Panetta, F.D., Groves, R.H. and Shepherd, R.C.H. (eds) *The Biology of Australian Weeds*. R.G. and F.J. Richardson, Melbourne, Australia, pp. 119–136.

Syrett, P., Briese, D.T. and Hoffmann, J.H. (2000) Measures of success in biological control of terrestrial weeds by arthropods. In: Wratten, S.D. and Gurr, G. (eds) *Measures of Success in Biological Control*. Kluwer, Amsterdam, pp. 189–230.

Thakur, M.L., Ahmad, M. and Thakur, R.K. (1992) Lantana weed (*Lantana camara* var. aculeata Linn) and its possible management through natural insect pests in India. *Indian Forester* 118, 466–488.

Thaman, R.R. (1974) *Lantana camara*: its introduction, dispersal and impact on islands of the tropical Pacific Ocean. *Micronesica* 10, 17–39.

Thorp, J.R. (2000) *Weeds of National Significance*. LANTANA (*Lantana camara*) Strategic Plan. National Weed Strategy Executive Committee, Launceston, TAS, Australia. Available at www.weeds.org.au/docs/lanstrat.pdf (accessed 12 July 2005).

Tilman, D. (1982) *Resource Competition and Community Structure*. Princeton University Press, Princeton, New Jersey.

Troup, R.S. (1921) *The Silviculture of Indian Trees*, vol I/II. Clarendon Press, Oxford, UK, 783 pp.

Turton, S.M. and Duff, G.A. (1992) Light environments and floristic composition across an open forest-rainforest boundary in north-eastern Queensland. *Australian Journal of Ecology* 17, 415–423.

Van Oosterhout, E., Clark, A., Day, M.D. and Menzies, E. (2004) Lantana Control Manual. Current Management and Control. Options for Lantana (*Lantana camara*) in Australian State of Queensland. Department of Natural Resources, Mines and Energy, Brisbane, QLD, Australia. Available at www.nrm.qld.gov.au/pests/wons/Lantana (accessed 23 November 2004).

Vazquez-Yanes, C. and Orozeo-Segovia, A. (1982) Seed germination of a tropical rain forest pioneer tree *Heliocarpus Donnell-smithii* in response to diurnal fluctuation of temperature. *Physiologia Plantarum* 56, 295–298.

Vinton, M.A. and Burke, I. (1995) Interaction between individual plant species and soil nutrient status in shortgrass steppe. *Ecology* 76, 1116–1133.

Vitousek, P.M. (1990) Biological invasions and ecosystem processes: towards an integration of population biology and ecosystem studies. *Oikos* 57, 7–13.

Vitousek, P.M. and Walker, L.R. (1989) Biological invasion by *Myrica faya* in Hawaii: plant demography, nitrogen fixation, and ecosystem effects. *Ecological Monographs* 59, 247–265.

Vitousek, P.M., Walker, L.R., Whiteaker, D., Müller-Dombois, D. and Matson, P.A. (1987) Biological invasion by *Myrica faya* alters ecosystem development in Hawaii. *Science* 238, 802–804.

Walters, M.B. and Reich, P.B. (1996) Are shade tolerance survival and growth linked? Low light and nitrogen effects on hardwood seedlings. *Ecology* 77, 841–853.

Wardle, D.A. and Bardgett, R.D. (2004) Human-induced changes in large herbivorous mammal density: the consequences for decomposition. *Frontiers in Ecology and Environment* 2, 145–153.

Waterhouse, D.F. and Norris, K.R. (1987) *Biological Control: Pacific Prospects*. Inkata Press, Melbourne, Australia.

Weeds of National Significance (2001) National Weed Strategy Executive Committee, Launceston, TAS, Australia (www.nrm.qld.gov.au/pests/wons/pdf/lantana.pdf).

Whiteaker, L. and Gardner, D.E. (1985) Cooperative National Park resource. Study Unit Technical Report 55, University of Hawaii, Honolulu, Hawaii.

Wilcove, D.S., Rothstein, D., Dubow, J., Phillips, A. and Losos, E. (1998) Quantifying threats to imperiled species in the United States. *Bioscience* 48, 607–615.

Williamson, M. (1993) Invaders, weeds and risk from genetically modified organisms. *Experientia* 49, 219–224.

Wilson, J.B. and Agnew, A.D.Q. (1992) Positive-feedback switches in plant communities. *Advances in Ecological Research* 23, 264–336.

Wilson, J.R.U., Dormontt, E.E., Prentis, P.J., Lowe, A.J. and Richardson, D.M. (2009) Something in the way you move: dispersal pathways affect invasion success. *Trends in Ecology and Evolution* 24, 136–144.

Winder, J.A. (1980) Factors affecting the growth of Lantana in Brazil. PhD thesis, Department of Agriculture and Horticulture, University of Reading, UK.

Witkowski, E.T.F. (1991) Effects of invasive alien acacias on nutrient cycling in the coastal lowlands of the Cape Fynbos. *Journal of Applied Ecology* 28, 1–15.

Wright, S.J. (2005) Tropical forest in a changing environment. *Trends in Ecology and Evolution* 20, 553–560.

Zuidema, P.A., Sayer, J.A. and Dijkman, W. (1996) Forest fragmentation and biodiversity: the case for intermediate-sized conservation areas. *Environmental Conservation* 23, 290–297.

# 4 Biology of *Chromolaena odorata*, *Ageratina adenophora* and *Ageratina riparia*: a Review

**R.S. Tripathi[1], A.S. Yadav[2] and S.P.S. Kushwaha[3]**

[1]*National Botanical Research Institute, Lucknow, Uttar Pradesh, India;* [2]*Department of Botany, R.R. Autonomous College, Rajasthan, India;* [3]*Indian Institute of Remote Sensing, Uttarakhand, India*

## Introduction

> Migration of species from one geographical region to another across the natural barriers such as high mountains, seas and oceans has been taking place since time immemorial. The movement of species through natural dispersal agents has been rather slow. However, with globalization there has been a phenomenal increase in trade, tourism, travel and other human activities, and this has caused both intentional and unintentional introduction of species from one country to another at a pace that was never witnessed before.
>
> (Tripathi, 2009)

Thus, over the last 100 years several invasive plant species have occupied new regions far away from their place of their origin, and some species of the genus *Eupatorium* (family *Asteraceae*) fall within this category. These species are native to Mexico, the West Indies and tropical South America. Ten species of *Eupatorium* (*E. adenophorum* Spreng., *E. cannabinum* L., *E. capillifolium* (Lam.), *E. chinense* L., *E. erythropappum* Robinson, *E. ligustrinum* DC., *E. nodiflorum* Wallich ex DC., *E. odoratum* L., *E. riparium* Regel. and *E. triplinerve* Vahl) have been reported from India (Rao and Rao, 1980; Uniyal, 1995). Of these, three species (*E. odoratum* L. [syn. *Chromolaena odorata* (L.) King & Robinson], *E. adenophorum* [syn. *Ageratina adenophora* (Spreng.) King & Robinson] and *E. riparium* [syn. *Ageratina riparia* (Regel.) King & Robinson]) are invasive and have become dominant in the vegetation of abandoned fields, tree plantations and degraded forest areas in the humid tropical regions of India. Besides invading India, these *Eupatorium* species have also become naturalized in several regions of Asia, Africa and the Pacific islands and are reported to be a major threat to the survival of indigenous species and the maintenance of natural and man-made ecosystems. In this chapter, we have reviewed the research work done on the biology of these three invasive alien species of *Eupatorium* in India and other regions of the world, which may be useful in understanding their biology and ecological impact on the biodiversity of the invaded regions. Throughout this chapter both former and current names are used for the genus: *Eupatorium/Chromolaena*; *Eupatorium/Ageratina*.

## Origins and Invasion

### *Eupatorium odoratum* L.

Originating from south Central America (King and Robinson, 1970; Fig. 4.1) and, as mentioned by Holm *et al.* in 1977, a serious weed at that time in around 23 countries, it has now spread to many more countries. It was reported to have arrived in South-east Asia around 1930 (McFadyen, 1989), and subsequently spread to East Timor some time after 1974, largely as a result of human movement – particularly the transportation of machinery and vehicles (McFadyen, 2004). It was first reported officially in Papua New Guinea in 1970 (Henty and Pritchand, 1973). *Eupatorium odoratum* affects a number of different land uses – oil palm, cocoa and coconut plantations, pastures, gardens, disturbed forests, roadsides and fringes of settlements and villages (McFadyen, 2002; Orapa *et al.*, 2004). The species has also invaded the tropical and subtropical regions of southern Africa (Gareeb *et al.*, 2004). In India, *E. odoratum* has become common on roadsides, open areas, pastures, abandoned gardens and forest clearings (Azmi 2000); dry deciduous forests and interior shrub jungles (Rao, 1977; Prashanthi and Kulkarni, 2005); and in abandoned jhum fallows and degraded forests in north-east India (Yadav and Tripathi, 1981). It has also infested tea, coffee and rubber plantations and vegetable fields, mostly in the humid parts of India. It is a dominant weed at lower altitudes in the north-eastern region, West Bengal, Kerala, Madras (Rao, 1977), Allahabad (Lata and Narain, 2008) and Dehradun (Raizada, 1976).

**Fig. 4.1**. *Chromolaena* (formerly *Eupatorium*) *odorata*.

This plant was introduced into Africa in 1937 and began to spread in the 1970s throughout central and West Africa, from Senegal in the west to the Central African Republic and Congo in the east (Obatolu and Agboola, 1993). It was introduced in Malaysia after World War I (Henderson, 1974). Zachariades *et al.* (2004) suggested that *E. odoratum* was introduced in South Africa along with imported agricultural or other products around 1900 from the West Indies, or that it could have been introduced as a garden plant by an individual. South Africa was the first African country to be affected through contaminated *Gmelina arborea* seeds in the 1940s and, from there, it spread to Ghana, Côte d'Ivoire and Cameroon (Bamikole *et al.*, 2004). It has invaded degraded forests in Ghana (Honu and Dang, 2000) and is a dominant weed and fallow species in slash–burn agricultural systems of the humid regions of Africa (Roder *et al.*, 1995; Weise, 1995; Weise and Tchamou, 1999).

### *Eupatorium adenophorum* Spreng. [syn. *Ageratina adenophora* (Spreng.) King & Robinson]

This native of Mexico (Fig. 4.2) was reported from Australia in 1920 (Blakeley, 1920) and has spread to New South Wales and

**Fig. 4.2**. *Eupatorium adenophorum* Spreng. [syn. *Ageratina adenophora* (Spreng.)].

Richmond Tweed regions (Auld, 1969a). It is also a serious weed of California (Fuller, 1961) and Hawaii (Hosaka and Thistle, 1954). In India, *E. adenophorum* was introduced as an ornamental plant in gardens in 1924 and then became naturalized. It is commonly distributed in degraded forests and abandoned fields above 550 m altitude in the north-eastern region of India (Khonglam and Singh, 1980; Yadav, 1980) and in Dehradun (Raizada, 1976).

### *Eupatorium riparium* Regel [syn. *Ageratina riparia* (Regel.) King & Robinson]

A native of Central and South America, this species (Fig. 4.3) has now colonized degraded forest and fallow lands at higher altitudes (above 1000 m) in Meghalaya (Khonglam and Singh, 1980; Rao and Rao, 1980; Yadav, 1980; Tripathi and Yadav, 1987), and has also been reported from Dehradun (Raizada, 1976).

## Biology

### *Eupatorium odoratum* L. [syn. *Chromolaena odorata* (L.) King & Robinson]

Also known as Siam weed, bitter bush or charismas weed (King and Robinson, 1970; Howard, 1989), *E. Odoratum* is a perennial herb that grows quite tall, sometimes attaining 3 m. In north-east India it shows vegetative growth until October after which sporadic flowering commences, attaining a peak in December followed by fruiting. The seeds (cypselae) are dispersed soon after maturation in February/March and seed germination commences in May. Reproduction in this weed occurs only through seeds (Yadav, 1980). The development of embryos and female gametophytes has been described by Ghosh (1969). Khonglam and Singh (1980) reported that *E. odoratum* is an allohexaploid of chromosome number 60 and is apomictic. On the other hand, Coleman (1989) stated that *E. odoratum* is an autohexaploid with a basic set of ten chromosomes, each represented six times.

### *Eupatorium adenophorum* Spreng. [syn. *Ageratina adenophora* (Spreng.) King & Robinson]

In north-east India, this species is seen as a many-stemmed perennial weed growing abundantly along roadsides and in open areas with high soil moisture content (Yadav, 1980). The mature shoots that dry up in April after seed-setting sprout from their tops again in May/June every year, while new shoots are produced in small numbers. The plant shows vegetative growth till November/December, attaining

**Fig. 4.3.** *Eupatorium riparium* Regel [syn. *Ageratina riparia* (Regel.) King & Robinson].

a height of 80–150 cm, flowering starts in the middle of December and seeds mature by March/April. Lu *et al.* (2008) observed that the pollen of *E. adenophorum* did not germinate in several kinds of media used, and young embryos developed into globular or heart shapes without meiosis and fertilization. The seeds (cypselae) are dispersed soon after maturation, and germination commences with the onset of rains in April/May. Reproduction through vegetative propagation is almost nil (Yadav, 1980). Auld (1969a) studied the distribution of this weed in Australia, and found that its growth is favoured by high rainfall, open areas and steep slopes. Auld and Martin (1975) observed that the seeds require light for germination; they also found a large number of non-viable and empty seeds that may be apomictic, as suggested by Holmgren (1919), Baker (1965) and Khonglam and Singh (1980).

Khonglam and Singh (1980) reported that *E. adenophorum* growing in Meghalaya is an autotriploid with chromosome number 51.

### *Eupatorium riparium* Regel. [syn. *Ageratina riparia* (Regel) King & Robinson]

Also called the mist flower, this species is a scrambling herb attaining 30–70 cm height in moist and shady habitats in north-east India (Yadav, 1980). Most of the old shoots of this species dry up after seed production in April. A large number of new shoots arise from the base of the old shoots in May/June with the onset of monsoon rains. Some of these shoots grow horizontally on the soil surface and, when they come in actual contact with the soil, produce new shoots from their nodes. These shoots become partially independent by producing adventitious roots from their base; this kind of vegetative propagation is quite common in *E. riparium* (Yadav, 1980). Khonglam and Singh (1980) found that *E. riparium* is also an autotriploid with chromosome number 51 and shows an apomictic form of reproduction.

## Reproductive Potential

All three species of *Eupatorium* described above, like other invasive species, exhibit high reproductive potential. The seeds (cypselae) are air-borne and are dispersed to long distances by high winds.

### *Eupatorium odoratum*

Each plant produced 27,000 seeds in a 6-year-old fallow (Yadav and Tripathi, 1981). However, it has also been reported that the number of fruits produced per plant ranges from 50,000 in the first year of growth to nearly 2,000,000 in fully established thickets (Olaoye and Egunjobi, 1974). It showed 34% seed viability and 39% seed germination in laboratory conditions (Yadav, 1980). A sizeable fraction of seeds (66%) was found to be non-viable. The seeds of *E. odoratum* require light for germination (Edward, 1974a; Yadav, 1980). Ramanujan and Ramakrishnan (2002) observed complete inhibition of seed germination in continuous darkness and maximum germination in alternating light/dark periods; they also reported that the optimum constant temperature for the germination of *E. odoratum* seed is 25°C; however, higher germination was observed at alternating 25/20°C temperatures.

### *Eupatorium adenophorum*

High reproductive potential is also evident in this species, which produced 3100 seeds per plant and 40,000 seeds/$m^2$ every year; seeds showed 78% viability and 75% germination under laboratory conditions (Yadav, 1980). Although a few seeds of *E. adenophorum* germinate in darkness, most germinate in continuous light (Ramanujam and Ramakrishnan, 2002). These workers also reported that the optimum constant temperature for germination of *E. adenophorum* seeds was 15–20°C, although higher germination was obtained at 20/15°C alternating temperatures.

### *Eupatorium riparium*

Production of 3100 seeds per plant and 193,600 seeds/m$^2$ under field situations was reported from Shillong, Meghalaya, with 85% viability and 71% germination under laboratory conditions (Yadav, 1980). Thus, a large proportion of *E. riparium* seeds are dormant compared with the other two species of *Eupatorium* discussed above; this characteristic enables it to survive under erratic climatic conditions at the higher elevations of hill slopes. According to Ramanujam and Ramakrishnan (2002), *E. riparium* exhibits seed germination in continuous darkness; however, it evidences higher germination under light. Like *E. adenophorum*, normal germination occurs at 15–20°C constant temperature and at 20/10°C alternating temperature. These authors further suggested that the differential germination behaviour of these three species of *Eupatorium* in relation to light and temperature could be related to their altitudinal distribution patterns. The higher germination level of seeds of the three *Eupatorium* species under alternating temperature conditions compared with constant temperature may be an adaptation to large fluctuations in temperature in surface soils compared with deeper layers, as the seed populations of these weeds are much larger in surface soils.

## Ecology

### Population ecology

The germination and growth behaviour of the populations of *E. odoratum* collected from seven localities from both the Old and New World have been studied by Edwards (1974a, b, 1975). Edwards (1974b, 1975) observed that these populations represent different ecotypes adapted to different climatic conditions. Edwards (1974a, 1977) also established edaphic ecotypes in *E. odoratum* growing under different soil moisture conditions. Ramanujam and Ramakrishnan (2002) have reported that the viability of seeds of *Eupatorium* species is related to the altitudinal extremity in the range of their distribution. They also observed that the germination behaviours of populations of the same species collected from different localities vary. These observations indicate that *Eupatorium* species may comprise populations adapted to varying environmental regimes.

In the north-eastern region of India, *E. odoratum* populations increase rapidly on abandoned jhum fallows, but in jhum fallows over 6 years old that support good growth of secondary successional plant communities, population size begins to decline (Yadav and Tripathi, 1999). Seedling populations of *E. odoratum* show exponential juvenile mortality and two peaks of mortality, one in the active growing season due to competition for light and mineral resources, and a second during the winter months due to low-temperature conditions and moisture stress (Yadav and Tripathi, 1981). Yadav and Tripathi (1981) observed that established plants showed good survival rates even in harsh environmental situations. Seedling mortality in *E. odoratum* increased with the age of the jhum fallow, and almost all newly recruited seedlings died in a 6-year-old fallow (Kushwaha *et al.*, 1981; Yadav and Tripathi, 1999). Kushwaha *et al.* (1981) studied the survival of *E. odoratum* shoots, and observed that the younger shoots showed high mortality compared with older ones.

*Eupatorium adenophorum* and *E. riparium* seedling populations show constant mortality throughout the year, with the death of entire seedling populations in established vegetation (Tripathi and Yadav, 1987). The main factor responsible for high seedling mortality at higher altitudes may be the accumulation of a thick layer of pine needles. The plants of both species show high survivorship, although the adults experienced a degree of mortality throughout the year. In burnt areas, survivorship of seedling populations and adults of both species was much greater compared with unburnt areas (Tripathi and Yadav, 1987). Yadav (1980) concluded that all three species of *Eupatorium* colonize new areas through seeds, but that in established communities seedling survivorship is almost negligible. Disturbances such as occasional fires and deforestation of new areas, due to the slash-and-burn agricultural system (jhuming) prevailing in north-east India, were helpful in the growth and maintenance of their populations.

### Effect of soil moisture

Soil moisture plays an important role in the maintenance of populations of *E. adenophorum* and *E. riparium*. Edwards (1974a) found a correlation between rainfall and growth of *E. odoratum*. This was also confirmed by Kushwaha and Ramakrishnan (1982), who reported a similar relation between above-ground dry weight yield of *E. odoratum* and available soil moisture. Biomass production per unit area decreased for both species, but the decrease was more pronounced in *E. riparium*, indicating that the two weeds differ in their moisture requirement and making their coexistence in nature a little easier. Likewise, the two species responded differently to moisture stress with respect to seed production. *Eupatorium riparium* showed decreased seed production per unit area under a low-moisture regime while *E. adenophorum* showed increased seed production under moisture-stress conditions. Yadav and Tripathi (1985) observed that *E. riparium* allocated more resources to reproductive parts as compared to *E. adenophorum* which is characterized by an increased biomass allocation to vegetative parts, indicating that *E. riparium* is an r-strategist while *E. adenophorum* is a K-strategist, as per the r–K continuum concept of McArthur and Wilson (1967).

The two species also reacted differently in other ways to moisture stress. Under a low-moisture regime, root allocation decreased in both species although *E. adenophorum* showed greater reduction in this respect. On the other hand, the allocation to reproductive parts increased in both species, although the increase was more pronounced in *E. adenophorum*. This shows that the reproductive effort made by *E. adenophorum* is greater than that by *E. riparium* under a moisture-stressed situation. *Eupatorium adenophorum* allocates a much greater proportion of its resources to vegetative structures such as stem and leaf under low-moisture conditions, which gives an indication that the competitive ability of this weed increases on relatively drier habitats. *Eupatorium riparium* on the other hand, showed increased vegetative and reproductive growth under a high-soil moisture regime, indicating that this species is better adapted to moist habitats. Yadav and Tripathi (1985) further observed that, in natural situations too, *E. riparium* grows more luxuriantly on the bottom of the hill slopes that are relatively more moist, while *E. adenophorum* grows more profusely at the top of the slopes where the soil is relatively dry.

### Light intensity and soil nitrogen

Besides soil moisture, the light and nutrient status of fallow lands are the chief environmental factors affecting the growth of three invasive species of *Eupatorium*. Kushwaha and Ramakrishnan (1982) reported that *E. odoratum* grew better in sun than in shade; it allocates a greater proportion of biomass to reproduction when grown in sun while it allocates more resources to roots and root sprouts in shade, indicative of shift in photosynthate allocation strategies as an adaptation to light availability. These authors also observed that dry matter production by *E. odoratum* increased with increased N/P/K levels in the soil.

Tripathi and Yadav (1982) studied the effects of soil nitrogen and light intensity on *E. adenophorum* and *E. riparium*. They observed that these weeds showed different growth behaviour at different density levels. At the highest density both species exhibited some mortality that tended to increase under a high soil nitrogen regime. Both weeds showed increased leaf area per pot under low light intensity and a low-soil nitrogen regime, while under a high-nitrogen regime *E. adenophorum* showed a decrease and *E. riparium* showed an increase in leaf area with reduced light intensity. This suggests that the latter prefers low light intensity while the former grows better under high light intensity if the soil is fertile. Based on their response to light intensity and soil nitrogen levels, it could be said that *E. riparium* is a shade-loving species, while *E. adenophorum* can grow well in both light and shade depending upon the soil fertility level. Biomass accumulation per plant of both the species declined with increased density under various treatments. The growth of *E. adenophorum* was, however, found to have suffered greater suppression with increased density compared with *E. riparium*, which may be attributed to the relatively higher growth rate of *E. adenophorum* resulting in more severe intraspecific competitive effect at high density.

Yield per pot of both species increased under high levels of soil nitrogen; however, the two species differed in their response to density stress under increased soil nitrogen. *Eupatorium riparium* showed greater yield with increased density, while maximum yield in *E. adenophorum* was obtained with mid-range density (12 plants per pot) and declined at the highest density. This suggests that, under high soil nitrogen, *E. riparium* can absorb greater density stress than *E. adenophorum*. Under a low-nitrogen regime, however, the latter showed better growth than the former, suggesting that the latter species requires a greater amount of nitrogen. In contrast, when investigating leaf area per pot in relation to light intensity, *E. adenophorum* gave a higher yield under low light intensity. This suggests that the growth performance of a species depends not only on the photosynthetic surface but also on its assimilation rate under a particular light condition (Tripathi and Yadav, 1982). Both weeds showed high rates of leaf death at low levels of soil nitrogen; however, *E. adenophorum* showed high leaf death at high light intensity and increased population density. It appears that any increase beyond a maximum limit of leaf area per pot attained by the two species results in the death of lower leaves. The adaptation of a species to a particular light intesntiy is also important in this context; *E. adenophorum*, being adapted to low light intensities, did not show increased leaf death with increase in leaf area per pot, which might have shaded the lower leaves.

The percentage of fertile plants in both weeds declined with increased density (Tripathi and Yadav, 1982), while reduced light intensity and increased soil nitrogen enhanced the number of fertile plants in both species. *Eupatorium adenophorum* showed higher seed production than *E. riparium* at higher light intensity, while the latter produced more seeds than the former under low light. In the case of *E. adenophorum* seed number per capitulum was found to be a function of density, while in *E. riparium* it was independent of density stress under various treatments. The number of seeds per capitulum of *E. adenophorum* was much greater than for the other species (Tripathi and Yadav, 1982). The effect of reduced light and increased soil nitrogen on seed output per unit area was same as on seed production per plant in both species, but the two species reacted differently to density stress (Tripathi and Yadav, 1982). In the case of *E. adenophorum* seed production declined with increased density, while seed output increased in *E. riparium* with increased density under high light intensity. Seed production of both species tended to decline with increased density, but the extent of reduction was very high in *E. adenophorum* on account of its high susceptibility to density. On the other hand, under low light intensity both species exhibited a tendency to increase seed production with increased density. Increased seed output in the case of *E. riparium* was comparatively high, showing its greater ability to absorb density stress. In both species the number of abortive capitula increased considerably under a high-soil nitrogen regime. The number of abortive capitula in *E. adenophorum* decreased with increased density, while their number in *E. riparium* increased with increased density. Both species produced large numbers of abortive capitula under conditions unfavourable to high seed production. It may be argued, therefore, that the initiation of capitula in both species depends on the factors that favour biomass accumulation, while further development does not require those conditions.

Density stress does not affect the resource allocation pattern of either species. Under high light intensity the allocation of resources to roots was higher in both species, which further increased under low soil nitrogen. However, under reduced light intensity both species allocated fewer resources to roots. It appears that under nutrient stress at high light intensity, both species allocated a high proportion of resources to their roots. At reduced light intensity when soil nitrogen levels were raised, the resource allocation to roots in *E. riparium* further declined while the other species did not respond in similar fashion. This indicates that *E. riparium* allocates more resources to roots under nutrient-stressed conditions as compared with the other species. Both species showed considerable increase in allocation to reproductive structures with increased soil nitrogen levels. In general, *E. adenophorum* allocated more resources to stem and leaves, which might increase its competitive ability and enable it to grow over a wide range of habitats, while *E. riparium* allocated a greater proportion of resources to its roots, leaves and reproductive structures, which suggests that it not only adopts a strategy that helps in efficient capture of below-ground resources and light energy, but by virtue of greater reproductive allocation it may also be in a better position to maintain and increase its population size (Tripathi and Yadav, 1982).

### Competitive interaction

Kushwaha *et al.* (1983) reported that *E. odoratum* is very sensitive to density stress and responded to it more through reduction in vegetative growth than mortality. Although the vegetative growth per plant was inversely proportional to density, the yield increased from 30 to 120 plants/m$^2$ density and then declined rapidly with further increase in density. Yadav (1980) observed that established plants of both *E. adenophorum* and *E. riparium* exhibited high survivorship with very low mortality. Biomass accumulation per plant of both weeds decreased with increased density of surviving plants, while biomass accumulation per unit area was observed to be independent of density stress. A similar response by both the weeds was observed with respect to reproductive growth with increased density stress; however, *E. riparium* can absorb high-density stress. One interesting characteristic difference observed between *E. adenophorum* and *E. riparium* is that the seed number per capitulum is a highly plastic character in case of the former, but this is not so in the latter species (Yadav, 1980). Dev and Ramakrishnan (1987a, b) studied inter-specific competition between the three species at different elevations and suggested that *E. odoratum* has a competitive edge over *E. adenophorum*, and *E. riparium* is more aggressive than *E. adenophorum*. Although *E. odoratum* is an aggressive competitor as compared with local weeds, its associated species *Imperata cylindrica* was found to be a superior competitor in a mixed culture experiment (Kushwaha *et al.*, 1983).

### Allelopathic interaction

Rai and Tripathi (1982) attributed the dominance of *E. riparium* as a noxious ruderal weed to its allelopathic effects. They observed that growth and population density of *Galinsoga ciliata* and *Galinsoga parviflora* was suppressed when grown in the vicinity of *E. riparium*. Rai and Tripathi (1984) also observed that *E. riparium* released toxic chemical substances into the soil through leaching during the active growing season, which reduced the soil microbial population. *Eupatorium adenophorum* exhibits strong allelopathic potential, which may be one of the attributes that enables it to dominate other plant species in vegetation (Tripathi *et al.*, 1981). This is in agreement with the findings of Li *et al.* (2007), who concluded that leaf extract of *E. adenophorum* inhibited seedling growth of *Chirita spinulosa* (Gesneriaceae), and the gesneriaceous species that grow on rocks appeared more susceptible to allelopathy of this weed than their allied species on trees. The root exudates and soil in which *E. odoratum* grows had allelochemicals that inhibited growth of crop plants – such as phenolics, alkaloids and amino acids (Ambika and Jayachandra, 1980, 1984). Kumar *et al.* (2007) reported that aqueous leaf extract of *E. odoratum* inhibited the germination of *Oryza sativa*. However, Ambika and Poornima (2004) demonstrated that allelochemicals released from *E. odoratum* enhanced crop productivity, and attributed this to the effect of allelochemicals that probably rendered the crop plants resistant to pathogens and water stress.

### Ecological role of *Eupatorium*

As all three invasive alien species of *Eupatorium* dominate vegetation, they are considered to suppress the growth and survival of indigenous species through both competition and their allelopathic effects on neighbouring plant species, and this is in agreement with the observations made by several workers. Rai and Tripathi (1984) reported that *E. riparium* reduces the soil microbial population, which may have adverse effect on soil fertility. Ambika and Jayachandra (1980, 1992), on the basis of their laboratory and field studies, suggested that the leachates and extracts of *E. odoratum* inhibited the growth of crop plants. Invasion by *E. odoratum* has been reported to be the chief factor in the poor regeneration of degraded forests (Swaine *et al.*, 1997). Honu and Dang (2000) suggested that the shade created by the canopy of *E. odoratum* is responsible for high mortality in tree seedlings and hence, and invasion by this weed leads to poor regeneration of trees in degraded forest areas in Ghana. The adverse effects of *E. odoratum* on biodiversity have also been reported by some workers (McFadyen, 2004; Prashanthi and Kulkarni, 2005). McFadyen (2004) suggested that *E. odoratum* has an adverse effect on biodiversity and decreases the livestock-carrying capacity of pastures. The effect of invasion by *E. odoratum* on the grass layer is also sufficiently severe to constitute a threat to the livestock and game industries (Tefera *et al.*, 2008). It has also been reported that *E. odoratum* is a strong competitor and suppresses the growth of other weeds in mixed food crops in tropical Africa (Akobundu *et al.*, 1992; Roder *et al.*, 1995; Akobundu *et al.*, 1999; Ikuenobe and Anoliefo, 2003). Some villages were abandoned due to non-productivity of the land infested by *E. odoratum* in the Philippines (Pancho and Plucknett, 1971). *Eupatorium adenophorum* inhibited movement of machinery and livestock in fields and decreased the livestock-carrying capacity of pastures in Australia (Auld and Martin, 1975). Both *E. adenophorum* and *E. riparium* suppressed the growth of the associated weeds by releasing allelochemicals in the environment (Tripathi *et al.*, 1981; Rai and Tripathi, 1984).

On the other hand, Ambika and Jayachandra (1984) observed that residues of *E. odoratum* promoted crop growth after six months of decomposition. They further concluded that allelochemicals make the crop plants resistant to pathogens and water stress, and these weeds may therefore be successfully exploited to increase crop productivity. In agroecological regions where the slash-and-burn cycle (jhum cycle) has been reduced to less than a 4-year period, *E. odoratum* appears to be a good alternative to natural bush growth. It is considered to be a good fallow plant, having characteristics such as fast growth rate by which it provides cover to the soil and suppresses the growth of low-yielding species. *Eupatorium odoratum* has been reported as a prolific producer of biomass (Olaoye, 1976; Jagdeesh *et al.*, 1990). Mulching by *E. odoratum* increased the yield of yam (Opara-Nadi and Lal, 1987) and cucumber plants (Eussen and Slamet, 1973). *Eupatorium odoratum* contributes to the maintenance of soil fertility through nutrient return to the soil, biomass production and litter fall, as well as providing favourable conditions for earthworm activity (Olaoye, 1976).

Ramakrishnan and Toky (1981) suggested that *E. odoratum*, along with *Imperata cylindrica*, holds the soil strongly and reduces the loss of nutrients through runoff in heavy rains. It is used as green manure in Cambodian rice fields and for black pepper cultivation (Garry, 1963; Litzenberger and Lip, 1968). It also increased crop production indirectly by preventing crop attack by nematodes and pathogens. However, the manure derived from this weed was found to be poisonous to fishes.

## Control of *Eupatorium* Populations

Manual removal of *E. odoratum* by stem cutting followed by digging out of its roots has been tried to clear agricultural land, but this is possible only for small areas. The forest department of Karnataka spends several lakhs of rupees annually to clear this weed in nurseries and young plantations, but the problem has remained (Prashanti and Kulkarni, 2005). Cultural methods such as mulching and cover crops have also been tried. Salgado (1972) observed that *Tephrosia purpurea* was grown as a cover crop in coconut plantations in Sri Lanka to prevent the establishment of *Chromolaena odorata*. Rai (1976) suggested that *Purераria phaseoloides* may be used as a cover crop in rubber plantations to reduce the impact of *E. odoratum*. Wu and Xu (1991) reported that signal grass (*Brachiaria decumbens*) competes with and decreases the incidence of *C. odorata* in pastures in China. Sheldrick (1968) suggested that a mixture of 2, 4-D and 2,4,5-T can be used as a herbicide for the control of *E. odoratum*.

Although these methods have reduced the impact of *E. odoratum*, they have their own limitations. Manual removal involves huge labour costs and it cannot be practised for long. Cultural methods are long-lasting but require the use of physical removal or chemical programmes to eliminate *E. odoratum* prior to implementation. Chemical control is effective but expensive, and may create other environmental problems. According to Goodall and Erasmus (1996), successful biological control remains the only viable solution for reducing the current and potential impact of *C. odorata* in southern Africa. McFadyen (2004) also reported that chemical control is not economically feasible; biological control is, therefore, the only method of controlling this weed in grasslands and abandoned fields.

Biological control of *E. odoratum* was first proposed by the Commonwealth Institute of Biological Control in the mid-1960s (Bennett and Rao, 1968). The moth *Pareuchaetes pseudoinsulata* was the first insect to be introduced from Trinidad into India and Ghana in the early 1970s (Bennett and Cruttwell, 1973), and has been successfully used in Indonesia, the Philippines and Guam (Bofeng *et al.*, 2004). It was effectively used to control *E. odoratum* in the Mariana islands (Muniappan *et.al.*, 2004). The leaf-mining fly *Calycomyza eupatorivora* and the stem-boring weevil *Lixus aemulus* have been tested for control of this weed in Africa. However, attempts to control it through *P. pseudoinsulata* became ineffective due to its poor establishment under field conditions in Karnataka (Prashanthi and Kulkarni, 2005). They evaluated the impact of *Aureobasidium pollulans* (deBary) Amaud on the flowering of *Eupatorium* and suggested that it may be used as a potential mycoherbicide to control this weed, as it considerably reduces seed production and thereby spread to new areas.

Bess and Haramoto (1959) successfully controlled *E. adenophorum* in Hawaii. However, Haseler (1966) and Auld (1969b) observed that only the rate of spread could be controlled by biological measures. Auld (1972) observed that a 0.6% solution of 2, 4-D amine was considered suitable for controlling *E. adenophorum*. A mixture of 2,4-D and 2, 4, 5-T esters can be used for the control of *E. adenophorum*. Auld (1969b) tried to control *E. adenophorum* by use of gall fly (*Procecidochares utilis*), a fungus (*Cercospora eupatorii*) and a crown-boring cerambycid (*Dihammus argentatus*), but despite the damage caused by these organisms the spread of this weed could not be reduced. In the case of *Eupatorium riparium* the best form of control is prevention. It can easily be removed manually, as the plants are small and restricted to moist, shaded sites.

## References

Akobundu, I.O., Ekeleme, F. and Agyakwa, C.W. (1992) Effect of alley farming on weed infestation and floral composition. In: Kang, B.T., Osinem, O.A. and Larbi, A. (eds) *Alley Farming Research and Development, Proceedings of the International Conference on Alley Farming.* International Institute of Tropical Agriculture, Ibadan, Nigeria, pp. 137–143.

Akobundu, I.O., Ekeleme, F. and Chikoye, D. (1999) Influence of fallow management system and frequency of cropping on weed growth and crop yield. *Weed Research* 39, 241–256.

Ambika, S.R. and Jayachandra (1980) Suppression of plantation crops by *Eupatorium* weed. *Current Science* 49, 874–875.

Ambika, S.R. and Jayachandra (1984) *Eupatorium odoratum* L. in plantations, an allelopath or a growth promoter? *Placrosysm – V. Proceedings of the Fiftieth Annual Symposium on Plantation Crops*, pp. 247–259.

Ambika, S.R. and Jayachandra (1992) Allelopathic effect of *Chromolaena odorata* (L) King and Robinson. In: Tauro, P. and Narwal, S.S. (eds) *Proceedings of the First National Symposium – Allelopathy in Agroecosystems.* Indian Society of Allelopathy, Haryana Agricultural University, Hissar, India, pp. 139–141.

Ambika, S.R. and Poornima, S. (2004) Allelochemicals from *Chromolaena odorata* (L.) King and Robinson for increasing crop productivity. In: Day, M.D. and McFadyen, R.E. (eds) *Chromolaena in the Asia–Pacific region.* Australian Centre for International Agricultural Research, Canberra, Australia, pp. 19–24.

Auld, B.A. (1969a) Distribution of *Eupatorium adenophorum* Spreng. on the far north–coast of New South Wales. *Journal and Proceedings, Royal Society of New South Wales* 102, 159–161.

Auld, B.A. (1969b) Incidence of damage caused by organisms which attack crofton weed in the Richmond Tweed region of New South Wales. *Australian Journal of Science* 32, 163.

Auld, B.A. (1972) Chemical control of *Eupatorium adenophorum* crofton weed. *Tropical Grasslands* 6, 55–60.

Auld, B.A. and Martin, P.M. (1975) The autecology of *Eupatorium adenophorum* Spreng. in Australia. *Weed Research* 15, 27–31.

Azmi, M. (2000) Status and biological control of *Chromolaena odorata* in Malaysia. *Proceedings of the Fifth International Workshop on Biological Control and Management of* Chromolaena odorata, Durban, South Africa. pp. 23–25.

Baker, H.G. (1965) The modes of origin of weeds. In: Baker, H.G. and Stebbins, G.L. (eds). *The Genetics of Colonizing Species.* Academic Press, New York, pp. 147–168.

Bamikole, M.A., Ikhatua, U.J. and Osemwenkhee, A.E. (2004) Converting bush to meat; the case of *Chromolaena odorata* feeding to rabbits. *Pakistan Journal of Nutrition* 3, 258–261.

Bennett, F.D. and Cruttwell, R.E. (1973) Insects attacking *Eupatorium odoratum* in the Neotropics. I. *Ammalo insulata* (Walk.) (Lep, Arctiidae), a potential biotic agent for control of *Eupatorium odoratum* L. (Compositae). *Technical Bulletin of the Commonwealth Institute of Biological Control* 16, 105–115.

Bennett, F.D. and Rao, V.P. (1968) Distribution of an introduced weed *Eupatorium odoratum* L. in Asia and Africa and possibilities of its biological control. In: *Proceedings of the 9th Commonwealth Forestry Conference, PANS* 14, 277–281.

Bess, H.A. and Haramoto, F.H (1959) Biological control of Pamakani *Eupatorium adenophorum* in Hawaii by a Tephritid gall fly, *Procecidochares utilis*. II. Population studies of the weed, the fly and the parasites of the fly. *Ecology* 40, 244–249.

Blakeley, W.F. (1920) Notes and exhibits at monthly meetings. *Proceedings of the Linnean Society of New South Wales* 45, 318.

Bofeng, I., Donnelly, G., Orapa, W. and Day, M.D. (2004) Biological control of *Chromolaena odorata* in Papua New Guinea. In: Day, M.D. and McFadyen, R.E. (eds) *Chromolaena in the Asia–Pacific Region.* Australian Centre for International Agricultural Research, Canberra, Australia, pp. 14–16.

Coleman, J.R. (1989) Embryology and cytogenetics of apomictic hexaploid *Eupatorium odoratum* L. (Compositae). *Revista Brasileira de Genetica* 12, 803–817.

Dev, P. and Ramakrishnan, P.S. (1987a) Co-existence of closely related *Eupatorium* species. I. *Eupatorium odoratum* L. versus *Eupatorium adenophorum* Spreng. and *Eupatorium adenophorum* Spreng. versus *Eupatorium riparium* Regel at different altitudes. *Proceedings of the Indian Academy of Science (Plant Science)* 97, 165–176.

Dev, P. and Ramakrishnan, P.S. (1987b) Co-existence of closely related *Eupatorium* species. II. *Eupatorium adenophorum* Spreng. and *Eupatorium riparium* Regel at different altitudes. *Proceedings of the Indian Academy of Science (Plant Science)* 97, 177–183.

Edwards, A.W.A. (1974a) The ecology of

*Eupatorium odoratum* L. II. Physiological studies of *E. odoratum. International Journal of Ecology and Environmental Sciences* 1, 97–105.

Edwards, A.W.A. (1974b) The ecology of *Eupatorium odoratum* L. V. Effect of moisture on growth of *E. odoratum. International Journal of Ecology and Environmental Sciences* 1, 61–69.

Edwards, A.W.A. (1975) The ecology of *Eupatorium odoratum.* III. Germination behaviour in seeds of *E. odoratum* from seven localities in the New and Old World Tropics. *Journal of the West African Science Association* 20, 23–32.

Edwards, A.W.A. (1977) The ecology of *Eupatorium odoratum* L. VI. Effect of habitat on growth. *International Journal of Ecology and Environmental Sciences* 3, 17–22.

Eussen, H.H. and Slamet, S. (1973) Mulching and crop growth. *Weeds in Indonesia* (*Indonesia*) 4, 15–17.

Fuller, T.C. (1961) New weed problems. Bulletin No. 50, California Department of Agriculture. Sacramento, California.

Gareeb, M., Pammenter, N.W., Bower, J.P. and Zachariades, C. (2004) Determining optimal growth conditions for the South African biotype of *Chromolaena odorata*. In: Day, M.D. and McFadyen, R.E. (eds) *Chromolaena in the Asia–Pacific Region.* ACIAR Technical Reports No. 55. Australian Centre for International Agricultural Research, Canberra, Australia, p. 18.

Garry, R. (1963) The changing fortunes and future of pepper growing in Cambodia. *Journal of Tropical Geography* 17, 133–142.

Ghosh, R. (1969) The development of female gametophyte and embryo in *Eupatorium odoratum* L. *Indian Journal of Science and Industry* 3, 59–60.

Goodall, J.M. and Erasmus, D.J. (1996) Review of the status and integrated control of the invasive alien weed, *Chromolaena odorata*, in South Africa. *Agriculture, Ecosystems and Environment* 56, 151–164.

Haseler, W.H. (1966) The status of insects introduced for the biological control of weeds in Queensland. *Journal of the Entomological Society of Queensland* 5, 1–4.

Henderson, M.R. (1974) *Malayan Wild Flowers, Part I.* The Malayan Nature Society, Kuala Lumpur.

Henty, E.E. and. Pritchand, P.H. (1973) Weeds of New Guinea and their control. Botany bulletin No. 7, Department of Forests, Lae, Papua New Guinea.

Holm, L.G., Plucknett, D.L., Pancho, J.V. and Herberger, J.P. (1977) *The World's Worst Weeds, Distribution and Biology.* University Press, Honolulu, Hawaii.

Holmgren, J. (1919) Zytologische studian uber die fortpflanging bei den Galtingen Eirageron and *Eupatorium.* K. Sevens. *I Vet. Akad. Hendl.* 59, 1.

Honu, Y.A.K. and Dang, Q.L. (2000) Responses of tree seedlings to the removal of *Chromolaena odorata* Linn. in a degraded forest in Ghana. *Forest Ecology and Management* 137, 75–82.

Hosaka, E.Y. and Thistle, A. (1954) Noxious plants of the Hawaiian ranges. University of Hawaii Extension Bulletin No. 1962.

Howard, R.A. (1989) Flora of the Lesser Antilles, Leeward and Windward Islands. Vol. 6. Arnold Arboretum, Harvard University, Jamaica Plain, Massachusetts, 658 pp.

Ikuenobe, C.E. and Anoliefo, G.O. (2003) Influence of *Chromolaena odorata* and *Mucuna pruriens* fallow duration on weed infestation. *Weed Research* 43, 199–207.

Jagdeesh, K.S., Geeta, G.S. and Reddy, T.K.R. (1990) Biogas production of anaerobic digestion of *Eupatorium odoratum* L. *Biological Wastes* 33, 67–70.

Khonglam, A and Singh, A. (1980) Altitudinal effect on chromosome associations in three species of *Eupatorium. Cell and Chromosome Newsletter* 3, 24–27.

King, R.M. and Robinson, H. (1970) Studies in the Eupatorieae (Compositae). 24. The genus *Chromolaena. Phytologia* 20, 196–209.

Kumar, M., Siangshaii, S. and Singh, B. (2007) Allelopathic influence of two dominant weeds on agricultural crops of Mizoram, India. *Pakistan Journal of Weed Science Research* 13, 83–92.

Kushwaha, S.P.S. and Ramakrishnan, P.S. (1982) Obsevations on growth of *Eupatorium odoratum* L. and *Imperata cylindrica* (L.) Beauv. Var. *major* under different light and moisture regimes. *Proceedings of the Indian National Science Academy* 48B, 689–693.

Kushwaha, S.P.S., Ramakrishnan, P.S. and Tripathi, R.S. (1981) Population dynamics of *Eupatorium odoratum* in successional environments following slash and burn agriculture. *Journal of Applied Ecology* 18, 529–535.

Kushwaha, S.P.S., Ramakrishnan, P.S. and Tripathi, R.S. (1983) Competition between *Imperata cylindrica* and *Eupatorium odoratum* at two levels of NPK treatment. *Proceedings of the Indian National Science Academy* 49B, 50–56.

Lata, K. and Narain, S. (2008) Addition to the family Asteraceae in flora of Allahabad. *Journal of Phytological Research* 21, 277–279.

Li, Y., Xu, H., Shi, L. and Li, Z. (2007) Allelopathic effects of *Eupatorium adenophorum* on five species of the family Gesneriaceae. *Biodiversity Science* 15, 486–491.

Litzenberger, S. and Lip, H.T. (1968) Utilizing

*Eupatorium odoratum* L. to improve crop yields in Cambodia. *Agronomy Journal* 53, 205–209.

Lu, H., Shen, J., Sang, W., Zhang, X. and Lin, J. (2008) Pollen viability, pollination, seed set and germination of crofton weed (*Eupatorium adenophorum*) in China. *Weed Science* 56, 42–51.

McArthur, R.H. and Wilson, E.O. (1967) *The Theory of Island Biogeography*. Princeton University Press, Princeton, New Jersey.

McFadyen, R.E.C. (1989) Siam weed, a new threat to Australia's north. *Plant Protection Quarterly* 4, 3–7.

McFadyen, R.E.C. (2002) *Chromolaena* in Asia and the Pacific: spread continues but control prospects improve. In: Zachariades, C. and Strathie, L. (eds). *Proceedings of the Fifth International Workshop on Biological Control and Management of* Chromolaena odorata, Durban, South Africa, pp. 13–18.

McFadyen, R.E.C. (2004) *Chromolaena* in East Timor, history, extent and control. In: Day, M.D. and McFadyen, R.E.C. (eds) *Chromolaena in the Asia–Pacific Region*. ACIAR Technical Reports No. 55, Australian Centre for International Agricultural Research, Canberra, Australia, pp. 8–10.

Muniappan, R., Englberger, K., Bamba, J. and Reddy, G.V.P. (2004) Biological control of *Chromolaena* in Micronesia. In: Day, M.D. and McFayden, R.E.C. (eds) *Chromolaena in the Asia–Pacific Region*. ACIAR Technical Reports No. 55, Australian Centre for International Agricultural Research, Canberra, Australia, pp. 11–12.

Obatolu, C.R. and Agboola, A.A. (1993) The potential of Siam weed (*C. odorata*) as a source of organic matter for soils in the humid tropics. In: Mulongoy, K. and Merckx, R. (eds) *Soil Organic Matter Dynamics and Sustainability of Tropical Agriculture*. John Wiley, Chichester, UK, pp. 87–93.

Olaoye, S.O.A. (1976) Litterfall and nutrient cycling in stands of *Eupatorium odoratum* Linn. (Siam weed) in Nigeria. *Annual Conference of the Weed Science Society of Nigeria* 6, 26–27.

Olaoye, S.O.A. and Egunjobi, J.K. (1974) Studies on the ecology of Siam weed (*Eupatorium odoratum* L.) in Nigeria; nature and distribution of the weed. *Nigerian Agricultural Journal* 11, 45–55.

Opara-Nadi, O.A. and Lal, R. (1987) Influence of method of mulch application on growth and yield of tropical root crops in South-eastern Nigeria. *Soil and Tillage Research (Netherlands)* 9, 217–230.

Orapa, W., Englberger, K. and Lal, S.D. (2004) *Chromolaena* and other weed problems in the Pacific Islands. In: Day, M.D. and McFadyen, R.E.C. (eds) *Chromolaena in the Asia–Pacific Region*. ACIAR Technical Report No. 55, Australian Centre for International Agricultural Research, Canberra, Australia, p. 13.

Pancho, J.V. and Plucknett, D.L. (1971) *Chromolaena odorata* (L.) R.M. King and H. Robinson – a new record of a noxious weed in the Philippines. *Philippines Journal of Animal Science* 8, 143–149.

Prashanthi, S.K. and Kulkarni, S. (2005) *Aureobasidium pullulans*, a potential myco-herbicide for biocontrol of Eupatorium [*Chromolaena odorata* (L.) King and Robinson] weed. *Current Science* 88, 18–21.

Rai, J.P.N. and Tripathi, R.S. (19820 Allelopathy as a factor contributing to dominance of *Eupatorium riparium* Regel. *Indian Journal of Ecology* 9, 14–20.

Rai, J.P.N. and Tripathi, R.S. (1984) Allelopathic effects of *Eupatorium riparium* on population regulation of two species of *Galinsoga* and soil microbes. *Plant and Soil* 80, 105–117.

Rai, S.N. (1976) *Eupatorium* and weedicides. *Indian Forester* 102, 449–454.

Raizada, M.B. (1976) *Supplement to Duthie's Flora of the Upper Gangetic Plain and of Adjacent Siwalik and Sub-Himalayan Tracts*. Bishen Singh Mahendra Pal Singh, Dehradun, India.

Ramakrishnan, P.S. and Toky, O.P. (1981) Preliminary observations on the impact of Jhum (shifting cultivation) on the forested ecosystem. In: *National Seminar on Resources, Development and Environment in the Himalayan Region*, pp. 343–354.

Ramanujam, P. and Ramakrishnan, P.S. (2002) Comparative studies on germination behaviour and seedling establishment of *Eupatorium* L. species. *My Forest* 38, 191–200.

Rao, R.R. (1977) Changing pattern in the Indian flora. *Bulletin Botanical Survey of India* 19, 156–166.

Rao, R.R. and Rao, M.K.V. (1980) The genus *Eupatorium* Linn. (Asteraceae) in north-east India. *Proceedings of the Indian National Science Academy* B 46, 587–592.

Roder, W., Phengchanh, S., Keoboualapha, B. and Maniphone, S. (1995) *Chromolaena odorata* in slash and burn rice systems of northern Laos. *Agroforestry Systems* 31, 79–92.

Salgado, M.L.M. (1972) *Tephrosia purpurea* (pila) for control of *Eupatorium* and as a green manure on coconut estate. *Ceylon Coconut Planters' Review* 6, 160–174.

Sheldrick, R.D. (1968) The control of Siam weed in Nigeria. *Journal of the Nigerian Institute for Oil Palm Research* 5, 7–19.

Swaine, M.D., Agyeman, V.K., Kyere, B., Orgle, T.K., Thomson, J. and Veenendaal, E.M. (1997) *Ecology of Forest Trees in Ghana.* ODA Forestry Series No. 7, London, pp. 64.

Tefera, S. Dlamini, B.J. and Dlamini, A.M. (2008) Invasion of *Chromolaena odorata* in the low veld region of Swaziland and its effects on herbaceous layer productivity. *International Journal of Agricultural Research* 3, 98–109.

Tripathi, R.S. (2009) Alien plant invasion, a hot ecological issue. *Environews* 15, 7–10.

Tripathi, R.S. and Yadav, A.S. (1982) Population regulation of *Eupatorium adenophorum* Spreng. and *E. riparium* Regel; effect of population density, soil nitrogen and light intensity. *Plant and Soil* 65, 35–49.

Tripathi, R.S. and Yadav, A.S. (1987) Population dynamics of *Eupatorium adenophorum* Spreng, and *E. riparium* Regel in relation to burning. *Weed Research* 27, 229–336.

Tripathi, R.S., Singh, R.S. and Rai, J.P.N. (1981) Allelopathic potential of *Eupatorium adenophorum* – a dominant ruderal weed of Meghalaya. *Proceedings of the Indian National Science Academy* B47, 458–465.

Uniyal, B.P. (1995) Eupatorieae Cass. In: Hajra, P.K., Rao, R.R., Singh, D.K. and Uniyal, B.P. (eds) *Flora of India, vol. 12 (Asteraceae).* Botanical Survey of India, Kolkata, India, pp. 345–365.

Weise, S.F. (1995) Distribution and significance of *Chromolaena odorata* (L.) R. King and H. Robinson across ecological zones in Cameroon. In: *Proceedings of the 3rd International Workshop on Biological Control and Management of* Chromolaena odorata, Abidjan, Côte d'Ivore, November 1993. Agricultural Experiment Station, University of Guam, Publication No. 202, pp. 29–38.

Weise, S.F. and Tchamou, N. (1999) Weeds in a changing slash and burn cropping environment in the humid forest zone of southern Cameroon. IITA Monograph, Yaounde, Cameroon.

Wu, R. and Xu, X. (1991) Effect of planting signal grass (*Brachiaria decumbens* Stapf.) on feificao (*Chromolaena odorata* King and Robinson) in Southern Yunnan, RRC. In: *Second International Workshop on Biological Control of* Chromolaena odorata, Bogor, Indonesia, February 4–8 1991.

Yadav, A.S. (1980) Studies on population dynamics of *Eupatorium odoratum* L., *E. adenophorum* Spreng. and *E. riparium* Regel. PhD thesis, North–Eastern Hill University, Shillong, India, 220 pp.

Yadav, A.S. and Tripathi, R.S. (1981) Population dynamics of the ruderal weed *Eupatorium odoratum* and its natural regulation. *Oikos* 35, 355–361.

Yadav, A.S. and Tripathi, R.S. (1982) A study on seed population dynamics of three weedy species of *Eupathorium. Weed Research* 22, 59–76.

Yadav, A.S. and Tripathi, R.S. (1985) Effect of soil moisture and sowing density on population growth of *Eupatorium adenonphorum* and *E. riparium. Plant and Soil* 88, 441–447.

Yadav, A.S. and Tripathi, R.S. (1999) Studies on the demography of *Eupatorium odoratum* in a secondary successional community. *Tropical Ecology* 40, 269–274.

Zachariades, C., Senger, I. and Barker, N.P. (2004) Evidence for a northern Caribbean origin for the southern African biotype of *Chromolaena odorata.* In: Day, M.D. and McFadyen, R.E.C. (eds) ACIAR Technical Reports No. 55, Australian Centre for International Agricultural Research, Canberra, Australia, pp. 25–27.

# 5 *Ageratum conyzoides*: an Alien Invasive Weed in India

**Shalinder Kaur[1], Daizy R. Batish[2], R. K. Kohli[2] and H.P. Singh[1]**

[1]*Department of Environment Studies, Panjab University, Chandigarh, India;* [2]*Department of Botany, Panjab University, Chandigarh, India*

## Introduction

Worldwide, invasion by exotic alien plants has caused a significant change in structure and composition of vegetation leading to homogenization of flora (Cushman and Gaffney, 2010). Due to increased globalization and burgeoning human population, there has been an unprecedented increase in movement (both intentional and unintentional) of species beyond their natural biogeographical range into new environments. In fact, the spread of invasive species is occurring at an exceptionally high rate throughout the world and is amongst one of the major threats to global biodiversity. It has greatly enhanced the interest of scientists in studying the magnitude of occurrence and impact of invasive species (Davis, 2009). Invasive plants damage ecosystems both economically and ecologically and, out of these, ecological impacts are more difficult to assess than economic effects (Pimentel *et al.*, 2005). Invasive plants affect native plants directly by competition for soil resources, light and space, as well as indirectly through alteration of ecosystem processes, services and ecological functioning such as soil nutrient cycling and pollination, etc. (Goodell, 2008; Weidenhamer and Callaway, 2010). Invasive species reduce the number of native species and may even lead to their extinction owing to competitive exclusion, niche displacement and hybridization (Mooney and Cleland, 2001). Furthermore, invasive plants reduce biodiversity via degradation of wildlife habitat, thus adversely affecting productivity, ecosystem properties and ecosystem development at the global level (Masters and Sheley, 2001; Kolb *et al.* 2002; Davies and Svejcar, 2008).

India is one of the richest centres of biodiversity due to its wide range of climatic conditions and geography (UNEP/WCMC, 2000). It hosts three biodiversity world hot spots, namely the Western Ghats, the Eastern Himalayas and the Indo-Burma region, regions showing high levels of endemism and species diversity. However, during the last few decades, these are under enormous threat due to various anthropogenic pressures such as tourism, industrialization and urbanization, in addition to the widespread introduction of invasive exotic species by humans. A number of exotic plant species either introduced deliberately or having entered accidentally have negatively affected ecosystem functions, reducing the effective habitat of endemic species and, in turn, have made the restoration of native species difficult.

In India there are a number of invasive weeds which have caused havoc with native biodiversity (Kohli *et al.*, 2009). It is very important to know and understand their biology, ecology and factors imparting invasiveness in order to manage them. This chapter provides details regarding *Ageratum conyzoides*, an invasive alien weed in India.

## *Ageratum conyzoides*

*Ageratum conyzoides* (family *Asteraceae*) is one such rapidly colonizing invasive alien species that has become a troublesome weed over a wide range of ecosystems in tropical and subtropical countries (Batish, 2008; Batish *et al.*, 2009a, b). It is one of about 30 species of the genus *Ageratum*, all of which originated in America (Okunade, 2002). The genus *Ageratum* is widely distributed throughout America, although most taxa have been reported in Mexico, Central America, the Caribbean and Florida (Ming, 1999). The literal meaning of *Ageratum* is non-ageing (referring to the longevity of its flowers or of the whole plant), with its origins in the Greek word *ageras*, whereas the species name *conyzoides* is derived from *konyz*, the Greek name for *Inula helenium* L., which it resembles (Kissmann and Groth, 1993). The English names of the plant, *goat weed* or *billy goat weed*, derive from its peculiar odour like that of male goat (Okunade, 2002). Johnson (1971) divided the taxon *A. conyzoides* into two subspecies, i.e. *latifolium* and *conyzoides*, of which *latifolium* is found throughout the Americas whereas *conyzoides* has a pantropical distribution.

The plant is now found as a weed of over 36 crops (including plantations) in 46 different countries (Holm *et al.*, 1977). It has been ranked as 19th of the world's worst weeds (Holm *et al.*, 1977). Waterhouse (1993) ranked *A. conyzoides* as the 15th most troublesome weed of South-east Asia and the Oceanic Pacific.

### Global distribution of *Ageratum conyzoides*

*Ageratum conyzoides*, a native of Central America and the Caribbean, is now found throughout the world (Xuan *et al.*, 2004). The alien range of *A. conyzoides* includes West Africa, Australia, Colombia, Costa Rica, Ecuador, Fiji, French Polynesia, the Guam Islands, the USA (Hawaiian Islands), Tonga, Vanuatu, Palau, Mauritius, Nicaragua, the Solomon Islands, Papua New Guinea, Samoa and South-east Asia (including China, India, the Philippines, Singapore, Thailand, Vietnam, Cambodia, Malaysia and Indonesia), Brazil and Korea (Kong *et al.*, 2004; Kohli *et al.*, 2006; Sankaran, 2007; Batish *et al.*, 2009a).

In Hawaii, *A. conyzoides* grows up to 1300 m above sea level (Wagner *et al.*, 1999). In Pacific islands the weed has been found growing in different habitats such as crops, pastures, plantations, wastelands and roadsides (Swarbrick, 1997; PIER 2008). In Fiji, *A. conyzoides* is cultivated but it has now become invasive and naturalized in grasslands, forests, forest clearings and along roadsides and trails up to an elevation of 950 m (Smith and Albert, 1991; Sankaran, 2007). It has been reported to be a weed of plantations and waste areas in Tonga (Yuncker, 1959) and Guam (Stone, 1970). In New Guinea, *A. conyzoides* is spread up to 2000 m above sea level in both waste and cultivated lands, in plantations, pastures and along roadsides (Henty and Pritchard, 1975).

In the Galapagos Islands the plant can be seen in moist uplands (McMullen, 1999). The species has been reported to be one of the most dominant weeds of upland crops throughout South-east Asia (Kato-Noguchi, 2001). In Central Sulawesi (Indonesia), *A. conyzoides* was recorded as an invasive weed in coffee and cacao plantations and in the submontane rain forests of Lore Lindu National Park (Siebert, 2002; Ramadhanil *et al.*, 2008).

In South Africa, *A. conyzoides* was introduced in 1949 as an ornamental plant and has now become invasive in many parts, including reserves and protected areas like Kruger National Park (Foxcroft *et al.*, 2008). It grows as an exotic herb in the riparian zones of South African rivers (Hood and Naiman, 2000). Nel *et al.* (2004) conducted a survey of riparian zones in South Africa and concluded that *A. conyzoides* is one of the most widespread and abundant invasive riparian weeds.

In China, the species was first recorded in *The Flora of Hong Kong* by Bentham in 1861, and later in late 19th century in southern Yunnan Province (Yan *et al.*, 2001). It was introduced as an ornamental plant, and has now invaded the ruderal habitats and croplands of central China, South China, Xizang and the lowlands, mountains, plains and hills of Yangtze Drainage (Yan *et al.*, 2001; Huang *et al.*, 2009).

In Australia, *A. conyzoides* has been reported to be a major invasive weed of crops, pastures and disturbed sites in northern Queensland and the Northern Territory (Holm *et al.*, 1977; Jessup, 2002). In Lord Howe Island (Australia), *A. conyzoides* originated as a cultivated plant but has now become a common weed (GISD, 2010). In Zimbabwe, the weed is invasive in nature and is commonly found in abandoned and disturbed sites, open woodlands and along stream and seepage zones at an altitude range of 700–1660 m (Hyde and Wursten, 2010). Akter and Zuberi (2009) reported *A. conyzoides* as an invasive weed in different habitats such as fallow lands, homesteads, roadsides and railway tracks in Bangladesh.

Additionally, the weed is also found/listed as invasive in several other countries (Table 5.1).

**Table 5.1.** Global distribution of *Ageratum conyzoides.*

| Region | Reference(s) |
|---|---|
| Continental Australia, Pacific offshore islands, Indian Ocean offshore islands, Christmas Island Group, Norfolk Island (Australia) | Holm *et al.* (1977); Orchard and Anthony (1994); Swarbrick (1997); Wagner *et al.* (1999) |
| Commonwealth of the Northern Mariana Islands, Lehua Island, Molokai Island, Maui Island, Hawaiian Islands, Guam Island, Virgin Island (America) | Fosberg *et al.* (1975); Wagner *et al.* (1999); PIER (2008); USDA-ARS (2009) |
| Chuuk Islands, Kosrae Island (Federal States of Micronesia) | Manner and Mallon (1989); Josekutty *et al.* (2002) |
| Society Islands, Polynesia, Marquesas Islands (French Polynesia) | Sachet (1983); Welsh (1998); Lorence and Wagner (2008) |
| Offshore islands, Bonin (Ogasawara) Islands (Japan) | Toyoda and Takeshi (2003); PIER (2008) |
| Nauru Island (Nauru) | Thaman *et al.* (1994) |
| New Caledonia Islands, Îles Loyauté (Loyalty Islands) (New Caledonia) | Gargominy *et al.* (1996) |
| Niue Island (Niue) | Yuncker (1943); Whistler (1988); Space *et al.* (2004) |
| Rock Islands, Angaur Island, Koror Island, Babeldaob Island, Malakal Island, Ngerkebesang Island (Palau) | Space *et al.* (2003, 2009) |
| Philippine Islands (Philippines) | Waterhouse (1993) |
| Savai'i Island, Upolu Island, Western Samoa Islands (Samoa) | Space and Flynn (2002); PIER (2008) |
| Tonga Islands, Tongatapu Island (Tonga) | Yuncker (1959) |
| Wallis and Futuna Islands (Wallis and Futuna) | Meyer (2007) |
| Kingdom of Cambodia (Cambodia) | Waterhouse (1993) |
| People's Republic of China (China) | Li-ying *et al.* (1997) |
| Republic of Indonesia (Indonesia) | Waterhouse (1993) |
| Republic of Singapore (Singapore) | Waterhouse (1993) |
| Taiwan Island (Taiwan) | Li-ying *et al.* (1997) |
| Kingdom of Thailand (Thailand) | Waterhouse (1993) |
| Socialist Republic of Vietnam (Vietnam) | Waterhouse (1993) |
| La Réunion Island (France) | MacDonald *et al.* (1991) |
| Maldive Islands (Maldives) | Fosberg (1957) |
| Mauritius and Rodrigues Islands (Mauritius) | Holm *et al.* (1977) |

Source: GISD (2010).

## Distribution in India

In India, *A. conyzoides* has been reported as existing prior to 1882 in *The Flora of British India* (Hooker, 1882). It was probably introduced as an ornamental plant in the 1860s (National Focal Point for APFISN, India, 2005), later attained a weedy habit and turned harmful to mankind. Its invasion and spread has caused ecological havoc to indigenous floristic composition in various regions of India, including the north-west Himalayas, eastern Himalayas, central India and Western Ghats (Yoganarnarasimham, 2000; Silori and Mishra, 2001; Kohli *et al.*, 2006; Reddy *et al.*, 2008).

The weed has been reported as one of the major invasive species, growing to an elevation of 2400 m in Himachal Pradesh (Kohli *et al.*, 2004; Dogra, 2008); in fact, around 50% of the area in Himachal Pradesh state is said to be infested by this obnoxious weed (Batta, 1988). It has been reported to be one amongst the ten most dominant herbs in forest grassland edge, weed-infested areas and low-lying wet grassland of the Jim Corbett tiger reserve, Terai and Bhabar regions (Rawat *et al.*, 1997) and in the wetlands of Samaspur Bird Sanctuary (Reddy *et al.*, 2009) in Uttar Pradesh, India. Sit *et al.* (2007) surveyed the eastern Himalayan region of West Bengal (India) and reported *A. conyzoides* as one of the most widely distributed weeds in various crop lands and palm gardens.

Negi and Hajra (2007) studied flora of the Doon Valley, north-west Himalaya and reported *A. conyzoides* as one of the invasive exotics. The species has been reported to occur in Veerapuli and Kalamalai forest reserve (Swamy *et al.*, 2000), tropical wet evergreen forests and the Anamalai Hills (Muthuramkumar *et al.*, 2006) of the Western Ghats of Tamil Nadu. *A. conyzoides* has been found to be a weed of disturbed sites in the tropical forest of Little Andaman Island in the Bay of Bengal (Rasingam and Parthasarathy, 2009). It has been found to be one of the dominant species in Agrakhal-Hindolakhal, Garhwal Himalaya, Uttaranchal (Bughani and Rajwar, 2005). In addition, *A. conyzoides* has been reported to be a major invasive weed on both the slopes and wetlands of Mothronwala swamp in the Doon Valley, Uttaranchal (Gupta *et al.*, 2006). Recently, *A. conyzoides* was found to be one of the most predominant weeds in Mandhala watershed in Himachal Pradesh (Rana *et al.*, 2010).

The plant has also been found as a major weed in the littoral and swamp forests of Assam (DOEF, 2010). In Arunachal Pradesh, *A. conyzoides* is one of the major weeds in West Siang (Singh *et al.*, 2002) and the most dominant weed under canopy of *Dendrocalamus hamiltonii* in tropical forests (Arunachalam and Arunachalam, 2002). In addition, the weed has also been reported in north-eastern and southern India (Rao, 2000) and the forests of the Gandhamardan Hills range, Orissa (Reddy and Pattanaik, 2009).

## Spread of *A. conyzoides* in different habitats

*Ageratum conyzoides* is a serious problem of cultivated lands in the hilly tracts of north-western India (Bansal, 1988), where it forms dense thickets in commonly grown crops such as chickpea, rice, maize and wheat, and adversely affects crop yields (Kohli *et al.*, 2006). Due to its enormous seed-producing capacity, fields left fallow are rapidly invaded and colonized by the weed. In Himachal Pradesh the weed starts appearing at the tassel stage in maize, produces flowers and sets seeds by the time crop is harvested (Kanwar and Kharwara, 1988).

A study conducted by Reddi *et al.* (1977) demonstrated that *A. conyzoides* is a major weed in ratoon sugarcane crop fields with a population of 250–400 plants/m$^2$. In maize fields, its population has been observed to be 1000 plants/m$^2$, thereby completely covering the ground surface (Anonymous, 1986–1987). Kanwar and Kharwara (1988) studied the population of *A. conyzoides* in various fields and wastelands in Himachal Pradesh (India). The population of *A. conyzoides* was greatest in kitchen gardens (50 plants/m$^2$) followed by maize fields (20 plants/m$^2$) and was least in meadows

(5 plants/$m^2$). *Ageratum conyzoides* was also found to be a common weed of habitats such as grasslands, forest/plantation areas, water channels and wastelands in Chandigarh (Arora, 1999), with maximum plant density and dominance in forests (Table 5.2).

In the hilly tracts of north-western India, particularly in the lower and middle Himalayas, weed infestation is increasing at an alarming rate and it has severely hampered the growth of native plants. Dogra (2008) carried out an extensive survey of areas ranging from the lower or Shivalik Himalayas, (300–1500 m) to higher ranges of altitude (1500–2400 m, middle Himalayas) to investigate the spread and invasion by *A. conyzoides* in north-western India. It was documented that the density of weed was greater in the lower Himalayas compared with the middle Himalayas. However, the upper Himalayas (2400–3800 m) were found to be free of the weed (Dogra, 2008). A survey conducted in Shimla and Kinnaur (Shivalik hills) showed that *A. conyzoides* constituted 21.42% of the total herb species (Rana and Sharma, 2009). Recently, it has been observed that infestation by *A. conyzoides*, along with *Cynodon dactylon* and *Parthenium hysterophorus*, constituted 75% of the total population in Mandhala watershed in Himachal Pradesh (Rana *et al.*, 2010).

The weed not only invades agricultural/arable field, but also severely infests disturbed sites. Due to its strong competitive ability and fast propagation, it has covered all terrain not under intensive use, care and managed efficiently, and causes significant ecological damage. Figures 5.1 and 5.2 show huge monocultures of *A. conyzoides* on wasteland and around a wheat field, respectively.

## Habitat and characteristic features

Physiognomically, *A. conyzoides* is an annual erect aromatic herb that shows considerable variation in shoot height (59–120 cm) in a stand at maturity stage (Table 5.3); however, the average height of plant is ~1 m (Kumar and Singh, 1988). The stem is erect, branched, cylindrical and decumbent, and covered with fine, white hairs. Leaves are opposite, ovate and triangular, and pubescent with the long petiole (1.5–2.0 cm or even up to 3.2 cm), covering an area of ~31 $cm^2$ and bearing trichomes (56 in number) on both surfaces (Kirtikar and Basu, 1984; Arora, 1999). Plants have a shallow tap root system with a radius of spread of ~8 cm. The plant bears blue–violet terminal inflorescence (Fig. 5.3) and a capitulum of homogamous disc florets (>70 capitula per plant) arranged in corymbose racemes (Kumar and Singh, 1988; Arora, 1999). The inflorescence changes colour from blue when young to white at maturity. The number of florets per head varies in the range 56–86 (Anonymous, 1988; Table 5.3).

The flowering period of the weed is long, and flowers retain their violet colour for a longer period of time. The fruit is a typical achene with pappus and easily spread by wind and animal hair. Seeds are minute (<1 cm), black and extremely light in weight, elliptical, bear a pappus, are produced in great numbers (Table 5.3) and are transported up to hundreds or thousands of miles. Arora (1999) reported that in Chandigarh, wild-growing plants produce on average ~5000 seeds. However, Rodriguez and Cepero (1984) reported a total of 94,772 seeds per plant. The weed completes its life cycle in about 10–12 weeks, and seeds are shed, resulting in tremendous increase in

**Table 5.2.** Distribution of *Ageratum conyzoides* in different habitats (Chandigarh, India).

| Habitat | Density (%) | Frequency (%) | Dominance (%) | Importance value index (IVI) (%) |
|---|---|---|---|---|
| Grassland | 59.98 | 22.24 | 57.96 | 140.18 |
| Forest | 81.73 | 15.79 | 83.67 | 181.19 |
| Water channel | 61.94 | 20.18 | 60.24 | 142.36 |
| Wasteland | 61.00 | 21.69 | 80.10 | 162.79 |

**Fig. 5.1.** Monoculture of *Ageratum conyzoides* on wasteland.

**Fig. 5.2.** *Ageratum conyzoides* around a wheat field.

its intensity next season (Bansal and Singh, 1986). The propagation of the weed through seeds is so rapid that it covers almost any terrain not under intensive use or abandoned. *Ageratum conyzoides* is a shade-tolerant plant and flourishes well in any type of garden soil such as clayey, sandy or loamy with wide range of pH. Soils rich in moisture, minerals and air are best suited to its growth.

In hilly tracts of subtropical to temperate environment, the weed is present throughout the year. However, in the plains, it emerges during the onset of winter and remains till early summer. In shady and moisture-rich areas the weed is also seen during the summer rainy season; however, in such areas it has a relatively shorter life span and reduced density (Kohli *et al.*, 2006).

## Invasive potential of *A. conyzoides*

*Ageratum conyzoides* has become an invasive weed in a wide variety of natural and man-

**Table 5.3.** Morphological features of *Ageratum conyzoides*.

| Parameter | Value |
|---|---|
| Shoot height (cm) | 82.60 ± 20.29 |
| Radius of spread (cm) | 8.01 ± 0.50 |
| Root length (cm) | 7.70 ± 0.82 |
| Basal area ($cm^2$) | 1.92 ± 0.68 |
| Shoot:root ratio | 10.72 ± 3.59 |
| Biomass (g) | 3.56 ± 0.20 |
| Trichomes (per $cm^2$) | |
| • leaf | 55.5 ± 11.7 |
| • stem | 32.0 ± 9.3 |
| Length of leaf petiole (cm) | 1.75 ± 0.61 |
| Leaf lamina | |
| • shape | Broadly ovate or rhomboid ovate to triangular |
| • length (cm) | 7.69 ± 3.24 |
| • width (cm) | 5.07 ± 2.64 |
| • perimeter (cm) | 20.81 ± 10.14 |
| • area ($cm^2$) | 30.54 ± 16.34 |
| • transpiration rate (μg/$cm^2$/s) | 79.36 ± 10.21 |
| • diffusion resistance (s/cm) | 0.046 ± 0.011 |
| • relative humidity (%) | 44.48 ± 5.96 |
| Peduncle length (cm) | 0.64 ± 0.11 |
| Flower colour | Sky blue to violet when young, white on maturity |
| Inflorescence | |
| • type | Capitulum made up of homogamous disc florets |
| • number | 69.00 ± 3.92 |
| • diameter of head (mm) | 6.1 ± 0.5 |
| • number of florets per head | 71.50 ± 10.13 |
| Seed | |
| • colour | Black, bearing pappus |
| • number | 4935.13 ± 77.08 |
| • length (cm) | 3.4 ± 0.5 |
| • width (cm) | 0.33 ± 0.01 |
| • weight (μg) | 5.06 ± 0.18 |

**Fig. 5.3.** Inflorescence of *Ageratum conyzoides*.

made ecosystems ranging from forests and grasslands to farmland. It is a prolific ephemeral herb with great morphological variations and easily adapts to different ecological conditions (Sauerborn and Kock, 1988). The weed has been recognized as one of the most serious pests of modern agriculture, one that has encroached on almost all other types of lands, i.e. wastelands, natural forest, plantations, vegetable gardens, overgrazed pastures, orchards, range/grasslands, riparian zones, ruderal/disturbed sites, shrub/shrublands, tea orchards, meadows, water courses, fresh landslides or areas with deep gullies and wetlands, and also along roadsides (Kanwar and Kharwara, 1988; Rao, 2000; Kunwar *et al.*, 2001; Kohli *et al.*, 2006; Sankaran, 2007; Batish *et al.*, 2009a, b; GISD, 2010).

### Features imparting invasiveness

Attributes for successful invasion of *A. conyzoides* include:

- fast growth and rapid spread due to potential to become well established;
- wide ecological amplitude;
- high reproductive potential (both sexual and vegetative);
- long flowering and fruiting periods;
- absence of natural predators/enemies/competitors;
- resistance to predators;
- unpalatable due to high phytotoxin content; and
- resource competition along with novel weapons such as allelopathy.

### Impacts of *A. conyzoides*

#### *Different ecosystems*

FORESTS. The shade-tolerant nature of *A. conyzoides* allows it to become established under tree canopies in developed forests. It has become an aggressive colonizer that can maintain dense populations within the understorey during its growing period (Arora, 1999). The tree canopy in forests and plantations provides favourable conditions such as shade for growth. The tree plantations of *Acacia catechu* and *Eucalyptus* spp., *Pinus* forests and mixed-culture tree plantations in the lower Himalayas of Himachal Pradesh have been observed to be occupied by *A. conyzoides* (Dogra, 2008). Native plant species growing in various plantations have been facing threats due to invasion by this exotic weed; this species was observed as the most frequently and densely growing under *Eucalyptus* and mixed-tree plantations (Dogra, 2008). Reddy and Pattanaik (2009) reported *A. conyzoides* among 64 exotic species posing a serious threat to the dense interior forests of the Gandhamardan Hill range, Orissa. The weed was found to encroach on the littoral and swamp forests of Assam, retarding/preventing natural succession and reforestation, so much so that some forests have lost their identity (DOEF, 2010).

GRASSLANDS. The fast-spreading stolons of *A. conyzoides* greatly enhance its potential to cover large areas of grassland. Once established, the weed displaces and replaces native grasses/plants and reduces grazing areas, resulting in fodder scarcity. The density, diversity and biomass of associated vegetation are severely affected in weed-infested grassland compared with weed-free areas in Chandigarh (Arora, 1999; Table 5.2). Furthermore, it reduces the carrying capacity of pastures, causes thinning of floral diversity and may lead to the disappearance of threatened and endemic species (Kohli *et al.*, 2004, 2006).

CULTIVATED LANDS. Agriculture ecosystems are most seriously affected by *A. conyzoides*, ecologically as well as economically. Due to the fast-spreading nature of *A. conyzoides*, it hinders field practices such as ploughing and field preparation, thereby increasing the maintenance cost of agricultural fields. It adversely affects the growth and production of major staple annual and perennial crops in India (Kohli *et al.*, 2006; Batish *et al.*, 2009a, b), and forms dense monocultures and interferes with the establishment of many crops such as maize, chickpea, rice and wheat (Kohli *et al.*, 2006; Batish *et al.*, 2009a, b). Additionally, it acts as a host for many crop diseases (Kashina *et al.*, 2003; Ekeleme *et al.*, 2005).

In the lower Shivalik range of the Himalayas this weed infests cultivated lands to such a serious extent that farmers may have to abandon their fields (Kohli and Batish, 1996; Batish *et al.*, 2004a). Due to this type of heavy infestation, nothing is left in areas where it has invaded – it is known as *Ujaroo* (the one that destroys everything) in this area (Negi, 1988).

Roder *et al.* (1995, 1998) reported *A. conyzoides* as one of the major weeds in rice fields in Asia, causing severe reduction in rice yield following invasion. It is considered as one of the worst weeds of paddy fields in Pokhra (Thapa and Kayastha, 1998) and Kirtipur in Nepal (Manandhar *et al.*, 2007). Manandhar *et al.* (2007) reported reductions in rice grain (25–47%) and straw (13–38%) yields due to infestation by *A. conyzoides*. It has been reported to be one among the problematic weeds associated with *Gladiolus* cultivation, adversely affecting productivity as well as product quality (Riaz *et al.*, 2007).

Many workers have attributed the observed loss in crop produce and quality to the allelopathic potential of the weed (Singh *et al.*, 2003; Batish *et al.*, 2006, 2009a, b; Manandhar *et al.*, 2007; Batish, 2008). However, Ekeleme *et al.* (2005) opined that shoot competition for light may also be one of the major causes of its interference with crops.

### Soil nutrients

*Ageratum conyzoides* directly or indirectly affects soil chemistry and composition, ecosystem functions and creates a novel environment for native species. Heavy infestations of the weed modify the soil environment through root exudation, affecting soil structure and mobilizing or chelating nutrients (Singh *et al.*, 2003). Furthermore, it causes depletion of soil nutrients because of resource competition, rendering it unfit for effective growth of crops and making sustainability difficult. Manandhar *et al.* (2007) reported a significant reduction in soil nitrogen and phosphorus in paddy fields due to weed infestation. On the other hand, Batish *et al.* (2009a, b) reported that weed residues enriched soil nutrient content despite the negative effects on crop growth.

### Native vegetation and biodiversity

The Himalayas in India is a global hot spot zone and has an exceptionally high level of biodiversity. However, the spread of *A. conyzoides* along with other exotics has remarkably altered the communities and ecosystem by depleting the indigenous species (Kohli and Batish, 1996; Kohli *et al.*, 2006; Dogra *et al.*, 2009a, b). Invasion of *A. conyzoides* is a matter of serious concern, as it causes a reduction in niche or habitat for growth and establishment of valuable plants. In fact, the, rapid spread of this exotic is a serious concern for foresters, farmers, ecologists, environmentalists, shepherds, orchardists, horticulturists and animal scientists.

Once established, it dominates large areas, disrupts nutrient cycles and alters the pattern of plant succession in the native ecosystems. Invasion by *A. conyzoides* results in negative interactions with native species for food and other natural resources, causing extinction of associated vegetation and thereby declining species richness. Kohli *et al.* (2004) observed a huge loss in plant diversity and density (~50–64%) of native flora after invasion by *A. conyzoides* in the Shivaliks, North India. The indices of richness and evenness were significantly less in weed-infested areas compared with weed-free areas, indicating the reduced numerical strength and uneven distribution of native flora. Weed-free areas exhibited higher plant diversity (Shannon's index), indicating heterogeneity in the communities. Compared with this, invasion by *A. conyzoides* resulted in a severe loss of plant diversity, and consequent homogeneity in communities. On the other hand, the index of dominance was higher in areas invaded by *A. conyzoides*, showing its dominance (occurrence of a single species) in the community (Kohli *et al.*, 2004). The homogenous nature of a community leads to instability that further aggravates the potential for invasion by opportunistic species, especially those with wider ecological amplitude. This changing pattern of vegetation in weed-occupied areas in comparison with weed-free areas may be attributed to the allelopathic nature of the weed, apart from other factors (Batish, 2008).

Dogra (2008) reported a decline in valuable indigenous medicinal plants such as *Achyranthes aspera*, *Trifolium repens*, *Centella asiatica*, *Zizyphus jujuba*, *Dichanthium annulatum*, *Murraya koenigii*, *Adhatoda vasica*, *Carissa* spp. and *Colebrookea* spp. in the Shivaliks of Himachal Pradesh (India) following invasion by *A. conyzoides*. Some species such as *Sonchus oleraceus*, *Sonchus asper*, *Vernonia cinerea*, *Abutilon indicum*, *Agave americana* and *Medicago lupulina* were not found in weed-infested areas, while they were found to be growing abundantly in weed-free areas (Dogra, 2008).

Batish (2008) revealed that *A. conyzoides* causes reduction in growth of other grasses and weeds measured in terms of dry biomass (up to 50%) and total number of species (up to 30%). Its dense monoculture-forming growth habit places it in the category of invasive species that can transform the integrity of community and ecosystems visually, structurally and chemically. A recent study showed that ~20% of the vegetational area in the lower Shivalik Hills of the Himalayas was occupied by *A. conyzoides*, along with other invasive plants; however, this weed accounted for ~30% reduction in plant species (Dogra *et al.*, 2009a, b).

Dogra *et al.* (2009b) investigated the impact of invasion by *A. conyzoides* along with other invasive species on the structure and composition of communities in the lower Himalayas. Following weed infestation, alpha diversity was reduced by ~40%. The number of abundant and very abundant species, index of evenness, and fresh and dry biomass of vegetation were also drastically reduced in areas infested by *A. conyzoides* compared with weed-free areas. All these observations show the homogenous nature of the ecosystem and loss of productivity in weed-infested areas (Dogra *et al.*, 2009a, b). Furthermore, fresh and dry biomass was also reduced in *A. conyzoides*-infested areas compared with weed-free areas, thus making those areas less productive (Dogra *et al.*, 2009a, b). *Ageratum conyzoides* has also been reported as being among the three most invasive weeds that have altered the community structure and population dynamics of native flora and fauna in the Shivalik Hills (Rana and Sharma, 2009). Recently, it has been reported that the undesirable effects of *A. conyzoides* – along with those of other plants – resulted in unevenness and instability of herb layers in the Mandhala watershed in Himachal Pradesh (Rana *et al.*, 2010).

Even the reserve parks/wildlife sanctuaries/protected areas have also been observed to be affected by this weed, resulting in disruption and loss of natural vegetation; for example, the Jim Corbett tiger reserve (Rawat *et al.*, 1997), the Veerapuli and Kalamalai forest reserve in the Western Ghats (Swamy *et al.*, 2000) and the Mudumalai wildlife sanctuary (Silori and Mishra, 2001) have been observed to be infested by *A. conyzoides*.

#### *Humans and animals*

Not only does *A. conyzoides* affect farmers and scientists, but it also has adverse effects on human and animal health. People in contact with this weed suffer from nausea, giddiness, irritation and asthma (Kohli and Batish, 1996; Negi and Hajra, 2007). Livestock do not feed on it as it causes ulceration and toxicity.

## Role of allelopathy in invasion potential of *A. conyzoides*

Allelopathy is a type of biotic interference in which plants release bioactive molecules into the surrounding environment and that negatively affects the growth of nearby vegetation. In fact, allelopathy has been adopted as a successful strategy for the spread, establishment, domination and colonization of alien environments by invasive weeds, thereby replacing native vegetation (Bais *et al.*, 2004; Kohli *et al.*, 2004, 2006; Hao and Qiang, 2005; Batish, 2008). A number of aggressive weeds have been reported to exhibit allelopathy as a mechanism of interference, which provides them competitive advantage over other plants (Qasem and Foy, 2001). *Ageratum conyzoides* has been reported to have adopted allelopathy as a strategy to out-compete native flora and suppress growth of crops. There is much evidence regarding the allelopathic nature of *A. conyzoides* providing it with selective advantage (Kong *et al.*, 1999; Batish *et al.*, 2009a, b). It has been demonstrated that root and shoot aqueous extracts of *A. conyzoides* are phytotoxic to seed germination and seed vigour in maize and soybean (Singh *et al.*, 1989) and to growth in wheat and rice (Jha and Dhakal, 1990; Prasad and Srivastava, 1991). Acetone extracts of *A. conyzoides* shoot residues inhibited the germination and growth of roots and shoots of *Amaranthus caudatus*,

*Digitaria sanguinalis* and *Lactuca sativa* (Kato-Noguchi, 2001).

Leaf debris and even soil entangled with the roots of *A. conyzoides* (rhizosphere soil) were also observed negatively to affect roots, shoot length and biomass accumulation in rice by ~18–30% (Batish *et al.*, 2009a). Apart from living tissues, leftover residues collected at the end of the growing season in the area of weed invasion also interfered with maize and rice (Batish *et al.*, 2004b). However, the allelopathic potential of *A. conyzoides* varies depending on organ, developmental stage and habitat (Hu and Kong, 1997).

Soil inhabited by the weed has been reported to be rich in non-volatile allelochemicals (Singh *et al.*, 2003) and to be phytotoxic to growth of other plants. Radicle and coleoptile lengths of crop plants were severely suppressed when grown in field soil previously infested with *A. conyzoides* (Kalia, 1998; Singh *et al.*, 2003). This may have been due to water-soluble phytotoxins present in *A. conyzoides* that interfere with the process of nutrient uptake (Batish, 2008; Batish *et al.*, 2009a, b).

Batish *et al.* (2006) demonstrated the allelopathic effect of *A. conyzoides* against chickpea and observed a severe reduction in the number of nodules, nodule weight and the leghaemoglobin content of the nodules of chickpea grown in soil amended with weed residues. Later, it was found that weed-infested soil and leaf extracts deleteriously affected the growth of rice, which was attributed to water-soluble phytotoxins, i.e. phenolics (Batish *et al.*, 2009a). Leaves were demonstrated to exhibit a greater inhibitory effect than stems and roots (Xuan *et al.*, 2004).

Even the extracts of isolated and purified precocene I and II, the major chromenes of *A. conyzoides*, also possessed allelopathic potential towards tomato, ryegrass and radish (Kong *et al.*, 1998a, b). Kong *et al.* (1999, 2002) demonstrated that volatile leaf oils from *A. conyzoides* inhibited seedling growth and reduced the chlorophyll content of *Cucumis sativus*, *Lolium ultiforum*, *Raphanus sativus*, *Phaseolus aureus*, *Triticum aestivum* and *Lycopersicon* sp. Greater inhibition was observed in response to volatile oils compared with pure constituents, thus demonstrating the synergistic effect of allelochemicals (Kong *et al.*, 1999). Furthermore, the allelopathic potential of the weed is enhanced when plants are grown under stress conditions such as nutrient deficiency (Kong *et al.*, 2002).

## Phytochemistry

Since *A. conyzoides* possesses diverse biological and physiological characteristics it has been explored intensively, particularly for its secondary metabolites (Gonzalez *et al.* 1991; Wiedentold and Roder, 1991). A wide range of secondary metabolites from different classes are found in *A. conyzoides*, including flavonoids, alkaloids, chromenes, phenolics and essential oils (Gonzalez *et al.*, 1991; Sharma and Sharma, 1995). Among these secondary metabolites, many are allelochemicals and inhibit the growth of other organisms (Pari *et al.*, 1998; Okunade, 2002). These allelochemicals are released either through leaching or volatilization into the soil or environment in bioactive concentrations, and retard the growth of other plants and organisms (Singh *et al.*, 2003; Batish *et al.*, 2009a, b).

### *Volatile compounds*

Almost every part of a plant contains volatile oils; the leaves and roots of *A. conyzoides* contain volatile oils in the range of 0.11–0.58% and 0.03–0.18%, respectively, depending on the season (Wandji *et al.*, 1996). As many as 51 constituents, including a number of mono- and sesquiterpenoids, have been identified from its volatile oils (Okunade, 2002). The chromenes 7-methoxy-2,2-dimethylchromene (precocene I), 6,7-dimethoxy derivative, ageratochromene (precocene II) and their derivatives (Burkill, 1985; Rastogi and Mehrotra, 1990; Kissmann and Groth, 1993; Kong *et al.*, 1998a, b) are the major constituents of leaves and flower oils, and possess biological activities (Bowers *et al.*, 1976). These allelochemicals are released from *A. conyzoides*, accumulate in the soil with the passage of time and were found to be responsible for growth suppression of weeds and reducing the population of soil-borne pathogenic fungi in intercropped citrus orchards (Kong *et al.*, 2004). Kong *et al.* (1999, 2002) demonstrated that fresh leaves and volatile oils of *A. conyzoides* exhibited adverse effect on crops, and they attributed this to the presence of precocenes and their derivatives, i.e. precocene I, precocene II, 3,3-dimethyl-tert-butylindone and β-caryophyllene, and several mono-terpenes and sesquiterpenes.

### *Non-volatile compounds*

In addition, *A. conyzoides* is very rich in poly-oxygenated flavonoids such as kaempferol, quercetin and their glucosides, and small quantities of triterpenoids and sterols (Okunade, 2002). A number of phenolic acids such as gallic, coumalic, protocatechuic, benzoic, sinapic, *p*-hydroxybenzoic and coumaric acid have been reported as active compounds in *A. conyzoides* (Xuan *et al.*, 2004). Batish *et al.* (2009a, b) reported the non-volatile components catechin and phenolic acids (gallic acid, coumalic acid, protocatechuic acid and *p*-hydroxybenzoic acid) from leaves and soil inhabited by the weed. However, the level of allelochemicals and hence allelopathic potential depends upon the growth stage and type of habitat (Kong *et al.*, 2004). Its seeds have also been known to yield fatty oils (Ambasta, 1992). These allelochemicals show a synergistic effect, and their allelopathic potential is intensified on exposure to various environmental stresses (Josep and Joan, 1997; Kong *et al.*, 2002; Batish *et al.*, 2009a, b).

## Uses of *A. conyzoides*

Although this weed is becoming a serious concern for farmers, horticulturalists, foresters, land managers, ecologists and environmentalist, the plant also has some useful aspects, including the following.

### *Medicinal value*

*Ageratum conyzoides* is an integral part of traditional medicine in many countries of world, particularly in tropical and subtropical regions. The extracts and metabolites of *A. conyzoides* have been used as a bacteriocide, antidysentric and antilithic (Borthakur and Baruah, 1987; Okunade, 2002) by traditional communities in India, South America and Africa. In Central Africa, Brazil and Congo it is used to cure pneumonia, wounds and burns (Ming, 1999). Its decoction is given to treat headache, fever and rheumatism (Kirtikar and Basu, 1984; Okunade, 2002). A crude material from leaves of species possesses *in vitro* antibacterial activity against *Staphylococcus aureus* and is also used for wound healing (Durodola, 1977). In Cameroon, aqueous extracts of the whole plant are known for their anti-diabetic properties (Tsabang *et al.*, 2001).

### *Insecticidal/bactericidal/antifungal value*

The oil obtained from its leaves and inflorescence has insect-repellent properties (Saxena *et al.*, 1994; Nogueira *et al.*, 2010). Traditional communities in Asia, South America and Africa use the aqueous extracts of the species as a bactericidal (Borthakur and Baruah, 1987; Ekundayo *et al.*, 1988). Leaf volatile oils of *A. conyzoides* have been reported to kill the maize grain weevil, *Sitophilus zeamais* (Bouda *et al.*, 2001). The weed has also been reported to have natural fungicidal potential (Pu *et al.*, 1990; Gravena *et al.*, 1993).

### *Herbicidal value*

*Ageratum conyzoides* has been reported as a natural herbicide for weed control in paddy fields (Xuan *et al.*, 2004). Application of *A. conyzoides* at 2 t/ha served as an effective herbicide in controlling paddy weeds such as *Echinochloa crus-galli* var. *formosensis*, *Monochoria vaginalis* var. *plantaginea* and *Aeschynomene indica*. Ramakrishna *et al.* (2006) reported that a mulch of *A. conyzoides* had good potential in checking the growth of weeds and enhancing crop yield.

### *Others*

A mulch of *A. conyzoides* has been reported to prevent loss of soil moisture and reduce erosion (Ramakrishna *et al.*, 2006). It can serve as good biological waste for the preparation of manure and biogas. Dried plant material from *A. conyzoides* is used for pest control (Vallador *et al.*, 1994). *Ageratum conyzoides* planted as a riparian wetland herb on the banks of the Rihand River in Renukoot plays a positive role in reducing erosion of organic carbon and cationic nutrients and helps in soil conservation (Kumar *et al.*, 1996). Recently, it has been observed that *A. conyzoides* helps in maintaining the texture and fertility of soil along the River Damodar riparian zone in Bhowra, as soil erosion was minimized to a great extent following its planting compared with bare areas (Srivastava *et al.*, 2009).

## Control/management of *A. conyzoides*

To date, a number of strategies including physical, chemical and ecological methods have been tried, though none has been found completely to control or manage the weed.

### *Physical methods*

These include manual uprooting, mowing, cutting with machete or burning of live plants. In general, these are of some use when the plant is at the vegetative stage. However, certain limitations are associated with these methods, for example the high cost of labour, ill-effects on workers' health, vegetative regeneration from stolons, etc. (Batish *et al.*, 2004a). Furthermore, at the maturity stage, uprooting of plants results in dispersal of seeds to other areas.

### *Chemical methods*

Pre-emergence application of commercial herbicides such as atrazine, diuron, methazole, metribuzin or simazine is an effective control strategy. For established infestations, the application of 2,4-D and 2,4,5-T provides excellent control of this weed (Rao, 2000). However, due to toxicological implications, synthetic herbicides are not recommended as a good method of controlling *A. conyzoides*.

### *Field and crop management*

To optimize weed control, farmers are required to know the time of emergence of weed seedlings relative to the crop, apart from any use of commercial herbicides (Ekeleme *et al.*, 2005). In northern Laos, increased fallow periods in slash-and-burn rice production systems have resulted in reduced weed infestation (Roder *et al.*, 1998).

### *Ecological methods*

The use of eco-friendly compounds such as natural plant products and other plant-based formulations offers a safer strategy for effective weed control. Volatile monoterpenes such as cineole and citronellol have been found to have considerable effects on germination, growth, chlorophyll content and cellular respiration of *A. conyzoides*, suggesting much potential for weed management (Singh *et al.*, 2002a). A study conducted by Batish *et al.* (1997) and Singh *et al.* (2002b) revealed that parthenin, a sesquiterpene lactone from *Parthenium hysterophorus*, negatively affected the growth and physiology of *A. conyzoides* and has potential for use as a novel agrochemical.

### *Role of community*

The methods outlined above cannot achieve successful weed control and management on their own; an integrated approach, incorporating all the above approaches, is essential. Furthermore, fallow lands likely to be invaded by the species should be put to some use such as planting of native grasses and fodder. To this end, local communities play a significant role (Anonymous, 2003). Local populations need to learn about environmental issues, the ecological hazards of *A. conyzoides* and the identification of the species so that it may be uprooted at an early stage before it spreads. Furthermore, organic compost and vermicompost may be useful to control and manage the weed (Batish *et al.*, 2004a).

## Conclusions

From the above discussion, we conclude that *Ageratum conyzoides* is one of the most harmful exotic weeds in a wide variety of habitats and adversely affects many facets of ecosystem function. The weed possesses many physiological, ecological and morphological features that help in its successful invasion. Therefore, efforts should be made to manage this weed through an integrated approach involving different management strategies in combination with community participation.

## References

Akter, A. and Zuberi, M.I. (2009) Invasive alien species in Northern Bangladesh, Identification, inventory and impacts. *International Journal of Biodiversity and Conservation* 1, 129–134.

Ambasta, S.P. (1992) *The Useful Plants of India.* National Institute of Science Communication, Delhi, India.

Anonymous (1986/87) *Annual Report of the All-India Coordinated Research Project on Weed Control.* Himachal Pardesh Krishi Vishavvidhalya, Palampur, India.

Anonymous (1988) The wealth of India – raw materials. Publications and Information Directorate, Council of Scientific and Industrial Research, New Delhi, India.

Anonymous (2003) *Report of Two-day Workshop on the Management of Weeds (Congress grass, Lantana, Neela Fulnu) through Community Participation*, 25–26 March 2003. State Council for Science, Technology and Environment, Shimla, India, pp. 2–3, 16.

Arora, V. (1999) Comparative assessment of eco-physiological functions between *Ageratum conyzoides L.* and *Parthenium hysterophorus L.* PhD thesis, Panjab University, Chandigarh. India.

Arunachalam, A. and Arunachalam, K. (2002) Evaluation of bamboos in eco-restoration of 'jhum' fallows in Arunachal Pradesh, ground vegetation, soil and microbial biomass. *Forest Ecology and Management* 159, 231–239.

Bais, H.P., Park, S.-W., Weir, T.L., Callaway, R.M. and Vivanco, J.M. (2004) How plants communicate using the underground information superhighway. *Trends in Plant Science* 9, 26–32.

Bansal, G.L. (1988) Weed biology of *Lantana* and *Ageratum* spp. In: Negi, G.C. and Singh, C.M. (eds) *Proceedings of Seminar on Control of Lantana and Ageratum Species.* Himachal Pradesh Agriculture University, Palampur, India. pp. 10.

Bansal, G.L. and Singh, C.M. (1986) Effect of herbicides on control and germination of *Ageratum conyzoides. Indian Journal of Weed Science* 18, 254–256.

Batish, D.R. (2008) *Tropical American Invasive Weeds in the Shiwalik Range of North Western Himalayas of India: An Assessment of Status and Impact.* Available online at http,//www.apafri.org/forestHealth08/DAY2_Forest%20Health%20Workshop/Daizy_APAFRI_DAIZY_Malaysia-1.pdf (accessed 18 October 2009).

Batish, D.R., Kohli, R.K., Singh, H.P. and Saxena, D.B. (1997) Studies on herbicidal activity of parthenin–a constituent of *Parthenium hysterophorus* towards billy-goat weed. *Current Science* 73, 369–371.

Batish, D.R., Singh, H.P., Kohli, R.K., Johar, V. and Yadav, S. (2004a) Management of invasive exotic weeds requires community participation. *Weed Technology* 18, 1445–1448.

Batish, D.R., Singh, H.P. Kaur, S., Arora, V. and Kohli, R.K. (2004b) Allelopathic Interference of residues of *Ageratum conyzoides. Journal of Plant Disease and Protection* XIX, 293–299.

Batish, D.R., Singh, H.P., Kaur, S. and Kohli, R.K. (2006) Phytotoxicity of *Ageratum conyzoides* residues towards growth and nodulation of *Cicer arietinum. Agriculture Ecosystems and Environment* 113, 399–401.

Batish, D.R., Kaur, S., Singh, H.P. and Kohli, R.K. (2009a) Role of root-mediated interactions in phytotoxic interference of *Ageratum conyzoides* with rice (*Oryza sativa*). *Flora* 204, 388–395.

Batish, D.R., Kaur, S., Singh, H.P. and Kohli, R.K. (2009b) Nature of interference potential of leaf debris of *Ageratum conyzoides. Plant Growth Regulation* 57, 137–144.

Batta, J.N. (1988) *Lantana camara* (Baraphulnu) – a threat to grazing and pasture land in Mid-Hill. In: Negi, G.C. and Singh, C.M. (eds) *Proceedings of Seminar on Control of Lantana and Ageratum.* Himachal Pardesh Krishi Vishavvidhalya, Palampur, India, pp.1–3.

Bentham, G. (1861) Flora Hongkongensis: A

description of the flowering plants and ferns. Lovell Reeve, London, 482 pp.

Borthakur, N. and Baruah, A.K.S. (1987) Search for precocenes in *Ageratum conyzoides* Linn. of North-East India. *Journal of the Indian Chemical Society* 64, 580–581.

Bouda, H., Tapondjou, L.A., Fomtem, D.A. and Gumedzoe, M.Y.D. (2001) Effect of essential oils from leaves of *Ageratum conyzoides*, *Lantana camara* and *Chromolaena odorata* on the mortality of *Sitophilus zeamais* (Coleoptera, Curculionidae). *Journal of Stored Products Research* 37, 103–109.

Bowers, W.S., Ohta, T. and Cleere, J.S. (1976) Discovery of insect antijuvenile hormones in plants. *Science* 193, 542–547.

Bughani, I. and Rajwar, G.S. (2005) Primary productivity and the impacts of the exotic weed *Eupatorium glandulosum* in a Montane grassland of Garhwal Himalaya. *The Environmentalist* 25, 31–38.

Burkill, H.M. (1985) *The Useful Plants of West Tropical Africa*. vol. 1. Royal Botanic Gardens, Kew, UK.

Cushman, J.H. and Gaffney, K.A. (2010) Community-level consequences of invasion, impacts of exotic clonal plants on riparian vegetation. *Biological Invasions* 8, 1073–1089.

Davies, K.W. and Svecjar, T.J. (2008) Comparison of medusahead-invaded and non-invaded Wyoming big sagebrush steppe in southeastern Oregon. *Rangeland Ecology and Management* 61, 623–629.

Davis, M.A. (2009) *Invasion Biology*. Oxford University Press, Oxford, UK, 244 pp.

DOEF (2010) *Assam's Rich Biodiversity*. Department of Environment and Forests, Government of Assam. Available at http//assamforest.in/environment/environment.php

Dogra, K.S. (2008) Impact of some invasive species on the structure and composition of natural vegetation of Himachal Pradesh. PhD thesis, Panjab University, Chandigarh, India.

Dogra, K.S., Kohli, R.K. and Sood, S.K. (2009a) An assessment and impact of three invasive species in the Shivalik hills of Himachal Pradesh, India. *International Journal of Biodiversity and Conservation* 1, 4–10.

Dogra, K.S., Kohli, R.K., Sood, S.K. and Dobhal, P.K. (2009b) Impact of *Ageratum conyzoides* L. on the diversity and composition of vegetation in the Shivalik hills of Himachal Pradesh (Northwestern Himalaya), India. *International Journal of Biodiversity and Conservation* 1, 135–145.

Durodola, J.I. (1977) Antibacterial property of crude extracts from a herbal wound healing remedy – *Ageratum conyzoides* L. *Planta Medica* 32, 388–390.

Ekeleme, F., Forcella, F., Archer, D.W., Akobundu, O. and Chikoye, D. (2005) Seedling emergence model for tropic ageratum (*Ageratum conyzoides*). *Weed Science* 53, 55–61.

Ekundayo, O., Sharma, S. and Rao, E.V. (1988) Essential oil of *Ageratum conyzoides*. *Planta Medica* 54, 55–57.

Fosberg, F.R. (1957) The Maldive Islands, Indian Ocean. *Atoll Research Bulletin* 58, pp. 37. Pacific Science Board, National Academy of Sciences, Washington, DC. Available at http//www.botany.hawaii.edu/faculty/duffy/ARB/051-58/58.pdf

Fosberg, F.R., Falanruw, M.V.C. and Sachet, M.-H. (1975) Vascular flora of the northern Marianas Islands. *Smithsonian Contributions to Botany* 22, 1–45.

Foxcroft, L.C., Richardson, D.M. and Wilson, J.R.U. (2008) Ornamental plants as invasive aliens, problems and solutions in Kruger National Park, South Africa. *Environmental Management* 41, 32–51.

Gargominy, O., Bouchet, P., Pascal, M., Jaffre, T. and Tourneu, J.-C. (1996). Conséquences des introductions d'espèces animales et végétales sur la biodiversité en Nouvelle-Calédonie. *Terre Vie* [Review of Ecology] 51, 375–401.

GISD (2010) *Global Invasive Species Database*. Available at http//www.issg.org.database

Gonzalez, A.G., Aguiar, Z.E., Grillo, T.A., Luis, J.G., Rivera, A. and Calle, J. (1991) Chromenes form *Ageratum conyzoides*. *Phytochemistry* 30, 1137–1139.

Goodell, K. (2008) Invasive exotic plant–bee interactions. In: James, R.R. and Pitts-Singer, T.L. (eds) *Bee Pollination in Agroecosystems*. Oxford University Press, Oxford, UK, pp. 166–183.

Gravena, S., Coletti, A. and Yamamoto, P.T. (1993) Influence of green cover with *Ageratum conyzoides* and *Eupatorium pauciflorum* on predatory and phytophagous mites in citrus. *Bulletin OILB-SROP* 16, 104–114.

Gupta, N., Anthwal, A. and Bahuguna, A. (2006) Biodiversity of mothronwala swamp, Doon Valley, Uttaranchal. *Journal of American Science* 2, 33–40.

Hao, J.H. and Qiang, S. (2005) Invasive alien weed–*Ageratum conyzoides*. *Weed Science* 4, 54–58 [in Chinese].

Henty, E.E. and Pritchard, G.H. (1975) *Weeds of New Guinea and Their Control*, 2nd edn. *Botany Bulletin* 7, 180, Department of Forests, Division of Botany, Lae, Papua New Guinea.

Holm, L.G., Plucknett, D.L., Pancho, J.V. and Herberger, J.P. (1977) *The World's Worst Weeds: Distribution and Biology*. An East-West Centre Book, University Press of Hawaii, Honolulu, Hawaii, 609 pp.

Hood, W.G. and Naiman, R.J. (2000) Vulnerability of riparian zones to invasion by exotic vascular plants. *Plant Ecology* 148, 105–114.

Hooker, J.D. (1882) *The Flora of British India*, vol. III. Reeve and Co., Ashford, UK.

Hu, E. and Kong, C. (1997) Allelopathy of *Ageratum conyzoides* L. I. Allelopathy of *A. conyzoides* aqueous extract and isolation and identification of its allelochemicals. *Chinese Journal of Applied Ecology* 8, 304–408 [in Chinese].

Huang, Q.Q., Wu, J.M., Bai, Y.Y., Zhou, L. and Wang, G.X. (2009) Identifying the most noxious invasive plants in China: Role of geographical origin, life form and means of introduction. *Biodiversity Conservation* 18, 305–316.

Hyde, M.A. and Wursten, B. (2010) *Flora of Zimbabwe: Species Information: Ageratum conyzoides.* Available at http//www.zimbabweflora.co.zw/speciesdata/species.php?species_id=158650

Jessup, L.W. (2002) *Queensland Herbarium. Flora of South-eastern Queensland. Volume 2. Changes to Names or Status of Taxa.* Queensland Herbarium Environmental Protection Agency, Brisbane Botanic Gardens, Toowong, Queensland, Australia.

Jha, S. and Dhakal, M. (1990) Allelopathic effects of various extracts of some herbs on rice and wheat. *Journal of the Institute of Agriculture and Animal Science* 11, 121–123.

Johnson, M.F. (1971) A monograph of the genus *Ageratum* L. (Compositae, Eupatorieae). *Annals of the Missouri Botanical Garden* 58, 6–88.

Josekutty, P.C., Wakuk, E.E. and Joseph, M.J. (2002) Invasive/weedy angiosperms in Kosrae. Federated States of Micronesia. In: *Invasive Species and Their Management. Micronesica Supplement* 6, 61–65.

Josep, P. and Joan, L. (1997) Effects of carbon dioxide, water supply and seasonality on terpene content and emission by *Rosmarinus officinalis*. *Journal of Chemical Ecology* 23, 979–993.

Kalia, P.K. (1998) Allelopathic properties of *Ageratum* – a cause of threat to plant diversity. PhD thesis, Panjab University, Chandigarh, India.

Kanwar, B.S. and Kharwara, P.C. (1988) Infestation of *Lantana* and *Ageratum* in Kangra Valley in Himachal Pradesh. In: Negi, G.C. and Singh, C.M. (eds) *Proceedings of Seminar on Control of Lantana and Ageratum*, Himachal Pardesh Krishi Vishavvidhalya, Palampur, India, pp. 4–10.

Kashina, B.D., Mabagala, R.B. and Mpunami, A.A. (2003) First report of *Ageratum conyzoides* L. and *Sida acuta* Burm F. as new weed hosts of tomato yellow leaf curl Tanzania virus. *Plant Protection Science* 39, 18–22.

Kato-Noguchi, H. (2001) Assessment of the allelopathic potential of *Ageratum conyzoides. Biologia Plantarum* 44, 309–311.

Kirtikar, K.R. and Basu, B.D. (1984) *Indian Medicinal Plants*, vol. 12. Periodical Book Agency, Delhi. India, pp. 1330.

Kissmann, G. and Groth, D. (1993) *Plantas Infestantes e Nocivas.* Basf Brasileira, São Paulo, Brazil.

Kohli, R.K. and Batish, D.R. (1996) Environmental threat from billy-goat weed *Ageratum conyzoides. Environmental Research Institute of Michigan (ERIM, Ann Arbor, Michigan)* 11, 925–929.

Kohli, R.K., Dogra, K.S., Batish, D.R. and Singh, H.P. (2004) Impact of invasive plants on the structure and composition of natural vegetation of northwestern Indian Himalayas. *Weed Technology* 18, 1296–1300.

Kohli, R.K., Batish, D.R., Singh, H.P. and Dogra, K.S. (2006) Status, invasiveness and environmental threats of three tropical American invasive weeds (*Parthenium hysterophorus* L., *Ageratum conyzoides* L., *Lantana camara* L.) in India. *Biological Invasions* 8, 1501–1510.

Kohli, R.K., Jose, S., Singh, H.P. and Batish, D.R. (2009) *Invasive Plants and Forest Ecosystems.* CRC Press, Boca Raton, Florida, pp. 437.

Kolb, A., Alpert, P., Enters, D. and Holzapfel, C. (2002) Patterns of invasion within a grassland community. *Journal of Ecology* 90, 871–881.

Kong, C., Xu, T. and Hu, F. (1998a) Allelopathy of *Ageratum conyzoides*. II. Releasing mode and activity of main allelochemicals. *Chinese Journal of Applied Ecology* 9, 257–260.

Kong, C., Xu, T. and Hu, F. (1998b) Study on interactions among allelochemicals of *Ageratum conyzoides. Acta Phytoecologica Sinica* 22, 403–408.

Kong, C., Hu, F., Xu, T. and Lu, Y. (1999) Allelopathic potential and chemical constituents of volatile oil from *Ageratum conyzoides. Journal of Chemical Ecology* 25, 2347–2356.

Kong, C., Hu, F. and Xu, X. (2002) Allelopathic potential and chemical constituents of volatiles from *Ageratum conyzoides* under stress. *Journal of Chemical Ecology* 28, 1573–1581.

Kong, C., Liang, W., Hu, F., Xu, X., Wang, P., Jiang, Y., Jiang *et al.* (2004) Allelochemicals and their transformations in the *Ageratum conyzoides* intercropped citrus orchard soils. *Plant and Soil* 264, 149–157.

Kumar, R., Ambasht, R.S., Srivastava, A.K. and Srivastava, N.K. (1996) Role of some riparian wetland plants in reducing erosion of organic carbon and selected cations. *Ecological Engineering* 6, 227–239.

Kumar, S. and Singh, C.M. (1988) Studies on control of *Ageratum conyzoides* under mid-hill conditions of Himachal Pradesh. In: Negi, G.C. and Singh, C.M. (eds) *Proceedings, Seminar on Control of Lantana and Ageratum species*, Himachal Pradesh Agricultural University, Palampur, India, pp. 74.

Kunwar, R.M., Gupta, V.N.G., Vetaas, O.R. and Chaudhary, R.P. (2001) Vegetation dynamics of herbaceous plants in lower Arun river basin of Makalu Barun buffer zone, east Nepal. In: Watanabe, T., Sicroff, S., Khanal, N.R. and Gautam, M.P. (eds) *Proceedings of International Symposium on the Himalayan Environments: Mountain Sciences and Ecotourism/ Biodiversity*, 24–26 November 2000, Kathmandu. Kathmandu, Hokkaido University, Tribhuvan University and the United Nations University, pp. 103–110.

Li-ying, L., Ren, W. and Waterhouse, D.F. (1997) *The Distribution and Importance of Arthropod Pests and Weeds of Agriculture and Forestry Plantations in Southern China*. ACIAR, Canberra, Australia, pp. 185.

Lorence, D.H. and Wagner, W.L. (2008) *Flora of the Marquesas Islands*. National Tropical Botanical Garden and the Smithsonian Institution. Available at http//botany.si.edu/pacificislandbiodiversity/ marquesasflora/index.htm

MacDonald, I.A.W., Thebaud, C., Strahm, W.A. and Strasberg, D. (1991) Effects of alien plant invasions on native vegetation remnants on La Reunion (Mascarenes Islands, Indian Ocean). *Environmental Conservation* 18, 51–61.

Manandhar, S., Shrestha, B.B. and Lekhak, H.D. (2007) Weeds of a paddy field at Kirtipur, Kathmandu. *Scientific World* 5, 100–106.

Manner, H.I. and Mallon, E. (1989) An annotated list of the vascular plants of Pulawat. *Atoll Micronesica* 22, 23–63.

Masters, R.A. and Sheley, R.L. (2001) Principles and practices for managing rangeland invasive plants. *Journal of Range Management* 54, 502–517.

McMullen, C.K. (1999) *Flowering Plants of the Galápagos*. Comstock Publication Association, Ithaca, New York, pp. 370.

Ming, L.C. (1999) *Ageratum conyzoides*, A tropical source of medicinal and agricultural products. In: Janick, J. (ed.) *Perspectives on New Crops and New Uses*. ASHS Press, Alexandria, Virginia, pp. 469–473.

Mooney, H.A. and Cleland, E.E. (2001) The evolutionary impact of invasive species. *Proceedings of National Academy of Sciences U.S.A.* 98, 5446–5451.

Muthuramkumar, S., Ayyappan, N., Parthasarathy, N., Mudappa, D., Raman, T.R.S., Selwyn, M.A. *et al.* (2006) Plant community structure in tropical rain forest fragments of the Western Ghats, India. *Biotropica* 38, 143–160.

National Focal Point for APFISN, India (2005) *Stocktaking of National Forest Invasive Species Activities*. Ministry of Environment and Forests, Delhi, India.

Negi, G.C. (1988) In: Negi, G.C. and Singh, C.M. (eds) *Proceedings of Seminar on Control of Lantana and Ageratum Species*, Himachal Pradesh Agricultural University, Palampur, India.

Negi, P.S. and Hajra, P.K. (2007) Alien flora of Doon Valley, Northwest Himalaya. *Current Science* 92, 968–978.

Nel, J.L., Richardson, D.M., Rouget, M., Mgidi, T.N., Mdzeke, N., Le Maitre, D.C. *et al.* (2004) A proposed classification of invasive alien plant species in South Africa, towards prioritizing species and areas for management action. *South African Journal of Science* 100, 53–64.

Nogueira, J.H.C., Gonçalez, E., Galleti, S.R., Facanali, R., Marques, M.O.M. and Flicio, J.D. (2010) *Ageratum conyzoides* essential oil as aflatoxin suppressor of *Aspergillus flavus*. *International Journal of Food Microbiology* 137, 55–60.

Okunade, A.L. (2002) *Ageratum conyzoides* L. (Asteraceae). *Fitoterapia* 73, 1–16.

Orchard, T. and Anthony, E. (1994) *Flora of Australia*, vol. 49, Oceanic Islands 1. Australian Government Publishing Service, Canberra, Australia.

Pacific Island Ecosystems at Risk (PIER) (2008) *Ageratum conyzoides*. Available at http//www.hear.org/Pier/species/ageratum_conyzoides.htm

Pari, K., Rao, P.J., Subrahmanyam, B., Rasthogi, J.N. and Devakumar, G. (1998) Benzofuran and other constituents of the essential oil of *Ageratum conyzoides*. *Phytochemistry* 49, 1385–1388.

Pimentel, D., Zuniga, R. and Morrison, D. (2005) Update on the environmental and economic costs associated with alien invasive species in the United States. *Ecological Economics* 52, 273–288.

Prasad, K. and Srivastava, V.C. (1991) Teletoxic effect of some weeds on germination and initial growth of groundnut (*Arachis hypogea*). *Indian Journal of Agricultural Science* 61, 493–494.

Pu, T.S., Liao, K.Y. and Chang, T. (1990) Investigations on predatious mite resources in citrus orchards in Guang Xi and their utilization. *Acta Phytophyarica Sinica* 17, 355–358.

Qasem, J.R. and Foy, C.L. (2001) Weed allelopathy,

its ecological impacts and future prospects: a review. *Journal of Crop Production* 4, 43–119.
Ramadhanil, Tjitrosoedirdjo, S.S and Setiadi, D. (2008) Structure and composition of understory plant assemblages of six land use types in the Lore Lindu National Park, Central Sulawesi, Indonesia. *Bangladesh Journal of Plant Taxonomy* 15, 1–12.
Ramakrishna, A., Tamb, H.M., Wani, S.P. and Long, T.D. (2006) Effect of mulch on soil temperature, moisture, weed infestation and yield of groundnut in northern Vietnam. *Field Crops Research* 95, 115–125.
Rana, J.C. and Sharma, S.K. (2009) Impact of climate change on floral composition in the Western Himalayan region and adaptive strategies thereof. In: *National Conference on Forestry Solutions: Strategies for Mitigation and Adaptations of the Impacts of Climate Change in Western Himalayan Mountain States*, Shimla, Himachal Pradesh, India, 19–21 November. Available at http//www.hpforestconference.org/ebooks/Rana's%20paper.pdf
Rana, J.C., Singh, A., Sharma, Y., Pradheep, K. and Mendiratta, N. (2010) Dynamics of plant bioresources in Western Himalayan region of India – watershed based study. *Current Science* 98, 192–203.
Rao, V.S. (2000) *Principles of Weed Science*. Science Publishers, Enfield, New Hampshire, 555 pp.
Rasingam, L. and Parthasarathy, N. (2009) Diversity of understory plants in undisturbed and disturbed tropical lowland forests of Little Andaman Island, India. *Biodiversity Conservation* 18, 1045–1065.
Rastogi, R.P. and Mehrotra, B.N. (1990) *Compendium of Indian Medicinal Plants*. Central Drug Research Institute/Publication and Information Directorate, Lucknow, India.
Rawat, G.S., Goyal, S.P. and Johnsingh, A.J.T. (1997) Ecological observations on the grasslands of Corbett Tiger Reserve, India. *Indian Forester* 123, 958–963.
Reddi, S., Soundra, G.H., Rajan, M.S. and Naidu, C. (1977) *Ageratum conyzoides* L. a new troublesome weed in sugarcane. In: *Proceedings ISWSI, Andhra Pradesh Agricultural University, Weed Science Conference*, Hyderabad, India. p. 261.
Reddy, C.S. and Pattanaik, C. (2009) An Assessment of Floristic Diversity of Gandhamardan Hill Range, Orissa, India. *Bangladesh Journal of Plant Taxonomy* 16, 29–36.
Reddy, C.S., Bagyanarayana, G., Reddy, K.N. and Raju, V.S. (2008) *Invasive Alien Flora of India*. NBII/USGS, Washington, DC. Available at http//www.gisinetwork.org
Reddy, C.S., Rangaswamy, M., Pattanaik, C. and Jha, C.S. (2009) Invasion of alien species in wetlands of Samaspur bird sanctuary, Uttar Pradesh, India. *Asian Journal of Water, Environment and Pollution* 6, 43–50.
Riaz, T., Khan, S.N., Javaid, A. and Farhan, A. (2007) Weed flora of *Gladiolus* fields in Lahore, Pakistan. *Pakistan Journal of Weed Science Research* 13, 113–120.
Roder, W., Phengchanh, S. and Keoboulapha, B. (1995) Relationships between soil, fallow period, weeds and rice yield in slash-and-burn systems of Laos. *Plant and Soil* 176, 27–36.
Roder, W., Keoboulapha, B., Phengchanh, S., Prot, J.C. and Matias, D. (1998) Effect of residue management and fallow length on weeds and rice yield. *Weed Research* 38, 167–174.
Rodriguez, G.S. and Cepero, G.S. (1984) Number of seeds produced by some weed species. *Centro Agricola* 11, 45–50.
Sachet, M.-H. (1983) Natural history of Mopelia Atoll, Society Islands. *Atoll Research Bulletin* 274, 37. Smithsonian Institution, Washington, DC. Available at http//www.botany.hawaii.edu/faculty/duffy/arb/273-281/274.pdf
Sankaran, K.V. (2007) Billy-goat weed (*Ageratum conyzoides*). *Newsletter of the Asia-Pacific Forest Invasive Species Network* (APFISN ) 13, 2.
Sauerborn, J. and Kock, W. (1988) Untersuchungen zur Keimungsbiologie von sechs tropischen Segetalaten. *Weed Research* 28, 47–52.
Saxena, R.C., Saxena, A. and Singh, C. (1994) Evaluation of growth disrupting activity of *Ageratum conyzoides* crude extract on *Culex quinquefasciatus* (Diptera, Culicidae). *Journal of Environmental Biology* 15, 67–74.
Sharma, P.D. and Sharma, O.P. (1995) Natural products chemistry and biological properties of *Ageratum* plant. *Toxicological and Environmental Chemistry* 50, 213–232.
Siebert, S.F. (2002) From shade- to sun-grown perennial crops in Sulawesi, Indonesia, Implications for biodiversity conservation and soil fertility. *Biodiversity and Conservation* 11, 1889–1902.
Silori, C.S. and Mishra, B.K. (2001) Assessment of livestock grazing pressure in and around the elephant corridors in Mudumalai Wildlife Sanctuary, south India. *Biodiversity and Conservation* 10, 2181–2195.
Singh, S.P., Pal, U.R. and Luka, K. (1989) Allelopathic effects of three serious weeds of Nigerian savanna on germination and seedling vigour of soybean and maize. *Journal of Agronomy and Crop Science* 162, 236–240.
Singh, H.P., Batish, D.R. and Kohli, R.K. (2002a)

Allelopathic effect of two volatile monoterpenes against billy goat weed (*Ageratum conyzoides* L.). *Crop Protection* 21, 347–350.

Singh, H.P., Batish, D.R., Kohli, R.K., Saxena, D.B. and Arora, V. (2002b) Effect of parthenin – a sesquiterpene lactone from *parthenium hysterophorus* – on early growth and physiology of *Ageratum conyzoides*. *Journal of Chemical Ecology* 28, 2169–2179.

Singh, H.P., Batish, D.R., Kaur, S. and Kohli, R.K. (2003) Phytotoxic interference of *Ageratum conyzoides* with wheat (*Triticum aestivum*). *Journal of Agronomy and Crop Science* 189, 341–346.

Singh, T.P., Singh, S., Roy, P.S. and Rao, B.S.P. (2002) Vegetation mapping and characterization in West Siang District of Arunachal Pradesh, India – a satellite remote sensing-based approach. *Current Science* 83, 1221–1230.

Sit, A.K., Bhattacharya, M., Sarkar, B. and Arunachalam, V. (2007) Weed floristic composition in palm gardens in plains of eastern Himalayan region of West Bengal. *Current Science* 92, 1434–1439.

Smith, T. and Albert, C. (1991) *Flora Vitiensis Nova: a New Flora of Fiji.* National Tropical Botanical Garden, Lawai, Kauai, Hawaii, vol. 5, 626 pp.

Space, J.C. and Flynn, T. (2002) Report to the Government of Samoa on Invasive Plant Species of Environmental Concern. USDA Forest Service, Honolulu, Hawaii, 83 pp.

Space, J.C., Waterhouse, B., Miles, J.E., Tiobech, J. and Rengulbai, K. (2003) Report to the Republic of Palau on Invasive Plant Species of Environmental Concern. USDA Forest Service, Honolulu, Hawaii, 177 pp.

Space, J.C., Waterhouse, B.M., Newfield, M. and Bull, C. (2004) Report to the Government of Niue and the United Nations Development Programme: Invasive plant species on Niue following Cyclone Heta. UNDP NIU/98/G31 – Niue Enabling Activity, United Nations, Rome, 80 pp.

Space, J.C., Lorence, D.H. and LaRosa, A.M. (2009) Report to the Republic of Palau: 2008 update on Invasive Plant Species. USDA Forest Service, Hilo, Hawaii.

Srivastava, N.K., Ram, L.C. and Masto, R.E. (2009) Role of selected riparian herbs in reducing soil erosion and nutrient loss under simulated rainfall. *Environment and Earth Science* 61, 405–417.

Stone, B.C. (1970) The flora of Guam. *Micronesica* 6, 1–659.

Swamy, P.S., Sundarapandian, S.M., Chandrasekar, P. and Chandrasekaran, S. (2000) Plant species diversity and tree population structure of a humid tropical forest in Tamil Nadu, India. *Biodiversity and Conservation* 9, 1643–1669.

Swarbrick, J.T. (1997) Environmental weeds and exotic plants on Christmas Island, Indian Ocean: A report to Parks Australia, Canberra, pp. 101.

Thaman, R.R., Fosberg, F.R., Manner, H.I. and Hassall, D.C. (1994) The flora of Nauru. *Atoll Research Bulletin* 392, 1–223. Smithsonian Institution, Washington, DC.

Thapa, C.B. and Kayastha, R.S. (1998) Enumeration of weed in the paddy fields of Pokhara Valley. *Prithwi* 2/3, 1–8.

Toyoda and Takeshi (2003) *Flora of the Bonin Islands.* Aboc & Co., Kamakura, Japan, pp. 522.

Tsabang, N., Nkongmeneck, B.A., Zapfack, L., Dongmo, Z., Nguenang, G.M., Lando, G. *et al.* (2001) Inventaire des plantes à vertus antidiabétiques dans la région de Yaoundé au Cameroun. *Revue de Médecines et Pharmacopées Africaines* 15, 87–94.

United Nations Environment Programme–World Conservation Monitoring Centre (UNEP–WCMC) (2000) *Biodiversity Profile of India.* Available at http//sea.unep-wcmc.org/nbp/index.html

USDA-ARS (2009) *Taxon,* Ageratum conyzoides *L.* Available at http//www.ars-grin.gov/cgi-bin/npgs/html/taxon.pl?103793 (accessed 21 October 2009).

Vallador, D., Fonollera, V., Nillama, N., Josue, A. and Dumalag, P. (1994) Screening of indigenous seed treatment materials for village level corn storage. *CMU Research Journal* 7, 2–15.

Wagner, W.L., Herbst, D.R. and Sohmer, S.H. (1999) *Manual of the Flowering Plants of Hawaii* (revised edn., Bernice P. Bishop Museum special publication). University of Hawaii Press/Bishop Museum Press, Honolulu, Hawaii, 1919 pp.

Wandji, J., Bissangou, M.F. and Ouambra, J.M. (1996) Essential oil of *Ageratum conyzoides*. *Fitoterapia* 67, 427–431 [in French].

Waterhouse, D.F. (1993) *The Major Arthropod Pests and Weeds of Agriculture in Southeast Asia.* Australian Centre for International Agricultural Research, Canberra, Australia, 141 pp.

Weidenhamer, J.D. and Callaway, R.M. (2010) Direct and indirect effects of invasive plants on soil chemistry and ecosystem function. *Journal of Chemical Ecology* 36, 59–69.

Welsh, S.L. (1998) *Flora Societensis: a Summary Revision of the Flowering Plants of the Society Islands.* E.P.S. Inc., Orem, Utah, 420 pp.

Whistler, W.A. (1988) Checklist of the weed flora of

Western Polynesia. Technical Paper No. 194, South Pacific Commission, Noumea, New Caledonia, 69 pp.

Wiedentold, H. and Roder, E. (1991) Pyrrolizidine alkaloids from *Ageratum conyzoides* L. *Planta Medica* 57, 578–579.

Xuan, D., Shinkichi, T., Hong, N.H., Khanh, T.D. and Min, C. (2004) Assessment of phytotoxic action of *Ageratum conyzoides* L. (billy goat weed) on weeds. *Crop Protection* 23, 915–922.

Yan, X., Zhenyu, L., Gregg, W.P. and Dianmo, L. (2001) Invasive species in China – an overview. *Biodiversity and Conservation* 10, 1317–1341.

Yoganarnarasimham, S.N. (2000) *Medicinal Plants of India—Tamil Nadu*, vol. 2. Regional Research Institute, Bangalore, India, 27 pp.

Yuncker, T.G. (1943) The flora of Niue Island. *Bishop Museum Bulletin* 178, Bishop Museum Press, Honolulu, Hawaii, 126 pp.

Yuncker, T.G. (1959) Plants of Tonga. *Bishop Museum Bulletin* 220, Bishop Museum Press, Honolulu, Hawaii, 343 pp.

# 6 Predicting the Geographical Distribution of an Invasive Species (*Chromolaena odorata* L. (King) & H.E. Robins) in the Indian Subcontinent under Climate Change Scenarios

**S.K. Barik and D. Adhikari**

*Centre for Advanced Studies in Botany, North-Eastern Hill University, Shillong, India*

## Introduction

Biological invasion, coupled with climate change, threatens the global environment as well as economics (Thuiller *et al.*, 2007; Tripathi, 2009; Walther *et al.*, 2009). Rising temperatures, rapid economic development and invasion by alien species can potentially affect ecosystems, rapidly disassemble communities and negatively impact native biodiversity (Sanders *et al.*, 2003; Lin *et al.*, 2007; Thuiller *et al.*, 2007; Kelly and Goulden, 2008; Walther *et al.*, 2009). Worldwide, alien invasive species cause an estimated annual economic loss of US$314 billion in the agriculture and forestry sectors, of which India's share is round US$116 billion (Pimentel *et al.*, 2001). Around 40% of the total plant species found in India are alien, of which 25% are considered to be invasive (Raghubanshi *et al.*, 2005). Invasive species such as *Chromolaena odorata*, *Lantana camara*, *Parthenium hysterophorous* and *Ageratum conyzoides* have caused serious ecological and health problems in this country (Tripathi *et al.*, 1981; Annapurna and Singh, 2003; Raghubanshi *et al.*, 2005; Negi and Hajra, 2007; Reddym, 2008). For early detection and control of these species in new areas it is essential to understand the environmental factors determining the species' distribution, ecological niche and actual and potential areas of invasion.

Ecological niche modelling (ENM) is a useful technique to depict the current as well as the future geographical distribution of species based on their ecological niche. ENM can therefore be used to predict the areas of invasion of alien plants as well as to assess the impact of climate change on their future distribution (Thuiller *et al.*, 2005; Hijmans and Graham, 2006). This method relates point distributional data of species to raster-based environmental grid data to construct a testable distributional hypothesis, usually in the form of a predictive distributional map. The resultant hypothesis is evaluated using various statistical tests to construct a distributional model for the species. Models generated by such techniques are transferable to varying geographical space and time (Peterson *et al.*, 1999) which makes it useful to track species' invasion in novel landscapes.

We selected *Chromolaena odorata* (Siam weed), an invasive shrub species belonging to the family *Asteraceae* native to continental America and now a cosmopolitan weed, for predicting its distribution and invasion in two different climate change scenarios in the Indian subcontinent. Characters such as high reproductive capacity, efficient dispersal mechanism, high competitive ability and wide ecological amplitude make the plant a troublesome weed and a successful invader (Joshi, 2001). The species is one of the world's 100 worst invasive alien species according to the Global Invasive Species Database (www.issg.org). In India, it was first introduced in 1845 in Calcutta as an ornamental plant (Muniappan and Bamba, 1999). It invaded north-eastern India during World War I and by 1924/1925 it spread to West Bengal and Orissa. From the eastern region it spread to Kerala in 1942, and then invaded rapidly to all the southern states (Singh, 2004). The species has also spread to Myanmar, Sri Lanka, Pakistan, Bhutan, Nepal, Bangladesh and Thailand. At present, the species is a serious weed with allelopathic effects in plantations, natural forests, pastures and natural reserves.

## Methods

### Species occurrence data

Distributional records for model building and evaluation in the native range were obtained from the web resource of Global Biodiversity Information Facility (http://www.gbif.org; Table 6.1). Overall, 458 records were collected for native range distributions in North and South America. These records were cleaned for spatial errors in Diva-GIS (Hijmans *et al.*, 2001) and duplicate records in the dataset were discarded using MS Excel. After data cleaning 400 occurrence records remained, which were used in model calibration and evaluation. Primary occurrence data in the Indian subcontinent were collected through field surveys in different parts of the subcontinent. Secondary occurrence data were obtained from the database of Indian Bioresource Information Network, Jeev Sampada (http://221.135.191.194/index.php) and other published literature (Singh, 2004; Shah and Khan, 2006; Negi and Hazra, 2007; Akter and Zuberi, 2009; Thomas *et al.*, 2009).

### Environmental data

For modelling the current distributional range of the species, raster data on eight environmental factors were used: data on physiographic factors – slope, aspect, topography and digital elevation model (http://www.usgs.org) and climatic factors – annual precipitation, mean annual temperature, mean annual minimum temperature and mean annual maximum temperature (http://www.worldclim.org), with a spatial resolution of 2.5 arc/min. The climatic variables, temperature and precipitation, along with the physiographic variables were chosen for modelling potential distribution as these have direct relevance to invasive species. The temperature variables describe the thermal tolerance of the species, the annual precipitation describes water availability and the physiographic factors describe the macro-habitat characteristics of the species. In order to avoid multicollinearity, derived bioclimatic variables were not used as these are strongly correlated to linear combinations of the four climatic variables listed above (Peterson and Nakazawa, 2008; Ficetola *et al.*, 2009).

To predict the possible impact of climate change on the future distributional range of invasive species, we used Hadley Centre coupled climate simulation models. These models were obtained from the worldclim website (http://www.worldclim.org), and are included in the third assessment report of IPCC (Intergovernmental Panel on Climate Change). Each climate model used

**Table 6.1**. Model calibration results for *Chromolaena odorata* derived from cross-validation runs. Area under the receiver operating characteristic curve (AUC) values for training and testing shown in the table are averages of the ten replicated runs.

| Total records | Training records | Testing records | AUC values | | |
|---|---|---|---|---|---|
| | | | Training | Testing | SD |
| 400 | 270 | 90 | 0.92 | 0.90 | 0.01 |

SD, standard deviation.

in the present study pertains to two time frames and two climate change scenarios – 2020 A2 & B2 and 2080 A2 & B2. Scenarios A2 and B2 represent two contrasting conditions of the future climate of the world that might develop in the 21st century based on a logical set of assumptions. Scenario A2 represents a future with a heterogeneous world with high human population growth, slow economic development and slow technological change, while B2 presents a conservative estimate of future climate changes and describes a world with intermediate human population and economic growth, emphasizing local solutions to economic, social and environmental sustainability (IPCC, 2007).

The resolution of environmental data is an important factor in determining model accuracy. This is related to preciseness in location of the species, size of sample, constraints in field survey, autocorrelational structure and computation time and resources. Therefore, resolution of environmental data affects the quality, relevance and dependability of the predictions (Guisan *et al.*, 2007). Spatial autocorrelation is a serious problem that could increase the significance value by up to 90-fold (Segurado *et al.*, 2006). Pixel resolution of raster-based environmental data is also associated with species range size, i.e. narrow-ranging species are associated more with fine-resolution microhabitat variables while wide-ranging species, i.e. the invasives, are associated with climatic variables (McPherson *et al.*, 2006; Luoto *et al.*, 2007). Considering the above facts, all environmental layers were re-sampled to a spatial resolution of 0.04° (~4 km) in order to match the species occurrence with environmental variables. The coarsening of environmental data might also reduce the possible bias of autocorrelation arising as a result of clumped sampling from the predictor space due to spatial clustering of some occurrence records.

## Model calibration

The fundamental niche of the species was modelled with the maximum entropy ecological niche modelling program, Maxent v3.3.1 (Phillips *et al.*, 2006). Maxent is one of the 'presence-only' groups of distributional modelling families that estimates the 'maximum entropy' probability distribution function to predict the geographical location of a species based on environmental variables. The program attempts to reconstruct the boundaries of the ecological niche by placing constraints on the probability distribution based on the environmental parameters of the grid-cell presence record (Phillips *et al.*, 2006). Maxent v3.3.1 facilitates replicated runs to allow cross-validation, bootstrapping and repeated subsampling in the model calibration process.

The niche model for *C. odorata* was constructed using 400 records from its native range in North and South America. Seventy-five per cent of the total records were used for model training and 25% were used for random testing. Other parameters were set to default as the program is already tuned to a wide range of datasets for plant and animal species (Phillips and Dudik, 2008). Preliminary trial runs changing the training and testing records yielded different prediction maps and AUC values in each run. To account for such variation and attain best fit we employed the cross-validation technique in Maxent by executing replicated runs. Due to constraints of computing time ten replicated runs were executed for the species, which generated average, maximum, minimum, median and standard deviation for the runs. Model quality was evaluated based on AUC values, and the model was graded as poor (AUC $<0.8$), fair ($0.8 <$ AUC $<0.9$), good ($0.9 <$ AUC $<0.95$) or very good ($0.95 <$ AUC $<1.0$), following Thuiller *et al.* (2006).

## Model transferability

The constructed niche model was transferred to the Indian subcontinent in order to anticipate current as well as future distributions. Prior to this, issues relating to transferability of ecological niche models addressed by several authors were reconsidered, and the inferences drawn from their studies were used to improve the generalizability of the niche model. Important issues on transferability of ecological niche models to novel landscapes – as well as under future climate change – have been discussed in detail by Peterson *et al.* (2007) and Phillips and Dudik (2008). It is inferred from their discussions that, for better transferability, species point data and sampled background data should be taken from entire area of the presence data of the species in its native range (Phillips, 2008). This inference is further strengthened by the proposal of Broennimann and Guisan (2008), which states that niche models trained using pooled data from all ranges reduce uncertainty and improve prediction of the extent of invasion. Estrada-Peña *et al.* (2007) opined that projections of the niche model derived from distributional data in the native range will not capture the full potential of the species if the native range distributional data are only a subset of the total climatic space occupied by the invasive species. However, Randin *et al.* (2006) suggested that, from a transferability perspective, a model with predictors covering the same or a wider range in the training region is more likely to give accurate predictions. In our study the native range (i.e. North and South America) is geographically much larger than the projected range (i.e. the Indian subcontinent; Fig. 6.1). Hence, we presume that the range of climatic gradients for *C. odorata* occurring in the invaded range is already present in its native range.

Nevertheless, transferability of the niche model is also sensitive to the number of predictor variables, model performance in the native range and use of derived climatic variables. Models generated using many predictor variables can overfit the data, and could reduce their transferability to novel landscapes (McCune, 2006; Randin *et al.*, 2006). The use of derived bioclimatic variables also reduces the transferability of models for species in the invasive range (Peterson and Nakazawa, 2008). Therefore, to enhance the generalizability of the model the number of environmental parameters was restricted to eight, which can be broadly classified into two categories – physiographic (slope, aspect, topography, digital elevation model) and climatic (total annual precipitation, mean annual temperature, mean annual maximum and mean annual minimum temperature).

### Model evaluation in the invaded landscape

The predictive capacity of the niche model transferred to the Indian subcontinent was evaluated by assessing the probability level at which the model was successful in correctly predicting the current distributional range of the selected species. Data on current distribution of the species in the projected range were collected through; (i) primary field surveys to collect GPS coordinates for species locations in north-eastern India; and (ii) occurrence records

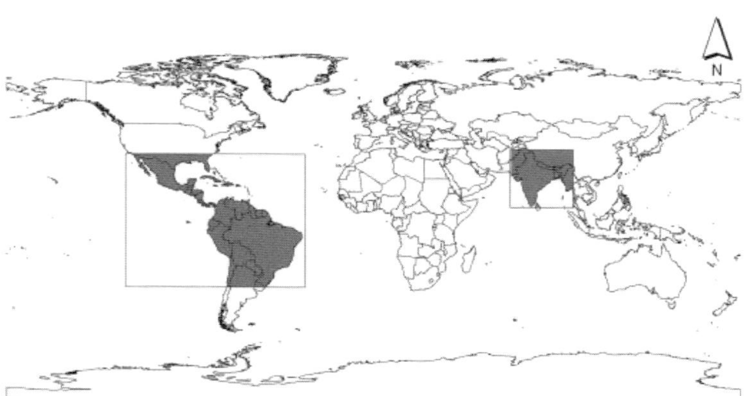


**Fig. 6.1.** Map showing the native range of *Chromolaena odorata* in North and South America and the projected geographical range in the Indian subcontinent.

from secondary sources, such as the database of the Indian Bioresource Information Network/Jeev Sampada and other published literature. The geographical coordinates of the place names collected from secondary sources were derived from the website http://www.fallinrain.com. Overall, 112 occurrence records were used for model validation, of which 49 were primary and 63 secondary records.

### Quantification of niche breadth

Niche breadth of a species is directly related to the extent of its geographic spread (Brown, 1984). To enable broad-scale comparison of niche model predictions under different climate change scenarios, we quantified the breadth of the fundamental niche of the selected species for different periods in the invaded range using ENMTools v1.0 (Warren *et al.*, 2008). The niche breadth metrics calculated using the ENMTools software are those of Levins (1968).

On the whole, we agreed with the following assumptions while modelling the distribution of the selected species under current as well as future climate change scenarios:

**1.** The selected species will retain the major ecological characteristics of its environmental niche following the 'niche conservatism' hypothesis (Peterson *et al.*, 1999).

**2.** Biotic factors, e.g. competition, will not constrain the species in realizing the full potential of its fundamental niche, as the species has a competitive edge over other species.

**3.** Anthropogenic activity will enable the species to overcome natural geographic barriers and fill the entire breadth of its fundamental niche.

## Results

The model calibration test for *C. odorata* in its native range yielded 'good' results ($AUC_{training}$ = 0.92 and $AUC_{testing}$ = 0.90). The standard deviation between the $AUC_{training}$ and $AUC_{testing}$ values for all species was <1.5 % (Table 6.1). Among all input environmental parameters, the climatic variables were more sensitive and contributed 62% to the structure of the modelled ecological niche of the species as compared with the physiographic variables, which contributed only 38%. The thermal variables – mean annual temperature (20.4%) and mean annual minimum temperature (29%) – governed the macro-scale niche characteristic (Fig. 6.2).

The niche model showed that the species' response peaked at a mean annual temperature range of 15–25°C, and there was a consistent increase in responsiveness with an increase in mean annual minimum temperature (Fig. 6.3).

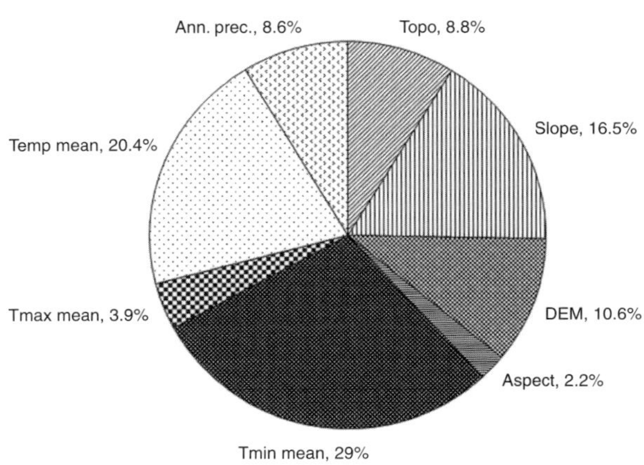


**Fig. 6.2.** Relative contribution of different environmental variables to the niche model of *C. odorata*. Ann. prec., annual precipitation; DEM, digital elevation model; Temp mean, mean annual temperature; Tmax mean, mean annual maximum temperature; Tmin mean, mean annual minimum temperature; Topo, topographical index.

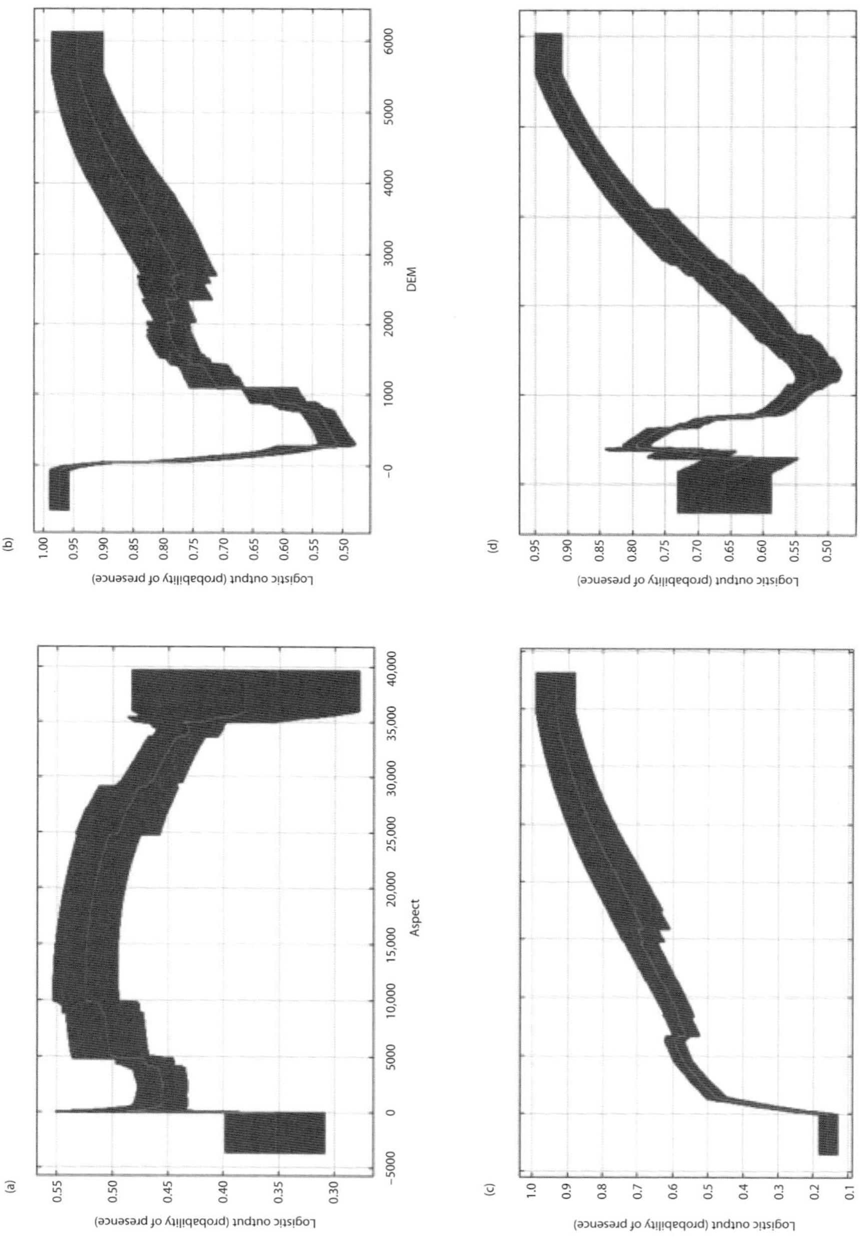
(a)
Logistic output (probability of presence)
Aspect
(b)
Logistic output (probability of presence)
DEM
(c)
Logistic output (probability of presence)
Slope
(d)
Logistic output (probability of presence)
Topo

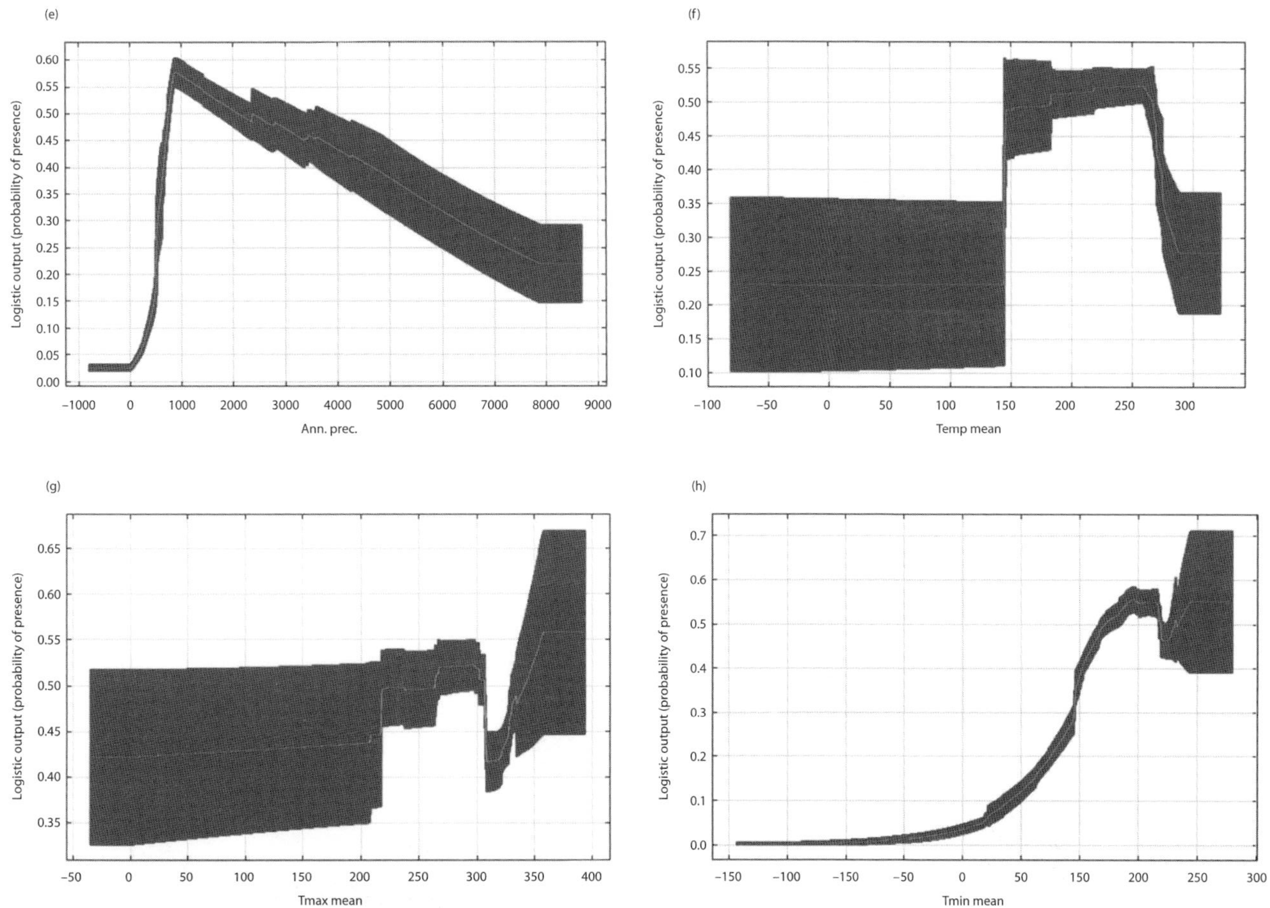


**Fig. 6.3.** Response curves showing how each environmental variable affects the niche model. These curves show how the logistic prediction changes as each environmental variable varies, keeping all other environmental variables at their average sample values. The curves show the mean response of the eight replicate model runs (pale line) and mean ± standard deviation (dark shaded area).

The niche model generated using native range distribution predicted most of the validation points in the northern part of the Indian subcontinent at >70% probability. However, some locations in central and southern India were predicted at a probability level of <30% (Fig. 6.4).

## Current and future invasive range in the Indian subcontinent

Overall, the ecological niche model shows higher suitability for species in north-eastern states, the central Himalayan provinces and the Western Ghats and Eastern Ghats (Fig. 6.4). The areas suitable for *C. odorata* include Jammu and Kashmir, Himachal Pradesh, Uttarakhand and West Bengal in the central Himalayas; and Arunachal Pradesh, Sikkim, Meghalaya, Nagaland, Mizoram, Manipur and Assam in the eastern Himalayas. The model also shows environmental suitability in Gujarat, Maharashtra, Karnataka, Kerala, Tamil Nadu, Andhra Pradesh and Orissa. Areas adjacent to the Indian sub-region showing high suitability for the species include Nepal, Myanmar and Pakistan.

The Western and Eastern Ghats, the central Himalayan region and the north-eastern part of the Indian subcontinent are shown to be more susceptible to invasion by *C. odorata* under current climatic conditions as well as predicted future climate warming (Fig. 6.5). Under the A2 scenario of climate change, species show a range expansion in the subcontinent in the 2020 time frame but a decrease in the 2080 time frame. As compared with the current distributional scenario for the species, under the B2 scenario, the species show a decline in its distributional range. This is inferred from the decreased niche breadth of the species in the Indian subcontinent (Fig. 6.6). In general, the climate models predict a slight

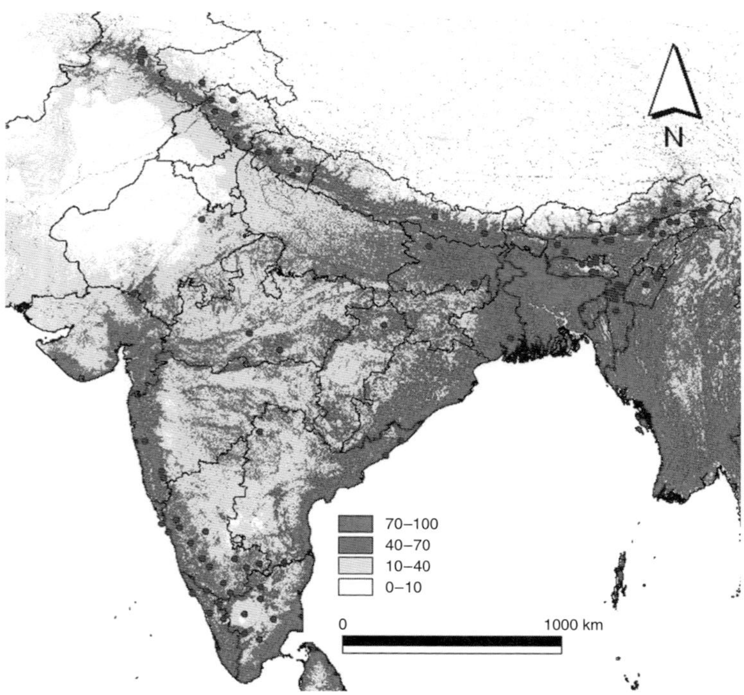


**Fig. 6.4.** Predicted potential distributional area of *C. odorata* in the Indian subcontinent. The grey colour ramp signifies different levels of probability of occurrence of the species under current climatic conditions, and the dots represent actual occurrences of the spcies.

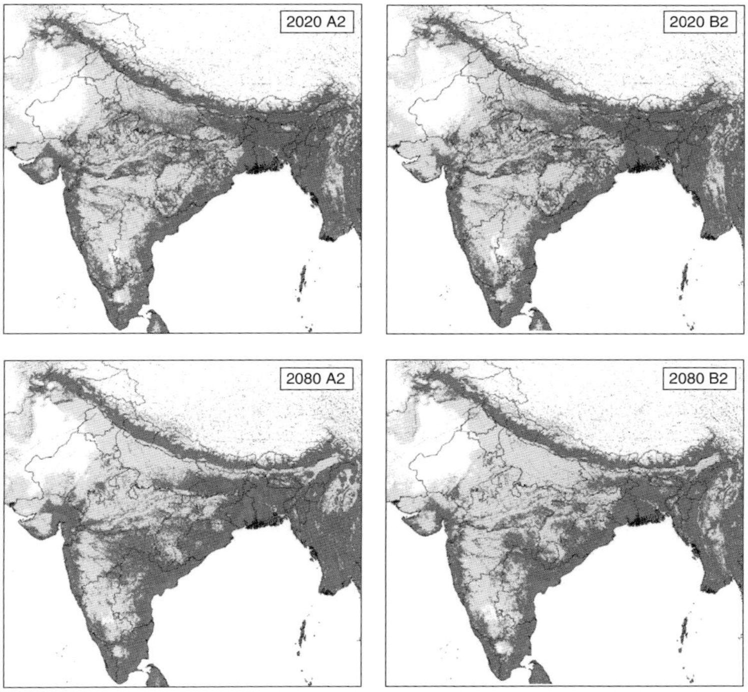


**Fig. 6.5.** Predicted potential distributional area of *C. odorata* in the Indian subcontinent under Hadley's climate change model.

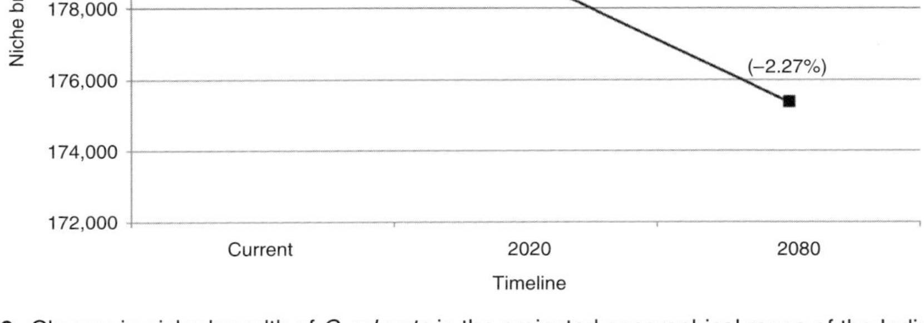


**Fig. 6.6.** Change in niche breadth of *C. odorata* in the projected geographical range of the Indian subcontinent under future climate change scenarios. The values in parentheses in the figure represent percentage change in niche breadth compared with current values; positive and negative values depict increase and decrease in niche breadth, respectively. The values for niche breadth in the figure do not have any units and simply reflect the geographical spread of the species in the Indian subcontinent.

northward expansion of the potential distributional range of the species under future climate change.

## Discussion

The ecological niche model in the present study showed good overall performance in its native range. The high AUC values for training and testing (>0.90) indicate that the niche model has a good ability to discriminate between presence and absence areas for the species. The niche model could successfully predict most of the validation points in the Indian subcontinent, showing its fair transferability (Fig. 6.4). However, some validation points in the southern and central parts of the subcontinent were predicted at a relatively low level of probability. Nevertheless, it can be said that the species has conserved its original niche properties in the northern part of the subcontinent. However, the lower degree of prediction probability for some of the validation points hints at the possibility of a niche shift for the species, which has to be studied further. It may be mentioned here that the process of niche shift in invasive plant species is facilitated by the processes of natural hybridization and genetic introgression that produces weedy races that can withstand harsher environmental conditions (Gopinathan and Babu, 1982). Some recent studies have also showed evidence for species shifting their climatic niches during biological invasion (Fitzpatrick *et al.*, 2007; Broennimann *et al.*, 2008; Da Mata *et al.*, 2010).

### Environmental niche of *C. odorata*

Analysis of the macro-scale niche property of *C. odorata* in the present study revealed that its distribution is sensitive mainly to thermal limits, i.e. mean minimum temperature and mean annual temperature followed by slope, elevation, precipitation and topography (Fig. 6.2). This observation is corroborated by micro-scale studies showing that seedling mortality of the species is increased due to low-temperature and moisture stresses, which is in fact the manifestation of thermal limit and precipitation (Yadav and Tripathi, 1981). The growth and development of the species is also favoured by high light intensity and a long photoperiod (Kushwaha *et al.*, 1981). The major role of temperature in determining the ecological niche of the species suggests that its future distribution will be positively affected by an increase in the minimum temperature level at higher altitudes.

### Implications for the Indian subcontinent

The A2 and B2 climate change scenarios of Hadley's climate model showed contrasting viewpoints of the future distributional range of *C. odorata*. Under scenario A2, which represents a harsher climate as compared with scenario B2, the species is predicted to increase its geographic range in the Indian subcontinent. The increased probability of occurrence and range expansion of *C. odorata* in the Indian subregion under climate change has implications for agriculture, commercial plantations and native biodiversity. *Chromolaena odorata* forms an important component of the weed community in these systems and competes with crop plants for light and nutrients, reducing crop yield (Tripathi and Yadav, 1982). Thus, expansion of the species range under both climate change scenarios is bound to affect crop productivity in the subcontinent. Studies on population dynamics of the species have concluded that it is more successful in the early successional stages (Kushwaha *et al.*, 1981), and hence activities such as shifting of agriculture in the mountain areas of the subcontinent might further accelerate the process of invasion under climate-warming scenarios and pose a serious problem for its management in culturable landscapes.

## References

Akter, A. and Zuberi, M.I. (2009) Invasive alien species in northern Bangladesh: Identification, inventory and impacts. *International Journal of Biodiversity and Conservation* 1, 129–134.

Annapurna, C. and Singh, J.S. (2003) Variation of *Parthenium hysterophorus* in response to soil quality: Implication for invasiveness. *Weed Research* 43. 190–198.

Broennimann, O. and Guisan, A. (2008) Predicting current and future biological invasions: both native and invaded ranges matter. *Biology Letters* 4, 585–589.

Brown, J.H. (1984) On the relationship between abundance and distribution of species. *American Naturalist* 124, 255–279.

Da Mata, R.A., Tidon, R., Cortes, L.G., Marco, P.D. and Diniz-Filho, J.A.F. (2010) Invasive and flexible: niche shift in the drosophilid *Zaprionus indianus* (Insecta, Diptera). *Biological Invasions* 12, 1231–1241.

Estrada-Peña, G., Pegram, R.G., Barré, N. and Venzal, J.M. (2007) Using invaded range data to model the climate suitability for *Amblyomma variegatum* (Acari: Ixodidae) in the New World. *Experimental and Applied Acarology* 41, 203–214.

Ficetola, G.F., Thuiller, W. and Schioppa, E.P. (2009) From introduction to the establishment of alien species: bioclimatic differences between presence and reproduction localities in the slider turtle. *Diversity and Distributions* 15, 108–116.

Fitzpatrick, M., Weltjin, J.F., Sanders, N.J. and Dunn, R.R. (2007) The biogeography of prediction error: why does the introduced range of the fire ant over-predict its native range? *Global Ecology and Biogeography* 16, 24–33.

Gopinathan, M.C. and Babu, C.R. (1982) Cytogenetics of *Galinsoga parviflora* Cav. and *G. ciliata* (Raf.) Blake, and their natural hybrids (Asteraceae). *New Phytologist* 91, 531–539.

Guisan, A., Graham, C.H., Elith, J. and Huettmann, F. (2007) Sensitivity of predictive species distribution models to change in grain size. *Diversity and Distributions* 3, 332–340.

Hijmans, R.J. and Graham, C.H. (2006) The ability of climate envelope models to predict the effect of climate change on species distributions. *Global Change Biology* 12, 2272–2281.

Hijmans, R.J., Cruz, J.M., Rojas, E. and Guarino, L. (2001) DIVA-GIS. A geographic information system for the management and analysis of genetic resources data. Manual (Internet). International Potato Center and International Plant Genetic Resources Institute, Lima, Peru. Available at http://www.diva-gis.org

IPCC (2007) Summary for policymakers. In: Solomon, S. *et al.* (eds) *Climate Change 2007: the Physical Science Basis. Contribution of Working Group I to the Fourth Assessment Report of the Intergovernmental Panel on Climate Change.* Cambridge University Press, Cambridge, UK and New York.

Joshi, C.M. (2001) Invasive Banmara (*Chromolaena odorata*): spatial detection and prediction. MSc thesis, International Institute for Aerospace Survey and Earth Sciences, The Netherlands, 53 pp.

Kelly, A.E. and Goulden, M.L. (2008) Rapid shifts in plant distribution with recent climate change. *Proceedings of National Academy of Sciences* 105, 11823–11826.

Kushwaha, S.P.S., Ramakrishnan, P.S. and Tripathi, R.S. (1981) Population dynamics of *Eupatorium odoratum* in successional environments following slash and burn agriculture. *Journal of Applied Ecology* 18, 529–535.

Levins, R. (1968) *Evolution in Changing Environments.* Monographs in Population Biology, vol. 2, Princeton University Press, Princeton, New Jersey.

Lin, W., Zhou, G., Cheng, X. and Xu, R. (2007) Fast economic development accelerates biological invasions in China. *PLoS ONE* 2, e1208.

Luoto, M., Virkkala, R. and Heikkinen, R.K. (2007) The role of land cover in bioclimatic models depends on spatial resolution. *Global Ecology and Biogeography* 16, 34–42.

McCune, B. (2006) Non-parametric habitat models with automatic interactions. *Journal of Vegetation Science* 17, 819–830.

McPherson, J.M., Jetz, W. and Rogers, D.J. (2006) Using coarse-grained occurrence data to predict species distributions at finer spatial resolutions – possibilities and limitations. *Ecological Modelling* 192, 499–522.

Mittermeier, R.A., Robles, G.P., Hoffmann, M., Pilgrim, J., Brooks, T., Mittermeier, C.G. *et al.* (2004) *Hotspots Revisited.* Garza Garcia N.L, Mexico: CEMEX.

Mooney, H.A. and Hobbs, R.J. (2000) *Invasive Species in a Changing World.* Island Press, Washington, DC.

Muniappan, R. and Bambam, J. (1999) Biological control of *Chromolaena odorata*: Successes and failures. In: Spencer, N.R. (ed.) *Proceedings of the Xth International Symposium on Biological Control of Weeds*, 4–14 July 1999, Montana State University, Bozeman, Montana, pp. 81–85.

Negi, P.S. and Hazra, P.K. (2007) Alien flora of Doon Valley, Northwest Himalaya. *Current Science* 92, 968–978.

Peterson, A.T. and Nakazawa, Y. (2008) Environmental data sets matter in ecological niche modelling: an example with *Solenopsis invicta* and *Solenopsis richteri. Global Ecology and Biogeography* 17, 135–144.

Peterson, A.T., Soberon, J. and Sanchez-Cordero, V. (1999) Conservatism of ecological niches in evolutionary time. *Science* 285, 1265–1267.

Peterson, A.T., Papes, M. and Eaton, M. (2007) Transferability and model evaluation in ecological niche modeling: a comparison of GARP and Maxent. *Ecography* 30, 550–560.

Phillips, S.J. and Dudik, M. (2008) Modeling of species distributions with Maxent: new extensions and a comprehensive evaluation. *Ecography* 31, 161–175.

Phillips, S.J., Anderson, R. and Schapire, R. (2006) Maximum entropy modelling of species geographical distributions. *Ecological Modelling* 190, 231–259.

Pimentel, D., McNair, S., Janecka, J., Wightman, J., Simmonds, C., O'Connell, C. *et al.* (2001) Economic and environmental threats of alien plant, animal and microbe invasions. *Agriculture, Ecosystems and Environment* 84, 1–20.

Raghubanshi, A.S., Rai, L.C., Gaur, J.P. and Singh, J.S. (2005) Invasive alien species and biodiversity in India. *Current Science* 88, 539–540.

Randin, C.F., Dirnbock, T., Dullinger, S., Zimmermann, N.E., Zappa, M. and Guisan, A. (2006) Are niche-based species distribution models transferable in space? *Journal of Biogeography* 33, 1689–1703.

Reddy, C.S. (2008) Catalogue of invasive alien flora of India. *Life Science Journal* 5, 84–89.

Sanders, N.J., Gotelli, N.J., Heller, N.E. and Gordon, D.M. (2003) Community disassembly by an invasive species. *Proceedings of the National Academy of Sciences* 100, 2474–2477.

Segurado, P., Araujo, M.B. and Kunin, W.E. (2006) Consequences of spatial autocorrelation for niche-based models. *Journal of Applied Ecology* 43, 433–444.

Shah, G.M. and Khan, M.A. (2006) Checklist of noxious weeds of district Mansehra, Pakistan. *Pakistan Journal of Weed Science Research* 12, 213–219.

Singh, S.P. (2004) *Some Success Stories in Classical Biological Control of Agricultural Pests in India.* Asia-Pacific Association of Agricultural Research Institutions, FAO Regional Office for Asia and Pacific, Bangkok, Thailand, 73 pp.

Thomas, S.G., Rehel, S.M., Varghese, A., Davidar, P. and Potts, S.G. (2009) Social bees and food plant associations in the Nilgiri Biosphere Reserve, India. *Tropical Ecology* 50, 79–88.

Thuiller, W., Richardson, D.M., Pysek, P., Midgley, G.F., Hughes, G.O. and Rouget, M. (2005) Niche-based modelling as a tool for predicting the risk of alien plant invasions at a global scale. *Global Change Biology* 11, 2234–2250.

Thuiller, W., Lavorel, S., Sykes, M.T. and Araujo, M.B. (2006) Using niche-based modelling to assess the impact of climate change on tree functional diversity in Europe. *Diversity and Distributions* 12, 49–60.

Thuiller, W., Richardson, D.M. and Midgley, G.F. (2007) Will climate change promote alien plant invasions? In: Nentwig, W. (ed.) *Ecological Studies*, vol. 193. Springer-Verlag, Berlin and Heidelberg, pp. 197–211.

Tripathi, R.S. (2009) Alien plant invasion: a hot ecological issue. *EnviroNews* 15, 7–10.

Tripathi, R.S. and Yadav, A.S. (1982) Population regulation of *Eupatorium adenophorum* Spreng. and *E. riparium* Regel: effect of population density, soil nitrogen and light intensity. *Plant and Soil* 65, 35–49.

Tripathi, R.S., Singh, R.S. and Rai, J.P.N. (1981) Allelopathic potential of *Eupatorium adenophorum* – a dominant ruderal weed of Meghalaya. *Proceedings of the Indian National Science Academy Part B* 47, 458–465.

Walther, G.R. *et al.* (2009) Alien species in a warmer world: risks and opportunities. *Trends in Ecology and Evolution* 24, 686–693.

Warren, D.L., Glor, R.E. and Turelli, M. (2008) Environmental niche equivalency versus conservatism: quantitative approaches to niche evolution. *Evolution* 62, 2868–2883.

Yadav, A.S. and Tripathi, R.S. (1981) Population dynamics of the ruderal weed *Eupatorium odoratum* and its natural regulation. *Oikos* 36, 355–361.

# 7 Impacts of Cultivation of *Kappaphycus alvarezii* on Coral Reef Environs of the Gulf of Mannar and Palk Bay, South-eastern India

**J.K. Patterson Edward[1] and J.R. Bhatt[2]**

[1]*Suganthi Devadason Marine Research Institute, Tuticorin, Tamil Nadu, India;* [2]*Ministry of Environment and Forests, Paryavaran Bhavan, CGO Complex, New Delhi, India*

## Introduction

*Kappaphycus alvarezii* is among the largest tropical red algae, with a high growth rate, resilient morphology and extremely successful vegetative regeneration, making it a potentially destructive invasive species (Doty Ex Silva, *Kappaphycus alvarezii* (http://www.hawaii.edu/reefalgae/invasive_algae/rhodo/kappaphycus_alvarezii.htm). It is tough, fleshy and firm, growing up to 2 m. Its thalli are coarse, with axes and branches 1–2 cm diameter; heavy, with major axes relatively straight, lacking secondary branches near apices. Frequently and irregularly branched, most branches primary, secondary branches intercalated between primary branches or mostly lacking. Colour is green to yellow orange (Doty, 1996).

Commercial cultivation of *K. alvarezii* (Doty) was developed jointly by Marine Colloids Corporation and Dr Maxwell Doty of the University of Hawaii Botany Department (Parker, 1974). This occurred during the latter half of the 1960s in the Philippines, using local varieties selected from the wild (Doty, 1973; Parker, 1974). Subsequently, these selected and cultivated varieties were introduced to numerous parts of the world for the purpose of research or development of a commercial cultivation industry, though only a fraction of these countries are commercial producers today.

According to Woo *et al.* (1989), 'Observations have shown that the alga is able to coalesce into the tissues of the coral, providing a strong means for attachment, and thus allowing the alga to persist in high wave energy environments'. Woo (1999) documented fragments of *Kappaphycus striatum* weighing 0.05 g that were capable of net growth in the field, suggesting that fragments created by physical disturbance can be carried by waves and currents to new locations where they can possibly establish growth and perpetuate. The alga spreads mainly by fragmentation (pieces of seaweed float to new locations) and can overgrow and kill coral by smothering, shading it from sunlight and abrasion. It causes shifts from diverse coral reef to a seaweed-dominated, low-diversity reef and changes the bottom structure of the reef, reducing access to crevices and holes. Thus the habitat loss may impact commercial and recreational fisheries (http://www.hawaiiinvasivespecies.org/pests/kappaphycus.html).

*Kappaphycus* is the source of sulfated polysaccharides, commonly called 'carrageenan', which are used in the food and pharmaceutical industries. The largest producer is the Philippines, where cultivated

 *Invasive Alien Plants: An Ecological Appraisal for the Indian Subcontinent* (eds J.R. Bhatt *et al.*)

seaweed produces about 80% of the world supply. Carrageenans are natural ingredients that have been used for decades in the food industry, and are generally regarded as safe; they are widely used as thickening, gelling and stabilizing agents (Van de Velde and De Ruiter, 2002). The different types of carrageenan are obtained from different species of the Rhodophyta. Kappa (K)-carrageenan is predominantly obtained by extraction of *K. alvarezii*, known in the trade by its earlier name of *Eucheuma cottonii* (or simply Cottonii) (Rudolph, 2000).

## Methodology

The extent of invasion by *Kappaphycus* in coral reef was assessed from October 2008 to October 2010 using the Line Intercept Transect (LIT) method (English *et al.*, 1997). Twenty meter transects were laid parallel to a series of islands at 0.5–2.0 m depth during high tide, the number of transects depending on the size of the reef. Six transects in Shingle Island, 15 in Krusadai Island and 6 in Poomarichan Island were laid in each assessment, which was carried out regularly at quarterly intervals to identify changes in benthic community structure. Underwater photographs were taken for documentation. Thirty local fishermen were interviewed for their views on recent trends in fish catch.

## Observations

### Cultivation of *K. alvarezii* in South Palk Bay (Mandapam region)

*Kappaphycus alvarezii* is cultivated by around 140–175 fisherfolk (some belonging to self-help groups) in three coastal villages (T. Nagar, Muniakadu and Thonithurai) in the Mandapam region of South Palk Bay, South-eastern India (Fig. 7.1) for commercial purposes, and the produce is immediately purchased by various buyers. Many fisherfolk involved in cultivation carry out this activity on a part-time basis along with regular fishing activities.

Initially, over 400 fisherfolk were involved in this activity and they were trained in cultivation and initial processing. Those who were given training also received bank loans.

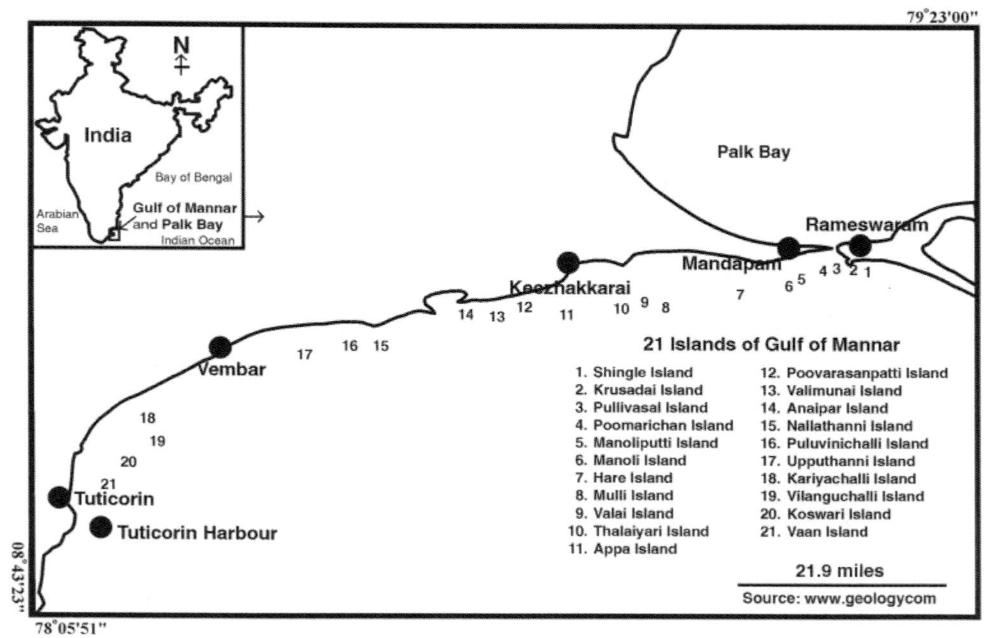


**Fig. 7.1.** Map showing Palk Bay and the Gulf of Mannar in south-eastern India.

The state government also provided aid through the self-help groups of this region. *Kappaphycus* is cultivated on rafts, and generally about 100 rafts are owned by three or four cultivators. *Kappaphycus* is cultivated in two to three cycles a year in Palk Bay, mostly two, and the most favourable months for cultivation are from March to August. Cultivation covers a 5 km stretch of coastline in the Mandapam region of the Palk Bay, about 50–100 m from the shore with a maximum depth of 2 m. The harvest period is around 45–60 days. The buying rate for the fresh product is 2 Rs./kg, and for the dried product 20 Rs./kg. Buyers prefer the dried product; approximately 500 kg of *wet* algae yields 50 kg of dried material.

The entire *Kappaphycus* cultivation enterprise in this region is conducted on luxuriant seagrass beds and corals, which in former times were very productive fishing areas for the locals. The fishing villages are located around 50–100 m from the seashore. Underwater surveys reveal that marine resources such as seagrasses, corals and associated fisheries are affected due to the cultivation of *Kappaphycus*, which alters the light source, aeration and nutrient content of the water. Loss of fish production has occurred in and around the cultivation area. Prior to *Kappaphycus* cultivation, local fisherfolk mostly practised hand-line and gill net fishing, but now there is no fishing activity near the algal cultivation area as, according to local people, the fish catch has declined in these areas. More research is required to substantiate the local claims.

### Impact of *K. alvarezii* invasion on coral reefs and seagrass beds

It was observed from a survey in 2008 that *Kappaphycus* had invaded the coral reef colonies of Shingle, Krusadai and Poomarichan Islands. The source of this spread was from the ongoing cultivation in South Palk Bay via sea currents, which was also evidenced by the underwater observations in the present study of considerable quantities of *Kappaphycus* fragments on the dense seagrass beds along the Pamban Pass.

Where *Kappaphycus* invaded coral colonies these were destroyed because of the shadowing and smothering effects of the attached fragments, which attach firmly and form a thick mat on the coral colony, penetrating deeply up to 5–10 cm and preventing sunlight reaching zooxanthallae for essential photosynthesis (Figs 7.2 to 7.7). Table 7.1 provides data on the impact of *Kappaphycus* on coral on three islands:

**Fig. 7.2.** Bioinvasion of *Kappaphycus alvarezii* on branching coral (*Acropora* sp.) on Krusadai Island.

**Fig. 7.3.** Bioinvasion of *Kappaphycus alvarezii* on massive coral reef (*Porites* sp.) on Krusadai Island.

**Fig. 7.4.** Branching coral colony destroyed by *Kappaphycus* invasion.

**Fig. 7.5.** Branching corals on Shingle Island smothered and destroyed by *Kappaphycus* invasion.

**Fig. 7.6.** A view of coral reef invaded by *Kappaphycus* on Krusadai Island during low tide.

**Fig. 7.7.** Aerial view of coral reef invaded by *Kappaphycus* on Krusadai Island.

areas affected, coral colonies and coral species, as of October 2010. Invasion also adversely affected other reef-associated fauna, especially native fish species. During underwater observations for the present study, it was noted that *Kappaphycus* was not preferred by native herbivorous fishes such as surgeon fish (acanthurids) and parrot fish (scarids), and this provides an opportunity for the invading alga to grow very rapidly without any predators. However, more data is required to endorse this view. Figure 7.8 illustrates the sequence of *K. alvarezii* bioinvasion and coral mortality based on monitoring observations in the affected reef area over a period of around 2 years.

*Kappaphycus* cultivation, as practised using rafts on the luxuriant seagrass beds, reduces the penetration of light essential for healthy growth of seagrass. It was observed during an underwater survey in the cultivation area that this activity

**Table 7.1.** Details of sites and coral reefs invaded by the alga *Kappaphycus alvarezii* on the islands of the Gulf of Mannar, India (October 2010).

| Island | GPS coordinates of affected reef area | Affected (dead) coral colonies (*n*) | Affected coral colony size (*n*): Small (<20 cm) | Intermediate (20–80 cm) | Larger (>80 cm) | Affected reef area (km$^2$) | Affected coral species |
|---|---|---|---|---|---|---|---|
| Shingle Island | Lat. 09°14, 543′ N; Long. 079°13, 994′ E | 116 | 36 | 51 | 29 | 0.20 | *Acropora cytherea, A. nobilis* |
| Krusadai Island | Lat. 09°14, 764′ N; Long. 079°13, 045′ E; Lat. 09°14, 736′ N; Long. 079°13, 031′E; Lat. 09°14, 683′ N; Long. 079°13, 007′ E | 298 | 53 | 203 | 42 | 0.42 | *A. cytherea, A. formosa, A.nobilis; Montipora digitata* |
| Poomarichan Island | Lat. 09°014, 439′ N; Long. 079°010, 434′ E | 54 | 12 | 26 | 16 | 0.16 | *Porites solida; A. cytherea* |

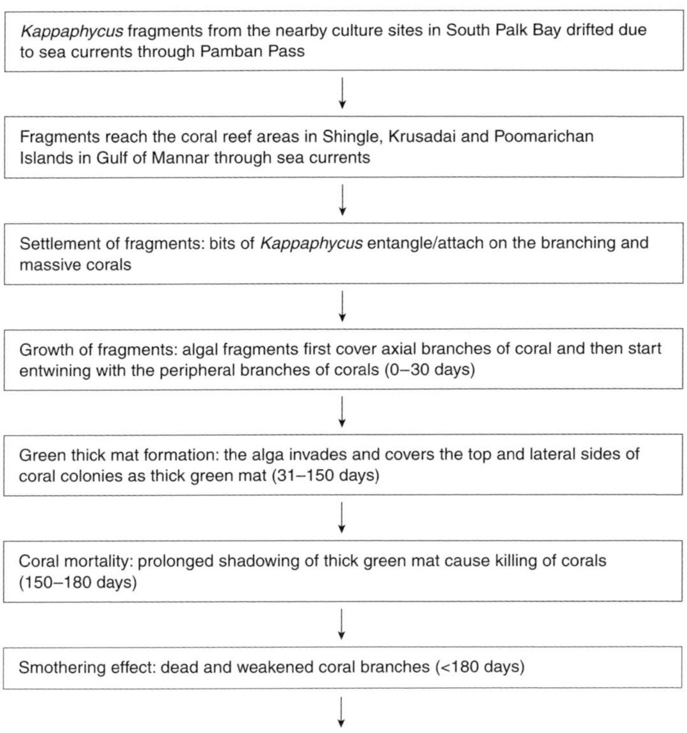


**Fig. 7.8.** Flow chart describing steps in the bioinvasion by *K. alvarezii* of coral reefs in the Gulf of Mannar, India.

ultimately affects the seagrass beds, resulting in stunted growth with reduced shoot density, a turbid environment (Figs 7.9 and 7.10) and less associated biodiversity – in particular, fish. Fisherfolk working on the catch reveal that native herbivorous fish species are gradually declining in the area as they do not appear to prefer *K. alvarezii* as their food source.

**Fig. 7.9.** View of *Kappaphycus* raft on seagrass bed in South Palk Bay, with associated turbidity.

**Fig. 7.10.** *Kappaphycus* cultivation: impact on coral in South Palk Bay.

## Discussion

A large number of species, either intentionally or accidentally, are transported between biogeographical regions or introduced to new areas. Humans are responsible for accelerating the rate and geographical scale of invasions, and the effect of introductions has been dramatic in many occasions (Mack *et al.*, 2000). In many areas, introduced species are among the greatest threats to native biodiversity and the functioning of natural ecosystems. The ability of invasive species to become established in a community and the effect of the invasion on the recipient biota have provided information on general ecological issues such as community assembly theory (Wilson *et al.*, 2000), evolutionary stability of ecological interactions (Bowers *et al.*, 1992; Richardson *et al.*, 2000; Trowbridge and Todd, 2001), factors limiting species distributions (Richardson and Bond, 1991) and the importance of species identities to ecosystem functioning (Vitousek and Walker, 1989). The identification of invasive species and following their pattern of spread is regarded as important for the development of policies to prevent the introduction of undesirable species (Reichard and Hamilton, 1997; Rejmánek, 2000). The study of invasion processes and their effects also allows observations of community responses to perturbation on a scale not usually attainable with experimental manipulations. Despite the growing concern for their possible effect on the coastal ecosystem, marine invasion processes and the impact of introduced species on native biota are still poorly understood (Grosholz, 2002).

Over the last 30 years, strains of *K. alvarezii* and *K. striatum* – mainly from the Philippines – have been introduced into more than 20 tropical countries for the purposes of mariculture (Russell, 1983; Doty, 1986; Glenn and Doty, 1990; Areces, 1995; Paula *et al.*, 1998; Oliveira and Paula, 2003). However, introductions have been made without observation of minimum protocols of quarantine (Oliveira and Paula, 2003). At present, there are no risk assessments or environmental impact procedures in place for the intentional introduction of *Kappaphycus*.

The best-documented case of the impacts of non-indigenous marine algae in the tropics comes from Hawaii. *Eucheuma striatum* (this species was later split into *Kappaphycus striatum* and *K. alvarezii*) and *Eucheuma denticulatum* were intentionally introduced into the fringing coral reef surrounding the Hawaii Institute of Marine Biology (HIMB) at Coconut Island (Moku o Loe), Kaneohe Bay, Oahu, Hawaiian Islands throughout the 1970s, for experimental research and cultivation (Russell, 1983). Subsequent research has demonstrated that these algae have spread rapidly throughout the bay and can be found in a variety of reef habitats overgrowing and killing corals (Conklin and Smith, 2005). Regrowth of the algae following their removal was rapid at most sites, probably due to the experimentally demonstrated ability of the algae to regrow from minute attachment points and the low palatability of the algae to native herbivorous fishes (Conklin and Smith, 2005).

Russell (1983) conducted a number of studies to determine the potential for spread and the impacts of *E. striatum* (*Kappaphycus* spp.) on the native reef community. Surveys conducted by Rodgers and Cox (1999) in 1996 determined that *Kappaphycus* spp. had indeed spread from the initial sites of introduction at HIMB to reefs as far as 6 km away, at an estimated rate of 250 m/year. Smith *et al.* (2002) conducted surveys assessing the distribution of non-indigenous marine algae throughout the main Hawaiian Islands, and found that *Kappaphycus* spp. had still not spread outside of Kaneohe Bay but had continued to spread northward in the bay since the Rodgers and Cox (1999) study. Also, Conklin and Smith (2005) noted the presence of *Kappaphycus* at a number of sites where they were absent 5 years previously.

Pereira and Verlecar (2005) reported that the fast-growing marine alga *K. alvarezii* introduced to India for seaweed aquaculture in 1995 has already established growth in many parts of the GOM Marine Biosphere Reserve. This species had already exhibited its invasion, shadowing and smothering effects over coral colonies and fears were expressed by scientists that it could switch to sexual reproduction by spores under favourable environmental conditions in the future (Chandrasekaran *et al.*, 2008). Namboothri and Shankar (2011) have also described harmful effects of *Kappaphycus alvarezii* on corals. The SDMRI Reef Research Team conducted a preliminary underwater survey in the shallow seaward areas of the Krusadai and Shingle Islands in the first week of October 2008, and reported to Tamilnadu Forest Department that *K. alvarezii* had invaded coral areas of around 6 $m^2$ in size, was firmly attached to the branching corals *Acropora formosa* and *Acropora nobilis* and was spreading to other nearby coral colonies (SDMRI Report, 2008). Regular follow-up monitoring revealed that within about 24 months the invasive algae had spread over a further 0.7 $km^2$ of reef area, affecting over 450 coral colonies in Shingle, Krusadai and Poomarichan Islands, and impacts were also noticed on other coral species such as *Acropora cytherea*, *Montipora digitata* and *Porites solida*. It is now appropriate to look carefully into the protection and conservation of not only coral reefs, but also the associated biodiversity of the Gulf of Mannar from the impact of *Kappaphycus* invasion, through well-planned and comprehensive management strategies involving surveying, monitoring, research, eradication and control.

## Conclusion

In general, native algal culture is often economically beneficial and ecologically benign. Krishnan and Narayanakumar (2010) documented the detailed history of *Kappaphycus alvarezii* cultivation initiatives in Tamil Nadu along with socio-economic benefits to cultivators, especially self-help groups and production projections. The present study is neither in argument nor in conflict with their work but points out the harmful invasive nature of the exotic alga, *K. alvarezii*. There is a livelihood benefit in exotic *Kappaphycus* cultivation to a very limited number of people. However, over 250,000 fisherfolk on both Gulf of Mannar and Palk Bay coasts in Tamil Nadu, South-eastern India depend on the reef- and seagrass-associated fishery resources for their livelihoods. As per the UNEP-WCMC (2006) estimate, the total economic value of healthy coral reefs ranges from US$100,000 to 600,000/km$^2$/year. The estimated value of 1 km$^2$ coral reef area in the Gulf of Kachchh, India is around Rs.7.95 million per year (Dixit *et al.*, 2010), the services including fisheries, tourism, biodiversity and protection against salinity and coastal erosion, wherein fisheries is most important followed by protection against coastal erosion.

It is therefore the impact on coral reefs and seagrass beds that will ultimately have considerable impact on the dependent fisherfolk's livelihood. Losses are largely environmentally caused because of the disappearance of native species and the instability of coastal areas due to loss of habitats such as coral reefs and seagrass beds.

Considering the deleterious impact of *Kappaphycus* on corals in other parts of the world, especially the Hawaiian Islands, it is now time to initiate effective long-term remedial measures to eradicate existing *Kappaphycus* in coral areas of the Gulf of Mannar to prevent further spread/invasion. The Government of Tamil Nadu issued orders in December 2005 (G.O. Ms. No.229, E & F (EC.3) Department, dated 20 December 2005) that *K. alvarezii* can be cultivated in the waters north of Palk bay and south of the Tuticorin coast. However, cultivation is now practised on seagrass beds on the south side of the Palk Bay, which is very close to the coral reef areas. Following a report from SDMRI, the Tamil Nadu Forest Department – along with team of scientists from the Indian Institute of Science – inspected and assessed the impacted coral reef areas in November 2008 and the issues related to *Kappaphycus* invasion brought to the notice of the Chief Secretary to the Government of Tamil Nadu and subsequently discussed.

The National Biodiversity Authority of India has signed an agreement with Pepsico (Hindu Business Line Report, 2010), and the latter has provided Rs. 37 lakh royalty monies to be shared among the local fishing community in Tamil Nadu for acquisition of *Kappaphycus* to be used by the multinational firm under the Access and Benefit Sharing (ABS) scheme. In the light of the observations of deleterious and irrevocable damage caused by the *Kappapycus*, there is an impending need to revise this agreement in order to protect and conserve ecologically sensitive habitats such as coral reefs and seagrass beds – and the livelihoods of over 250,000 fisher folk who are solely dependent on associated fishery resources. The agreement needs to be suitably modified to incorporate the basic principles of benefit-sharing, such as conservation and local development (socio-economic), in particular on the basis of economic valuation of damages caused by *Kappaphycus*.

The observations from the present study underscore the need for urgent reconsideration of the cultivation of *Kappaphycus* in the Gulf of Mannar and Palk Bay, both of which are exceptionally rich in species diversity, including both endemic and threatened species. Further to in-depth detailed research, conservation and management strategies and comprehensive guidelines should be set in place towards the introduction of exotic species in any marine environment. A coordinated approach among various departments, scientists and conservation managers is key prior to the introduction of new exotic species.

Long-term continuous monitoring and remedial measures are required in this region in order to check further impacts of *Kappaphycus* on corals and seagrass beds: damage pattern, duration, species involved, season, depth, etc.; impacts on associated fisheries and other biological resources; impacts on abiotic factors such as physical, chemical and nutrient; impact on socio-economic/ecological factors. Research is also necessary to improve understanding of the short- and long-term consequences in an already impacted area on fish catch.

## Acknowledgements

The authors thank the Principal Chief Conservator of Forests, Chief Wildlife Warden and Wildlife Warden, Gulf of Mannar Marine National Park for permission to conduct this study.

## References

Areces, A.J. (1995) Cultivo comercial de carragenofitas del género *Kappaphycus* Doty. In: Alveal, K., Ferrario, M.E., Oliveira, E. and Sar, E. (eds) *Manual de Métodos Ficológicos.* Universidad de Concepción, Concepción, pp. 529–549.

Bowers, M.D., Stamp, N.E. and Collinge, S.K. (1992) Early stage of host range expansion by a specialist herbivore, *Euphydras phaeton* (Nymphalidae). *Ecology* 73, 526–536.

Chandrasekaran, S., Arun Nagendran, P., Pandiaraja, P. and Krishnankutty, N.B. (2008) Bioinvasion of *Kappaphycus alvarezii* on corals in the Gulf of Mannar, India. *Current Science* 94, 1167–1172.

Conklin, E.J. and Smith, J.E. (2005) Abundance and spread of the invasive red algae, *Kappaphycus* spp., in Kane'ohe Bay, Hawaii and an experimental assessment of management options. *Biological Invasions* 7, 1029–1039.

Dixit, A.M., Kumar, P., Kumar, L., Pathak, K.D. and Patel, M.I. (2010) Economic valuation of coral reef systems in Gulf of Kachchh. Final Report. World Bank-aided Integrated Coastal Zone Management Project, submitted to Gujarat Ecology Commission, 156 pp.

Doty, M.S. (1973) Farming the red seaweed, *Eucheuma*, for Carrageenans. *Micronesica* 9, 59–73.

Doty, M.S. (1986) The production and use of *Eucheuma.* In: Doty, M.S., Caddy, J.F. and Santelices, B. (eds) *Case Studies of Seven Commercial Seaweed Resources.* FAO Fish Technical Paper 281, The Food and Agriculture Organization of the United Nations, Rome, pp. 123–161.

English, S., Wilkinson, C. and Baker, V. (1997) *Survey Manual for Tropical Marine Resources.* Australian Institute of Marine Science, Townsville, Australia, pp. 390.

Glenn, E.P. and Doty, M.S. (1990) Growth of the seaweeds *Kappaphycus alvarezii, K. striatum and Eucheuma denticulatum* as affected by environment in Hawaii. *Aquaculture* 84, 245–255.

Grosholz, E. (2002) Ecological and evolutionary consequences of coastal invasions. *Trends in Ecology and Evolution* 17, 22–27.

Hindu Business Line Report (2010) Delhi likely to host 11th Conference of Parties in 2012 (16 October).

Krishnan, M. and Narayanakumar, R. (2010) Socio-economic dimensions of Seaweed Farming in India. *CMFRI special publication* (104), 1–78.

Mack, R.N., Simberloff, D., Lonsdale, W.M., Evans, H., Clout, M. and Bazzaz, F.A. (2000) Biotic invasions: causes, epidemiology, global consequences, and control. *Ecological Applications* 10, 689–710.

Namboothri, N. and Shanker, K. (2011) Corals or Cola: between the Devil & the Deep Blue Sea. *Current Conservation* Vol. 4 (1), 18–20.

Oliveira, E.C. and Paula, E.J. (2003) Exotic seaweeds: Friends or foes? In: Chapman, A.R.O., Anderson, R.J., Vreeland, R.J.V. and Davison, I. (eds) *Proceedings of the XVII International Seaweed Symposium.* Oxford University Press, New York, pp. 425–434.

Parker, H.S. (1974) The culture of the red algal genus *Eucheuma* in the Philippines. *Aquaculture* 3, 425–439.

Paula, E.J., Pereira, R.T.L. and Ostini, S. (1998) Introdução de espécies exóticas de *Eucheuma* e *Kappaphycus* (Gigartinales, Rhodophyta) para fins de maricultura no litoral brasileiro: abordagem teórica e experimental. In: Paula, E.J., Cordeiro-Marino, M., Pupo Santos, D., Fujii, M. , Plastino, E.M. and Yokoya, N. (eds) *IV Congresso Latino Americano de Ficologia, II Reunião Ibero-Americana de Ficologia e VII Reunião Brasileira de Ficología.* Sociedade Brasileira de Ficología, São Paulo, Brazil, pp. 340–357.

Pereira, N. and Verlecar, X.N. (2005) Is Gulf of Mannar heading for marine bioinvasion? *Current Science* 89, 1309–1310.

Reichard, S.H. and Hamilton, C.W. (1997) Predicting invasions of woody plants introduced into North America. *Conservation Biology* 11, 193–203.

Rejmánek, M. (2000) Invasive plants: approaches and predictions. *Australian Ecology* 25, 497–506.

Richardson, D.M. and Bond, W.J. (1991) Determinants of plant distribution: evidence from pine invasions. *American Naturalist* 137, 639–668.

Richardson, D.M., Pysek, P., Rejmánek, M., Barbour, M.G., Panetta, F.D. and West, C.J. (2000) Naturalization and invasion of alien plants: concepts and definitions. *Diversity and Distributions* 6, 93–107.

Rodgers, S.K. and Cox, E.F. (1999) The distribution

of the introduced rhodophytes, *Kappaphycus alvarezii*, *Kappaphycus striatum* and *Gracilaria salicornia* in relation to various physical and biological factors in Kane'ohe Bay, O'ahu, Hawaii. *Pacific Science* 53, 232–241.

Rudolph, B. (2000) In: Martin, R.E., Carter, E.P. and Flich, G.J. (eds) *Seaweed Product: Red Algae of Economic Significance*. Technomic, Lancaster, Pennsylvania, pp. 515–529.

Russell, D.J. (1983) Introduction of *Eucheuma* to Fanning Atoll, Kiribati, for the purpose of mariculture. *Micronesica* 18, 35–44.

SDMRI Report (2008) Report on the preliminary survey conducted in Krusadai and Shingle Islands to check the presence of seaweed, *Kappaphycus alvarezii*, 5 pp.

Smith, J.E., Hunter, C.L. and Smith, C.M. (2002) Distribution and reproductive characteristics of non-indigenous and invasive marine algae in the Hawaiian Islands. *Pacific Science* 56, 299–315.

Trowbridge, C.D. and Todd, C.D. (2001) Host–plant change in marine specialist herbivores: ascoglossan sea slugs on introduced macroalgae. *Ecology Monographs* 71, 219–243.

UNEP-WCMC (2006) In the front line: shoreline protection and other ecosystem services from mangroves and coral reefs. UNEP-WCMC, Cambridge, UK, 33 pp.

Van de Velde, F. and De Ruiter, G.A. (2002) In: Steinbuchel, A., Debates, S. and Banadme, E.J. (eds) *Polysaccharides II, 6, Biopolymers*, Wiley-VCH, Weinheim, Germany, pp. 245–274.

Vitousek, P.M. and Walker, L.R. (1989) Biological invasion by *Myrica faya* in Hawai'i: plant demography, nitrogen-fixation, ecosystem effects. *Ecology Monographs* 59, 247–265.

Wilson, J.B., Steel, J.B., Dodd, M.E., Anderson, B.J., Ullmann, I. and Bannister, P. (2000) A test of community reassembly using the exotic communities of New Zealand roadsides in comparison to British roadsides. *Journal of Ecology* 88, 757–764.

Woo, M. (1999) Ecological impacts and interactions of the introduced red alga *Kappaphycus striatum* in Kaneohe Bay, Oahu. MS thesis, University of Hawai'i at Ma¯noa, Honolulu, Hawaii.

Woo, M., Smith, C. and Smith, W. (1989) Ecological interactions and impacts of invasive *Kappaphycus striatum* in Kaneohe Bay. In: *Proceedings, First National Conference, Tropical Reef Marine Bioinvasions*, 186 pp.

# 8 Biology of *Mikania micrantha* H.B.K.: a Review

**R.S. Tripathi[1], M.L. Khan[2] and A.S. Yadav[3]**

[1]*National Botanical Research Institute, Lucknow, Uttar Pradesh, India;* [2]*Department of Forestry, North-Eastern Regional Research Institute of Science and Technology (Deemed University), Nirjuli, Arunachal Pradesh, India;* [3]*Department of Botany, R.R. Autonomous College, Alwar, Rajasthan, India*

## Introduction

Invasion of alien plant species is one of the most serious global problems. Alien plant species invade and adversely affect both natural and semi-natural ecosystems (Higgins *et al.*, 1996; Mgidia *et al.*, 2007). *Mikania micrantha* is one of the 100 worst alien species (Lowe *et al.*, 2001), is among the ten worst exotic species in South-east and South Asia, and one of the 16 exotic species in China (Zhang, L.Y. *et al.*, 2004). It is commonly known as mile-a-minute weed (Waterhouse, 1994), is an extremely fast-growing, perennial vine and is regarded as one of the world's most notorious invaders (Holm *et al.*, 1977; Cronk and Fuller, 1995). Once established, it kills nearby plant species by reducing light beneath its canopy (Huang *et al.*, 2000) (Fig. 8.1). Invasive species such as *M. micrantha* pose a serious threat to biodiversity (Wang *et al.*, 2004; Zhang, L.Y. *et al.*, 2004), and hence it is essential to understand the biology of this weed and its impact on biological communities. A review of research work carried out on this invasive species is presented in this chapter.

## Distribution and History of Invasion

*Mikania micrantha* H.B.K. (family: *Asteraceae*) is native to Central and South America (Wirjahar, 1976; Holm *et al.*, 1977). There are around 425 species of *Mikania* (King and Robinson, 1987), most of these native to tropical America and many are serious weeds (Wang *et al.*, 2001). The widespread distribution of *M. micrantha* may be attributed to humans (Holmes, 1981, 1982). The weed first invaded Hong Kong via the sea route at the beginning of the last century, and from there it spread to the coast of Guangdong province in the 1980s (Xie *et al.*, 2000; Wang *et al.*, 2001; Ye and Zhou, 2001; Zhang, L.Y. *et al.*, 2004; Li *et al.*, 2006). *Mikania micrantha* was introduced into Taiwan for soil conservation in the 1970s and it now infests nurseries, orchards, lawns, plantations and disturbed forests between latitudes 22–24° N and longitudes 111–117° W, and between 50 and 200 m altitude (Zhang, L.Y. *et al.*, 2004). It also grows in forests, along rivers and streams, in disturbed situations and along roadsides (Kong *et al.*, 2000; Zan *et al.*, 2000; Feng *et al.*, 2002). It has now also spread to Mauritius, India, Sri Lanka, Bangladesh, South-east Asia and the Pacific (Deng *et al.*, 2004). It was introduced into India during World War II to camouflage airfields or as ground cover for tea plantations. It has spread to moist tropical and subtropical regions (Fig. 8.2a) and the north-east (Fig. 8.2b) of India. The species was introduced into Indonesia as ground cover and it then it spread to the Pacific, South-east Asia and

**Fig. 8.1.** *Mikania micrantha* covering the canopy of a small tree in north-east India.

New Guinea (Waterhouse, 1994). It has invaded agricultural lands and plantation crops such as tea, teak, rubber and oil palm in moist tropical forest zones of Asia, particularly South-east Asia (Choudhury, 1972; Parker, 1972; Holm *et al.*, 1977).

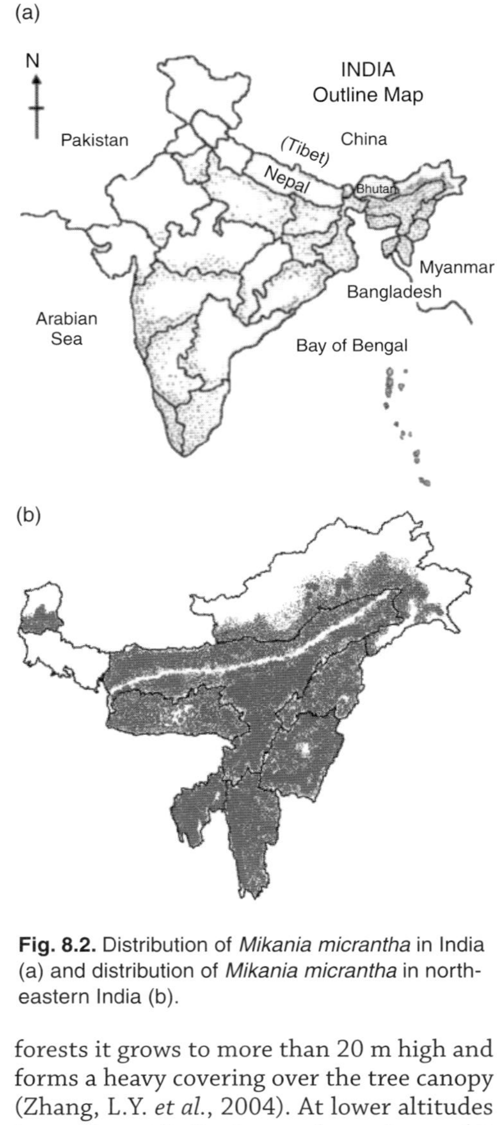


**Fig. 8.2.** Distribution of *Mikania micrantha* in India (a) and distribution of *Mikania micrantha* in north-eastern India (b).

## Biology

*Mikania micrantha* has been described by Robinson and King (1985) as being a perfect genus. It is easily delimited by the capitulum of four flowers (disc florets) surrounded by four phyllaries (King and Robinson, 1987). It is a multi-branched, perennial, scrambling vine with a 5-ribbed stem, internodes pubescent or glabrous and 7.5–21.5 cm long. The leaves are opposite, cordate or triangular with an acute apex and broad base, 4–13 cm long. Inflorescence is axillary, panicled corymbs; capitulum is cylindrical and 1.5 mm in diameter (Fig. 8.3). In open areas *M. micrantha* has cushion-like growth, with twining roots up to 30 cm thick, whereas in forests it grows to more than 20 m high and forms a heavy covering over the tree canopy (Zhang, L.Y. *et al.*, 2004). At lower altitudes in north-east India, too, we have observed it climbing both small and tall trees, covering their canopy completely. It has small, compact florets that are protandrous.

During anthesis, the stigma and style grow through the tube formed by syngenesious anthers and the stigma protrudes out above the anthers along with pollen grains. Thus, the stigma plays an important role in presenting pollen grains to pollinators (Hong *et al.*, 2007). *Mikania micrantha* is also reported to be self-

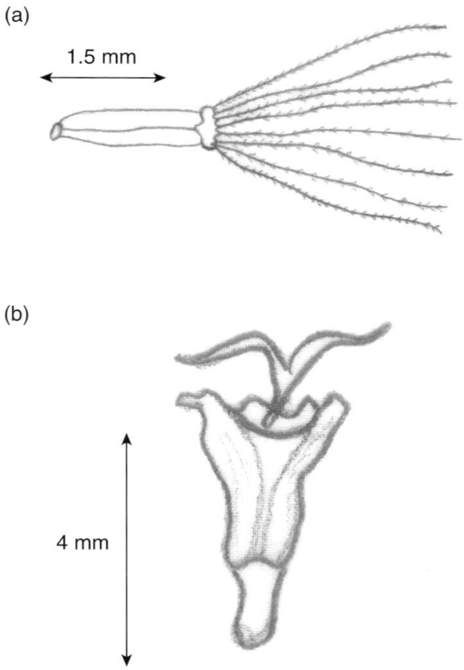


**Fig. 8.3.** *Mikania micrantha* cypsella (a) and flower (b).

incompatible, and requires insect pollination (Hong *et al.*, 2007). In China the maximum mean growth rate of *M. micrantha* has been reported to be 47 cm per week in mid-October during the rainy season (Zhang, L.Y. *et al.*, 2004). In Hong Kong it grows vegetatively from March to August, produces flowers in September/October and seeds in November–February (Hu and But, 1994), whereas in Donguan in China it grows vegetatively from March to October and flowers during October–December (Cao *et al.*, 2003). In north-east India, flowering takes place from October to late November and the plants die in December and January (Swamy and Ramakrishnan, 1987a). Seed dispersal takes place from October to April via wind, water and animals. *Mikania micrantha* has the chromosome number $n$ = 17–19 (King and Robinson, 1987). Twelve polymorphic microsatellite loci have been isolated and characterized from an AC-enriched genomic library of this species (Hong *et al.*, 2008).

## Reproductive Potential

*Mikania micrantha* perpetuates through seeds and by rooting at stem nodes (Ismail and Mah, 1993). There are three types of propagule in this species: (i) seeds; (ii) ramets; and (iii) perennating rosettes derived from a rosette caudex or ramet caudices (Swamy and Ramakrishnan, 1987a). A ramet recruitment pattern is a function of successional development of fallow following agricultural changes, their population decreasing in older fallows (Swamy and Ramakrishnan, 1987a). *Mikania micrantha* produces enormous numbers of seeds, i.e. 170,000/$m^2$ (Kuo *et al.*, 2002), which are small ($8.92 \times 10^{-5}$ g) (Hu and But, 1994). Seeds are dispersed by wind, and are the chief means of invasion of newly disturbed areas (Zhang, L.Y. *et al.*, 2004). Following establishment, local spread of this weed in north-east India is by vegetative propagules (Swamy and Ramakrishnan, 1987a; Wen *et al.*, 2000).

## Physiology

In China, *M. micrantha* shows higher biochemical activity as compared with *Mikania cordata*, which is a native species. The higher assimilation rate, light saturation point and $CO_2$ saturation point suggest that the former is adapted to high light intensity. Its greater water use efficiency is also observed to be related to its higher photosynthetic activity (Deng *et al.*, 2004). It has been reported to show lower maximum light saturated photosynthetic rate, photosynthetic nitrogen use efficiency and relative growth rate under low irradiance (Zhang and Wen, 2009). These workers also reported maximum growth rate under a combination of high light intensity and high soil moisture content. Likewise, in its native region, *M. micrantha* is abundant in open areas and shows less vigorous growth under 25% shade (Ipor, 1991).

Hu and But (1994) observed that the *M. micrantha* flower takes 5 days to develop from a flower bud, 5 days from flower to anthesis and 5–7 days are required for seed maturation. After 6–8 days of seed production, 96.3% germination is achieved at ambient temperature (Hu and But, 1994). The optimum temperature for germination is reported to be 25–30°C (Yang *et al.*, 2003). These authors also suggested that the seeds of *M. micrantha* have an after-ripening requirement, as the autumn and spring seeds exhibited 70 and 80% germination, respectively. *Mikania micrantha* showed a net photosynthetic rate of 21.56 μ mol $CO_2$/$m^2$/s under full light, which is reported to be higher than that for other associated species studied (Wen *et al.*, 2000). They also suggested that, although *M. micrantha* is a heliophytic species with higher quantum use efficiency and lower maximum compensation point, it can also tolerate shady situations.

## Biochemistry

Bakir *et al.* (2004) reported the isolation of mikanolide (I), a sesquiterpene dilactone, from different species of *Mikania*. It is used as a medicine by locals in Jamaica and is reported to have antibacterial, anti-tumour, antimicrobial, cytotoxic and phytotoxic activities (Aguinaldo *et al.*, 1995; Facey *et al.*, 1999). But *et al.* (2009) isolated a new sesquiterpene, 3β-acetoxy-1, 10-epoxy-4-germacrene-12, 8; 15, 6-diolide, along with six known constituents: (i) 1,10-epoxy-4-germacrene 12, 8; 15, 6-diolide; (ii) dihydromikanolide; (iii) potassium mikanin 3-sulfate; (iv) alpinetin; (v) mikanin; and (vi) ergosta-7, 22-dien-3β-ol. But *et al.* (2009) further reported that compound (i) exhibited moderate activity against respiratory syncytial virus and parainfluenza type 3 virus, and potassium mikanin 3-sulfate is the main compound of *M. micrantha* showing inhibitory activity against parainfluenza type 3 virus. The sesquiterpenes, mikanolide and dihydromikanolide show antibacterial and antimicrobial activity. Essential oils and sesquiterpenoid lactones of the germacrane type have been reported from *M. micrantha* (But *et al.*, 2009). Essential oils from *Mikania micrantha* showed sufficient repellent effect and oviposition-deterrent activity for the diamondback moth, *Plutella xylostella* (Zhang, M.X. *et al.*, 2004).

## Allelopathic Effects

Allelopathy has been considered as a mechanism for the successful invasion by alien plant species (Chen *et al.*, 2009). Weng (1964) reported that extracts of *Mikania* plants significantly decreased the biomass and nitrogen content of tomato seedlings and legume cover crops. Leaf or root extracts of *M. micrantha* significantly reduced germination and fresh weight of *Asytasia intrusa*, *Chrysopogon aciculatus* and *Paspalam conjugatum* (Ismail and Mah, 1993). However, Ismail and Chong (2002) reported that seed germination, radicle elongation and biomass production of *Lycopersicon esculentum* and *Brassica chinensis* were inhibited by aqueous extracts of *M. micrantha*, but *Zea mays* and *Vigna sesquipedelis* seedlings were not affected. Similar observations were made when debris from *M. micrantha* was added to the soil used for growing the four tested species. They also identified caffeic acid, P-hydroxybenzaldehyde, resorcinol and vanillic acid as allelochemicals in *M. micrantha* leaf extract. Chen *et al.* (2009) reported that the soil in the vicinity of *M. micrantha* inhibited seed germination and growth of tested species. They also suggested that water-soluble allelochemicals from *Mikania micrantha* improved soil nutrient availability, which may help this weed successfully invade and establish in new areas.

Shao *et al.* (2005) isolated and identified three sesquiterpenoids from *M. micrantha*: dihydromikanolide, deoxymikanolide and 2, 3-epoxy-1-hydroxy-4, 9-germacradiene-12, 18: 15, 6 diolide. Seed germination and seedling growth in tested species were inhibited by these sesquiterpenoids, with deoxymikanolide exhibiting the strongest phytotoxicity. Wu *et al.* (2009) also reported the inhibitory effect of leaf and root extracts from *M. micrantha* on woody species. The leaf extract of this weed showed a higher allelopathic effect on seed germination as compared with the root extract, whereas the root extract showed greater inhibitory effect on the roots of tested species as compared with the leaf extract. Therefore, the root extract may promote the invasiveness of *M. micrantha* due to its more direct effective allelopathic inhibition of the roots of associated species.

## Ecology

Ye and Zhou (2001) reported that *M. micrantha* grows in moist soils, whether acidic or alkaline (pH 4.15–8.35), and from less fertile to highly fertile soils, in southern China. The wide ecological amplitude of this weed helps it invade and spread in different regions of Asia and the Pacific. It can also grow in dry soils and in shade in China. It grows best where annual temperature is >21°C and soil moisture >15% (Huang *et al.*, 2000). However, *M. micrantha* was found to grow at >600 m altitude on Wutong Mountain in Shenzhen, China (Huang *et al.*, 2000). It can also invade aquatic habitats such as ponds, covering and killing aquatic plants such as *Eichhornia crassipes* (Zhang, L.Y. *et al.*, 2004).

Swamy and Ramakrishnan (1987b) studied the population dynamics and reproductive potential of *M. micrantha* in north-east India. In natural situations, reproduction through rametes, which develop from the rosettes, was greater than through seeds. Ramet population growth was highest during the rainy season as a result of high recruitment. Seedling recruitment was recorded in freshly disturbed fallows, and in the 8-year old jhum fallow where ground-level herbaceous vegetation was poor due to shading by shrubs and small trees. In established plant communities recruitment occurred only through ramets. The high recruitment and death rates of ramets and seedlings occurred during the rainy season and reached a maximum in October, levelling off subsequently (Swamy and Ramakrishnan, 1987a). The maximum risk of death occurred during the peak growth period from June to September, and death due to senescence occurred during November and December. The percentage of surviving rosettes and expected turnover rates declined with the age of the jhum fallows. Seed output per square metre also declined with age, the lowest being in the 8-year-old fallow (Swamy and Ramakrishnan, 1987a). Net population of *M. micrantha* increased with ageing of fallow for 3 years, but declined sharply in 6-year-old and 12-year-old fallows (Swamy and Ramakrishnan, 1987b). They also found that biomass allocation was mainly to the ramet root system in younger fallows and to the rosette root system in older fallows.

Swamy and Ramakrishnan (1987a) evaluated the effect of fire on the population behaviour of *M. micrantha*. It responded to burning through both enhanced recruitment and mortality of seedlings and ramets. Seedling recruitment occurred in burnt plots of 2-, 4- and 8-year-old fallows. The highest seedling recruitment was recorded in 4-year-old fallows and lowest in 8-year-old jhum fallows in any given year. The number of seedlings established was very low, as heavy mortality of seedlings was experienced during the growing season. Almost all the seedlings established produced rosettes after perennation (Swamy and Ramakrishnan, 1987a).

Swamy and Ramakrishnan (1987a) reported that the number of rosettes of *M. micrantha* was lower in burnt plots. Figures for rosette formation were higher in unburnt as compared with burnt 2-year-old fallows. Although vegetative recruitment was higher in unburnt plots, the population growth rate declined in 4- and 9-year-old fallows in the unburnt plots. The net rosette population of *M. micrantha* consistently increased in burnt fallow plots over a 1-year period, whereas such an increase was observed only in 2-year-old unburnt fallow plots. The proportion of ramets that flowered was higher in the burnt plots than in the unburnt. Seed production was highest in 4-year-old fallows, declining sharply in 8-year-old fallows. Rosette perennation and seedling recruitment in younger fallows was enhanced by fire, but inhibited by it in 8-year-old fallows. Burning also promoted the reproductive potential of *M. micrantha*, but this declined rapidly in older fallows. Swamy and Ramakrishnan (1987a) concluded that frequent disturbances in slash-and-burn agriculture with short cycles of 4–5 years would enhance both clonal and seed reproduction in *M. micrantha*. This would result in increased weed potential and the succession would be maintained at the weed stage, whereas long slash-and-burn cycles would effectively suppress plant growth.

Swamy and Ramakrishnan (1988) reported higher biomass and nutrient allocation to seeds in burnt than in unburnt 8-year-old fallow plots. Nutrient use efficiency was observed to be inversely related to uptake efficiency, and may be a compensatory mechanism for survival. The vigour of *M. micrantha* was maintained in burnt fallow plots but decreased with age in unburnt plots, which indicates that it is adapted to survival after burning (Swamy and Ramakrishnan, 1988).

## Ecological Impact

Swamy and Ramakrishnan (1987c) suggested that *M. micrantha* conserves large quantities of potassium in its biomass and therefore exerts considerable influence on nutrient cycling in secondary succession after slash-and-burn agriculture in north-eastern India. Growth of *M. micrantha* had a profound effect on soil by increasing numbers of aerobic bacteria and decreasing those of anaerobic bacteria (Li *et al.*, 2006). They observed maximum enzymatic activity, influenced by soil pH and organic matter, in *M. micrantha*-infested communities, intermediate in ecotones and lowest in the native communities. Chen *et al.* (2009) suggested that, although water-soluble allelochemicals of *M. micrantha* inhibit seed germination and seedling growth of some plants, they also improve the availability of soil nutrients by increasing levels of carbon, nitrogen, nitrate and ammonia in the soil.

## Control Methods

The dense growth habit typical of *M. micrantha* can be suppressed by the use of herbicides such as glyphosate, 2, 4-D, napropamide and atrazine (Palit, 1981). According to Hu and But (1994), seed germination of *M.* micrantha was effectively inhibited by 0.4% bentazone, and 0.2% picloram/0.4% bentazone was most effective against the growth of 25-, 45- and 60-day-old seedlings. Although spray application of 2, 4-D and glyphosate killed above-ground parts of this weed, it had fully regenerated by 3 months, while injection of 25% hexazinone and spraying of 75% sulfometuron methyl killed the whole plant (Zan *et al.*, 2001). Kumar (2005) reported that treatment with 0.8 kg gramaxone/1.0 kg 2,4-D amine/ha significantly reduced *M. micrantha* growth within 2–3 days of spraying, and there was no regeneration of the weed for up to 80 days afterwards.

Several natural enemies of *M. micrantha* have been identified. Among these, the tea mosquito bug (*Helopeltis theivora*) inflicted serious damage but none could be considered as having the potential to control this weed (Abraham *et al.*, 2002). *Cuscuta campestris*, a holoparasite, is reported to have reduced cover, biomass and nutrient levels of *M. micrantha* and significantly increased native species richness by increasing soil pH and water and nutrient levels (Yu *et al.*, 2009). Shao *et al.* (2002) reported a growth rate for *Cuscuta* of 10 cm/day, which is higher than that for *M. micrantha*. According to Han *et al.* (2002), the thread-like stem of *Cuscuta* coils around the petiole, stem and branches of *M. micrantha*, a haustorium is formed where it touches the latter and this extracts water and nutrients from the weed, eventually killing it (Fig. 8.4). A fungus, *Puccinia spegazzinii*, has also been reported to have potential to control *M. micrantha* as it can cause extensive damage to its leaves, petioles and branches, which may result in death. This fungus is reported to be specific to *M. micrantha* and also has a wide range of environmental tolerance (Cock *et al.*, 2000).

Biological and chemical control of *M. micrantha* are both very costly, and may adversely affect other associated species, and therefore manual removal is the most effective method. The cutting of vines near the ground once a month for three consecutive months in summer and autumn, and again in winter and spring, can eliminate 90% of *M. micrantha* (Kuo *et al.*, 2002). In

**Fig. 8.4.** *Mikania micrantha* engulfed by *Cuscuta* sp. in north-eastern India.

Guangdong province of China, the most effective method was found to be the pulling out of *M. micrantha* before flowering at the end of October (Zhang, L.Y. *et al.*, 2004). To prevent reinvasion of this weed, bare areas should be covered with fast-growing native species.

## Conclusions

*Mikania micrantha*, an invasive alien species, is native to Central and South America. It has invaded the moist tropical regions of the Pacific, South-east Asia, China, India and other countries of the world. It invades new areas by seed dispersal and, once established, it spreads locally by ramets produced by perennating rosette caudices and ramet caudexes. It drastically suppresses the growth of associated plant species, mainly by reducing the availability of light, as it very rapidly forms a dense cover over host plants, and also by secreting allelochemicals in their vicinity. Consequently, it adversely affects the biodiversity of the invaded region. Several measures have been attempted in controlling this weed, but to date little success has been achieved. It has been reported that some biochemical substances obtained from this weed exhibit antibacterial and antimicrobial activity, and show inhibitory action against parainfluenza type 3 and respiratory syncytial viruses. Thus, *Mikania micrantha* has great potential as a medicinal plant which may be exploited in the future for human welfare. Exploitation of *Mikania micrantha* to this end would also help in controlling its spread to new regions.

## References


Abraham, M., Abraham, C.T. and Joy, P.J. (2002) Natural enemies on *Mikania micrantha* H.B.K. in Kerala. *Journal of Tropical Agriculture* 40, 39–41.

Aguinaldo, A.M., Abe, F., Yamanchi, T. and Padolina, W.G. (1995) Germacranolides of *Mikania cordata*. *Phytochemistry* 38, 1441–1443.

Bakir, M., Facey, P.C., Hassan, I., Mulder, W.H. and Porter, R.B. (2004) Mikanolide from Jamaican *Mikania micrantha*. *Acta Crystallographica Section C: Crystal Structure Communications* 60, 798–800.

But, P.P.H., He, Z.D., Ma, S.C., Chan, Y.M., Shaw, P.C., Ye, W.C. *et al.* (2009) Antiviral constituents against respiratory viruses from *Mikania micrantha*. *Journal of Natural Products* 72, 925–928.

Cao, H.L., Feng, H.L., Deng, X., Zhang, L.Y. and Ye,

W.H. (2003) Study on invasion, harm, prevention and control of *Mikania micrantha*. In: Xu, R.M. and Ye, W.H. (eds) *Biological Invasion: Theory and Practice*. Science Press, Beijing, pp. 236–248.

Chen, B.M., Peng, S.L. and Ni, G.Y. (2009) Effect of the invasive plant *Mikania micrantha* H.B.K. and soil nitrogen availability through allelopathy in South China. *Biological Invasions* 11, 1291–1299.

Choudhury, A.K. (1972) Controversial *Mikania* (climber) – a threat to forests and agriculture. *Indian Forester* 98, 178–186.

Cock, M.J.W., Ellison, C.A., Evans, H.C. and Ooi, P.A.C. (2000) Can failure be turned into success for biological control of mile-a-minute weed (*Mikania micrantha*)? In: Spencer, N.R. (ed.) *Proceedings of the Xth International Symposium on Biological Control of Weeds*, Montana State University, Bozeman, Montana, pp. 155–167.

Cronk, Q.C.B. and Fuller, J.L. (1995) *Plant Invaders: The Threat to Natural Ecosystems*. Chapman and Hall, London.

Deng, X., Ye, W.H., Feng, H.L., Yang, Q.H., Cao, H.L., Xu, K.Y. *et al.* (2004) Gas exchange characteristics of the invasive species *Mikania micrantha* and its indigenous congener *M. cordata* (Asteraceae) in South China. *Botanical Bulletin of Academia Sinica* 45, 213–220.

Facey, P.C., Pascoe, K.O., Parter, R.B. and Jones, A.D. (1999) Investigation of plants used in Jamaican folk medicine for antibacterial activity. *Journal of Pharmacy and Pharmacology* 51, 1455–1460.

Feng, H.L., Cao, H.L., Liang, X.D., Zhou, X. and Ye, W.H. (2002) The distribution and harmful effects of *Mikania micrantha* in Guangdong. *Journal of Tropical and Subtropical Botany* 10, 263–270.

Han, S., Li, K., Luo, L., Liu, W., Chen, Q., Peng, T. *et al* (2002) *Mikania micrantha* destroyed by parasitic weed dodder, *Cusuta chinensis*, in Guangdong. *Natural Enemies of Insects* 24, 7–14.

Higgins, S.I., Richardson, D.M. and Cowling, R.M. (1996) Modeling invasive plant spread: the role of plant–environment interactions and model structure. *Ecology* 77, 2043–2045.

Holm, L.G., Plucknett, D.L., Pancho, J.V. and Herberger, J.P. (1977) *The World's Worst Weeds. Distribution and Biology*. University Press of Hawaii, Honolulu, Hawaii.

Holmes, W.C. (1981) *Mikania* (Compositae) of the United States. *Sida* 9, 147–158.

Holmes, W.C. (1982) Revision of the Old World *Mikania* (Compositae). *Botanisches Jahres Beitraegefuer Systematik* 103, 211–246.

Hong, L., Shen, H., Ye, W.H., Cao, H.L. and Wang, Z.M. (2007) Self incompatibility in *Mikania micrantha* in South China. *Weed Research* 47, 280–283.

Hong, L., Niu, H., Shen, H., Ye, W. and Cao, H. (2008) Development and characterization of microsatellite markers for the invasive weed *Mikania micrantha* (Asteraceae). *Molecular Ecology Resources* 8, 193–195.

Hu, Y. and But, P.P.H. (1994) A study on life cycle and response to herbicides of *Mikania micrantha*. *Acta Scientiarum Naturalium Universitatis Sunyatseni* 33, 88–95.

Huang, Z.L., Cao, H.L., Liang, X.D., Ye, W.H., Feng, H.L. and Cai, C.X. (2000) The growth of damaging effects by *Mikania micrantha* in different habitats. *Journal of Tropical and Subtropical Botany* 8, 131–138.

Ipor, I.B. (1991) The effect of shade on the growth and development of *Mikania micrantha* H.B.K. *Malaysian Applied Biology* 20, 57–63.

Ismail, B.S. and Chong, T.V. (2002) Effects of aqueous extracts and decomposition of *Mikania micrantha* H.B.K. debris on selected agronomic crops. *Weed Biology and Management* 2, 31–38.

Ismail, B.S. and Mah, L.S. (1993) Effects of *Mikania micrantha* H.B.K. on germination and growth of weed species. *Plant and Soil* 157, 107–113.

King, R.M. and Robinson, H. (1987) *The genera of the Eupatorieae (Asteraceae) – Monographs in Systematic Botany*, Vol. 22. Missouri Botanical Garden, St. Louis, Missouri.

Kong, G.H., Wu, Q.G. and Hu, Q.M. (2000) Appearing of exotic weed *Mikania micrantha* H.B.K. in China. *Journal of Tropical and Subtropical Botany* 8, 27.

Kumar, S. (2005) *Forest invasive species*. Indian Council of Forestry Research and Education, Dehradun, India.

Kuo, Y.L., Chen, T.Y. and Lin, C. C. (2002) Using a consecutive cutting method and allelopathy to control the invasive vine, *Mikania micrantha* H.B.K. *Taiwan Journal of Forest Science* 17, 171–181.

Li, W.H., Zhang, C.B., Jiang, H.B., Xin, G.R. and Yang, Z.Y. (2006) Changes in soil microbial community associated with invasion of the exotic weed, *Mikania micrantha* H.B.K. *Plant and Soil* 281, 309–324.

Lowe, S., Browne, M., Boudjelas, S. and Depoorter, M. (2001) *100 of the World's Worst Invasive Alien Species*. A selection from the Global Invasive Species Database. IUCN/SSC Invasive Species Specialist Group (ISSG), Auckland, New Zealand.

Mgidia, T.N., Le Maitre, D.C., Schonegevela, L., Nel, J.L., Rouget, M. and Richardson, D.M.

(2007) Allien plant invasions-incorporating emerging invaders in regional prioritization: a pragmatic approach for Southern Africa. *Journal of Environmental Management* 84, 173–187.

Palit, S. (1981) *Mikania* – a growing menace in plantation forestry in West Bengal. *Indian Forester* 107, 96–101.

Parker, C. (1972) The *Mikania* problem. *Pest Articles and News Summaries* 18, 312–315.

Robinson, H. and King, R.M. (1985) Comments on the generic concepts in the *Eupatorieae. Taxon* 34, 11–16.

Shao, H., Peng, S., Wei, X., Zhang, D. and Zhang, C. (2005) Potential allelochemicals from an invasive weed *Mikania micrantha* H.B.K. *Journal of Chemical Ecology* 31, 1657–1668.

Shao, W., Han, S., Huang, S., Liu, W., Li, K., Peng, T. *et al.* (2002) Advances in controlling exotic weed *Mikania micrantha* H.B.K. *Guangdong Agricultural Sciences* 1, 43–45.

Swamy, P.S. and Ramakrishnan, P.S. (1987a) Effect of fire on population dynamics of *Mikania micrantha* H.B.K. during early succession after slash and burn agriculture (jhum) in North-East India. *Weed Research* 27, 397–403.

Swamy, P.S. and Ramakrishnan, P.S. (1987b) Contribution of *Mikania micrantha* during secondary succession following slash and burn agriculture (jhum) in North-East India. II nutrient cycling. *Forest Ecology and Management* 22, 239–249.

Swamy, P.S. and Ramakrishnan, P.S. (1987c) Weed potential of *Mikania micrantha* H.B.K. and its control in fallows after shifting agriculture (jhum) in North-East India. *Agriculture, Ecosystems and Environment* 18,195–204.

Swamy, P.S. and Ramakrishnan, P.S. (1988) Effect of fire on growth and allocation strategies of *Mikania micrantha* under early successional environments. *Journal of Applied Ecology* 25, 635–658.

Wang, B.S., Liao, W.B. and Miao, R.H. (2001) Revision of *Mikania* from China and the key of four relative species. *Acta Scientiarum Naturalium Universitatis Sunyatseni* 40, 72–75.

Wang, B.S., Wang, Y.J., Liao, W.B., Zhan, Q.J., Li, M.G. and Peng, S.L. (2004) *The Invasion Ecology and Management of Alien Weed Mikania micrantha* H.B.K. Science Press, Beijing, pp. 76–85.

Waterhouse, D.F. (1994) *Biological Control of Weeds: South East Asian Prospects.* ACIAR, Canberra, Australia.

Wen, D.Z., Ye, W.H., Feng, H.L. and Cai, C.X. (2000) Comparison of basic photosynthetic characteristics between exotic invader weed *Mikania micrantha* and its companion species. *Journal of Tropical and Subtropical Botany* 8, 139–146.

Weng, W.P. (1964) Evidence for the presence of growth inhibitory substances in *Mikania cordata* (Brum. f.) B.L. Robinson. *Journal of Rubber Research Institute Malaya* 18, 231–242.

Wirjahar, S. (1976) Autoecological study of *Mikania* spp. In: Kataoko, T., Furuya, S., Hyakutake, H., Ishizuka, K., Matsunaka, S. and Takabayashi, M. (eds) *Proceedings of the Fifth Asian Pacific Weed Science Society Conference*, Tokyo, pp. 70–73.

Wu, A.P., Yu, H., Gao, S.Q., Huang, Z.Y., He, W.M., Miao, S.L. *et al.* (2009) Differential belowground allelopathic effects of leaf and root of *Mikania micrantha. Trees – Structure and Function* 23, 11–17.

Xie, Y., Li, Z., William, P.G. and Li, D. (2000) Invasive species in China – an overview. *Biodiversity and Conservation* 10, 1317–1341.

Yang, Q.H., Feng, H.L., Ye, W.H., Cao, H.L., Deng, X. and Xu, K.Y. (2003) An investigation of the effects of environmental factors on the flowering and seed setting of *Mikania micrantha* H.B.K. (Compositae). *Journal of Tropical and Subtropical Botany* 11, 123–126.

Ye, W.H. and Zhou, X. (2001) The plant killer *Mikania micrantha* in South China. *Aliens* 13, 7.

Yu, H., He, W.M., Liu, J., Miao, S.L. and Dong, M. (2009) Native *Cuscuta campestris* restrains exotic *Mikania micrantha* and enhances soil resources beneficial to natives in the invaded communities. *Biological Invasions* 11, 835–844.

Zan, Q.J., Wang, Y.J., Wang, B.S., Liao, W.B. and Li, M.G. (2000) The distribution and harm of the exotic weed *Mikania micrantha. Chinese Journal of Ecology* 19, 58–61.

Zan, Q.J., Wang, Y.J., Liang, Q.Y., Wang, B.S. and Liao, W.B. (2001) Effectiveness of four herbicides on the harmful weed *Mikania micrantha. Ecological Science* 20, 32–36.

Zhang, L.L. and Wen, D.Z. (2009) Structural and physiological responses of two invasive weeds, *Mikania micrantha* and *Chromolaena odorata* to contrasting light and soil water conditions. *Journal of Plant Research* 122, 69–79.

Zhang, L.Y., Ye, W.H., Cao, H.L. and Feng, H.L. (2004) *Mikania micrantha* H.B.K. in China – an overview. *Weed Research* 44, 42–49.

Zhang, M.X., Ling, B., Chen, S.Y., Liang, G.W. and Pang, X.F. (2004) Repellent and oviposition deterrent activities of the essential oil from *Mikania micrantha* and its compounds on *Plutella xylostella. Entomologia Sinica* 11, 7–45.

# 9 *Anthemis cotula* L.: a Highly Invasive Species in the Kashmir Himalaya, India

**Zafar A. Reshi, Manzoor A. Shah, Irfan Rashid and Nazima Rasool**

*Department of Botany, University of Kashmir, Srinagar, Jammu and Kashmir, India*

## Introduction

The rapidly increasing alien species invasions across biogeographical regions in the contemporary world, with huge socio-economic and environmental costs, underscore the urgency of detailed investigation of the process of invasion, species invasiveness and community invasibility (Moles *et al.*, 2008). Different hypotheses proposed to explain the mechanism of plant invasion (reviewed and integrated by Catford *et al.*, 2008) suggest that different factors operate, either separately or synergistically, at various spatio-temporal scales and stages to cause successful invasion. While the process of invasion is highly complex and largely context-specific, detailed understanding of the invasiveness of a large number of flagship invasive species in different parts of world can aid in arriving at some broad generalizations about the 'syndrome' of traits that contribute to invasion, predictive modelling of spatial spread and effective management.

Despite considerable progress in certain facets of invasion biology over the years, basic information about many invasive species is lacking in respect of key attributes that promote their invasion, the analysis and documentation of which is a challenging task for invasion ecologists (Pysek and Richardson, 2007). Thus, the question *what makes a plant species invasive* merits special attention in the rapidly progressing field of invasion ecology. It is in this context that a series of studies has been conducted over the last few years (Allaie *et al.*, 2005a, b, 2006; Rashid *et al.*, 2007a, b; Shah and Reshi, 2007; Shah *et al.*, 2008a, b, 2009) on *Anthemis cotula* L. (*Asteraceae*) to identify the key attributes that contribute to the invasiveness of this species in the Kashmir Himalaya, India. This chapter is an attempt to consolidate the information generated hitherto in the form of a unified conceptual and operational framework to portray the success story of *A. cotula* invasion and its impact on above- and below-ground communities. The information generated would be of immense importance in formulating effective management strategies to combat the menace of extensive growth and spread of this non-native species at the expense of native floristic elements in the Kashmir Himalaya.

## *Anthemis cotula* L.: a Smart Species with Smart Strategies

The genus *Anthemis* comprises around 100 species of aromatic herbs belonging to the sunflower family (Asteraceae), of which *Anthemis cotula* L., commonly known as either mayweed chamomile or stinking mayweed, is a monocarpic annual species native to southern Europe–western Siberia

(Erneberg, 1999) where it grows in both natural and agricultural habitats. Its European distribution extends to southern Norway, central Sweden and southern Finland, whereas its southern extent includes the Canary Islands, Egypt and western Asia. It grows as a common weed of arable lands and farmyards in the Scottish lowlands, Ireland, England and Wales (Kay, 1971). Besides the Kashmir Himalaya, the species has been introduced to the USA, Canada, Argentina, Australia and New Zealand (Hulten, 1968; USDA-ARS, 2005) and is a noxious weed in Colorado and Nevada (USDA, 2002).

In the Kashmir Himalayan region, *A. cotula* is a highly invasive species of disturbed habitats, which is attributed to its protracted recruitment pattern aided by habitat disturbance; high population size even after seedling mortality (Allaie *et al.*, 2005b); allelopathic activity of its aqueous leaf leachate (Allaie *et al.*, 2006); mycorrhizal mutualism (Shah and Reshi, 2007; Shah *et al.*, 2008a, b); herbivore-induced, over-compensatory growth (Rashid *et al.*, 2006); and profuse achene production and synchrony between their germination and favourable environmental conditions (Rashid *et al.*, 2007b). It flowers from mid May to late August, producing capitula of 15–30 mm diameter containing white neuter ray florets and 70–200 yellow hermaphroditic disc florets (Kay, 1971). It is strongly self-incompatible and pollinated by syrphids and other Diptera (Kay, 1971).

Many allelochemicals, such as linear sesquiterpene lactones, sterols, taraxasterol, 7-methoxy-6-acetyl-2,2-dimethylchromene (Baruah *et al.*, 1985) and flavonoids (Glennie and Harborne, 1971) have previously been isolated from *A. cotula* L. While a number of parasites and pathogens are reported to limit the population size of this species in its native range (Shah and Reshi, 2007), no incidence of pests has been recorded in its introduced Kashmir Himalayan range. In the Valley of Kashmir, part of the recently declared Himalayan biodiversity hot spot, this fast-spreading species is emerging as a major threat to native biodiversity and ecosystem processes.

## Phenotypic Variability: Jack-and-Master Strategy

Phenotypic variability among different traits of *A. cotula* is summarized in Table 9.1. Vegetative and reproductive characters, excepting floret mass per plant, differed significantly between populations and habitats ($P < 0.001$). In riparian populations of *A. cotula*, all traits, except height, exhibited higher values (Table 9.2). Fitness measured as number of disc florets per plant was invariably high though being highest (19,235.37) in riparian habitats and lowest (2429.09) in terrestrial habitats with low disturbance. Thus, *A. cotula* is not only able to maintain fitness across a broad range of environments, a characteristic related to the concepts of a ‘general purpose genotype’ and of fitness homeostasis, but is also able to increase fitness in favourable environments (Jack-and-master strategy). Riparian habitats act as havens and, along with roads and trails, serve as dispersal corridors for invasion of upland sites by *A. cotula* (Rashid *et al.*, unpublished data).

## Germination Ecology: Sensing the Right Environment

Studies (Rashid *et al.*, 2007) carried out in the Kashmir Himalaya have revealed that fresh achenes of *A. cotula* exhibit dormancy as the percentage of germination of fresh achenes either under continuous light or dark conditions was very low (Fig. 9.1); in contrast, achenes of 1–4 years of age under continuous light conditions showed a relatively high percentage of germination. In comparison with controls, application of different nitrate ($NO_3^-$) concentrations applied in the present not only improved germination under continuous light in achenes of all ages but also had a significant effect on the germination of achenes of different ages under continuous dark conditions (Fig. 9.1). Treatments of 2.0, 2.5 and 3.0 mM $NO_3^-$ had more or less similar effects on achene germination under continuous light conditions, but the highest concentration also improved germination

**Table 9.1.** Vegetative and reproductive characters of *Anthemis cotula* (mean ± SE) among different populations.

| Habitat type | Site | Height/ plant | Lateral branches/ plant (*n*) | Root mass/ plant (g) | Shoot mass/ plant (g) | Capitula/ plant (*n*) | Disc florets/ capitulum (*n*) | Disc florets/ plant (*n*) | Floret mass/ plant (g) |
|---|---|---|---|---|---|---|---|---|---|
| Terrestrial open habitats with low disturbance | LD1 | 35.80 ± 1.28 | 1.50 ± 0.22 | 0.84 ± 0.12 | 5.55 ± 0.20 | 21.20 ± 1.07 | 109.50 ± 1.45 | 2,311.30 ± 96.85 | 1.30 ± 0.20 |
| | LD2 | 50.80 ± 1.38 | 1.10 ± 0.10 | 0.35 ± 0.09 | 1.97 ± 0.30 | 12.10 ± 1.02 | 96.00 ± 1.01 | 1,156.10 ± 90.69 | 0.52 ± 0.16 |
| | LD3 | 19.80 ± 2.93 | 1.20 ± 0.13 | 0.41 ± 0.04 | 1.69 ± 0.19 | 12.70 ± 1.27 | 94.90 ± 1.68 | 1,196.50 ± 113.21 | 0.58 ± 0.08 |
| | LD4 | 52.80 ± 1.71 | 1.00 ± 0.00 | 0.92 ± 0.07 | 4.49 ± 0.22 | 15.20 ± 0.92 | 113.80 ± 0.71 | 1,730.40 ± 105.85 | 0.54 ± 0.08 |
| | LD5 | 49.20 ± 2.03 | 1.40 ± 0.22 | 1.10 ± 0.14 | 9.42 ± 0.42 | 23.30 ± 2.58 | 115.60 ± 1.07 | 2,698.90 ± 306.54 | 1.57 ± 0.12 |
| | LD6 | 76.60 ± 4.21 | 1.20 ± 0.13 | 1.51 ± 0.12 | 7.35 ± 0.15 | 19.20 ± 1.86 | 126.30 ± 1.14 | 2,415.30 ± 225.94 | 1.30 ± 0.12 |
| | LD7 | 26.30 ± 1.37 | 2.00 ± 0.30 | 1.09 ± 0.10 | 8.77 ± 0.24 | 29.00 ± 2.13 | 120.70 ± 1.14 | 3,501.10 ± 258.66 | 1.07 ± 0.08 |
| | LD8 | 27.70 ± 1.20 | 1.50 ± 0.17 | 1.07 ± 0.08 | 8.86 ± 0.22 | 27.30 ± 2.17 | 125.20 ± 2.02 | 3,389.90 ± 237.88 | 1.17 ± 0.06 |
| | LD9 | 36.10 ± 3.43 | 1.50 ± 0.22 | 0.37 ± 0.04 | 3.10 ± 0.31 | 26.70 ± 1.99 | 129.90 ± 1.86 | 3,462.30 ± 249.02 | 1.23 ± 0.13 |
| Terrestrial open habitats with high disturbance | HD1 | 93.50 ± 4.90 | 1.30 ± 0.15 | 1.18 ± 0.18 | 9.16 ± 1.04 | 41.60 ± 3.60 | 120.10 ± 1.68 | 4,971.10 ± 408.74 | 1.97 ± 0.47 |
| | HD2 | 72.20 ± 2.76 | 1.60 ± 0.22 | 0.88 ± 0.03 | 15.17 ± 0.33 | 38.10 ± 2.41 | 130.00 ± 1.21 | 4,932.60 ± 281.29 | 2.77 ± 0.13 |
| | HD3 | 43.00 ± 2.44 | 1.30 ± 0.15 | 1.29 ± 0.05 | 8.29 ± 0.12 | 35.30 ± 2.01 | 122.60 ± 1.09 | 4,323.50 ± 242.37 | 2.00 ± 0.03 |
| | HD4 | 62.60 ± 2.85 | 1.30 ± 0.15 | 0.99 ± 0.10 | 14.29 ± 0.51 | 43.90 ± 1.68 | 114.30 ± 0.72 | 5,013.90 ± 181.85 | 2.33 ± 0.13 |
| | HD5 | 53.50 ± 2.66 | 2.00 ± 0.30 | 1.09 ± 0.08 | 27.80 ± 1.59 | 52.10 ± 3.09 | 132.20 ± 1.28 | 6,871.30 ± 381.59 | 3.23 ± 0.26 |
| | HD6 | 50.90 ± 2.00 | 1.70 ± 0.21 | 1.23 ± 0.19 | 14.00 ± 0.43 | 34.60 ± 2.42 | 129.00 ± 1.23 | 4,449.70 ± 291.05 | 2.62 ± 0.20 |
| | HD7 | 85.50 ± 4.23 | 1.30 ± 0.15 | 2.00 ± 0.20 | 21.91 ± 2.10 | 53.00 ± 2.36 | 128.70 ± 1.10 | 6,807.20 ± 274.95 | 3.81 ± 0.38 |
| | HD8 | 35.00 ± 1.53 | 2.10 ± 0.35 | 1.66 ± 0.17 | 11.63 ± 0.26 | 30.50 ± 2.25 | 134.80 ± 1.31 | 4,101.50 ± 289.66 | 1.54 ± 0.15 |
| Riparian habitat | RH1 | 74.20 ± 3.53 | 10.50 ± 0.56 | 4.62 ± 0.19 | 46.09 ± 0.70 | 138.10 ± 2.81 | 136.20 ± 1.09 | 18,801.00 ± 369.36 | 11.78 ± 0.34 |
| | RH2 | 24.40 ± 1.12 | 10.00 ± 0.65 | 2.00 ± 0.07 | 16.68 ± 1.04 | 158.90 ± 3.94 | 138.50 ± 1.35 | 22,021.10 ± 648.36 | 10.69 ± 0.58 |
| | RH3 | 30.20 ± 1.00 | 10.40 ± 0.45 | 1.61 ± 0.18 | 17.07 ± 1.46 | 147.30 ± 4.81 | 114.80 ± 1.29 | 16,884.00 ± 492.62 | 9.39 ± 0.99 |

*n*, number.
g, grass.

**Table 9.2.** Vegetative and reproductive characters of *Anthemis cotula* (mean ± SE) from different habitats.

| | Habitat types | | |
|---|---|---|---|
| Trait | Terrestrial open habitats with low disturbance | Terrestrial open habitats with high disturbance | Riparian habitats |
| Height (cm) | 41.68 ± 1.91 | 62.03 ± 2.40 | 42.93 ± 4.31 |
| Lateral branches/plant (*n*) | 1.38 ± 0.07 | 1.58 ± 0.08 | 10.30 ± 0.32 |
| Root mass/plant (g) | 0.85 ± 0.05 | 1.29 ± 0.06 | 2.74 ± 0.26 |
| Shoot mass/plant (g) | 5.69 ± 0.32 | 15.28 ± 0.78 | 26.62 ± 2.63 |
| Capitula/plant (*n*) | 20.74 ± 0.85 | 41.14 ± 1.22 | 148.10 ± 2.70 |
| Disc florets/capitulum (*n*) | 114.66 ± 1.34 | 126.46 ± 0.83 | 129.83 ± 2.10 |
| Disc florets /plant (*n*) | 2,429.09 ± 112.87 | 5,183.85 ± 152.10 | 19,235.37 ± 487.47 |
| Floret mass/plant (g) | 1.03 ± 0.05 | 2.53 ± 0.12 | 10.62 ± 0.43 |

under continuous dark conditions (Fig. 9.1). Similarly, the effects of different concentrations of ammonium ($NH_4^+$) on achene germination are shown in Fig. 9.2. Relative to controls, treatments increased germination under continuous light but not under continuous dark conditions.

Achene response to various concentrations of gibberellic acid ($GA_3$) (Fig. 9.3) and kinetin (Fig. 9.4) revealed that 1.0 mM $GA_3$ had the most significant effect on germination under both light and dark conditions. Germination of even fresh achenes was very significantly improved in comparison with controls. Though the application of kinetin marginally improved percentage germination under continuous dark conditions, the effect across treatments was not significant. These results suggest that achenes have an elaborate mechanism

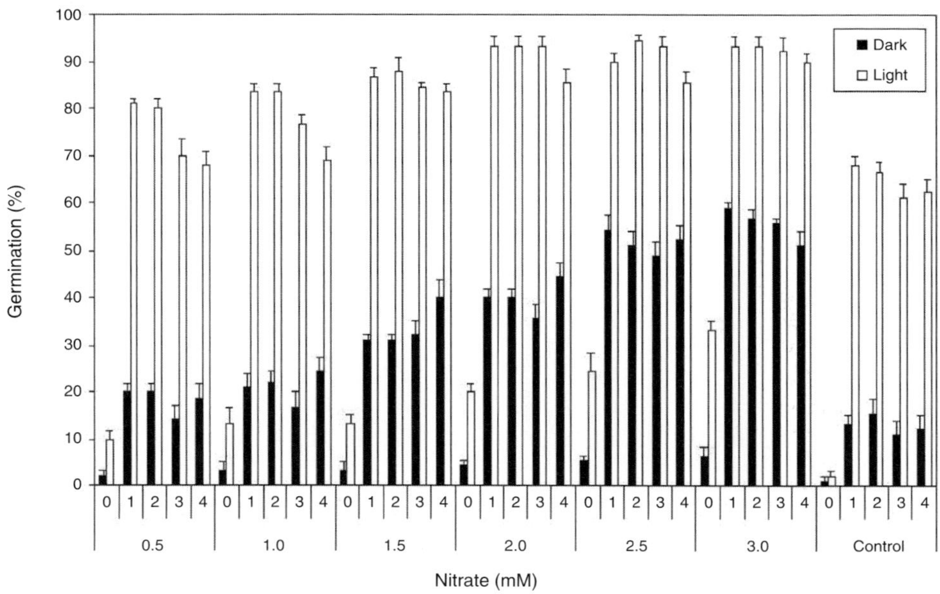


**Fig. 9.1.** Effect of different nitrate concentrations on achene germination under light and dark conditions. 0–4, age of achenes (years); error bars indicate SE of mean.

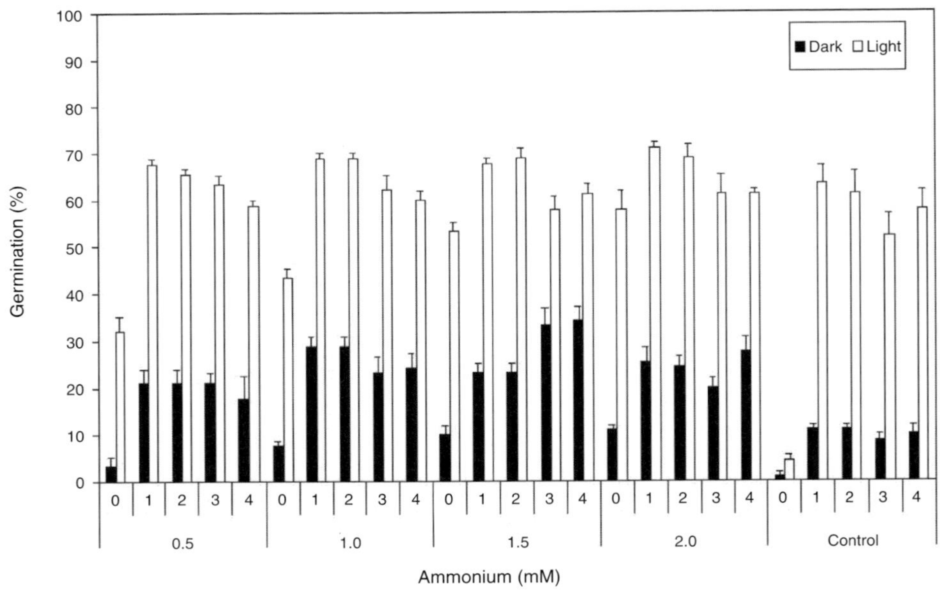


**Fig. 9.2.** Effect of different ammonium concentrations on achene germination under light and dark conditions. 0–4, age of achenes (years); error bars indicate SE of mean.

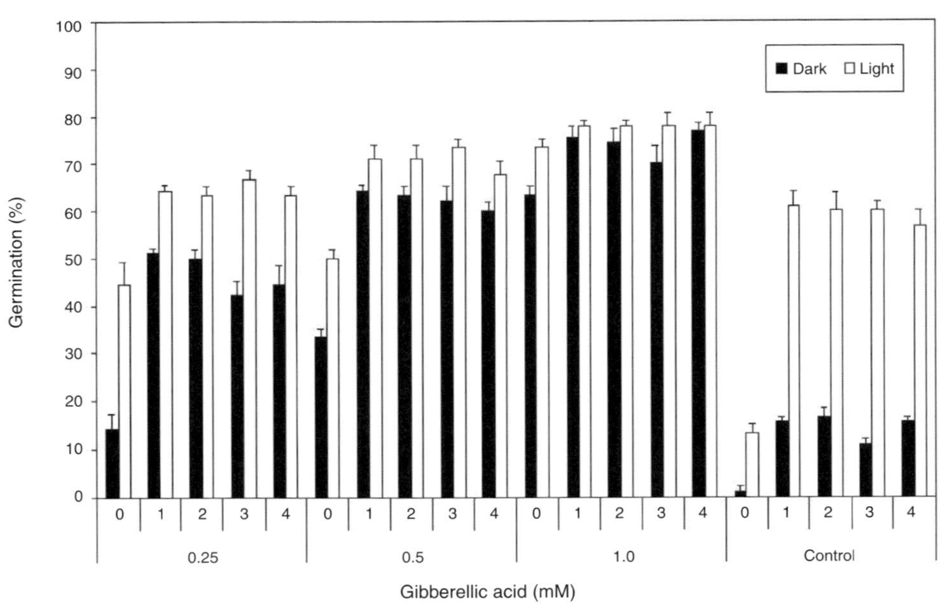


**Fig. 9.3.** Effect of different gibberellic acid ($GA_3$) concentrations on achene germination under light and dark conditions. 0–4, age of achenes (years); error bars indicate SE of mean.

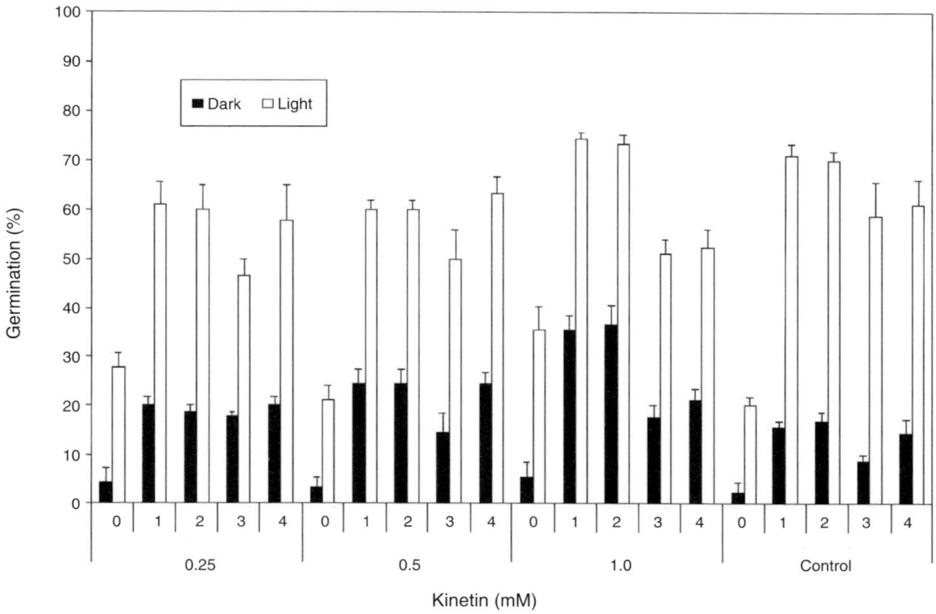


**Fig. 9.4.** Effect of different kinetin concentrations on achene germination under light and dark conditions. 0–4, age of achenes (years); error bars indicate SE of mean.

for sensing habitat conditions that helps them synchronize their germination with favourable environmental conditions – a strategy that aids the species in ensuring recruitment, survival and its spread in the Kashmir Himalaya (Rashid *et al.*, 2007b).

## Seedling Emergence Time: First Come, First Succeed

Being a winter annual, seedling emergence in *A. cotula* starts in early autumn (September) and lasts till early spring (March). During this entire period, recruitment occurs in two main pulses (Fig. 9.5). The major pulse occurs in autumn prior to the onset of winter (December–February), the minor pulse occurring during spring (March). The recruitment pattern is largely determined by the climatic factors of temperature and soil moisture, aside from the physiological state of the achenes. Data revealed an irregular mortality pattern in the populations investigated – this was higher during autumn and thereafter declined (Fig. 9.5). Hence established populations of *A. cotula* in the Kashmir Himalaya consist of individuals of both pre- and post-winter seedling cohorts. The number of seedlings surviving during the study period at each site is presented as survivorship curves in Fig. 9.6. Pre-winter cohorts exhibited a Deevey (1947) Type II curve, while post-winter cohorts showed a Type I survivorship curve (Rashid *et al.*, 2007a).

The individuals of two cohorts differed significantly in respect to vegetative parameters such as stem height, number of primary branches per plant, number of secondary branches per primary branch and reproductive attributes (i.e. number of capitula and achenes per plant (Rashid *et al.*, 2007a). Individuals of pre-winter seedling cohorts at all the study sites (Table 9.3) showed significantly higher performance than those of post-winter cohorts. The number of primary branches per plant, which in turn influenced the number of capitula and hence the number of repro-

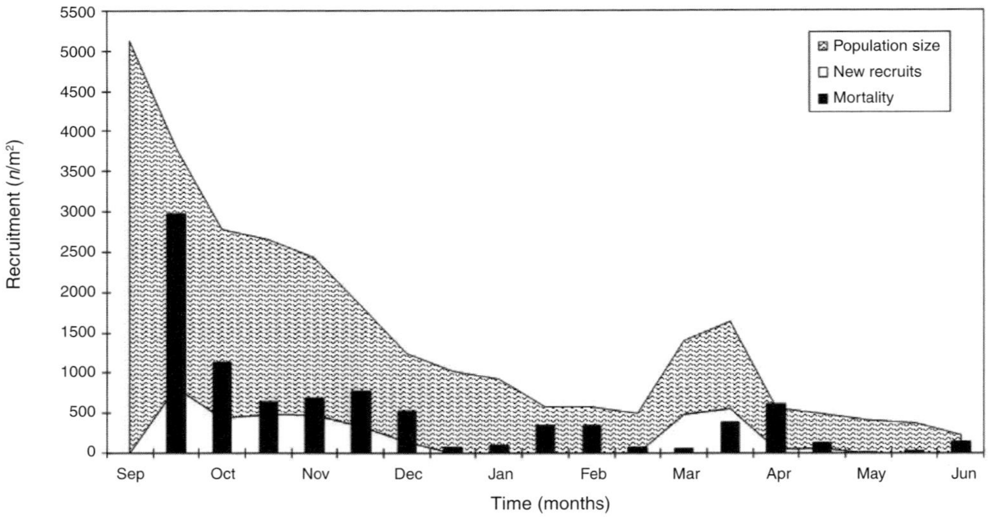


**Fig. 9.5.** Recruitment and mortality patterns in a natural population of *Anthemis cotula*. Vertical bars indicate mortality.

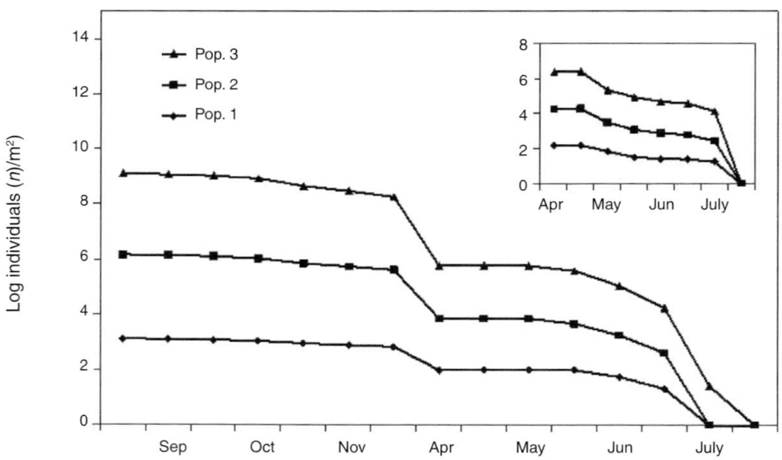


**Fig. 9.6.** Survivorship curves for pre- (main) and post-winter (inset) cohorts of *Anthemis cotula* in the Kashmir Himalaya (September 2003 to July 2004).

ductive propagules and invasiveness of the species, was higher in pre-winter individuals.

Thus, the existence of an ecological trade-off in the life history of *A. cotula* is clearly discernible, with individuals of pre-winter cohorts contributing to fecundity and those of post-winter cohorts ensuring survival and continuance of the species in invaded habitats. These results, therefore, conclusively demonstrate the importance of early seedling emergence in the successful colonization and spread of this invasive species in the Kashmir Himalaya (Allaie *et al.*, 2005a).

**Table 9.3.** Comparison of vegetative and reproductive attributes of pre- and post-winter cohorts of *Anthemis cotula* across different study sites (mean ± SD based on three replicates).

| Attribute | Cohort | Population A | Population B | Population C | Significance (*P*) |
|---|---|---|---|---|---|
| Stem height (cm) | Pre-winter | 91.80 ± 6.92 | 87.87 ± 5.94 | 87.47 ± 5.07 | <0.0001 |
| | Post-winter | 42.00 ± 5.06 | 36.67 ± 5.62 | 37.56 ± 2.70 | |
| Primary branches/ plant (*n*) | Pre-winter | 14.13 ± 1.86 | 14.67 ± 1.74 | 13.87 ± 1.32 | <0.0001 |
| | Post-winter | 1.89 ± 0.72 | 2.22 ± 0.89 | 1.89 ± 0.72 | |
| Secondary branches/ primary branch (*n*) | Pre-winter | 9.87 ± 0.58 | 9.87 ± 0.75 | 9.67 ± 0.66 | <0.0470 |
| | Post-winter | 8.67 ± 0.72 | 9.44 ± 0.58 | 9.33 ± 0.58 | |
| Branches/plant (*n*) | Pre-winter | 132.47 ± 18.35 | 135.27 ± 19.39 | 122.33 ± 14.82 | <0.0001 |
| | Post-winter | 14.00 ± 3.61 | 17.44 ± 8.46 | 16.00 ± 6.35 | |
| Capitula/plant (*n*) | Pre-winter | 1,095.20 ± 137.23 | 1,146.27 ± 253.61 | 1,034.67 ± 213.37 | <0.0001 |
| | Post-winter | 99.11 ± 27.28 | 152.44 ± 76.38 | 120.56 ± 58.53 | |
| Achenes/capitulum (*n*) | Pre-winter | 156.07 ± 2.77 | 154.27 ± 3.20 | 157.73 ± 4.16 | <0.0001 |
| | Post-winter | 125.56 ± 3.31 | 131.33 ± 2.91 | 125.44 ± 6.66 | |
| Achenes/plant (*n*) | Pre-winter | 173,432.60 ± 22,659.44 | 180,632.80 ± 38,974.10 | 166,912.27 ± 31,910.43 | <0.0001 |
| | Post-winter | 13,005.00 ± 3428.71 | 20,509.44 ± 10,515.13 | 15,590.33 ± 7,551.83 | |
| Wt of 1000 achenes (g) | Pre-winter | 0.489 ± 0.013 | 0.496 ± 0.030 | 0.510 ± 0.028 | <0.033 |
| | Post-winter | 0.448 ± 0.004 | 0.473 ± 0.037 | 0.477 ± 0.034 | |

## Allelopathy of *A. cotula*: a Novel Weapon for Successful Invasion

Studies pertaining to the effect of aqueous leaf leachate from *A. cotula* on germination of different field crops and *A. cotula* itself at different time periods have indicated the allelopathic potential of this species (Allaie *et al.*, 2006; Table 9.4). In regard to *Brassica campestris*, germination decreased with increased concentration of the leachate (Table 9.4). A germination rate of only 32.2% was recorded when a leachate of 5000 ppm concentration was used as compared

**Table 9.4.** Effect of different concentrations of aqueous leaf leachate on seed/fruit germination (average ± SD) of field crops and *Anthemis cotula*.

| | | Seed germination (%) | | | | | | |
|---|---|---|---|---|---|---|---|---|
| Concentration | Species | 1[a] | 2 | 3 | 4 | 5 | 6 | 7 |
| 1000 ppm | *B. campestris* | 49.9 ± 2.73 | 58.9 ± 5.10 | 61.6 ± 1.96 | 61.6 ± 1.96 | 61.6 ± 1.96 | 61.6 ± 1.96 | 61.6 ± 1.96 |
| | *V. radiata* | 63.3 ± 11.5 | 76.7 ± 5.77 | 80.0 ± 0.00 | 83.3 ± 5.77 | 83.3 ± 5.77 | 83.3 ± 5.77 | 83.3 ± 5.77 |
| | *T. aestivum* | 93.3 ± 11.5 | 100.0 ± 0.0 | 100.0 ± 0.0 | 100.0 ± 0.0 | 100.0 ± 0.0 | 100.0 ± 0.0 | 100.0 ± 0.0 |
| | *A. cotula* | 0.0 ± 0.00 | 4.4 ± 1.91 | 14.4 ± 1.91 | 25.5 ± 1.90 | 28.9 ± 5.10 | 35.5 ± 1.91 | 39.9 ± 3.35 |
| 2500 ppm | *B. campestris* | 25.5 ± 5.69 | 48.8 ± 6.93 | 49.9 ± 5.77 | 51.1 ± 5.08 | 52.2 ± 3.81 | 54.4 ± 5.10 | 57.7 ± 3.87 |
| | *V. radiata* | 40.0 ± 17.3 | 60.0 ± 26.46 | 73.3 ± 25.1 | 76.7 ± 20.8 | 76.7 ± 20.8 | 76.7 ± 20.8 | 76.7 ± 20.82 |
| | *T. aestivum* | 86.7 ± 5.77 | 100.0 ± 0.0 | 100.0 ± 0.0 | 100.0 ± 0.0 | 100.0 ± 0.0 | 100.0 ± 0.0 | 100.0 ± 0.00 |
| | *A. cotula* | 0.0 ± 0.00 | 1.1 ± 1.90 | 8.8 ± 3.87 | 18.9 ± 9.64 | 24.4 ± 10.8 | 27.7 ± 7.68 | 35.5 ± 10.18 |
| 5000 ppm | *B. campestris* | 0.0 ± 0.00 | 8.9 ± 7.69 | 21.1 ± 8.40 | 24.4 ± 13.8 | 24.4 ± 13.8 | 24.4 ± 13.8 | 32.2 ± 8.40 |
| | *V. radiata* | 30.0 ± 0.00 | 43.3 ± 11.5 | 63.3 ± 20.8 | 63.3 ± 20.8 | 63.3 ± 20.8 | 63.3 ± 20.8 | 63.3 ± 20.82 |
| | *T. aestivum* | 63.3 ± 11.5 | 73.3 ± 11.5 | 76.7 ± 5.77 | 80.0 ± 10.0 | 80.0 ± 10.0 | 80.0 ± 10.0 | 80.0 ± 10.00 |
| | *A. cotula* | 0.0 ± 0.00 | 0.0 ± 0.00 | 2.2 ± 1.90 | 8.8 ± 3.87 | 16.6 ± 3.35 | 19.9 ± 3.35 | 23.3 ± 6.70 |
| Control | *B. campestris* | 54.4 ± 1.96 | 56.6 ± 3.35 | 60.0 ± 0.00 | 61.6 ± 1.96 | 61.6 ± 1.96 | 61.6 ± 1.96 | 62.2 ± 1.96 |
| | *V. radiata* | 73.3 ± 11.5 | 83.3 ± 20.8 | 93.3 ± 5.77 | 93.3 ± 5.77 | 93.3 ± 5.77 | 93.3 ± 5.77 | 93.3 ± 5.77 |
| | *T. aestivum* | 96.7 ± 5.77 | 100.0 ± 0.0 | 100.0 ± 0.0 | 100.0 ± 0.0 | 100.0 ± 0.0 | 100.0 ± 0.0 | 100.0 ± 0.00 |
| | *A. cotula* | 2.2 ± 3.51 | 7.7 ± 1.96 | 17.7 ± 1.96 | 30.0 ± 0.00 | 30.0 ± 0.00 | 30.0 ± 0.00 | 30.0 ± 0.00 |

[a] time in days.

with 62.2% in controls, while the lowest leachate concentration showed only a marginal depressing effect. The seeds of *Vigna radiata* also exhibited a declining trend in germination with increased concentration of leaf leachate (Table 9.4); while under control conditions 93.3% germination was recorded, only 63.3% of seeds germinated when 5000 ppm leachate concentration was used. Seed germination in this species also declined at lower leachate concentration (1000 ppm), but the effect was less pronounced (Table 9.4). As in the two previous crops, germination in *Triticum* was also adversely affected by *A. cotula* leaf leachate (Table 9.4); while leachate concentrations of 1000 and 2500 ppm had no effect on caryopsis germination (100% germination in controls), only 80.0% of grains germinated in response to 5000 ppm leachate concentration. The effect of leaf leachate was also tested on achenes of *A. cotula* itself. The data revealed a promotory effect of 1000 and 2500 ppm leachate concentrations on achene germination in comparison with controls; the highest leachate concentration (5000 ppm), however, had an adverse negative effect with only 23.3% of achenes germinating compared with controls (30.0%) (Table 9.4). Inhibition of seed/grain germination was recorded in all the investigated crop species, although the extent of inhibition varied across species and treatments. Such an observation assumes significance in view of the reported occurrence of this species in cropping systems in the native area (Gealy *et al.*, 1994) and its incursion in the crop fields of the Kashmir Valley is likely to occur, following which significant interference with emergence, growth and development of crop plants is expected because of the observed effect of its leaf leachate on seed/grain germination of the species bioassayed.

## Mycorrhizal Mutualism: Hidden Friends in the Dark Half

Studies investigating the extent of AMF (arbuscular mycorrhizal fungi) association in *A. cotula* vis-à-vis its co-occurring species showed the highest RLC (root length colonized) (88%) in *A. cotula*, while that of its co-associates ranged from a low of 25% (*Avena fatua*) to a high of 78% (*Trifolium pratense*) (Shah *et al.*, 2008a, b; Table 9.5). The survey of 14 populations of *A. cotula*, spread across different areas in the Kashmir Valley, showed positive results for AMF colonization (Shah and Reshi, 2007). However, the extent of colonization was variable across populations (Table 9.6). Plant samples drawn from populations at Mirzabagh, Zukura and Shikarghat (all falling within Srinagar district) showed very high percentage RLC (85.1, 84.6 and 80.6%, respectively), while very low results were found (11.1, 14.0 and 8.5%) for populations distant from Srinagar (Uri, Drass and Bandipora, respectively) (Table 9.6).

Studies on the effect of *Glomus mossae*, a predominant AM symbiont associated with *A. cotula* under field conditions, on its growth and invasiveness revealed that it significantly enhanced both morphological and reproductive attributes of the host species (Shah and Reshi, 2007). While morphological traits such as stem length, shoot biomass and number of branches were favourably influenced by mycorrhizal inoculation (Table 9.7), beneficial effects were statistically significant in fitness attributes, such as number of capitula per plant and number of seeds (achenes) per capitulum ($P < 0.01$). However, length and

**Table 9.5.** Presence (percentage RLC) of AMF in *Anthemis cotula* and selected associated species.

| Species | RLC by AMF (%) |
|---|---|
| *Anthemis cotula* L. | 88 |
| *Avena fatua* L. | 25 |
| *Cerastium glomeratum* Thuill. | 35 |
| *Lactuca sativa* L. | 31 |
| *Medicago polymorpha* Bartal. | 62 |
| *Plantago major* L. | 72 |
| *Poa bulbosa* L. | 55 |
| *Sonchus asper* Hill. | 65 |
| *Sorghum halepense* Pers. | 25 |
| *Trifolium pratense* L. | 78 |

RLC, root length colonization; AMF, arbuscular mycorrhizal fungi.

**Table 9.6.** Percentage RLC in different populations of *Anthemis cotula* in the Kashmir Valley.

| Sampling site[a] | RLC (%) | Sampling site[a] | RLC (%) |
|---|---|---|---|
| Mirzabagh (13) | 85.1 | Lalbazar (14) | 50.0 |
| Zukura (17) | 84.6 | Hazratbal (12) | 41.1 |
| Shikarghat (12) | 80.6 | Ferozpur (30) | 35.2 |
| Nagbal (19) | 67.5 | Baramulla (50) | 34.8 |
| Narbal (18) | 64.5 | Uri (80) | 11.1 |
| Gondhasibat (16) | 54.1 | Bandipur (55) | 8.5 |
| Pampore (9) | 50.0 | Drass (140) | 14.0 |

[a] Distances from Srinagar (km) given in parentheses.
RLC, root length colonization.

**Table 9.7.** Effect of *Glomus mosseae* on vegetative and reproductive attributes of *Anthemis cotula* (statistical analysis based on mean ± SE of log-transformed data, apart from number of branches).

| Attribute | Control (mean ± SE) | *Glomus* treated (mean ± SE) | F-ratio (attribute × treatment) |
|---|---|---|---|
| Root length (cm) | 15.1 ± 1.053 | 13.866 ± 1.374 | 0.518 |
| Stem length (cm) | 36.633 ± 2.451 | 45.066 ± 3.547 | 3.849 |
| Root biomass (g) | 0.32 ± 0.055 | 0.176 ± 0.028 | 5.668 |
| Shoot biomass (g) | 2.133 ± 0.135 | 2.693 ± 0.386 | 1.820 |
| Root:shoot ratio | 0.236 ± 0.088 | 0.066 ± 0.02 | 6.062 |
| Branches/plant (*n*) | 18.400 ± 1.907 | 21.766 ± 1.299 | 2.172 |
| Capitula/plant (*n*) | 4.400 ± 0.115 | 24.000 ± 4.932 | 53.945[a] |
| Achenes/capitulum (*n*) | 39.200 ± 2.666 | 59.066 ± 3.644 | 20.398[b] |

[a] $P < 0.001$; [b] $P < 0.01$.

dry mass of roots were higher for uninoculated control plants. On the basis of the extent of increase in growth due to mycorrhizal inoculation, mycorrhizal dependency in *A. cotula* was estimated at 116.95%.

Assessment of percentage RLC in monospecific patches of *A. cotula* established under field conditions at five study sites revealed >80% *Arum*-type AMF root length colonization, with characteristic intercellular hyphae and arbuscles (Shah *et al.*, 2008a). The degree of mycorrhizal RLC was, however, less when *A. cotula* was grown in association with its neighbours, namely *Conyza canadensis*, *Galinsoga parviflora* and *Sisymbrium loeselii*, and the extent of this reduction was largely determined by the identity of the neighbour (Table 9.8). Maximum reduction (percentage RLC) by AMF in *A. cotula* was shown by *S. loeselii* at the study site of Bandipore (8.5%), with Lalbazar (50% colonization) showing the lowest. In comparison with *S. loeselii*, two other species showed reduced RLC to a lesser extent. When *G. parviflora* was the neighbouring plant, percentage RLC ranged from a maximum of 70.2 to a minimum of 39.5. Likewise, root length colonization in *A. cotula* when growing in association with

*C. canadensis* ranged from a maximum of 74.5% to a minimum of 54.5% (Table 9.8).

Consistent with our field observations, the effects of mycorrhizal inoculation were investigated in a pot trial using local and foreign AMF versus untreated controls in regard to different attributes of *A. cotula* in combination with each of its co-associates (Table 9.9). Not only was the differential impact of local and foreign AMF noted, but neighbour-specific effects were also observed. The effect of local and foreign AMF associates (*Glomus mosseae*, *Glomus fasciculatum*, *Gigaspora margarita*) on the vegetative and reproductive attributes of *A. cotula* was more positive when local AMF were used (Shah *et al.*, 2008). Analysis of the results indicates that mycorrhizal mutualists of *A. cotula* in monocultures significantly influenced stem length, stem dry mass and number of branches and capitula per plant, but influenced root dry mass relatively less significantly. In the presence of *C. canadensis* and *G. parviflora*, local AMF showed greater effects than foreign AMF on stem length and number of branches and capitula per plant in *A. cotula*, but a relatively insignificant influence on root length and dry mass per plant. Neither local nor foreign AMF in the presence of *S. loeselii*, however, showed any significantly positive influence on any of the attributes of *A. cotula* except for stem dry mass, but both had a markedly negative effect on root length and dry mass.

The response of *A. cotula* to mycorrhizal inoculation in the presence of *C. canadensis* was statistically significant in regard to stem length, stem dry mass and capitula number, not significant for root dry mass (Tables 9.9 and 9.10). When *G. parviflora* was the co-associate, mycorrhizal inoculation of *A. cotula* significantly influenced capitula number, stem dry mass and stem length. Root dry mass, however, was least influenced and root length was not influenced at all. Contrarily, when grown with *S. loeselii*, the most significantly influenced attributes of *A. cotula* were root dry mass and root length, with no effect on number of capitula and branches per plant. Two-way analysis of variance for various growth and fitness attributes of *A. cotula* with mycorrhizae and neighbour identity as fixed factors showed that root length and drymass did not differ between different treatments or their interaction (Table 9.10). However, *A. cotula* showed significantly higher stem length and dry mass and produced significantly more branches and capitula per plant in response to mycorrhizal inoculation (Table 9.10). Moreover, there was a significant mycorrhiza × neighbour identity interaction for *A. cotula* in respect of stem length and dry mass, as well as number of branches and capitula per plant (Table 9.10).

Whilst the root:shoot ratio of *A. cotula* under control conditions (0.36) and in plants inoculated with foreign AMF (0.35) was almost identical, it was reduced to 0.21 in response to inoculation with local AMF (Fig. 9.7). However, in the presence of *S. loeselii*, the root:shoot ratio of *A. cotula* decreased from 0.72 with no AMF to 0.16 and 0.47 under local and foreign AMF

**Table 9.8.** Percentage RLC[a] for *Anthemis cotula* at different sites with *Sisymbrium loeselii* (A+S), *Conyza canadensis* (A+C) or *Galisoga parviflora* (A+G) as neighbouring plant species.

| | A+S | | A+C | | A+G | |
|---|---|---|---|---|---|---|
| Sampling site | *A. cotula* | *S. loeselii* | *A. cotula* | *C. canadensis* | *A. cotula* | *G. parviflora* |
| Mirzabagh | 28 ± 2.2 | 24 ± 0.3 | 70.8 ± 0.7 | 62.0 ± 1.3 | 67.8 ± 1.7 | 36.0 ± 2.4 |
| Lalbazar | 50 ± 3.1 | 0 | 74.5 ± 1.5 | 72.3 ± 1.7 | 70.2 ± 2.5 | 35.1 ± 0.3 |
| University campus | 41 ± 1.5 | 13 ± 2.0 | 73.3 ± 1.6 | 59.4 ± 1.8 | 63.3 ± 2.1 | 52.4 ± 0.5 |
| Bandipore | 8.5 ± 0.8 | 0 | 54.5 ± 1.6 | 40.6 ± 1.7 | 47.5 ± 2.4 | 21.6 ± 3.1 |
| Naranag | 30 ± 3.2 | 0 | 65.2 ± 2.4 | 45.7 ± 1.9 | 39.5 ± 2.1 | 17.3 ± 1.2 |

RLC, root length colonization.

[a] In all monoculture patches of *A. cotula* RLC was >80%.

**Table 9.9.** Effect of AMF (local versus foreign), neighbouring plants (*Conyza* (A+C), *Galinsoga* (A+G) and *Sisymbrium* (A+S)) and their interaction on different attributes of *Anthemis cotula*. Figures indicated by different letters in the same column are significantly different at $P$ <0.05 (post-ANOVA Tukey tests).

| Treatments and their interactions | | Shoot length (cm) | Root length (cm) | Shoot dry mass (g) | Root dry mass (g) | Capitula/ plant (*n*) | Branches/ plant (*n*) | Root: shoot ratio |
|---|---|---|---|---|---|---|---|---|
| Mycorrhizae | Control | 13.29[a] | 11.21[a] | 0.41[a] | 0.19[a] | 4.33[a] | 7.92[a] | 0.51 |
| | Local | 25.33[c] | 10.75[a] | 1.23[c] | 0.27[a] | 15.58[c] | 13.00[b] | 0.21 |
| | Foreign | 15.98[b] | 10.98[a] | 0.54[b] | 0.20[a] | 8.42[b] | 9.17[a] | 0.38 |
| CD | | 1.67 | 1.63 | 0.10 | 0.13 | 1.83 | 1.55 | |
| Competition | A | 19.94[c] | 10.61[a] | 0.93[c] | 0.24[a] | 9.56[b] | 15.00[c] | 0.31 |
| | A+C | 25.00[d] | 10.33[a] | 0.98[c] | 0.27[a] | 14.33[c] | 10.00[b] | 0.34 |
| | A+G | 15.67[b] | 10.00[a] | 0.64[b] | 0.22[a] | 11.44[b] | 11.44[b] | 0.37 |
| | A+S | 12.20[a] | 12.97[b] | 0.36[a] | 0.16[a] | 2.44[a] | 3.67[a] | 0.45 |
| CD | | 1.92 | 1.88 | 0.15 | 0.11 | 2.11 | 1.79 | |
| No AMF | A | 13.33[a] | 10.67[a] | 0.41[a] | 0.15[a] | 4.00[a] | 12.00[b] | 0.36 |
| | A+C | 15.00[a] | 11.00[a] | 0.55[a] | 0.28[a] | 8.00[a] | 8.00[a] | 0.50 |
| | A+G | 11.00[a] | 9.00[a] | 0.34[a] | 0.15[a] | 2.67[a] | 7.33[a] | 0.45 |
| | A+S | 13.83[a] | 14.17[a] | 0.33[a] | 0.24[a] | 2.67[a] | 4.33[a] | 0.72 |
| Local AMF | A | 29.67[c] | 10.67[a] | 1.85[c] | 0.39[a] | 14.67[b] | 19.00[c] | 0.21 |
| | A+C | 38.00[d] | 12.00[a] | 1.71[c] | 0.29[a] | 28.00[c] | 15.00[b] | 0.16 |
| | A+G | 21.00[b] | 10.00[a] | 1.00[b] | 0.32[a] | 18.00[b] | 15.00[b] | 0.32 |
| | A+S | 12.67[a] | 10.33[a] | 0.37[a] | 0.06[a] | 1.67[a] | 3.00[a] | 0.16 |
| Foreign AMF | A | 16.83[a] | 10.50[a] | 0.53[a] | 0.19[a] | 10.00[a] | 14.00 | 0.35 |
| | A+C | 22.00[b] | 8.00[a] | 0.67[a] | 0.24[a] | 7.00[a] | 7.00[a] | 0.35 |
| | A+G | 15.00[a] | 11.00[a] | 0.58[a] | 0.20[a] | 13.67[b] | 12.00[b] | 0.34 |
| | A+S | 10.10[a] | 14.40[a] | 0.38[a] | 0.17[a] | 3.00[a] | 3.67[a] | 0.47 |
| CD | | 3.32 | 3.25 | 0.19 | 0.25 | 3.65 | 3.10 | |

AMF, arbuscular mycorrhizal fungi; CD, critical difference.

treatments, respectively (Fig. 9.7). In the presence of *C. canadensis* it showed a sharp decline (from 0.50 to 0.16) under local AMF against a marginal decline (from 0.50 to 0.35) under foreign AMF treatment. The association of *A. cotula* with *G. parviflora* showed decrements in root:shoot ratio, from 0.45 (without AMF) to 0.32 and 0.34 under local and foreign AMF, respectively (Fig. 9.7). Mycorrhizal dependency showed higher values with local than foreign AMF, both in mono- and mixed cultures. Whilst the highest mycorrhizal dependency values for *A. cotula* were obtained with local AMF in monocultures (0.75), negative values were obtained when *S. loeselii* was the co-associate irrespective of local (–0.24) or foreign AMF (–0.04). When grown with *G. parviflora*, *A. cotula* showed a mycorrhizal dependency of 0.62 on local and 0.38 on foreign AMF but, when grown with *C. canadensis*, these values were 0.58 and 0.09, respectively.

**Table 9.10.** Results of two-way ANOVA analysis for various growth and fitness attributes of *Anthemis cotula* with mycorrhizae and neighbour identity as fixed factors.

| Mean effect and interaction | DF | Shoot length (cm) | | Root length (cm) | | Shoot dry mass (g) | | Root dry mass (g) | | Capitula/plant (*n*) | | Branches/plant (*n*) | |
|---|---|---|---|---|---|---|---|---|---|---|---|---|---|
| | | *F* | *p* | *F* | *p* | *F* | *p* | *F* | *p* | *F* | *p* | *F* | *p* |
| Mycorrhizae | 2 | 227.29 | 0.000 | 0.31 | 0.76 | 327.44 | 0.000 | 1.33 | 0.28 | 152.29 | 0.000 | 45.92 | 0.000 |
| Neighbour identity | 3 | 130.45 | 0.000 | 8.09 | 0.000 | 102.66 | 0.000 | 1.75 | 0.18 | 90.29 | 0.000 | 109.95 | 0.000 |
| Mycorrhizae × neighbour identity | 3 | 38.96 | 0.000 | 5.11 | 0.000 | 52.11 | 0.000 | 2.29 | 0.07 | 39.97 | 0.000 | 10.53 | 0.000 |
| Error (mean square) | 24 | 2.11 | – | 2.06 | – | 0.01 | – | 0.01 | – | 2.56 | – | 1.83 | – |

DF, degree of freedom; *F*, *F*-test; *p*, probability.

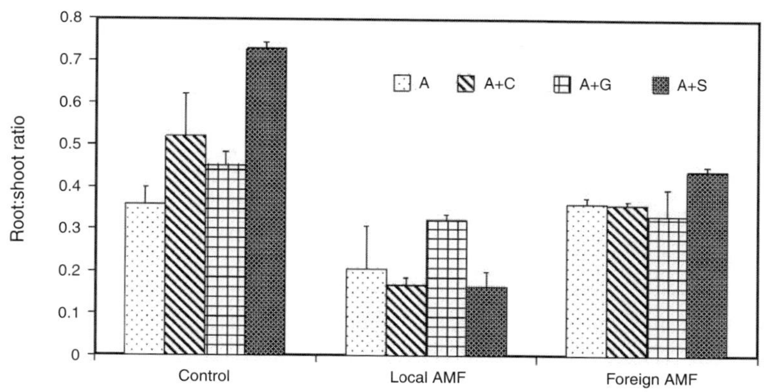


**Fig. 9.7.** Root:shoot ratio of *Anthemis cotula* with and without local and foreign AMF in the presence and absence of *Conyza canadensis* (A+C), *Galinsoga parviflora* (A+G) and *Sisymbrium loeselii* (A+S).

## Has *A. cotula* Been Released from the Native Enemies?

In its native range, *A. cotula* is attacked by a number of herbivores and pathogens. The above-ground herbivores associated with *A. cotula* in its native range of southern Europe–western Siberia are dominated by aphids (Aphidae), spittlebugs (Cercopidae), bugs (Heteroptera), moths, slugs and snails. In western Europe the species is attacked by around 68 insect pathogens of which 13 specifically use this species as their host (personal communication, S. Benvenuti). In addition, a stem-mining agromyzid fly (*Napomyza* sp.) and two insect species, namely *Cochylidia implicitana* (Tortricidae) and *Homeosoma nimbella* (Pyralidae), are also reported to feed on its shoots, flowers, fruits and seeds. We surveyed five European populations of *A. cotula* at different sites in Gothenburg, Sweden, which revealed an abundance of black aphids associated with this species. However, notwithstanding intensive surveys of innumerable populations since 2000, no parasite or pathogen was found associated with this plant species in the Kashmir Himalayan region (Shah and Reshi, 2007).

## Impact on Native Biodiversity: Guilty until Proved Innocent

The analysis of multiple assembly rule models (Table 9.11) reveals that the community invaded by *A. cotula* shows a random co-occurrence pattern, in contrast to the non-random (segregated) pattern exhibited by natural communities. Thus, invasion causes shift from segregated (non-random) to random or weakly aggregated co-occurrence patterns, leading to community disassembly and alterations in the interaction of surviving species in the community.

A comparative study (Shah *et al.*, 2009), dealing with the diversity and abundance of AMF in the rhizospheres of *A. cotula* at different study sites in the Kashmir Himalaya in comparison with AMF communities from nearby uninvaded sites with relatively diverse native plant assemblages, revealed that nine AMF species were recovered from the rhizospheres of *A. cotula*. In contrast, uninvaded sites near *A. cotula* yielded 20 AMF species (Table 9.12). Variation in the number of spores recovered from invaded and uninvaded soils by *A. cotula* is presented in Fig. 9.8, showing significant differences in the density of AMF spores across sites. However, AMF spore density, unlike their species richness, was higher in the invaded rhizospheric soils than nearby uninvaded soils. The number of AMF spores per gram of soil in the rhizosphere of *A. cotula* was highest at Bandipore and lowest at Nagbal.

## Conclusion

The series of studies conducted on *A. cotula* suggests that many key attributes of this species synergistically promote its invasion in the Kashmir Himalaya (Fig. 9.9). The main conclusions drawn from these studies are presented below.

**1.** Fine-tuning of the life history strategy of *A. cotula* to the prevailing environmental conditions facilitates its invasion of ruderal habitats in the Kashmir Himalaya. A protracted recruitment pattern aided by habitat disturbance and favourable moisture, temperature, light and nutrient regimes, and high population size even after seedling mortality, are the key demographic attributes contributing to the establishment and spread of this species.
**2.** The germination ecology of the invasive alien *A. cotula* helps it synchronize its successful recruitment with favourable habitat conditions, thereby ensuring seedling survival, establishment, growth and fitness.
**3.** An ecological trade-off in the life history of *A. cotula* is clearly discernible, with

**Table 9.11.** Summary of null model algorithms used for four co-occurrence indices.

| Index | Row constraints | Column constraints | Observed index | Simulated index | *P* (observed ≥ expected) |
|---|---|---|---|---|---|
| C-score | Fixed | Fixed | 3.62678 | 3.59015 | 0.18840 |
| | Fixed | Equiprobable | 3.62678 | 3.84207 | 0.89320 |
| CHECKER | Fixed | Fixed | 81.000 | 73.8690 | 0.12820 |
| | Fixed | Equiprobable | 81.000 | 80.86 | 0.55240 |
| COMBO | Fixed | Fixed | 10.000 | 10.000 | 1.00 |
| | Fixed | Equiprobable | 10.000 | 10.000 | 1.00 |
| V-ratio | Fixed | Equiprobable | 1.71173 | 0.98996 | 0.7660 |
| | Proportional | Proportional | 1.71173 | 1.73822 | 0.46220 |

**Table 9.12.** Arbuscular mycorrhizal species composition in rhizospheric soils invaded (I) by *Anthemis cotula* in comparison with nearby uninvaded (U) sites.

| AMF taxon | U | I |
|---|---|---|
| Glomaceae | | |
| *Glomus caledonium* Trappe & Gerd. | + | – |
| *G. lamellosum Dalpe* Koske & Tews | + | – |
| *G. etunicatum* Becker & Gerdemann | + | – |
| *G. intraradices* Schenck & Smith | + | – |
| *G. mosseae* Gerd. & Trappe | + | + |
| *G. claroideum* Schenck & Smith | + | – |
| *G. dimorphicum* Boyetchko & Tewari | – | + |
| *G. fasciculatum* Walker & Koske | – | – |
| *G. diaphanum* Morton & Walker | – | + |
| *G. luteum* | – | + |
| Gigasporaceae | | |
| *Gigaspora decipiens* Hall & Abbott | + | – |
| *Gi. margarita* Becker & Hall | + | + |
| *Scutellospora erythropa* Walker & Sanders | + | – |
| *S. verrucosa* Walker & Sanders | + | + |
| *S. heterogama* Walker & Sanders | + | + |
| *S. dipurpurascens* Morton & Koske | + | – |
| *S. calospora* Walker & Sanders | + | – |
| *S. pellucida* Walker & Sanders | + | – |
| *S. laevis* | – | – |
| *Scutellospora* sp. | – | – |
| Acaulosporaceae | | |
| *Acaulospora spinosa* Walker & Trappe | + | + |
| *A. laevis* Gerdemann & Trappe | + | + |
| *A. foveata* Trappe & Janos | – | – |
| *A. paulineae* Blaszkowski | + | – |
| *Acaulospora* sp. | + | – |
| *Entrophospora infrequens* Ames & Schneider | + | – |
| Archaeosporaceae | | |
| *Archaeospora trappei* Morton & Redecker | – | – |
| Paraglomaceae | | |
| *Paraglomus occultum* Morton & Redecker | + | – |

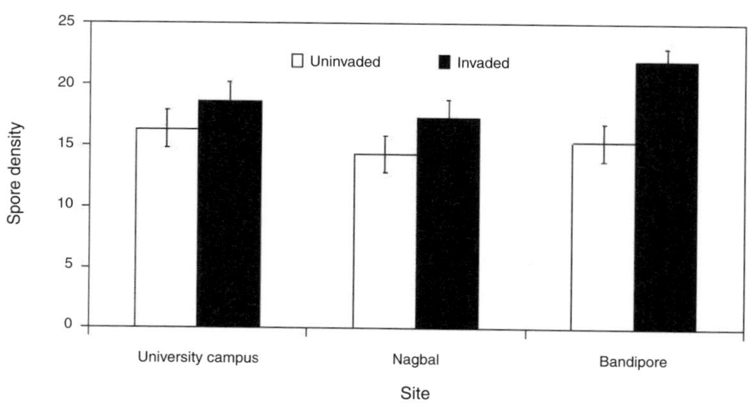


**Fig. 9.8.** AMF spore number (± SD) per gram of rhizospheric soil at three sites invaded by *Anthemis cotula* in comparison with nearby uninvaded sites.

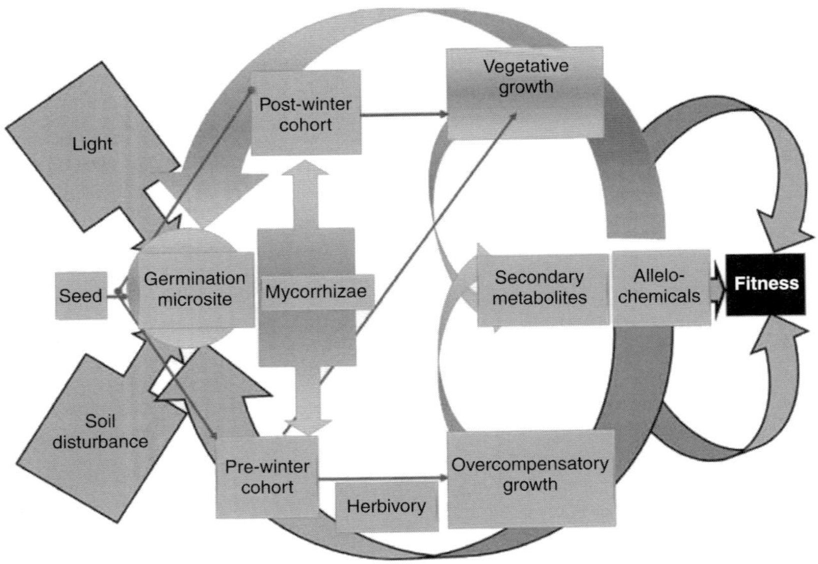


**Fig. 9.9.** Schematic representation of the attributes that synergistically promote the invasion of *Anthemis cotula.*

individuals of pre-winter cohorts contributing to fecundity and those of post-winter cohorts ensuring the survival and continuance of the species in the invaded habitats. Thus, early seedling emergence is critically important in the successful colonization and spread of this invasive species in the Kashmir Himalaya.

**4.** In view of allelopathic inhibition of seed/grain germination in all the crop species investigated, the potential spread of *A. cotula* in cropping systems of the invaded area is likely to interfere with the emergence, growth and development of the crop plants.

**5.** *Anthemis cotula* is highly mycorrhizal under field conditions, although the extent of colonization varies among its populations in the Kashmir Himalaya. Moreover, *Glomus mosseae* significantly promoted its invasiveness. Avoidance of native enemies and absence of AMF in the invaded range contribute to the invasive success of *A. cotula*.

**6.** Mycorrhizal source (local versus foreign) and neighbouring plant species play a significant role in *A. cotula* invasion.

**7.** Invasion by *A. cotula* causes a shift in community assembly and AMF spore density, composition and diversity in rhizospheric soils in comparison to non-rhizospheric soils.

## Acknowledgements

Thanks are due to the Head, Department of Botany, University of Kashmir, Srinagar for providing all the facilities required to conduct these studies.

## References

Allaie, R.R., Reshi Z., Rashid, I. and Wafai, B.A. (2005a) Demographic plasticity in relation to growth and resource allocation pattern in *Anthemis cotula*–an alien invasive species in Kashmir Himalaya, India. *Applied Ecology and Environmental Research* 4, 63–74.

Allaie, R.R., Reshi, Z. and Wafai, B.A. (2005b) Demographic studies on alien invasive *Anthemis cotula* L. in Kashmir Himalaya. *Trends in Life Science* 20, 49–56.

Allaie, R.R., Reshi, Z., Rashid, I. and Wafai, B.A. (2006) Effect of aqueous leaf leachate of

*Anthemis cotula*–an alien invasive species on germination behaviour of some field crops. *Journal of Agronomy and Crop Science* 192, 186–191.

Baruah, R., Bohlmann, F. and King, R. (1985) Eriolanolides, eudesmanolides and a rearranged sesquiterpene from *Eriophyllum* species. *Planta Medica* 72, 182–187.

Catford, J.A., Jansson, R. and Christer, N. (2008) Reducing redundancy in invasion ecology by integrating hypotheses into a single theoretical framework. *Diversity and Distributions* 15, 22–40.

Erneberg, M. (1999) Effects of herbivory and competition on an introduced plant in decline. *Oecologia* 118, 203–209.

Gealy, D.R., Squier, S.A. and Ogg, A.G. Jr (1994) Soil environment and temperature affect germination and seedling growth of mayweed chamomile (*Anthemis cotula*). *Weed Technology* 8, 668–672.

Glennie, C.W. and Harborne, J.B. (1971) Flavone and flavonol 5 glucosides. *Phytochemistry (Oxford)* 10, 1325–1329.

Hulten, E. (1968) *Flora of Alaska and Neighboring Territories.* Stanford University Press, Stanford, California.

Kay, Q.O.N. (1971) Biological flora of the British Isles: *Anthemis cotula* L. *Journal of Ecology* 59, 623–636.

Moles, A.T., Gruber, M.A.M. and Bonser, S.P. (2008) A new framework for predicting invasive plant species. *Journal of Ecology* 96, 13–17.

Pysek, P. and Richardson, D.M. (2007) Traits associated with invasiveness in alien plants: Where do we stand? In: Nentwig, W. (ed.) *Biological Invasions.* Ecological Studies, vol. 193, Springer-Verlag, Berlin and Heidelberg, pp. 97–125.

Rashid, I., Reshi, Z., Shah, M.A. and Wafai, B.A. (2006) Does herbivory promote invasiveness of *Anthemis cotula* in Kashmir Himalaya, India? In: *International Symposium on Biology, Ecology and Management of World's Worst Plant Invasive Species*, pp. 83–84.

Rashid, I., Reshi, Z. and Wafai, B.A. (2007a) Effect of seedling emergence time on the performance of mayweed (*Anthemis cotula* L.): an alien invasive species in Kashmir Himalaya. *Proceedings of the 21st Asian-Pacific Weed Science Society (APWSS) Conference*, 2–6 October, pp. 408–412.

Rashid, I., Reshi, Z., Allaie, R.R. and Wafai, B.A. (2007b) Germination ecology of invasive alien *Anthemis cotula* L. helps it synchronise its successful recruitment with favourable habitat conditions. *Annals of Applied Biology* 150, 361–369.

Shah, M.A. and Reshi, Z. (2007) Invasion by alien *Anthemis cotula* L. in a biodiversity hotspot. Release from native foes or relief from alien friends? *Current Science* 92, 21–22.

Shah, M.A., Reshi, Z. and Rashid, I. (2008a) Mycorrhizal source and neighbour identity differently influence *Anthemis cotula* L. invasion in the Kashmir Himalaya, India. *Applied Soil Ecology* 40, 330–337.

Shah, M.A., Reshi, Z. and Rashid, I. (2008b) Mycorrhizosphere-mediated mayweed chamomile invasion in the Kashmir Himalaya, India. *Plant and Soil* 312, 219–225.

Shah, M., Zafar, A., Reshi, A. and Rasool, N. (2009) Plant invasion induces shift in Glomalean spore diversity. *Tropical Ecology* 51, 317–323.

USDA (US Department of Agriculture)/NRCS (Natural Resource Conservation Service) (2002) The PLANTS Database, Version 3.5, National Plant Data Center, Baton Rouge, Louisiana (http://plants.usda.gov).

USDA-ARS (2005) National Genetic Resources Program. Germplasm Resources Information Network (GRIN), National Germplasm Resources Laboratory, Beltsville, Maryland (http://www.arsgrin).

# 10 A Brief Appraisal of Genus *Potamogeton* L. in the Kashmir Valley

**Aijaz Hassan Ganie, Zafar A. Reshi and B.A. Wafai**

*Department of Botany, University of Kashmir, Srinagar, Jammu and Kashmir, India*

## Introduction

The genus *Potamogeton* (Potamogetonaceae) is distributed worldwide (Chambers *et al.*, 2008) and is represented by 69 species and 50 natural hybrids (Weigleb and Kaplan, 1998). The number of species existing in different continents of the world and those shared by various continents are summarized in Table 10.1. The largest number of species (29) of the genus inhabit aquatic ecosystems in Asia, followed by North America (28 species) and Europe (22 species). Nine species are common to Asia and North America, 14 to Asia and Europe and 11 are shared by North America and Europe. The high species diversity of the genus in the northern hemisphere indicates that this region could be the centre of diversity and origin of the genus (Lindqvist *et al.*, 2006). A relatively higher number of species of this genus has also been recorded in central, eastern and western USA, Canada, temperate Europe, northern China and Japan (Weigleb, 1988).

Our surveys in the Kashmir Valley of India resulted in the identification of 11 species and two presumably natural interspecific hybrids from different freshwater habitats (Table 10.2). Of these *Potamogeton amblyphyllus*, *Potamogeton berchtoldii*, *Potamogeton trichoides* and *Potamogeton wrightii* were reported for the first time in the Kashmir Valley. However, several species such as *Potamogeton alpines*, *Potamogeton fluitans*, *Potamogeton filiformis*, *Potamogeton octandrus*, *Potamogeton zizi*, *Potamogeton polygonifolius* and *Potamogeton tepperi*, though previously reported from the Kashmir Valley (Stewart, 1972; Kak, 1979) were not recorded despite extensive field surveys.

The distribution of species reported from

**Table 10.1.** Worldwide intercontinental distribution of *Potamogeton* spp. The bottom value in each column represents the total number for that continent.

| Continent | Africa | Australia | Europe | N. America | S. America | Asia |
|---|---|---|---|---|---|---|
| Asia | 9 | 4 | 14 | 9 | 5 | 29 |
| S. America | 3 | 2 | 4 | 7 | 15 | |
| N. America | 5 | 3 | 11 | 28 | | |
| Europe | 9 | 4 | 22 | | | |
| Australia | 5 | 13 | | | | |
| Africa | 14 | | | | | |

**Table 10.2.** Number of *Potamogeton* species reported by various workers from Jammu and Kashmir State.

| Region | Number of species reported | Reference |
|---|---|---|
| Kashmir Valley | 12 | Stewart (1972) |
| Kashmir Valley | 14 | Kak (1979) |
| Kashmir Valley | 11 | Present study |
| Jammu region | 4 | Sharma and Kachroo (1981) |
| Lower Ladakh | 7 | Klimes and Dickore (2005) |

**Table 10.3.** Global distribution of *Potamogeton* spp. recorded from the Kashmir Valley.

| | Status in other continents | | | | |
|---|---|---|---|---|---|
| Species | Africa | Europe | N. America | S. America | Australia |
| *P. amblyphyllus* C.A. May | – | – | – | – | – |
| *P. berchtoldii* Fieb. | – | + | + | – | – |
| *P. crispus* L. | + | + | + | + | + |
| *P. lucens* L. | + | + | + | – | – |
| *P. natans* L. | – | + | + | – | – |
| *P. nodosus* Poir. | + | + | + | + | + |
| *P. pectinatus* L. | + | + | + | + | + |
| *P. perfoliatus* L. | + | + | + | – | + |
| *P. pusillus* L. | + | + | + | + | – |
| *P. trichoides* Chan. | + | + | – | – | – |
| *P. wrightii* Morong. | – | – | – | – | – |

the Kashmir Valley is presented in Table 10.3. Of these 11 species, *Potamogeton crispus*, *Potamogeton nodosus* and *Potamogeton pectinatus* are present in all six continents of the world, while *P. amblyphyllus* and *P. wrightii* are restricted to Asia. Phytogeographical analysis (Table 10.4) reveals that Kashmir shares a large number of *Potamogeton* species with Central Asian countries and Russia, indicating thereby that the genus might have been introduced from Central Asia. Of the six species of *Stuckenia* lineage (Potamogetonaceae) reported by Kaplan (2008) from Central Asia, five are also present in the Kashmir Himalaya, which further strengthens the view that these species might have been introduced into Kashmir from Central Asia.

## Growth Form and Habitat Preference

All *Potamogeton* species are rooted and have floating or submerged leaves. On the basis of leaf morphology, the species can be divided into four groups: floating broad-leaved, submerged broad-leaved, submerged linear-leaved and submerged narrow–filiform-leaved species (Fig. 10.1). The floating-leaved species are heterophyllous while the submerged-leaved species are always homophyllous.

*Potamogeton* species show varied habitat requirements, with *P. crispus*, *P. nodosus* and *P. pectinatus* inhabiting all types of water bodies in the Kashmir Valley, ranging from lotic to lentic systems and oligotrophic to eutrophic water bodies. These species are

**Table 10.4.** Regional distribution of *Potamogeton* spp. growing in the Kashmir Valley.

| | Status in neighbouring countries | | | | |
|---|---|---|---|---|---|
| Species | Pakistan/ Afghanistan | Central Asia | Russia | China | Japan |
| *P. amblyphyllus* | + | + | + | + | – |
| *P. berchtoldii* | –? | + | + | – | – |
| *P. crispus* | + | + | + | + | + |
| *P. lucens* | + | + | + | + | – |
| *P. natans* | + | + | + | + | + |
| *P. nodosus* | + | + | + | – | – |
| *P. pectinatus* | + | + | + | + | – |
| *P. perfoliatus* | + | + | + | + | + |
| *P. pusillus* | + | + | + | + | + |
| *P. trichoides* | –? | + | + | – | – |
| *P. wrightii* | + | – | – | + | + |

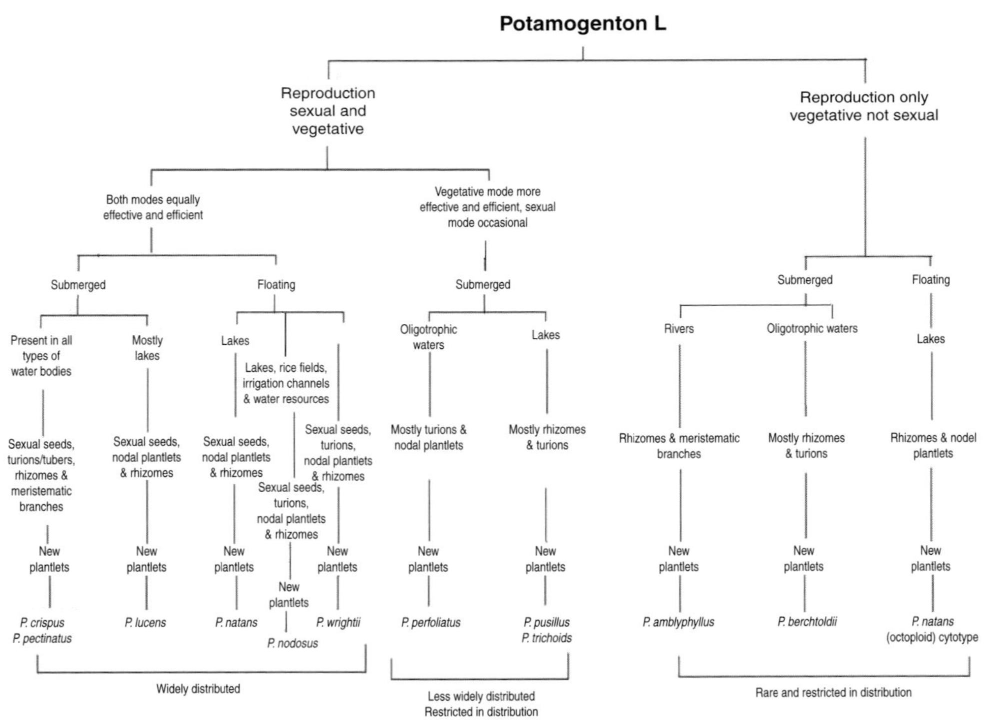


**Fig. 10.1.** Growth forms, modes of reproduction and habitat preferences of *Potamogeton* spp. in the Kashmir Valley.

thus highly invasive in the freshwater ecosystems of the Kashmir Valley, and *P. crispus* is known to grow in water bodies that are contaminated with high levels of pollutants, such as phenols (Hafez *et al.*, 1998), heavy metals (Samecka-Cymerman and Kempers, 2007), and in eutrophic lakes (Jain *et al.*, 2003). Similarly, *P. pectinatus* shows tolerance to high concentrations of several metals (Rai *et al.*, 2003; Tripathi *et al.*, 2003; Singh *et al.*, 2005). On the other hand, *P. perfoliatus* and *P. berchtoldii* inhabit very clear waters of springs and spring streams and are hence of rare occurrence in Kashmir Valley.

The above species also display plastic responses in their morphological characteristics, such as leaf shape, size and clonal architecture; and in resource allocation with respect to flow, depth and water temperature. Those species inhabiting both standing and running water and water bodies of different depths are more plastic in their morphological characters (Fig. 10.1). Consequently, the ability to tolerate multiple stressors through morphological adjustments is a major feature that determines survival, colonization and ecological amplitude of the species (Chapin *et al.*, 1987; Bazzaz, 1996; Sultan *et al.*, 1998).

## Modes of Reproduction

*Potamogeton*, like other aquatic plants, displays a wide variety of reproductive systems including both sexual as well as asexual modes of reproduction (Fig. 10.1). Sexual reproduction is accomplished through seeds, while vegetative propagation takes place by means of turions (axillary and underground), rhizomes, nodal plantlets, meristematic branches and plant fragments. Perusal of the data (Fig. 10.1) reveals that those species of *Potamogeton* operating multiple reproductive strategies are widespread while those with limited reproductive options are less widely distributed. Besides, species also employ habitat-specific reproductive strategies. For example, *P. crispus* and *P. pectinatus* produce sexual seeds in lentic water bodies, while the vegetative mode of reproduction is dominant in both lentic as well as lotic systems. Turions are the most important propagules in lentic water bodies but, in lotic waters, rhizomes and meristematic branches are the primary propagules of proliferation. Nodal plantlets in these species are produced at the time of senescence, and these contribute to the establishment of new populations and are mostly operative in habitats with standing waters. At the end of the growing season rhizomes and older stems give rise to branches with apical meristematic tips (meristematic branches), which grow continuously in a horizont plane and recline downwards in standing waters (*P. natans*, *P. nodosus*); in running water these branches grow along the surface of the sediment (*P. crispus*, *P. pectinatus*), giving rise to new plantlets at their nodes.

## Invasion Status

The genus *Potamogeton* in the Kashmir Valley is represented by widespread and highly invasive species, as well as by very rare and scarce species. Invasive species include *P. crispus*, *P. lucens*, *P. natans* and *P. nodosus*. In particular, *P. crispus* does not show any habitat preference, instead being widespread in all aquatic ecosystems.

## References


Bazzaz, F.A. (1996) *Plants in Changing Environments: Linking Physiological, Population and Community Ecology*. Cambridge University Press, Cambridge, UK.

Chambers, P.A., Lacoul, P., Murphy, K.J. and Thomaz, S.M. (2008) Global diversity of aquatic macrophytes in fresh water. *Hydrobiologia* 595, 9–26.

Chapin, F.S., Bloom, A.J., Field, C.B. and Waring, R.H. (1987) Plant response to multiple environmental factors. *Bioscience* 37, 49–57.

Hafez, N., Abdulla, S. and Ramadan, Y.S. (1998) Accumulation of phenols by *Potamogeton crispus* from aqueous industrial waste. *Bulletin of Environmental Contamination and Toxicology* 60, 944–948.

Jain, Y., Li, B., Wang, J. and Chen, J. (2003) Control

of turion germination in *Potamogeton crispus. Aquatic Botany* 75, 59–69.

Kak, A.M. (1979) Potamogeton in North western Himalayas. *Journal of the Bombay Natural History Society* 80, 312–319.

Kaplan, Z. (2008) A taxonomic revision of Stuckenia (Potamogetonaceae) in Asia with notes on diversity and variation of the genus on a worldwide scale. *Folia Geobotanica* 43, 159–234.

Klimes, L. and Dickore, B. (2005) A contribution to vascular plant flora of Lower Ladakh (Jammu and Kashmir, India). *Willdenowia* 35, 125–153

Lindqvist, C., Laet, J.D., Haynes, R.R., Aagesen, L., Keemer, B.R. and Albert, V.A. (2006) Molecular phylogenetics of an aquatic plant lineage, Potamogetonaceae. *Cladistics* 22, 568–588.

Rai, V.N., Tripathi, R.D., Vajpayee, P., Panday, N., Ali, M.B. and Gupta, D.K. (2003) Cadmium accumulation and its phytotoxicity in *Potamogeton pectinatus* L. (Potamogetonaceae). *Bulletin of Environmental Contamination and Toxicology* 70, 566–575.

Samecka-Cymerman, A. and Kampers, A.J. (2007) Heavy metals in aquatic macrophytes from two small rivers polluted by urban, agricultural and textile industry sewages in SW Poland. *Archives of Environmental Contamination and Toxicology* 53, 198–206.

Sharma, B.M. and Kachroo, P. (1981) *Flora of Jammu*, vol. I. Bishen Singh, Dehra Dun, India.

Singh, N.K., Pande, G.C., Rai, U.N., Tripathi, R.D., Singh, H.B. and Gupta, D.K. (2005) Metal accumulation and ecophysiological effects of distillery effluent on *Potamogeton pectinatus. Bulletin of Environmental Contamination and Toxicology* 74, 857–863.

Stewart, R.R. (1972) *An Annotated Catalogue of the Vascular Plants of West Pakistan and Kashmir.* Fakhri Printing Press, Karachi, Pakistan.

Sultan, S.E., Wilczek, A.M., Bell, D.L. and Hand, G. (1998) Physiological response to complex environments in annual Polygonum species of contrasting ecological breadth. *Oecologia* 115, 564–578.

Tripathi, R.D., Rai, U.N., Vajpayee, P., Ali, M.B., Khan, E., Gupta, D.K. *et al.* (2003) Biochemical response of *P. pectinatus* L. exposed to higher concentration of zinc. *Bulletin of Environmental Contamination and Toxicology* 71, 255–262.

Weigleb, G. (1988) Notes on pondweeds – outlines for a monographic treatment of the genus *Potamogeton* L. *Feddes Report* 99, 249–266.

Weigleb, G. and Kaplan, Z. (1998) An account of the species of *Potamogeton* L. (Potamogetonaceae). *Folia Geobotanica* 33, 241–316.

# 11 Remote Sensing of Invasive Alien Plant Species

**S.P.S. Kushwaha**

*Forestry and Ecology Division, Indian Institute of Remote Sensing, Indian Space Research Organisation, Dehradun, India*

## Introduction

Invasive alien species (IAS) are plants, animals, pathogens and other organisms that are non-native to an ecosystem and which may cause economic or environmental harm or adversely affect human health. In particular, they impact adversely upon biodiversity, including decline or elimination of native species through competition, predation or transmission of pathogens, and the disruption of local ecosystem and ecosystem functions (CBD, 2006). Invasive alien species, introduced and/or spread outside their natural habitats, have affected native biodiversity in almost every ecosystem type on earth and are one of the greatest threats to biodiversity. Since the 17th century, invasive alien species have contributed to nearly 40% of all animal extinctions for which the cause is known. The problem continues to grow, at great socio-economic, health and ecological cost around the world. Invasive alien species exacerbate poverty and threaten development through their impact on agriculture, forestry, fisheries and natural systems, which are an important basis of peoples' livelihoods in developing countries. This damage is aggravated by climate change, pollution, habitat loss and human-induced disturbance.

The international Convention on Biological Diversity (CBD) and its members (there were 191 parties, as of October 2008) recognize that there is an urgent need to address the impact of invasive alien species. Article 8(h) of the CBD states that 'each contracting party shall, as far as possible and as appropriate, prevent the introduction of, control or eradicate those alien species which threaten ecosystems, habitats or species'. The CBD sets global priorities, guidelines, collects information and helps coordinate international action on invasive alien species that threaten ecosystems, habitats or species, and these can be accessed on the CBD website (Decision VI 23). The website also provides further information on invasive species and relevant decisions of the Conference of the Parties (COPs) (CBD, 2006).

The Global Invasive Species Programme (GISP) promotes the ecosystem approach as being essential for the effective management of biological invasions. Invasive alien species threaten biodiversity as well as food security, human health, trade, transport and economic development. Biodiversity is impacted negatively in different ways, including competition, predation and herbivory, parasitism and pathogenesis, hybridization, facilitation (i.e. by changing nutrient, air or water levels) and ecosystem processes but, in all cases, the impact of invasions – and indeed the reason for the invasion phenomenon, is the interaction of the invasive species and the recipient ecosystem. For this reason, the GISP actively promotes the ecosystem approach for the management of invasive species so that the target is not just the reduction or eradication of the invading species but involves the recipient ecosystem and communities in a multi-sector and multi-tier approach. A good example can be found in the UNEP/GEF-funded project entitled 'Removing barriers to invasive plant management in Africa', currently being executed by CAB International (CABI) and the International Union for Conservation of Nature (IUCN), who have adopted the ecosystem approach in managing invasive species at the lowest level with the involvement of all stakeholders, notably, the local community (Principle 2). A great deal of emphasis has also been attached to the importance of ensuring inter-sector cooperation, including the formation of inter-ministerial bodies, which is consistent with the 5th operational guidance point of CBD Decision V/6.

Native to the Amazon basin, the water hyacinth (*Eichhornia crassipes*) has invaded tropical habitats worldwide, spreading to more than 50 countries on five continents. Water hyacinth blocks waterways, decimates aquatic wildlife and the livelihood of local people and creates ideal conditions for disease and its vectors. Native to the Indian subcontinent, the ship rat (*Rattus rattus*) has caused extinction of and catastrophic decline in native birds on islands and has spread throughout the world. Deadly new disease organisms such as avian influenza A (H5N1) attack both humans and animals, in both temperate and tropical countries. Invasive species have caused serious damage to aquatic ecosystems: native to the Caspian and Black Seas, the zebra mussel (*Dreissena polymorpha*) affects fisheries, mollusc diversity and electric power generation in the Great Lakes of North America and in the Mississippi basin (CBD, 2006).

Almost every ecosystem on earth has serious problems with invasive alien species, with invasions into natural systems posing a key threat to global biodiversity and ecosystem functioning as well as incurring economic costs (Mooney and Cleland, 2001; Pimentel *et al.*, 2005). Over the past century invasive species have attracted the attention of ecologists, biologists, conservationists and natural resources managers due to their rapid colonization of forestlands, croplands and wastelands resulting in the depletion of biodiversity and damage to ecosystems. Czech and Krausman (1997) and Wilcove and Chen (1998) considered species invasion as a major cause of biodiversity loss. Biological invasion is also considered as one of the major cause of species extinction (Drake *et al.*, 1989). Invasions can alter ecosystem structure and function, in addition to carbon sequestration in grasslands. Mooney and Hobbs (2000) indicated that a rapid increase in the levels of invasive species are homogenizing the world's flora and fauna.

## Remote Sensing of Invasive Alien Species

Information on the spatial distribution of invasive species is critical in planning for their management and control. Conventional invasive plant mapping has utilized ground-based manual mapping with the global positioning system (GPS). The shortfalls of ground-based methods include limited area coverage and high inputs of time and cost. Mapping vegetation with remote sensing covers large spatial areas, and maps can be updated at required intervals determined by those in charge. Remote sensing and geographic information system (GIS) technologies offer potentially valuable tools for mapping and monitoring of invasive species, as well as providing data inputs for predicting areas susceptible to invasion. Cost-effective, large-scale and long-term documentation and monitoring of IAS are recognized as fundamental research needs (Johnson, 1999) that are increasingly being addressed. Over the past decade the number of publications on remote-sensing applications for invasion biology has grown to over 100 (Fig. 11.1). Although the use of remote sensing and GIS techniques for mapping invasive species and invaded ecosystems is increasing rapidly, literature on the means and methods for this remains in short supply and is often sketchy. Most of the IUCN's (International Union for Conservation of Nature) worst invasive species fall under the category of Class IV species, in which straightforward application of remote sensing is almost impossible.

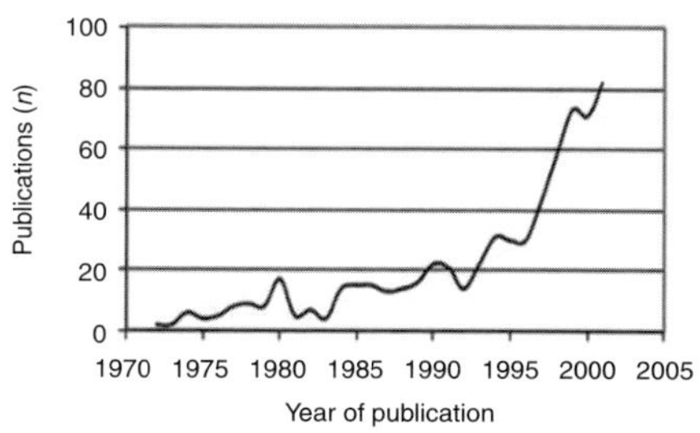


**Fig. 11.1.** Publications involving remote sensing, 1970–2005 (Joshi, 2006).

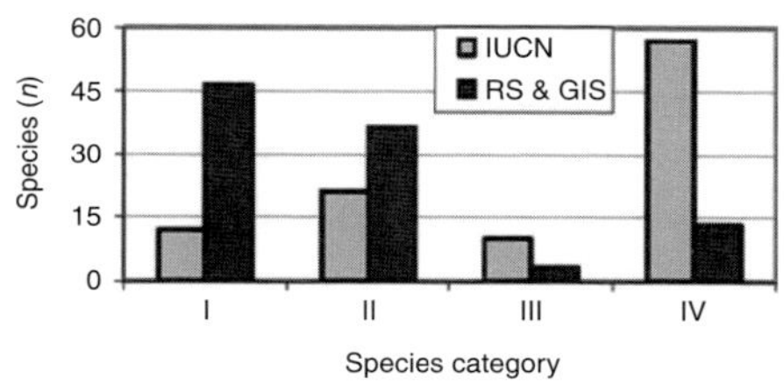


**Fig. 11.2.** The world's 100 worst invasive species as categorized by RS (remote sensing) and GIS (global positioning system) (Joshi, 2006).

Recent remote sensing and GIS applications for detection of invasive species have dealt mainly with species belonging to Class I (Fig. 11.2; Joshi, 2006).

Remote sensing, in comparison with field-based techniques, allows an entire region to be mapped simultaneously. Image-derived locations of IAS provide a permanent geo-referenced record that can be input to GIS databases for control activities, and repeated remotely sensed data acquisitions allow analysis of the abundance and distribution patterns to be monitored over time. In contrast, in areas where an IAS has not yet occurred, prevention is the most efficient way of dealing with problematic species. Consequently, using remote sensing techniques to map points at risk is valuable, such as mapping airstrips in remote locations or seaports. Alternatively, once an IAS has established in a region, mapping potential pathways with imagery can help prevent spread into new areas. For example, roads, hiking and horse trails, forest footpaths and off-road vehicle trails can be mapped, these being routes by which seeds are dispersed via vehicle tyres and footwear. Similarly, boat-launching sites can be identified since turbines on boat engines can spread fragments of aquatic invasive plants as well as small, invasive aquatic animals such as snails (CBD, 2006).

Walsh *et al.* (2008) examined the spatial structure and composition of guava (*Psidium guajava*) on Isabella island in the Galapagos Islands of Ecuador through remote-sensing and image-processing techniques: (i) hyper-spatial and multi-spectral 2.44 m × 2.44 m resolution *QuickBird* spring and summer data were classified using a pixel-based approach; and (ii) *Hyperion* hyper-spectral data were used to examine the spectral characteristics of guava and to link that knowledge to the classified *QuickBird* data to spectrally 'unmix' the 30 m × 30 m *Hyperion* data pixel into the fraction (percentage) of guava within each 30 m pixel using a linear modelling approach.

Remote sensing is also a valuable tool for monitoring the degree of infestation, as well as detecting impediments to water-borne transportation caused by aquatic plant infestations. Frequent large-scale remote sensing-based monitoring provides field managers with real-time information on the routing and scheduling of ships for vegetation eradication cost analysis. Traditional survey methods such as hand-mapping and GPS-assisted mapping (Cooksey and Sheley, 1997) specify sufficiently high accuracies (~80%) for small management areas (Cooksey and Sheley, 1998), but might be technically, financially and logistically impractical for many managers. Remote sensing has been used for decades to measure and map the biophysical characteristics of vegetation (Tucker *et al.*, 1984; Treitz *et al.*, 1992; Anderson *et al.*, 1996; Lawrence and Ripple, 2000). Both spatial and spectral resolutions of remotely sensed data impact mapping accuracy of individual species. Moderate-resolution satellite imagery is more suited to mapping at the ecosystem level because the spatial resolution is generally too coarse to distinguish individual species, unless growing gregariously (Dewey *et al*., 1991; Everitt and Escobar, 1996; Sohn and McCoy, 1997).

Jakubauskas *et al*. (2000) used visible and near-infrared bands of *IKONOS* (4 m spatial resolution) and *ASTER* data (15 m spatial resolution) for mapping and monitoring of water hyacinth. Their results showed that even *ASTER* data could detect the distribution of water hyacinth on the River Rio Grande. It was also shown that reflectance from water hyacinth at the near-infrared region of the electromagnetic spectrum was directly correlated with the density of the weed (Fig. 11.3). Kimothi *et al*. (2010) used IRS P6 LISS-IV and *Cartosat-1* data and found that merged product facilitated up to 96.4% of the differentiation and mapping of *Lantana camara*, compared with 92.9% using LISS-IV and 65% with Cartosat-1 alone, and that April imagery provided maximum contrast for differentiation of this species in the Doon Valley in India (Fig. 11.4).

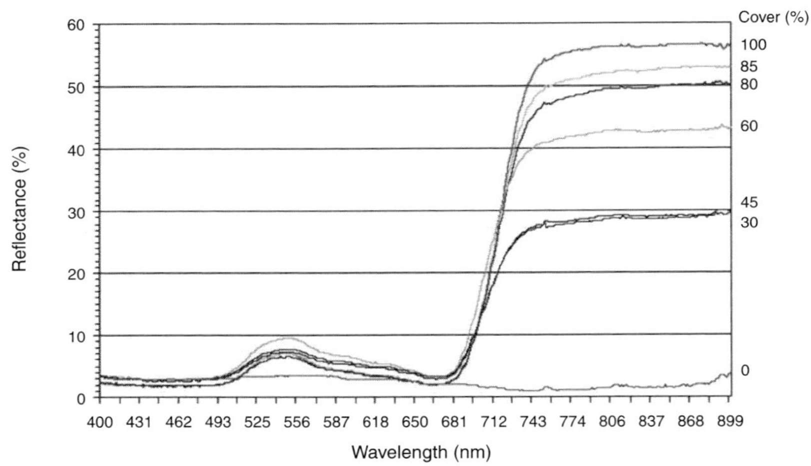


**Fig. 11.3.** Spectral reflectance of water hyacinth (*Eichhornia crassipes*) as a function of percentage cover (Jakubauskas *et al*., 2000).

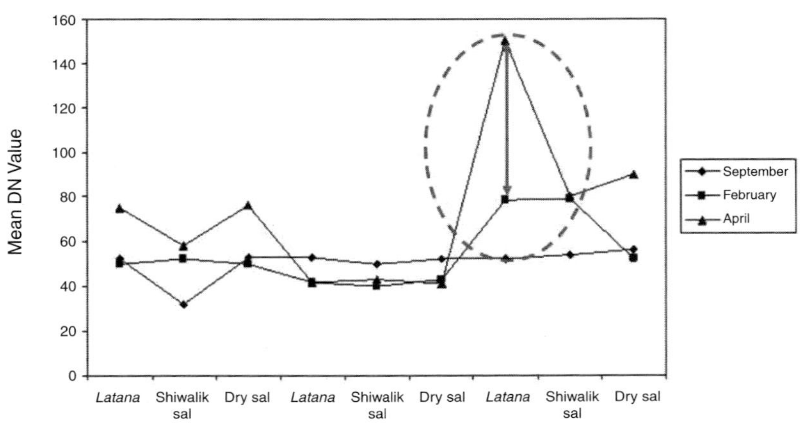


**Fig. 11.4.** Mean reflectance value for *Lantana camara* and its associates at different seasons (Kimothi *et al*., 2010).

Identifying invasive plants in a heterogeneous landscape is often difficult with multispectral imagery (Dewey *et al.*, 1991), because healthy vegetation exhibits similar spectral responses in the visible and near-infrared portions of the electromagnetic spectrum due to similar cellular chemistry and architecture (Woolley, 1971; Price, 1994). In a recent study in the Doon Valley (India) by the author using December *Landsat ETM*$^+$ data, it was noticed that differences in mean reflectance value for *Lantana camara* vis-à-vis associated species were not sufficiently significant for discrimination of *Lantana* from the other species (Table 11.1). An alternative to multispectral imagery is hyperspectral imagery (Ustin *et al.*, 2002).

The continuous nature of spectra inherent to hyperspectral imagery might be utilized to differentiate vegetation into taxonomic levels on account of the vast amount of information stored in the data, especially at the near-infrared band (Jakubauskas *et al.* (2000). Researchers have used hyperspectal imagery, for example, to map leafy spurge in the following US areas: Theodore Roosevelt National Park (O' Neill *et al.*, 2000), undesirable woody vegetation encroaching on grassland in the Niobrara Valley (Wylie *et al.*, 2000), flowering leafy spurge in north-eastern Wyoming (Parker and Hunt, 2002, 2004) and Idaho (Glenn *et al.*, 2005), and hairy cress in Idaho (Mundt *et al.*, 2005).

A particular advantage in using hyperspectral imagery for invasive plant mapping is its potential for determining the relative components or unmixed pixels, which can be especially valuable for determining percentage cover of species for detection of sub-pixel infestations (Glenn *et al.*, 2005; Mundt *et al.*, 2005). Underwood *et al.* (2003) utilized hyperspectral imagery for mapping water hyacinth and Brazilian waterweed (*Egeria densa*) in the Sacramento–San Joaquin delta in the Central Valley of California. Lawrence *et al.* (2006) mapped leafy spurge (*Euphorbia esula*) and spotted knapweed (*Centaurea maculosa*) using 128-band hyperspectral imagery (of 5 m and 3 m resolution) and assessed the accuracy of the resulting maps; Breiman–Cutler Classification

**Table 11.1.** Spectral reflectance of *Lantana camara* and its commonly associated species in the Doon Valley, India.

| | Band 1 | | Band 2 | | Band 3 | |
|---|---|---|---|---|---|---|
| Land use/cover category | Mean | SD | Mean | SD | Mean | SD |
| Khair (*Acacia catechu*) plantation | 55.0 | 3.94 | 38.2 | 4.05 | 58.6 | 3.86 |
| Pine (*Pinus roxburghii*) plantation | 48.6 | 0.84 | 27.5 | 0.53 | 58.7 | 2.00 |
| Crop (mainly wheat and sugar cane) | 67.6 | 5.74 | 40.5 | 3.81 | 94.4 | 2.99 |
| Dense sal (*Shorea robusta*) forest | 44.8 | 1.14 | 24.9 | 0.74 | 67 | 3.53 |
| Dry deciduous forest | 54.6 | 2.22 | 35.3 | 1.95 | 86.1 | 2.08 |
| *Eucalyptus* plantation (Pathri) | 52.9 | 1.52 | 31.1 | 0.88 | 90 | 2.16 |
| *Eucalyptus* plantation (Motichur) | 56.3 | 1.89 | 30.9 | 1.20 | 112.7 | 3.06 |
| Grassland (Motichur) | 59.3 | 4.24 | 44.9 | 6.44 | 77.4 | 2.07 |
| *Lantana* (Mohand) | 60.1 | 2.02 | 35.0 | 2.58 | 105.6 | 4.86 |
| *Lantana* (Motichur) | 59.0 | 0.94 | 32.3 | 1.57 | 112.5 | 5.62 |
| *Lantana* (Timli) | 51.5 | 1.27 | 28.2 | 0.79 | 106.3 | 4.60 |
| Open sal forest | 55.0 | 2.49 | 35.4 | 3.72 | 79.4 | 8.09 |
| Shisham (*Dalbergia sissoo*) plantation | 50.1 | 1.52 | 29.8 | 0.79 | 71.1 | 4.79 |
| Tea garden | 51.8 | 1.55 | 30.7 | 0.95 | 102.8 | 9.67 |
| Teak (*Tectona grandis*) plantation | 48.8 | 1.14 | 29.8 | 1.23 | 65.1 | 2.47 |

(BCC) was used to classify the imagery using the *RandomForest* package in the *R*-statistical program. BCC builds multiple classification trees by repeatedly taking random subsets of the observational data and uses random subsets of the spectral brands to determine each split in the classification trees. Overall classification accuracy of 86 and 84% was recorded for leafy spurge and spotted knapweed, respectively.

## Discussion and Conclusions

As will be evident from the above, remote sensing, GIS and GPS technologies have already been applied with varying degrees of success to mapping the distributions of several plant and animal species, their ecosystems, landscapes, bio-climatic conditions and the factors facilitating invasions (Stow *et al.*, 1989, 2000; McCormick, 1999; Rowlinson *et al.*, 1999; Haltuch *et al.*, 2000; Los *et al.*, 2002). Selecting a season for data acquisition, when contrast between invasive and native species is high, is the most important decision besides choosing the appropriate technique for classification. Using *Landsat ETM⁺* digital image data, the author found that the near-infrared band provided the highest spectral contrast (Fig. 11.5), suggesting that reflectance in near-infrared could be used to discriminate *Lantana camara* from its associates. Furthermore, the cover density of invasives also determines whether a species could be detected. The moist and dry deciduous forests in India represent ~65% of the forest cover in that country. By April the trees in these forests have shed their leaves completely, paving the way for the detection and mapping of invasives such as *Lantana camara*. Operationally, it is a lot easier to detect and map larger patches of invasives than smaller ones, due to differences in size. Spatial, spectral and radiometric remote-sensing imagery of higher resolution is certainly helpful in improving the identification, mapping and monitoring of invasive species.

## References


Anderson, G.L., Everitt, J.H. and Escobar, D.E. (1996) Mapping leafy spurge (*Euphorbia esula*) infestations using aerial photography and geographic information systems. *Geocarto International* 11, 81–89.

CBD (2006) What are invasive alien species? Available at https://www.cbd.int/idb/2009/about/what/ (accessed 16 October 2011).

Cooksey, D. and Sheley, R.L. (1997) Noxious weed


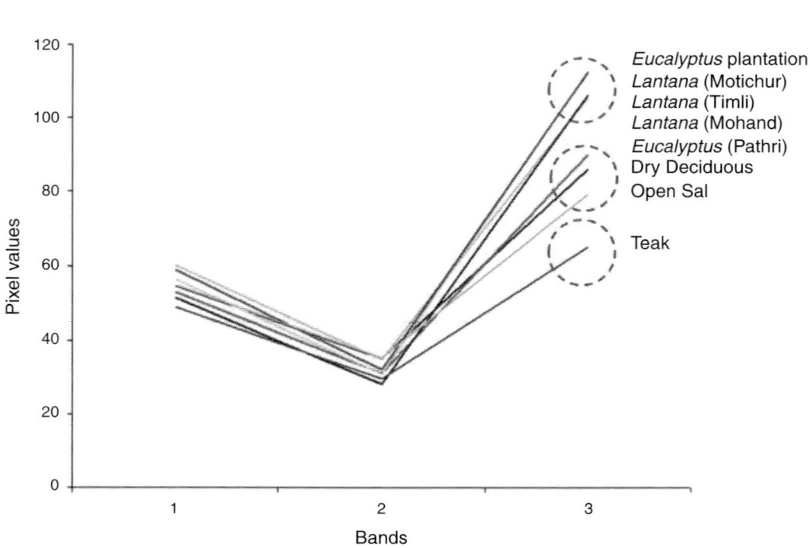


**Fig. 11.5.** Mean pixel values of *Lantana camara* and its associates in Doon Valley, India.

survey and mapping system. *Rangelands* 19, 20–23.

Cooksey, D. and Sheley, R.L. (1998) Mapping noxious weeds in Montana. EB148, Montana State University Extension Service, Bozeman, Montana, 23 pp.

Czech, B. and Krausman, P.R. (1997) Distribution and causation of species endangerment in the United States. *Science* 277, 116–117.

Dewey, S.A., Price, K.P. and Ramsey, D. (1991) Satellite remote sensing to predict potential distribution of Dyers weed (*Isatis tinctoria*). *Weed Technology* 5, 479–484.

Drake, J.A., Mooney, H.A., Castri, F.D., Groves, R.H., Kruger, F.J., Aejmamek, M. *et al.* (1989) *Biological Invasions: A Global Perspective.* John Wiley & Sons, Chichester, UK.

Everitt, J.H. and Escobar, D.E. (1996) Use of spatial information technologies for noxious plant detection and distribution on rangelands. *Geocarto International* 11, 63–80.

Glenn, N.F., Mundt, J.T., Weber, K.T., Prather, T.S., Lass, L.W. and Pettingill, J. (2005) Hyperspectral data processing from repeat detection hyperspectral imagery using mixture tuned matched filtering. *Remote Sensing of the Environment* 82, 446–456.

Haltuch, M.A., Berkman, P.A. and Garton, D.W. (2000) Goegraphic information system (GIS) analysis of ecosystem invasion: Exotic mussels in Lake Erie. *Limnology and Oceanography* 45, 1778–1787.

Jakubauskas, M.E., Kindscher, K., Fraser, A., Debinski, D.M. and Price, K.P. (2000) Close-range remote sensing of aquatic macrophyte vegetation cover. *International Journal of Remote Sensing* 21, 3533–3538.

Johnson, D.E. (1999) Surveying, mapping and monitoring noxious weeds on rangelands. In: Sheley, R.L. and Petroff, J.K. (eds) *Biology and Management of Noxious Rangeland Weeds.* Oregon State University Press, Corvallis, Oregon, pp. 19–35.

Joshi, C.M. (2006) Mapping cryptic invaders and invasibility of tropical forest ecosystems: *Chromolaena odorata* in Nepal. PhD thesis, ITC Dissertation 133, ISBN: 90-8504-470-7, Wageningen University, The Netherlands, 196 pp.

Kimothi, M.M., Anitha, D., Vasistha, H.B., Soni, P. and Chandola, S.K. (2010) Remote sensing to map the invasive weed, *Lantana camara* in forests. *Tropical Ecology* 51, 67–74.

Lawrence, R.L. and Ripple, W.J. (2000) Fifteen years of revegetation of Mount St. Helens: a landscape-scale analysis. *Ecology* 81, 2742–2752.

Lawrence, R.L., Wood, S.D. and Sheley, R.L. (2006) Mapping invasive plants using hyperspectral imagery and Breiman Cutler classifications (*RandomForest*). *Remote Sensing of the Environment* 100, 356–362.

Los, S.O., Tucker, C.J., Anyamba, A., Cherlet, M., Collatz, G.J., Giglio, L. *et al.* (2002) *Environmental Modelling with GIS and RS.* Taylor & Francis, London.

McCormick, C.M. (1999) Mapping exotic vegetation in the Everglades from large-scale aerial photographs. *Photogrammetric Engineering and Remote Sensing* 65, 179–184.

Mooney, H.A. and Cleland, E.E. (2001) The evolutionary impact of invasive species. *Proceedings of the National Academy of Sciences USA* 98, 5446–5451.

Mooney, H.A. and Hobbs, R.J. (2000) *Invasive Species in a Changing World.* Island Press, Washington, DC, 457 pp.

Mundt, J.T., Glenn, N.F., Wever, K.T., Prather, T.S., Lass, L.W. and Pettingill, J. (2005) Discrimination of hoary cress and determination of its detection limits via hyperspectral image processing and accuracy assessment techniques. *Remote Sensing of the Environment* 96, 509–517.

O'Neill, M., Ustin, S.L., Hager, S. and Root, R. (2000) Mapping the distribution of leafy spurge at Theodore Roosevelt National Park using AVIRIS. In: Green, R.O. (ed.) *AVIRIS Airborne Geoscience Workshop*, Pasadena, California.

Parker, W.A. and Hunt, E.R. Jr (2002) Estimation of leafy spurge from hyperspectral imagery using mixture tuned matched filtering. *Remote Sensing of the Environment* 82, 446–456.

Parker, W.A. and Hunt, E.R. Jr (2004) Accuracy assessment of detection of leafy spurge with hyperspectral imagery. *Journal of Range Management* 57, 106–112.

Pimentel, D., Zuniga, R. and Morrison, D. (2005) Update on the environmental and economic costs associated with alien-invasive species in the United States. *Ecological Economics* 52, 273–288.

Price, J.C. (1994) How unique are spectral signatures? *Remote Sensing of the Environment* 49, 181–186.

Rowlinson, L.C., Summerton, M. and Ahmad, F. (1999) Comparison of RS data sources and techniques for identifying and classifying alien invasive vegetation in riparian zones. *Water SA* 25, 497–500.

Sohn, Y. and McCoy, R.M. (1997) Mapping desert shrub rangeland using spectral unmixing and modeling spectral mixtures with TM data. *Photogrammetric Engineering and Remote Sensing* 63, 707–716.

Stow, D.A., Hope, A.S. and George, T.H. (1989) Reflectance characteristics of Arctic tundra vegetation from airborne radiometry. *International Journal of Remote Sensing* 14, 1239–1244.

Stow, D.A., Hope, A., Richardson, D., Chen, D., Garrison, C. and Service, D. (2000) Potential of colour-infrared digital camera imagery for inventory and mapping of alien plant invasions in South African shrublands. *International Journal of Remote Sensing* 21, 2965–2970.

Treitz, P.M., Howarth, P.J. and Suffling, R.C. (1992) Application of detailed ground information to vegetation mapping with high spatial resolution imagery. *Remote Sensing of the Environment* 42, 65–82.

Tucker, C.J., Holben, B.N. and Goff, T.E. (1984) Intensive forest clearing in Rondonia, Brazil as detected by satellite remote sensing. *Remote Sensing of the Environment* 15, 255–264.

Underwood, E., Ustin, S. and DiPietro, D. (2003) Mapping non-native plants using hyperspectral imagery. *Remote Sensing of the Environment* 86, 150–161.

Ustin, S.L., DiPietro, D., Olmstead, K., Underwood, E. and Scheer, G.J. (2002) Hyperspectral remote sensing for invasive species detection and mapping. *Proceedings of the IEEE International Geoscience and Remote Sensing Symposium* (IGARSS'02), Toronto, Canada, vol. 3, pp. 1658–1660.

Walsh, S.J., McCleary, A.L., Mena, C.F., Shao, Y., Tuttle, J.P., González, A. *et al.* (2008) *QuickBird* and *Hyperion* data analysis of an invasive plant species in the Galapagos Islands of Ecuador: Implications for control and land use management. *Remote Sensing of the Environment* 112, 1927–1941.

Wilcove, D.S. and Chen, L.Y. (1998) Management costs for endangered species. *Conservation Biology* 12, 1405–1407.

Woolley, J.T. (1971) Reflectance and transmittance of light by leaves. *Plant Physiology* 47, 656–662.

Wylie, B.K., Meyer, D.J., Choate, M.J., Vierling, L. and Kozak, P.K. (2000) Mapping woody vegetation and eastern red cedar in the Nebraska Sand Hills using AVIRIS. In: Green, R.O. (ed.) *AVIRIS Airborne Geoscience Workshop*, Pasadena, California.

# 12 Invasive Alien Weeds of the Western Ghats: Taxonomy and Distribution

**R.R. Rao[1] and Kavitha Sagar[2]**

[1] *INSA Honorary Scientist, Bangalore, India;* [2] *Vidya Herbs, Bangalore, India*

## Introduction

Alien weeds have been established on Indian soil ever since the late 15th century, following Portuguese settlement in that country. While the vast majority of these have become naturalized and seem to be the permanent denizens of our flora, a few aliens – particularly those that have established in the last century or so – have proved to be invasives that have rapidly spread like wildfire in all biogeographic zones and have taken a heavy toll of native biodiversity. Although most invasive weeds originated from Mexico and the tropical American Region, they are not truly invasives in their native habitats but have acquired the invasive character in India. While it is important to look for causes that make a species invasive, it is equally important to obtain baseline data on their correct taxonomic identification, distribution, habitat, economic uses (if any) and such other field-related data. The present chapter is focused mainly through a taxonomic perspective on some of these issues in regard to invasive aliens in the Western Ghats region, and is based on available information in published literature as well as on our own field explorations and studies of invasive species in various parts of that region.

## The Western Ghats: a Rich Genetic Estate

The Western Ghats biogeographic zone, which forms the 'Malabar' Province according to some biogeographers (Clarke, 1898; Anonymous, 1985–1995; Chatterjee, 1940) is a narrow stretch of land running from the hills south of the Tapti River in the north to Kanyakumari in the south, along India's west coast. This zone stretches for approximately 1500 km, encompassing a wide variety of climatic conditions. The Ghats rise up abruptly in the west to a highly dissected plateau up to 2900 m in height, and descend to dry Deccan plains below 500 m in the east. This extreme climatic and altitudinal gradient has resulted in a variety of forest types, from evergreen to semi-evergreen, from moist deciduous to dry deciduous formations. At higher ranges, stunted montane communities have also developed (Gamble, 1967). The development of tropical rain forests in the southern Western Ghats and the 'Sholas' in the Nilgiri and Anamalai regions are the most outstanding features of the Western Ghats. The entire Western Ghats biogeographic region is a major genetic estate with an enormous biodiversity of ancient lineage. Although the Western Ghats cover only 5% of India's total geographical area, the region contains over 30% of the country's plant species of which around 12,000 species, from lower groups to flowering plants, are estimated to occur here.

## Characteristics of Invasive Species

What invasive species are and why some species acquire invasive characters are issues that require elaborate discussion and involve phytogeographic, edaphic and ecological considerations. In brief, invasive species are those that, following introduction to a new environment, spread like wildfire and out-compete all native and other species in the area, some even by producing root leachates inhibiting the growth of all neighbouring flora. According to Reddy *et al.* (undated), invasive species possess characteristic features such as: (i) 'pioneer species' in varied landscapes; (ii) tolerant of a wide range of soil and weather conditions; (iii) generalist in distribution; (iv) producing copious amounts of seed that disperse easily; (v) growing aggressive root systems; (vi) short generation time; (vii) high dispersal rates; (viii) long flowering and fruiting periods; (ix) broad native range; and (x) abundant in native range. Furthermore, invasive weeds exhibit physiological features such as: (i) early germination; (ii) fast growth and early maturity; (iii) adaptability to adverse environmental conditions; (iv) drought tolerance; (v) seed dormancy; (vi) easily dispersible seeds; (vii) deep root systems; (viii) tall, leafy shoots; (ix) extended longevity; (x) mass germination of seedlings; (xi) mass reproductive output; and (xii) low herbivore pressure (Newsome and Noble, 1986; Rotherham, 1990).

## Factors Influencing the Introduction of Invasive Species

Among various factors responsible for the introduction of invasive aliens on Indian soil, some of the most important are: (i) wanton introduction as horticultural plants (*Lantana*, *Eichhornia*); (ii) as cover crops in plantations, especially rubber plantations (*Pueraria*, *Mikania*); (ii) import of food grains contaminated by seeds of alien weeds (*Parthenium*); and (iv) plantation workers moving from one country to other and unintentionally carrying with them alien weed seeds (*Mikania*). Furthermore, after their introduction the spread of these weeds is favoured by numerous anthropogenic disturbances such as: (i) deforestation and destruction of natural habitats; (ii) shifting agriculture; (iii) sale and introduction of impure seeds; (iv) construction of roads and long-distance railway lines; (v) construction of irrigation channels and canals; and (vi) nomadic tribes and labourers shifting from one region to another. Added to this, natural factors such as recurring floods, and migratory birds visiting different aquatic habitats and dispersing aquatic weed seeds to newer habitats, have greatly favoured the spread of alien weeds to different bio-geographic zones in the country. In the Western Ghats the growth of the human population, extension of plantations, clearance of forests or destruction of original habitats for human settlement are among key factors responsible for the spread of alien invasives.

## The Menace of Invasive Weeds in the Western Ghats

Invasive weeds, being very aggressive and having high competitive ability, once introduced to a new environment can alter the entire ecosystem composition of the area. These species, which form almost pure populations, have tremendous capacity to eliminate the entire local flora, some even by means of root and leaf leachates to the environment. Comparison of the wayside monsoon flora of Mysore district before and after the introduction of an invasive weed such as *Parthenium hysterophorus* (Ramaswamy *et al.* 1972–73; Rao, 1973) reveals the disappearance or fragmentation of as many as 50 native species from that district alone within a short span of 35 years (Rao and Razi, 1981; Rao, 1996). Similarly, weeds such as *Mikania micrantha*, *Puereria phaseoloides* and others have eliminated much of the indigenous flora in the Western Ghats (Table 12.1).

**Table 12.1.** Some wild species in India whose diversity has been affected by invasive aliens of the family *Asteraceae.*

| Species | Family |
|---|---|
| *Abutilon indicum* | *Malvaceae* |
| *Acalypha indica* | *Euphorbiaceae* |
| *Alternanthera echinata* | *Amaranthaceae* |
| *Alysicarpus vaginalis* | *Papilionaceae* |
| *Amaranthus spinosus* | *Amaranthaceae* |
| *Andrographis paniculata* | *Acanthaceae* |
| *Anisomeles indica* | *Lamiaceae* |
| *Aristolochia bracteata* | *Aristolochaceae* |
| *Asystasia dalzelliana* | *Acanthaceae* |
| *Boerhaavia diffusa* | *Nyctaginaceae* |
| *Cassia occidentalis* | *Caesalpiniaceae* |
| *C. tora* | *Caesalpiniaceae* |
| *Chenopodium album* | *Chenopodiaceae* |
| *Chrozophora rottlerii* | *Euphorbiaceae* |
| *C. ambrosioides* | *Euphorbiaceae* |
| *Cleome viscosa* | *Capparidaceae* |
| *Convolvulus arvensis* | *Convolvulaceae* |
| *Corchorus tridens* | *Tiliaceae* |
| *Crotolaria medicaginea* | *Papilionaceae* |
| *Cucumis callosus* | *Cucurbitaceae* |
| *Desmodium triflorum* | *Papilionaceae* |
| *Digera alternifolia* | *Amaranthaceae* |
| *Euphorbia hirta* | *Euphorbiaceae* |
| *E. microphylla* | *Euphorbiaceae* |
| *E. orbiculata* | *Euphorbiaceae* |
| *E. prostrata* | *Euphorbiaceae* |
| *Evolvulus alsinoides* | *Convolvulaceae* |
| *Gomphrena serrata* | *Amaranthaceae* |
| *Hedyotis aspera* | *Rubiaceae* |
| *Hybanthus enneaspermus* | *Violaceae* |
| *Indigofera cordifolia* | *Papilionaceae* |
| *I. linifolia* | *Papilionaceae* |
| *I. trita* | *Papilionaceae* |
| *Ipomoea muricata* | *Convolvulaceae* |
| *Leucas aspera* | *Lamiaceae* |
| *Mollugo cerviana* | *Molluginaceae* |
| *Phyllanthus amarus* | *Euphorbiaceae* |
| *Physalis minima* | *Solanaceae* |
| *Polycarpaea corymbosa* | *Caryophyllaceae* |
| *Polycarpon prostratum* | *Caryophyllaceae* |
| *Polygala arvensis* | *Polygalaceae* |
| *P. chinensis* | *Polygalaceae* |
| *Portulaca oleracea* | *Portulacaceae* |
| *Rungia repens* | *Acanthaceae* |
| *Scoparia dulcis* | *Scrophulariaceae* |
| *Solanum nigrum* | *Solanaceae* |
| *Tephrosia purpurea* | *Papilionaceae* |
| *Tragia involacrata* | *Euphorbiaceae* |
| *Trianthema portulacastrum* | *Aizoaceae* |
| *Tribulus terrestris* | *Zygophyllaceae* |
| *Viola serpens* | *Violaceae* |

While the above weeds have become a huge problem to both agriculture and wild native flora in terrestrial ecosystems, alien weeds such as *Eichhornia crassipes* have taken a heavy toll of aquatic diversity. Biological invasion is considered the second largest current cause of loss of biodiversity, second only to habitat destruction (Vitousek *et al.*, 1997). The problem of biological invasion has been recognized by SCOPE (Scientific Committee On Problems of the Environment) as a central problem in the conservation of biological communities. Heywood (1989) remarks that 'Invasion of natural communities in many parts of the world by introduced plants, especially woody species constitutes one of the most serious threats to their survival'. Over the past two decades, the potential for non-native species dramatically to alter ecosystem structure and function has become increasingly recognized. Although all invasive aliens have shown a severe negative impact on native biodiversity, the following 12 species can be prioritized as the worst weeds in the Western Ghats, having spread very extensively and depleted the native flora (Table 12.2).

## Description of Alien Invasive Weeds of the Western Ghats

In the following, the families and species therein are arranged in alphabetical order. The nomenclature of all species has updated according to the latest ICBN (International Code of Biological Nomenclature). Species citation is kept to a minimum, covering only the original citation and important synonyms in the region. Only a few regional floras are cited. Vernacular/common names, where available, are provided immediately following botanical names, and this is followed by a brief description of the species and its phenology. Notes on distribution, nativity and other remarks are provided at the end.

### *Amaranthaceae*

#### *Alternanthera ficoidea*

*Alternanthera ficoidea* (L.) R..Br. ex Griseb. Fl. Brit. W. Ind. Isl. 67. 1867; Murthy & Singh in Curr. Sci. 32: 597.1966; Rao & Razi Fl. Mysore 401. 1981. *Gomphrena ficoidea* L. Sp. Pl. 235.1753. *Gomphrena polygonoides* L. Sp. Pl. 225. 1753. *A. polygonoides* (L.) R. Br. Prodr.417. 1914.

Common names: alligator weed, joyweed, Joseph's coat.

Herbs; stems decumbent to erect, moderately branched; leaves oblanceolate or narrowly elliptic, acute and mucronate at apex, attenuate at base and narrowed along poorly defined petiole, subglabrous to pilose; flower heads axillary, sessile, subglobose to short cylindric; bracts and bracteoles subequal, scarious with midrib ending in sharp point, tepals acuminate and sharp-tipped, richly pilose with scabrulous trichomes, becoming tough.

Flowers and fruits: June–July.

**Table 12.2.** Twelve worst invasive alien weeds in the Western Ghats.

| Species | Family |
|---|---|
| *Ageratina adenophora* | *Asteraceae* |
| *Alternanthera paronychioides* | *Amaranthaceae* |
| *Cassia uniflora* | *Caesalpiniaceae* |
| *Chromolaena odorata* | *Asteraceae* |
| *Eichhornia crassipes* | *Pontederiaceae* |
| *Hyptis suaveolens* | *Lamiaceae* |
| *Ipomoea fistulosa* | *Convolvulaceae* |
| *Lantana camara* | *Verbenaceae* |
| *Mikania micrantha* | *Asteraceae* |
| *Parthenium hysterophorus* | *Asteraceae* |
| *Prosopis chilensis* | *Mimosaceae* |
| *Ricinus communis* | *Euphorbiaceae* |

Distribution: a native of tropical America, now naturalized throughout the Western Ghats, commonly found in ponds and ditches.

#### *Alternanthera paronychioides*

*Alternanthera paronychioides* A. St.-Hilaire, Voy. Distr. Diam. 2: 439. 1833. *Illecebrum ficoideum* Jacquin, Select. Stirp. Amer. Hist., 88. 1763, non Linnaeus 1762.

Common name: smooth joyweed.

Perennial, creeping, herbs; stems prostrate, villous, hairy; leaves sessile; blade elliptic, ovate–rhombic, apex acute or obtuse, villous; inflorescence axillary, sessile; heads white, globose, 0.5–1.0 cm diam.; bracts less than half as long as tepals; flowers: tepals monomorphic, white, lanceolate, apex acuminate, without rigid, spinose tips, hairs not barbed; stamens 5; anthers 3–5, globose; pseudostaminodes ligulate, shorter than filaments, margins entire or dentate; utricles included within tepals, stramineous, orbiculate to rounded–obovate, apex truncate; seeds lenticular, 1.2–1.5 mm.

Distribution: native of tropical America, now naturalized throughout the Western Ghats, found abundantly in waste places, open sandy disturbed places, fallows, marshy places.

#### *Alternanthera philoxiroides*

*Alternanthera philoxiroides* (Mart.) Griseb. Wiss. Gnett. 24:36. 1879; Madhusoodanan & Ajit in J. Econ. Tax. Bot. 17(3): 651.1993.

Much-branched perennial herbs forming dense mats; leaves opposite, elliptic to obovate-lanceolate with acute base; heads white, axillary, globose; bracts ovate–lanceolate, glabrous; perianths 5, dorsally compressed, white, glabrous, subequal, oblong–lanceolate with mucronate tips; stamens 5, pseudo-staminodes longer than the filaments; ovary short-stalked.

Very abundant in marshy, shallow waters in eutrophicated lakes, ponds and ditches.

Flowers and fruits: generally throughout the year.

Distribution: neotropical aquatic weed very extensively distributed in the aquatic systems of Kerala, Karnataka and Tamil Nadu.

## *Araceae*

#### *Pistia stratiotes*

*Pistia stratiotes* L. Sp. Pl. 963.1753; Hk. f. Fl. Brit. India 4:497.1885; Cooke Fl. Bombay 3: 328. 1967 (Rep.ed.2); Gamble Fl. Madras 1097. 1967 (Rep.ed.2); Subram. Aq. Ang. -75, f. 42. 1974 (Rep.ed); Rao & Razi Fl. Mysore 585. 1981; Matthew Mat. Fl. Tamilnadu Carnatic 368. 1981 & III. Fl. Tamilnadu Carnatic t. 768. 1982; Murthy & Yoganar. Fl. Coorg 492. 1990.

Common/vern. names: water lettuce; Kan: *Anthara thavare*.

Aquatic, free-floating, stoloniferous, rosulate herbs; leaves obovate, cuneate at base, rounded or obtuse at apex, densely white, pubescent beneath; spathe oblong; spadix equalling spathe; staminate flowers whorled, placed above the pistillate flowers; stamens 6–8, connate into a synandrium; ovary unilocular, many-ovuled; style short, stout, decurved.

Flowers and fruits: December–February, fruits throughout the year.

Distribution: native of South America, now invasive in many ponds in Karnataka and Tamil Nadu.

## *Asteraceae*

#### *Acanthospermum hispidum*

*Acanthospermum hispidum* DC. Prodr. 5: 522.1836;Tadulingam et al., Handb. South Ind. Weeds. 231.1955; Gamble Fl. Madras 495. 1967 (Rep.ed.2); Rao & Razi Fl. Mysore 541.1981; Matthew Mat. Fl. Tamilnadu Carnatic 238. 1982; Rao *et al.*, Fl. Ind. Enum. Aster.1.1988; Murthy & Yoganar. Fl. Coorg. 237.1990; Hajra *et al.*, Fl. India 12: 361.1995.

Common/vern. names: bristly starburr/ Kan: *Kadle mullu*; Tam: *Kombumul*, *Mul Thulasi*; Mal: *Kattu Nerinji*.

Erect hairy herbs; stems diffusely

branched; leaves elliptic to ovate, 2–10 cm long, with glands on lower surface, margins serrate to sub-entire, gradually narrowed to base, sessile; flowers yellow, in heads; ray florets 4–7; burs cuneate, strongly compressed; the ribs bearing 1–2 rows of hooked prickles, the 2 apical prickles stout; achenes clustered star-like.

Flowers and fruits: June–September.

Uses: the species finds use in local medicines as an anthelmintic, abortifacient, and for the treatment of boils, hypertension, dermatology disorders (Tadulingam *et al.*, 1955).

Distribution: native of Brazil, abundant in wide range of habitats in the Western Ghats – roadsides, pastures, waste ground, along railways, etc. in Tamil Nadu, Karnataka and Goa.

### *Ageratina adenophora*

*Ageratina adenophora* (Spreng.) King & Robinson in Phytologia 19: 211. 1970; Rao *et al.*, Fl. Ind. Enum. Aster. 2.1988. *Eupatorium adenophorum* Spreng. Syst. 3: 420.1826; Hajra *et al.*, Fl. India 12: 350.1995. *E. glandulosum* Kunth Nov. Gen. sp. 4: 122. t. 346. 1820, *non* Michaux.

Common names: catweed, crofton weed, sticky agrimony.

Perennial, subshrubs; stems erect, often purple, branched, densely glandular puberulent; leaves darker on upper surface, rhombic–deltate, margins coarsely serrate, petioles usually 1–6 cm; inflorescence erect; involucral bracts glandular puberulent, florets white; achenes reddish-brown.

Flowers and fruits: April–June.

Distribution: native of Mexico, extensively distributed in open forests, hill slopes in Western Ghats particularly Tamil Nadu, Karnataka and Kerala.

### *Ageratum conyzoides*

*Ageratum conyzoides* L. Sp. Pl. 839. 1753; Clarke Comp. Indicae 30, 1876; Hk.f. Fl. of Brit. India 3: 243.1882; Gamble Fl. Madras 476. 1967 (Rep.ed.2); Matthew Indian J. Forestry (Addl.ser. 2) 48. 1981; Rao & Razi Fl. Mysore 540. 1981; Rao *et al.*, Fl. Ind. Enum. Aster. 2.1988; Murthy & Yoganar. Fl. Coorg 238. 1990; Hajra *et al*; Fl. India 12. 348.1995.

Vern. names: Kan: *Oorala gida*, *Mooguthi gida*; Tam: *Vaadaichedi*; Mal: *appa*, *muryanpacha*.

Erect, branched, annual herbs; stem terete; hairy; leaves simple, lower opposite, ovate, acute or obtuse, crenate and hairy on both surfaces; heads homogamous, in terminal corymbs; involucre of bracts linear lanceolate, acute, nerved on the back, margins scarious; florets white or pale blue, corolla tubular, 5-lobed; pappus of 5 scales, as long as corolla lobes; stamens 5; anthers syngenecious; ovary inferior and 1-celled; achenes, long, angled, black, sometimes glandular.

Flowers and fruits: all seasons except dry months.

Uses: traditional communities in the Western Ghats use this species as a bacteriocide, antidysenteric and antilithic, and for pneumonia, burns, wounds, rheumatism and fever; it also has antibacterial, insect-repellent and analgesic properties (http://www.rain-tree.com/ageratum.htm).

Distribution: a native of South America abundant throughout the Western Ghats and listed as one of the 12 worst weeds of that area. The species is highly variable, particularly in the colour of its flowers – both white- and blue-flowered populations are seen. Very common both in plains and on hills, mostly in disturbed areas. Also very abundant along the bunds in agricultural fields near moist areas, plantations, orchards and roadsides throughout the Western Ghats region.

### *Bidens biternata*

*Bidens biternata* (Lour.) Merr. & Sherff in Bot. Gaz. 88: 293. 1929; Murthy & Yoganar. Fl. Coorg 240. 1990; Hajra *et al*; Fl. India 12: 367.1995. *Cereopsis biternata* Lour. Fl. Cochinch. 508. 1790. *Bidens pilosa auct non.* L.; Hk.f. Fl. Brit. India 3: 309.1882, p.p.; Gamble Fl. Madras 499. 1967 (Rep.ed.2);

Matthew Mat. Fl. Tamilnadu Carnatic 239. 1981; Rao *et al.*, Fl. Ind. Enum. Aster.13.1988.

Erect, branched, glabrous, annual herbs; stem quadrangular, grooved, hairy; branches opposite; leaves trifoliate, imparipinnate, acuminate, dentate, lower surface white, translucent crystals; heads heterogamous, radiate, in lax corymbose panicles; involucral bracts hairy at base, narrowly lanceolate; bracts biseriate, outer green, with dark, thick mid-nerves; inner dark, small, ovate, with green or whitish scarious margins; ray florets ligulate, white or yellow, strap-shaped, few, sessile; disc florets yellow. Troublesome as the mature achenes adhere to the clothes of the passer-by.

Distribution: a troublesome invasive weed from tropical America, abundantly found throughout the Western Ghats, particularly along roadsides on hills.

### *Chromolaena odorata*

*Chromolaena odorata* (L.) King & Robinson in Phytologia 20: 204. 1970; Rao *et al.*, Fl. Ind. Enum. Aster. 24.1988. *Eupatorium odoratum* L. Syst. 10(2): 1205. 1759; Matthew Fl. Tamilnadu Carnatic 1: 776. 1983; Clarke Comp. Ind. 30. 1876; Hk.f.Fl. Brit. India 3: 244.1882; Hajra *et al.*, Fl. India 12: 354. 1995 (Fig. 12.1e).

Common names: bitter bush, chromolaena, jack in the bush, triffid weed.

Large bushy herbs/undershrubs with long, rambling (but not twining) branches; stems terete, pubescent; leaves opposite, flaccid–membranous, velvety–pubescent, deltoid–ovate, acute, 3-nerved, very coarsely toothed; base obtuse or subtruncate, shortly decurrent; blade 5–12 × 3–6 cm, capitula in sub-corymbose axillary and terminal clusters; peduncles 1–3 cm long, bracteate; bracts slender, 10–12mm long; involucre of about 4–5 series of bracts, pale with green nerves, acute, the lowest ones about 2 mm long, upper ones 8–9 mm long, all acute, distally ciliate, flat, appressed except the extreme divergent tip; florets all alike (disc-florets), pale purple to dull off-white, the styles extending about 4 mm beyond the apex of the involucre, radiately spreading; receptacle very narrow; florets about 20–30; corolla slender, trumpet-shaped; pappus of dull white hairs, 5 mm long; achenes glabrous or nearly so.

Flowers and fruits: December–April.

Uses: ornamental plant; sometimes encouraged for use in shifting slash-and-burn agriculture to compete with *Imperata cylindrica*, which is harder to control. Luxuriant vegetative growth coupled with spreading root system extracts large amounts of nutrient elements from soil and acts as nutrient pump; succulent biomass as a source of green manure in wetland cultivations; capsules made from this weed help in reducing the desire for smoking (http://www.oztion.com); as smoking has been banned in many parts of the world this property begs further attention. Aqueous extract of leaves is good remedy for soft tissue wounds, burns and skin infections; eupolin, a compound isolated from *E. odoratum*, helps repair cell damage.

Forms dense stands preventing establishment of other species, due to both competition and allelopathic effects. When dry, *C. odorata* usable as fuel, which may promote wild bushfires; may also cause skin complaints and asthma in allergy-prone individuals.

Distribution: native of South and Central America, found very extensively in forests, forest-cleared areas, agricultural land and plantations throughout the Western Ghats, from plains to altitudes of 1000 m. Often forms pure patches as forest undergrowth. One of the 12 worst invasive weeds in the Western Ghats.

### *Conyza bonariensis*

*Conyza bonariensis* (L.) Cronq. In Bull. Torrey Bot. Club. 70(6): 632. 1943; Matthew Fl. Tamilnadu Carnatic 1: 777. 1983; Rao *et al.*, Fl. Ind. Enum. Aster.27.1988; Hajra *et al.*, Fl. India 12: 104.1995. *Erigeron bonariensis* L. Sp. Pl. 863. 1753. *E. linifolius* Willd. Sp. Pl. 3: 1955. 1804; Hk. f. Fl. Brit. India 3: 254. 1882. *Conyza ambigua* DC. Fl. Fr. Suppl. 6: 468. 1915; Gamble Fl. Madras 480.1967 (Rep.ed.2).

**Fig. 12.1.** Alien invasive weeds. A, B, *Ipomoea cairica* spreading over tall trees; C, *Ipomoea fistulosa*; D, *Hyptis suaveolens* along roadsides; E, *Prosopis chilensis* on waste ground in northern Karnataka; F, *Pueraria phaseoloides* on rubber plantations in Kerala.

Common name: asthma weed.

Erect, stout, hirsute, annual herbs with corymbose branches overtopping the main stem; leaves alternate, simple, sessile, linear lanceolate, toothed or entire, acute, hispidly villous; heads racemose, terminal and in upper axils, heterogamous; bracts bi-seriate, linear, narrow; corolla of female florets filiform, short, yellow; bisexual flowers tubular, yellow; stamens 5; filaments short; anthers obtuse, not tailed; achenes compressed, sparsely silky; pappus reddish-white.

Flowers and fruits: January–May.

Distribution: native of South America (Argentina). Established and naturalized quite early in the Western Ghats and was known as early as 1875.

### *Erigeron karvinskianus*

*Erigeron karvinskianus* DC. Prodr. 5: 285.1836; Matthew Exot. Fl. Kod. 145. 1969; Rao & Razi Fl. Mysore 544. 1981; Rao *et al.*, Fl. Ind. Enum. Aster. 37.1988; Hajra *et al.*, Fl. India 12: 122.1995. *E. mucronatus* DC. Prodr. 5: 285. 1836; Tadulingam *et al.*, Handb. South Ind. weeds 219. 1955; Gamble Fl. Madras 480. 1967 (Rep.ed.2).

Common/vern. names: pink & white Nilgiri bank daisy; Tam: *Thaer chedi.*

Much-branched, perennial herbs; branches slender, decumbent, sparsely pubescent throughout; leaves dimorphic; lower stem-leaves obovate–attenuate; ciliate at base, lobes mucronate; upper leaves oblong to oblanceolate; attenuate; heads heterogamous, terminal, white or pink, solitary, involucre of 3 rows of lanceolate, green bracts, 2–4 mm long; disc florets 5 mm long, bisexual; pappus as long as the corolla, barbed; corolla-lobes yellow; stamens 5; achenes scabrid.

Flowers and fruits: February–May.

Distribution: native of Mexico. Perhaps first introduced at Ooty along with seeds of ornamental plants from Australia and has since then established as gregarious wayside weed, on old walls, road cuttings and many other places in the Western Ghats (Gamble, 1967). Naturalized on all hill stations of the Western Ghats.

One of the highly variable weeds and, because of this, this taxon has been named differently by various authors.

### *Galinsoga parviflora*

*Galinsoga parviflora* Cav. Icon. Deser. Pl. 3: 41. t. 281. 1796; Hk. f. Fl. Brit. India 3: 311. 1882; Gamble Fl. Madras 500. 1967 (Rep. ed.2); Rao & Razi Fl. Mysore 545. 1981; Matthew Mat. Fl. Tamilnadu Carnatic 243. 1981; Rao *et al.*, Fl. Ind. Enum. Aster.39.1988; Hajra *et al;* Fl. India 12: 388.1995.

Common names: potato weed, gallant-soldiers, small-flowered galinsoga.

Soft, erect, annual herbs; branches slender, glabrous; stems erect, branched, slender and striate (parallel-ridged); leaves opposite, ovate, acute, cuneate at base, undulate, puberulous, palmately nerved, margins fringed with short hairs like an eyelash, upper leaves smaller, narrower; heads heterogamous, small; involucral bracts bi-seriate, ovate; florets white; ray florets fertile; ligule 1–2 mm long, obtuse; tube with spreading hairs; stylar arms spreading; disc florets bisexual, pappus up to 2 mm long.

Flowers and fruits: December–April.

Uses: leaf juice used as astringent, for stings, cuts and wounds.

Distribution: native of South America, abundant in many places in Western Ghats.

### *Galinsoga quadriradiata*

*Galinsoga quadriradiata* Ruiz et Pavon Syst. Veg. 1: 198.1798; Matthew Ill. Fl. Tamilnadu Carnatic 5. 371. 1982; Rao *et al.*, Fl. Ind. Enum. Aster.40.1988; Hajra *et al.*, Fl. India 12: 388.1995. *G. ciliata* (Rafin.)Blake in Rhodora 24: 35.1922; Babu in Bull. Bot. Surv. India 11: 184. 1969. *Adventina ciliata* Rafin. New. Fl. Am. 1: 67. 1836.

Erect annual hirsute herbs; stems 10–80 cm high, many-branched, with coarse hairs; leaves opposite, dark green, up to 6 × 4 cm, simple, ovate, acute–acuminate, broadest at the base; leaf margin dentate; both surfaces are bristly hairy; heads less than 1 cm across consisting of 4 or 5 white, 3-toothed ray florets and many yellow disc florets; fruit an achene, 1.5 mm long, hairy, tapering from the base to the apex, with a white pappus that resembles a crown; pappus scales entire.

The species is often confused with *G. parviflora*, which is paler green, has short hairs and trilobed pappus scales and involucral bracts with broader and rounded apex. Both species often grow sympatrically and even hybridize (Reinhard *et al.*, 2003).

Flowers and fruits: Dec–May.

Distribution: cosmopolitan weed, native of Mexico, adapted to a warm climate and heavy, nitrogen-rich and clayey soils. It is found in gardens, greenhouses, public plantations, cultivated plots, roadsides, railway lines and waste places.

### Mikania micrantha

*Mikania micrantha* Kunth. In H.B.K. Nov. Gen. Sp. 4: 134. 1820; Rao & Razi Fl. Mysore 547. 1981; Rao *et al.*, Fl. Ind. Enum. Aster.54.1988; Hajra *et al.*, Fl. India 12: 357.1995. *M. scandens auct non* Willd. 1804; Clarke Comp. Ind. 34.1876 *non Willd.*; f. Fl. Brit. India 3. 244. 1882. (Fig. 12.2e).

Common/vern. names: American rope, Chinese creeper: Hin: *titaiya baur.*

Fast-growing, perennial, creeping or twining plants; stems branched, pubescent to glabrous, ribbed; leaves opposite, thin, cordate, triangular or ovate, blade 4–13 cm long; heads terminal and lateral, in corymbose panicles, involucral bracts 4, oblong–obovate, acute, green and with one additional smaller bract; flowers 4 in each head; corollas white; achenes linear–oblong, black, 5-angled, glabrous, pappus soft, white and bristly.

Flowers and fruits: June and February.

Uses: natives use the crushed leaves for curing wounds.

Distribution: Central and South American native, introduced into India after World War II to camouflage air fields; it grows very abundantly in agricultural fields, plantations, cleared forest areas and even near forest margins, covering the entire photosynthetic surface of smaller trees and shrubs.

Very aggressive climbing/twining weed; completely chokes the native vegetation, including forest trees; cuts off the photosynthetic surface of native flora; deprives much-needed sunlight for saplings of forest species; the species has adopted an efficient mode of reproduction and seed dispersal; resistant to wildfires; avoided by grazing animals. It has a vigorous capacity for both vegetative and sexual reproduction.

### Parthenium hysterophorus

*Parthenium hysterophorus* L. Sp.Pl. 988. 1753; Rao in J. Bombay Nat. Hist. Soc. 54: 218.1956; Rao & Razi Fl. Mysore 547. 1981; Matthew Mat. Fl. Tamilnadu Carnatic 244. 1981; Rao *et al.*, Fl. Ind. Enum. Aster.56.1988; Hajra *et al.*, Fl. India 12: 403. 1995 (Fig. 12.2a, b).

Common/vern. names: Santa Maria feverfew, congress grass, ragweed parthenium, white topweed; Kan: *congress gida.*

Erect, profusely branched, 1–2 m tall; stems hairy, longitudinally grooved; leaves alternate, pinnately or bipinnately dissected, capitula 4–5 mm in diam, heterogamous, numerous; female florets urceolate; males developed in pairs along with 5 female ones on the ray; disc florets all male; achenes obovate, black, crowned by the persistent remnants of corolla and appendages of the styles.

Flowers and fruits: throughout the year.

Distribution: native of tropical America, Mexico, first reported from Karnataka, now naturalized throughout the Western Ghats, grows in agricultural areas, range/grasslands, disturbed areas, suburb/shrublands, urban areas. This is one of the worst weeds of the Western Ghats, colonizing in almost pure populations, vast areas along railway lines, roadsides, fallows, cleared forest areas and even in protected areas; has the capacity to fragment the native species to extinction (Rao, 1996). The species also causes health hazards for both human and cattle populations.

### Synedrella vialis

*Synedrella vialis* (Less.) A Gray in Proc. Amer. Acad. 17: 217.1882; Ahuja and Pataskar in Ind. For. 95.267.1969; Hajra *et al.*, (eds) Fl. India 12: 415. 1995. *Calyptocarpus vialis* Les., Syn. Comp. 221. 1832 (Fig. 12.1d).

Annual herbs branched from base; scabrid hairy, rooting at nodes; leaves ovate, lanceolate, acute, cuneate at base, serrate, up to 4 × 3 cm, hairy on both surfaces, 3-nerved from base; heads axillary, solitary, sessile; involucral bracts 4-herbaceous, ciliate along margins; ray florets ligulate, 5–9, female; corolla yellow, 2–3-lobed; achenes dorsally compressed, crowned with 2 spines; disc florets hermaphrodite, tubular; achenes triquetrous, crowned by 3 spines.

**Fig. 12.2.** Alien invasive weeds. A, *Parthenium hysterophorus* in Bandipur sanctuary; B, *Parthenium hysterophorus* along roadsides; C, *Cassia uniflora*; D, *Cassia uniflora* forming dense populations along roadsides in North Karnataka; E, *Mikania micrantha* spreading over other vegetation; F, *Ricinus communis* in fallows along railway lines.

Distribution: native of South America, Mexico and West Indies. Earlier reported from Maharashtra and Uttar Pradesh. The species is extensively spreading in Karnataka along moist, shady roadsides, below trees, in gardens and parks along water courses. etc. This species has already replaced a large number of native ground flora, including grasses, in many places where we have noticed a vast spread of compact, green carpets of this invasive species. Although this is one of the major invasive species in Karnataka, the weed has not received due attention by biologists.

### *Taraxacum javanicum*

*Taraxacum javanicum* Soest. Wentia 10: 57. f. 27. 1963; Matthew Mat. Fl. Tamilnadu Carnatic 246. 1981. *T. officinale auct. non* Wigg. Prim. Fl. Holst. 56. 1780; Hk.f. Fl. Brit. Ind. 3. 401. 1882; Gamble Fl. Madras 514. 1967 (Rep.ed.2); Fyson Fl. Hill Stat. 356. 1932; Cooke Fl. Bombay 2: 127 1967 (Rep.ed.2); Ramaswamy & Razi Fl. Bangalore 632. 1973; Rao *et al.*, Fl. Ind. Enum. Aster. 81.1988; Hajra *et al.*, (eds) Fl. India 12: 252. 1995.

Common/vern. names: dandelion, bitter-wort, blowball; Sind: *Bathoo*; Tam: *Pathri.*

Glabrous, scapigerous herbs; leaves radical, obovate–spathulate in outline, lobes denticulate; involucral bracts in 2 whorls; outer whorls of subulate, inner whorl lanceolate, erect; disc flat; florets ligulate, fertile, bisexual; stamens 5, syngenesious; filaments long, 2 mm, anthers 2-celled, each with elongated bases; ovary inferior, 1-celled with basal ovule; style 1 cm long; stigma bilobed, lobes recurved; achenes fusiform, 4–5 angled, beaked, 3 mm long; pappus capillary, 6 mm long.

Flowers and fruits: November–May.

Note: extremely variable species. De Candolle (Prodr. 7: 145–149. 1838) describes different species that have since been reduced to varieties or even synonyms.

Distribution: mainly at higher elevations in the Western Ghats. Abundant on grass tufts, in unweeded gardens.

### *Tithonia diversifolia*

*Tithonia diversifolia* (Hemsl.) A. Gray in Proc. Amer. Acad. 19: 5. 1883; Rao *et al.*, Fl. Ind. Enum. Aster.84.1988; Rao and Razi Fl. Mysore 549. 1981; Hajra *et al.*, Fl. India 12: 416.1995. *Mirasalia diversifolia* Hemsl. Biol Centr. Amer. Bot. 2: 168. t. 47. 1881.

Shrubs or undershrubs, 2.5–5.0 m tall, with subterranean stolons; stems brown to green, round, glabrous to villous; leaves large, alternate, deltoid, 3–5-lobed, acuminate with attenuate base; serrate on the margins, shortly hispid hairy above; heads large, heterogamous, usually solitary, sparsely tuberculate; peduncles hispid; involucral bracts many, 3–4-seriate, oblong–ovate, occasionally scarious; ray florets 7–14; corolla yellow, linear, bi- or trifid at the apex; achenes 1-seriate, white to yellow; pappus absent; disc florets numerous, corolla yellow; anthers black.

Flowers and fruits: September–May.

Uses: very attractive flower heads used for wounds and bruises.

Distribution: Native of America. The species is an escape from garden now runs wild in waste places throughout Western Ghats. The species forms huge populations wherever established.

### *Xanthium indicum*

*Xanthium indicum* Koenig. in Roxb. Fl. Ind. 3: 601. 1832; Matthew Ill. Fl. Tamilnadu Carnatic t. 397. 1982; Murthy & Yoganar. Fl. Coorg 258. 1990; Hajra *et al.*, Fl. India 12: 427.1995. *X. strumarium* L. Sp. Pl. 987.1753. p.p.; Hk.f. Fl. Brit. India 3:303.1882 (excl. Syn. *X. strumarium* Boiss.) p.p.; Gamble Fl. Madras 494. 1967 (Rep.ed.2); Rao *et al.*, Fl. Ind. Enum. Aster. 91.1988.

Common/vern. names: cocklebur or bur weed; Sanskrit: *Sarpakshi*; Hin: *Kuthua, shankeshvar*; Kan: *Maralu ummathi*; Tam: *Uttarachedi, Harlummutta.*

Unarmed, erect, annual herbs; leaves alternate, triangular–cordate or ovate, 3–5-lobed, irregularly serrate, hispid or scabrid; heads many, white or green, in terminal and axillary racemes; male heads globose, female heads ovoid; covered with hooked prickles, end in 2 strongly hooked beaks; achenes obovoid, enclosed in the hard involucre.

Flowers and fruits: January–February.

Distribution: South American weed, distributed throughout the Western Ghats; common on waste ground, roadsides, rubbish heaps, fallows.

## Butomaceae

### Limnocharis flava

*Limnocharis flava* (L.) Buchenau Abh. Naturwiss. Verine Bremen 2: 2. 1868; Biswas & Calder Handb. Water and Marsh Pl. India. 80. 1937; Van Steenis Fl. Males. 1, 5(1): 120. 1954; Bahadur & Raizada Indian For. 94 (8): 641–644.1968; Dassanayake Rev. Handb. Fl. Ceylon 14: 237. 2000. Karthikeyan *et al.*, in Curr. Sci. 87(2): 140. 2004. *Alisma flava* L. Sp. Pl. 232. 1753.

Common names: velvetleaf, yellow burhead.

Tropical America. First introduced at Bogor Botanic Garden in 1866 as an ornamental. In India first recorded in Kerala around 1935, along with imported paddy from SE Asian countries. Grows in wet, waterlogged, exposed lands and rice fields. Completely chokes the aquatic vegetation; propagates by seeds, a single fruit producing ~1000 seeds and a single plant producing 1,000,000 seeds (Karthigeyan *et al.*, 2004).

## Calceolariaceae

### Calceolaria gracili

*Calceolaria gracili* Kunth in Humb. *et al.*, Nov. Gen. SP. 2: 339. 1818; Matthew Fl. Tamilnadu Carnatic 1072. 1981. *C. mexicana* Benth. Pl. Hartw. 47. 1839; Hk. f. Fl. Brit. India 4: 246.1885; Gamble Fl. Madras 684. 1967 (Rep.ed.2); Rao & Razi Fl. Mysore 518. 1981.

Common names: gold purse.

Erect flaccid herbs; leaves pinnate or pinnatisect; flowers in irregular cymes; calyx 4-partitite; corolla deep yellow, upper lip very small, lower lip large, inflated; perfect stamens 2; capsule septicidal at the top; seeds many in each locule.

Flowers and fruits: November–December.

Distribution: native of South America, abundant on roadsides and degraded forests in the high ranges of the Western Ghats, particularly in plantations. According to Gamble (1967), this species 'is a pretty yellow-flowered weed from Mexico now naturalized on the higher parts of the Nilgiris and Pulneys, also on the hills of N. Coimbatore and the Bababudan Hills of Mysore'.

## Caesalpiniaceae

### Cassia occidentalis

*Cassia occidentalis* L. Sp. Pl. 377.1753; Hk. f. Fl. Brit. India 2: 262.1879; De Witt. in Webbia 11: 256. 1956; Cooke Fl. Bombay 1: 445. 1967 (Rep.ed.2); Gamble Fl. Madras 284. 1967 (Rep.ed.2); Matthew Mat. Fl. Tamilnadu Carnatic 203. 1981; Rao & Razi Fl. Mysore 445. 1981; Murthy & Yoganar. Fl. Coorg 170. 1990. Sanjappa Leg. India 19. 1991. *Senna occidentalis* (L.) Link. Handbuch. 2: 140. 1831.

Common/vern. names: fedegoso, stinking weed, coffee senna; Kan: *Ane togace*, *Dodda togace*; Hin: *Banar*, *Kasandi*; Tam: *paayavarai*; Mal: *Ponnaveeram*.

Foetid-smelling, diffusely branched undershrubs, 0.6–1.5 m high; leaves alternate, pinnately compound; stipules lanceolate, long-pointed and deciduous; petiole dialated at the base with solitary gland; leaflets opposite, 4–6 pairs, ovate–lanceolate, glabrous on short stalks; flowers yellow in terminal and axillary racemes; calyx of 5, oblong, membranous, oblong, obtuse, unequal, imbricate sepals; corolla 5, yellow, obovate rounded, veined; stamens 6–7; ovary superior, 1-celled, bent towards one side; legumes compressed, linear, glabrous, brown, seeds 12–30.

Flowers and fruits: February–December.

Uses: antidermatophytic, insecticidal, anthelmintic, antimalarial, antihepatotoxic, fertility regulator, antidiabetic, molluscicidal (Caceres *et al.*, 1993).

Distribution: native of West Indies, a pantropical invasive alien weed, abundantly appears in waste places, meadows and sometimes on hills and distributed throughout the Western Ghats.

### *Cassia tora*

*Cassia tora* L.Sp.Pl. 376.1753; Hk.f. Fl. Brit. India 2: 263.1879; Sanjappa Leg. India 22. 1991; Cooke Fl. Bombay 1: 448. 1967 (Rep. ed.2); Gamble Fl. Madras 284.1967 (Rep. ed.2); Matthew Mat. Fl. Tamilnadu Carnatic 203. 1981; Rao & Razi Fl. Mysore 446. 1981; Murthy & Yoganar. Fl. Coorg 171. 1990. *Senna tora* (L.) Roxb. Fl. Ind. 2: 340.1832 (Fig. 12.2c, d).

Common/vern. names: sickle senna, sickle pod; Kan: *gandu togache*; Hin: *chakunda*; Tel: *Tantipu*; Tam: *Tagarai.*

Foetid-smelling subshrubs, up to 1 m tall; leaves 5–7 cm long with a conical gland between each of the two lowest pairs of leaflets; leaflets 3 pairs, opposite, obovate–oblong, cuneate–rotund, margin entire, apex obtuse; flowers yellow, in axillary pairs; calyx lobes 5, ovate, pubescent; petals 5, golden yellow, obvate, subequal; stamens 10, upper 3 staminodes to 2.5 mm, antheriferous ones 7; two lower ones large with erect filaments; ovary subsessile, curved, pubescent; pods sub-tetragonal, short–stipitate, flat, compressed, 9 × 4–6 mm long, glabrous, septate between seeds; seeds 20–30, oblong, longitudinal.

Flowers and fruits: April–December.

Uses: germicide, laxative and mucilaginous activity, used as coffee substitute and has a naturant and anodyne action, useful in treating skin diseases such as ringworm and psoriasis (Kirtikar and Basu, 1981); also has antiplatelet aggregation properties and antifungal and antibacterial activity (Caceres *et al.*, 1993).

Distribution: South American native; grows gregariously on plains, river banks, fallow fields, margins of forests, waste places, railway lines and roadsides, particularly abundant forming almost pure populations in central and northern parts of Karnataka.

### *Cassia uniflora*

*Cassia uniflora* Mill., Gard. Dict. Ed. 8, no. 5. 1768; Sanjappa Leg. India 24. 1991. *Senna uniflora* (Mill.) H. S. Irwin & Barneby Mem. New York Bot. Gard. 35:258. 1982; *Cassia sericea* Swartz, Fl. Ind. Occ. 2: 724. 1798; Singh Bull. Bot. Surv. India 21: 203. 1981 (Fig. 12.2c, d).

Erect pubescent herbs; leaflets 4-pairs with 3 glands in between the three basal leaflets, obovate, soft pubescent; corymbs in terminal axils; flowers yellow; pods 3–4 cm long hairy, 6-10 seeded.

Flowers and fruits: June–October.

Distribution: native of Topical America. First reported from India as a new record by Singh (1979). Today this is one of the most aggressive weeds, occupying huge areas especially along roadsides, forming almost pure populations in parts of Karnataka, Andhra Pradesh, Kerala and Tamilnadu.

## *Convolvulaceae*

### *Ipomoea cairica*

*Ipomoea cairica* (L.) Sweet Hort. Brit. 287. 1827; Gamble Fl. Madras 645. (Rep.ed.2); 1967; Matthew Fl. Tamilnadu Carnatic 1024.1983; Murthy & Yoganar. Fl. Coorg 299. 1990. *Convolvulus cairicus* L., Syst. Nat. 10: 922. 1759. *I. pulchella* Wight, Icon. Pl. Ind. Orient. t. 156. 1839, *non* Roth. 1821; Cooke Fl. Bombay 2: 320 1967 (Rep. ed.2). *I. palmata* Forsk., Fl. Aegypt.-Arab. 43. 1775; Hk.f. Fl. Brit. India 4: 214. 1885. (Fig. 12.3a, b).

Common/vern. names: railway creeper; Kan: *Kothihejje balli.*

Twining, glabrous, perennial herbs; mature stems semi-woody, rough with lenticels; leaves alternate, 7-lobed; leaflets entire, palmately pedate, the mid-lobe marginally the largest, elliptic–lanceolate, acute, mucronate, others smaller and more obtuse, petioles with stipular leaves at their base; peduncles short, 1–3 flowers, curved downwards, sepals ovate, obtuse, chartaceous, glabrous outside, dotted with whitish specks inside and with membranous margins; corolla large, dull violet, funnel-shaped, up to 6 cm long; capsule ovoid; seeds tawny, pubescent, margins with long, white, silky hairs; stamens 5; ovary 2-celled; ovules 4, stigma capitate, bi-globose.

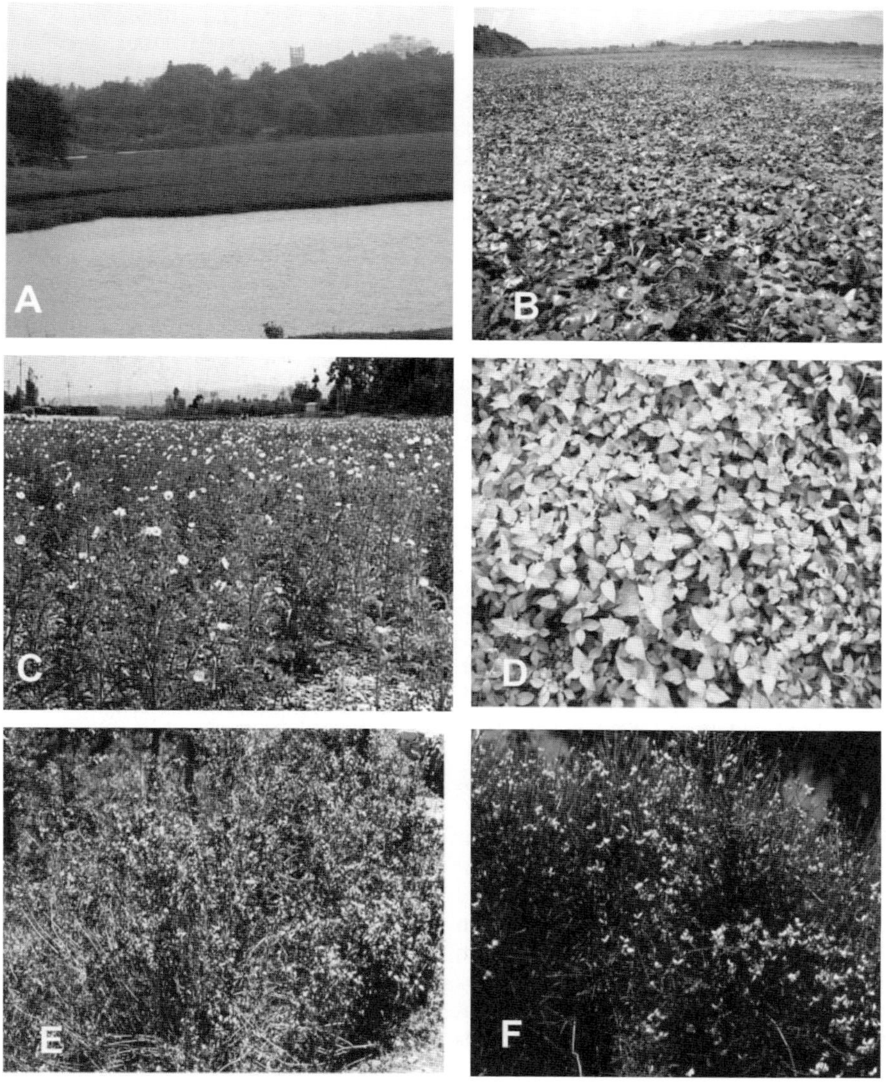


**Fig. 12.3.** Alien invasive weeds. A, *Eichhornia crassipes* forming a dense, floating carpet, Hebbal lake, Bangalore; B, *Eichhornia crassipes*; C, *Argemone mexicana* on waste ground; D, *Synedrella vialis*; E, *Chromolaena odorata*; F, *Cytisus scoparius* in Nilgiris.

Flowers and fruits: all seasons.

Distribution: Native of America, found abundantly in wayside thickets in Tamilnadu and Karnataka. One of the ornamental climbers but has the capacity to climb large trees or shrubby species in the area, completely covering the surrounding vegetation; has become invasive in recent years.

### *Ipomoea fistulosa*

*Ipomoea fistulosa* Mart. ex Choisy in DC. Prodr. 9: 349. 1845; Rao & Razi Fl. Mysore 502. 1981. *I. carnea non* Jacq. 1760; Gamble Fl. Madras 645. 1967(Rep.ed.2); Matthew Fl. Tamilnadu Carnatic 1024. 1983 (Fig. 12.3c).

Vern. name: Tam: *Vaelikkaathaan*.

Suberect shrubs, ~2 m high; stems fistulose; leaves alternate, entire; flowers large, in axillary cymes, pale–rose; sepals 5, equal, imbricate; corolla campanulate, limb plicate, slightly lobed; bands usually defined by 2 prominent lines; stamens 5, unequal, included; filaments filiform, anthers straight; ovary 2-celled; ovules 4; style filiform; stigma capitate; capsule dehiscent.

Flowers and fruits: throughout the year.

Distribution: native of South America, found abundantly throughout the Western Ghats, particularly on the plains; very extensively colonizes marshy areas, bunds of lakes and ponds and even fallow lands. Forms almost pure populations. The flower colour varies from white to light pink.

## Euphorbiaceae

### Croton bonplandianum

*Croton bonplandianum* Baill. Adansonia 4:339.1864; Rao & Razi 468. 198; Matthew Fl. Tamilnadu Carnatic 1420. 1983. *C. sparsiflorus* Morung in Ann. N. Y. Acad. Sci. 7: 221.1893; Gamble Fl. Madras 920. 1967 (Rep.ed.2).

Vern. names: Kan: *Alpabedi soppu*; Tam: *Nai milakai, Aathuppondu*; Tel: *kukka mirapa.*

Erect, diffusely branched, monoecious, annual herbs or undershrubs, tender parts with white, warty outgrowths; leaves 3-ribbed, often with 2 sessile glands, simple, alternate, ex-stipulate, crowded towards the top; petiole short; lamina lanceolate, long, acute closely serrate and tapering; inflorescence terminal, racemes, erect, androgynous spike, 10–20 cm long, few female flowers at its base, small, male flowers clustered above; calyx lobes 5, ovate acute; corolla lobes minute, obovate in male and absent in female; disc annular; stamens 15–20, with slender filaments; ovary scaly, 3-loculed, capsule 4 mm long, dehiscent; seeds smooth, oblong.

Flowers and fruits: April–June.

Distribution: native of South America, found abundantly on wastelands, rocky slopes, cultivated fields, roadsides, railway lines. A drought-resistant weed. In many places the weed has been suppressed by other invasive weeds such as *Parthenium*, *Chromolaena*, *Lantana*, etc. This is one of the dominant weeds in the *Prosopis*-infected areas in North Karnataka.

### Ricinus communis

*Ricinus communis* L.Sp. Pl. 1007, 1753; Hk. f. Fl.Brit. Ind. 5: 457.1887; Cooke Fl. Bombay 3: 125. 1967 (Rep.ed.2); Gamble Fl. Madras 933. 1967(Rep.ed.2); Rao & Razi Fl. Mysore 473. 1981; Matthew Fl. Tamilnadu Carnatic 1471. 1983 (Fig. 12.2f).

Common name: castorbean.

Vern. names: Kan: *haralu*; Hin: *Arend*; Tam: *Sittamunuk*; Tel: *Amadam.*

Tall erect, monoecious shrubs/small trees, 10–13 m tall; stems succulent, herbaceous; leaves alternate, peltate, palmately compound, 6–10-lobed, glabrous; flowers in terminal racemes, male flowers below, 1.5 cm across, female flowers above, 6 cm across; stamens numerous; filaments connate repeatedly branched; ovary globose, echinate, 3-locular; styles 3; capsule trilobed 2 cm across, prickly; seeds oblong shiny, carunculate.

Flowers and fruits: throughout the year.

Distribution: native of Africa, cultivated but mostly runs wild, rather abundantly along roadsides, fallows, railway lines, etc. forming dense thickets under the shade of which native species disappear. Common throughout the Western Ghats on the plains. Very abundant in vacant lands, even in urban areas such as Bangalore.

## Fabaceae

### Cytisus scoparius

*Cytisus scoparius* (L.) Link. Enum. Hort. Berol. 2: 241.1823; Fyson Fl. S. Ind. Hill Stat.138. t. 104. 1932; Sanjappa Leg. India 133. 1991; Matthew Ill. Fl. Palni Hills t. 173. 1996. *Spartium scoparius* L., Sp. Pl. 709. 1753. *Sarothamnus scoparius* (L.) Wimm. ex. W.D.J. Koch Syn. Fl. germ. Helv. 152. 1837;

Matthew Rec. Bot. Surv. India 20(1): 72.1969 (Fig. 12.1f).

Erect, much-branched, grooved shrubs; leaves trifoliate at base, simple upwards, adpressed–pubescent, elliptic, acute; leaflets elliptic–oblong to obovate; flowers yellow, axillary, solitary; pedicel 1 cm, glabrous; calyx abaxial, 2-toothed, glabrous; petals five, bright yellow; stamens 10, monadelphous, 6 longer, 4 shorter; ovary densely white pubescent with elongate adpressed hairs; style pubescent, curved; pod black, hairy along the sutures.

Flowers and fruits: June–October.

Distribution: native of Mediterranean region, naturalized in hill stations. The plant is said to be a garden introduction prior to 1900. Very abundant in the Palni region.

### *Leucaena latisiliqua*

*Leucaena latisiliqua* (L.) Gillis, Taxon 23: 190. 1974; Sanjappa Leg. India 67. 1991. *Mimosa latisiliqua* L. Sp. Pl. 519. 1753. *Leucaena latisiliqua* (Lamk.) de Wit in Taxon 10: 54. 1961; Rao & Razi Fl. Mysore. 451. 1981; Matthew Fl. Tamilnadu Carnatic 543. 1983; *Mimosa latisiliqua* Lamk. Encycl. 1: 12.1783. *Leucaena glauca* Benth. In Hk. Jour. Bot. 4: 416.1885; Hk. f. Fl. Brit. India 2: 290. 1878; Gamble Fl. Madras 297. 1967 (Rep.ed.2).

Common/vern. names: coffee bush, hedge acacia, horse tamarind; Hindi: *Balori*; Tam: *Naattucavindaal.*

Shrubs or small erect trees, up to 10 m tall; leaves bipinnate, pinnae about 5 (4–9) pairs; leaflets about 12 (11–17) pairs, opposite, lanceolate, acute; somewhat dull, greyish-green; flowers in globose, pedunculate heads; peduncles 5–6 cm long; corolla and stamens white; calyx 2.5 mm long; petals linear; stamens 10, anthers hairy; ovary faintly pubescent at apex; pods clustered, linear, flat, dark brown, beaked at apex, glossy brown, oval–oblong, flat, 6–10 cm long.

Flowers and fruits: November–March.

Distribution: native of tropical America, now pantropical, being a fast-growing species; cultivated for green manure, fodder, fuel and afforestation in barren areas, grows along river banks, coffee estates, etc. throughout the Western Ghats. Very successful and colonizes open areas in a short span of time.

### *Pueraria phaseoloides*

*Pueraria phaseoloides* (Roxb.)Benth. In J. Linn. Soc. Bot. Lond. 9: 125. 1865; Ramachandran & Nair Fl. of Cananore 152. 1988. Sanjappa Leg. India 234. 1991. *Dolichos phaseoloides* Roxb. Fl. Ind. 3: 316.1832 (Fig. 12.3f).

Vern. name: Tam: *Thottapayar.*

Gigantic climbers; branches densely pilose; leaves trifoliate; leaflets broadly rhomboid–ovate, cuneate, subacute, silky pubescent on both surfaces; flowers purple, up to 1.7 cm long, in axillary racemes, up to 30 cm long; pods long, linear, slender, beaked, appressed with brown hairs.

Flowers and fruits: February–March.

Distribution: probably a Malaysian weed, initially introduced in the rubber plantation as cover crop, but today the species has become very invasive in parts of Kerala and Karnataka near plantations. Being a giant climber, the weed spreads extensively killing the entire native flora.

### *Ulex europaeus*

*Ulex europaeus* L. Sp. Pl. 741. 1753; Gamble Manual Ind. Tim. 228. 1881; Sanjappa Leg. India 266. 1991.

Common name: gorse.

Profusely branched shrubs to 6–10 m tall; young branches usually terminating in a spine, younger parts glaucous, hirsute–tomentose; phyllodes 4–14mm long, usually spine-tipped; calyx yellow, densely villous, persistent; corolla yellow, 15–20 mm long; pods slightly compressed, densely villous; seeds 1–4, brownish-green, reniform.

*Ulex europaeus* is a major invasive weed in five countries. It is extremely competitive, displaces cultivated and native plants and alters soil conditions by fixing nitrogen and acidifying the soil. It creates an extreme fire hazard due to its oily, highly flammable foliage and seeds, and abundant dead material. It not only increases the risk of fire, but also produces a hotter fire than most weeds. This fire risk poses increased threat on the margins of native vegetation. Because of its various characteristics the soil is often bare between individual gorse plants, which increases erosion on the steep slopes where the gorse has replaced grasses or forbs. Spiny and mostly unpalatable when mature, gorse reduces pasture quality where it invades rangeland. It excludes grazing animals from rangelands and pasture. Gorse is a successful invasive species because it can: (i) fix nitrogen; (ii) acidify and (at least temporarily) impoverish soils by taking up bases; (iii) survive on a variety of soil types; (iv) produce copious amounts of heat-tolerant seeds with long-term viability; and (v) regenerate rapidly from seeds and stumps after disturbances such as brush clearing or fires (Hoshovsky, 1989).

Distribution: a weed of European origin, now completely naturalized in Nilgiri, particularly on hill slopes. Very attractive due to its bright yellow flowers.

## Lamiaceae

### Hyptis suaveolens

*Hyptis suaveolens* (L.) Poit., Ann. Mus. Natl. Hist. Nat. 7: 472. t. 29, f.2.1806; Mukerjee in Rec. Bot. Surv. India 14(1): 63. 1940; Tadulingam *et al.*, Handb. South Ind. Weeds 341. 1955; Gamble Fl. Madras 789. 1967(Rep.ed.2); Rao & Razi Fl. Mysore 511. 1981; Cooke Fl. Bombay 2: 560. 1967(Rep. ed.2); Hk.f. in Fl. Brit India 4: 630. 1885; Yoganar. Med. Pl. India (Tamilnadu) 282. 2000; Mohanan & Sivad., Fl. Agasthyamalai 532. 2002. *Ballota suaveolens* L. Syst. Nat. 10: 1100.1759 (Fig. 12.3d).

Common/vern. name: bush mint, pignut; Tam: *Kanathulasi.*

Tall, rigid, sweet-smelling herbs; stem 4-angled, rough hairy; leaves opposite, ovate, serrate, hairy; flowers collected in heads in the axils of leaves, small, blue; calyx campanulate and ribbed; corolla tube cylindric; lobes 5, the lower tip deflexed, upper spreading; stamens 4, didynamous; ovary superior, 2-celled, 4-partitite; style with a sub-entire or shortly bifid stigma; fruit of 4 dry ovoid or oblong nutlets, smooth or rugulose.

Flowers and fruits: throughout the year, more during June–October.

Uses: stimulant, carminative, sudorific and lactagogue; used for catarrhal conditions, uterine afflictions and parasitic cutaneous diseases; leaf juice used for colic disorders (Yoganarasimhan, 2000).

Distribution: native of North America, now naturalized in the Western Ghats; grows in waste places, roadsides, waysides, along hedges, railway lines and borders of fields. Abundant throughout South India in the plains regions.

## Melastomataceae

### Clidemia hirta

*Clidemia hirta* (L.) D. Don Mem. Wern. Hist. Soc. 4: 309. 1823; Mohanan & Henry Fl. Thiruvananthapuram 194. 1994. *Melastoma hirta* L. Sp. Pl. 390. 1753.

Bushy shrubs, stems, petioles, peduncles and calyx villous mixed with stellate hairs; leaves broadly elliptic to oblong lanceolate, 5-ribbed, lateral nerves parallel and connecting adjacent ribs, villous on both sides; flowers in dense axillary cymose peduncles; peduncles and pedicels villous; calyx campanulate; tube ending in a scaly ring appendages; petals white, ovate; stamens 1, attached at throat of calyx tube; oblong, connate with calyx tube at base, 5-celled; ovules many; stigma capitate, tubercled; fruit baccate, subglobose; seeds many, minute.

Flowers and fruits: April–July.

Distribution: hilly areas such as Pampa-Thriveny, forming almost pure population – a Malesian element, may have reached Kerala region through pilgrims.

## Mimosaceae

### Mimosa invisa

*Mimosa invisa* Martius ex Colla. Heb. Pedom. 2: 255. 1834: Sanjappa Leg. India 68. 1991. *Schrankia brachycarpa* Benth. In Hook. London J. Bot. 2: 130.1843.

Pubescent, trailing shrubs, with recurved prickles; leaves bipinnate; leaflets 15–20 pairs, oblong obtuse; flower heads globose in terminal racemes, calyx 3 mm long; petals 4, united at base, lobes oblong acute; stamens 8; pod compressed, oblong, spinous along margins.

Distribution: South American origin; found abundantly in Karkataka, Kerala and Tamilnadu in open, moist areas.

### Mimosa pudica

*Mimosa pudica* L., Sp. Pl. 518.1753; Baker in Hk. f. Fl. Brit. India. 2: 291. 1879; Gamble Fl. Madras 298. 1967 (Rep.ed.2); Cooke Fl. Bombay 1: 470. 1967 (Rep.ed.2); Rao & Razi Fl. Mysore 451. 1981; Matthew Fl. Tamilnadu Carnatic 545. 1983; Murthy & Yoganar. Fl. Coorg 177. 1990. *Mimosa octandra* Roxb., Pl. Corom. 2: 55 t. 200. 1805. *Mimosa mutabilis* Roxb., Fl. Ind. 2: 584. 1832.

Common/vern. names: sensitive plant; Kan: *nachike gida, muttidare muni*; Hin: *lajwati*; Tam: *total vadi, Thottaar sinungi*.

A diffuse, prickly undershrub; stems and branches sparingly prickly and clothed with long, weak bristles from bulbous bases; leaves bipinnate; leaflets 15–17 pairs, sensitive, coriaceous, linear oblong; ciliate; flowers 4-merous, in globose heads, polygamous; calyx very minute; corolla short, pink; stamens 4, much exserted, pink or red; bracteoles stiff, bristly; ovary many ovuled; style equal or slightly longer than stamens, filiform; stigma obscure; pods clustered, flat, bristly, jointed; seeds 2–5, compressed.

Flowers and fruits: September–October.

Distribution: native of South America, found as clumps in moist and ungrazed places such as river banks, bunds of arable lands or fallow fields, water courses, etc.

### Prosopis chilensis

*Prosopis chilensis* (Sw.) D. Don Prodr. 2: 447. 1825; Matthew Fl. Tamilnadu Carnatic 549. 1983; Sanjappa Leg. India 72. 1991 (Fig. 12.3e).

Vern. names: Kan: *Bellary jali, sarkari jali*; Tam: *Vaelikkaruvai*.

Armed shrubs or trees up to 8 m high; branchlets glabrous; spines axillary; leaves alternate or clustered, 2-pinnate; pinnae 1 or 2 pairs; leaflets 15–18 pairs, oblong, chartaceous, inequilateral, glabrous, base and apex obtuse; petiole glabrous, prolonged above as a bristle, eglandular; rachis with a solitary gland; petiolule 0; stipules spinescent; spikes axillary, solitary; bracteole linear; flowers 5-merous, bisexual; calyx-tube campanulate, 5-toothed; petals 5, creamy, pilose within, united below middle; stamens 10, free; filaments 4.5 mm long; anthers with deciduous gland; ovary stipitate, pubescent; pod drupaceous, straight or slightly curved, compressed, septate, mesocarp pulpy, indehiscent; seeds ovoid.

Flowers and fruits: December–March.

Distribution: native of South America, an aggressive colonizer in wastelands, along roadsides, railway lines, fallows, throughout the Western Ghats in the plain areas up to the foothills. The species has multiple uses in the arid regions, where it is a useful fuel. This species is also listed as one of the 12 worst weeds that have taken a heavy toll of native flora.

## Papaveraceae

### Argemone mexicana

*Argemone mexicana* L. Sp. Pl. 508.1753; et Gen. Pl. ed. 5: 225. 1754; Hk. f. Fl. Brit. India 1: 117. 1875; Tadulingam *et al.*, Handb.

South Ind. Weeds 69. 1954; Cooke Fl. Bombay 1: 29. 1967(Rep.ed.2); Gamble Fl. Madras 25. 1967(Rep.ed.2); Rao & Razi Fl. Mysore 390. 1981; Matthew Fl. Tamilnadu Carnatic 29. 1983 (Fig. 12.1c).

Common/vern. names: Mexican prickly poppy; Kan: *Dhatturada gida*, *arasina datoora*; Hin: *Bharband*, *Brahmadundi*; Tam: *Brahmadandu*; Mal: *Ponnmattam*; Tel: *Brahmadandu*.

Armed, erect, prickly, branching, annual herb; stems round, glaucous, usually very prickly; leaves simple, alternate, sessile and semi-amplexicaul; flowers yellow, solitary and terminal on pedicels which are prickly; bracts leaf-like; calyx 3, long valvate, prickly on back and spine tipped; corolla 6, subequal, bright yellow, obovate and rounded and easily shed; stamens numerous and free; seeds numerous, round, blackish-brown and reticulate.

Flowers and fruits: throughout the year.

Uses: oil from seeds used for skin diseases, scabies, dropsy and opthalmia. It is cathartic. The plant is suitable for the reclamation of alkaline soils. Dried and powdered plants are recommended as green manure as they contain sufficient quantities of nitrogen, phosphorus and potassium.

Distribution: native of the West Indies, now naturalized throughout the plain regions in Karnataka and Tamilnadu. A gregarious weed found in waste places, near habitations, along river banks, cultivated fields and roadsides.

## Pontederiaceae

### Eichhornia crassipes

*Eichhornia crassipes* (Mart.) Solmns.-Laub in A. DC., Monog. Phan. 4: 527. 1883; Fischer Fl. Madras 3: 1530 (1069) 1928; Murthy & Yoganar. Fl. Coorg 477. 1973; Tadulingam *et al.*, Handb. South Ind. Weeds. 409.1955; Gopal & Sharma Water hyacinth 6. 1981; Rao & Razi Fl. Mysore 586. 1981; Matthew Fl. Tamilnadu Carnatic 1649. 1983. *Pontederia crassipes* Mart., Nov. Gen. Sp. Pl. 9.t.4.1823 (Fig. 12.1a, b).

Common/vern. names: water hyacinth; Kan: *Antara tavare*, *Pishaachi thaavare*; Tam: *vengaya thamarai*; Mal: *kula vazha*; Tel: *anthara thamara*.

Aquatic free-floating, freshwater herb; roots elongate fibrous; leaves whorled, broadly ovate rhomboid; petioles elongate, spongy, with a fusiform bulbous middle; spikes terminal; flowers blue to violet, regular, 5 cm across; perianth tube greenish; lobes 3+3, obovate, oblanceolate, unequal, upper one lilac with a blue border, yellow median blotch; stamens 3+3, declinate, glandular pubescent; ovary 3-celled; ovules many; stigma globose.

Flowers and fruits: January–April.

Distribution: native of Brazil, originally introduced as an ornamental plant, now found abundantly in still, slow-flowing waters, tanks, rivers and other wetlands throughout plains of the Western Ghats. Once established, the weed rapidly multiplies forming a complete and dense mat on the surface of the water eliminating all native aquatic species. The species adversely affects aquatic systems by clogging of irrigation channels, choking up of navigational routes, smothering of rice paddies, loss of fishing areas and increasing the breeding habitat available to disease-transmitting mosquitoes.

## Salviniaceae

### Salvinia molesta

*Salvinia molesta* Mitch. Br. Fern Gaz. 10: 251(1973); Dixit Cens. Ind. Pterid. 174. 1984.

Common names: giant salvinia, water fern.

Aquatic free-floating, branched, gregarious, annual ferns; stem spongy, terete, branched with nodes and internodes; leaves modified into root-like organs, numerous, slender, covered by brown, septate hairs; normal leaves borne at the nodes in two opposite pairs, erect floating, sessile, obovate–oblong, upper surface with dense hairs; sporocarps borne in clusters on submerged leaves, 2 mm in diam, megaspores and microspores numerous, globular, covered with minute hairs.

Distribution: South American in origin, grows best in stagnant or slow-flowing water and prefers the small bays of dissected shorelines and the estuaries of small streams; common in wetlands of Kerala and Karnataka.

Note: the presence of numerous, cage-like, club-shaped hairs on the upper surface of the aerial leaves is very distinctive. This feature imparts additional buoyancy to the plant and serves to distinguish *S. molesta* from any other species of the genus. As mentioned by Madhusoodanan (1987), it is an introduced taxon and is a pentaploid hybrid between *Salvinia auriculata* Aubl. and *Salvinia biloba* (Mitchell, 1972).

## Typhaceae

### Typha angustata

*Typha angustata* Bory. & Chaub. Exped. Sci. Moree Bot. 1: 338.1833; Hk. f. Fl. Brit. Ind. 6: 489.1894; Cooke Fl. Bombay 3: 326 1967 (Rep.ed.2); Gamble Fl. Madras 1096. 1967(Rep.ed.2); Rao & Razi Fl. Mysore 581. 1981; Matthew Fl. Tamilnadu Carnatic 1684.1983.

Vern. names: Kan: *Jondu, anejondu*; Hin: *Pan-kanis, jangli-bajri*; Tel: *Jambu*; Tam: *Sambu*.

Aquatic or marshy robust herbs; leaves linear, erect, apex tapering, acute, base sheathing, exceeding flowering stem, semi-terete above the sheath; rhizome perennial, covered with distichous scales; flowers yellow, monoecious, minute, in terminal superposed dense cylindrical spikes, upper part of spike male, 10–20 cm long; rachis (separating male from female) 3 cm long, lower female flowers often intermixed with slender clavate bracts; stamens 2 or 3; anthers 4-celled, basifixed, linear; hairs around stamens linear, spathulate; female spike(s) 1–20 × 1.0–1.5 cm wide, cushion-like; gynophore densely hairy, white, 1-celled; fruit minute, membranous, fusiform, dehiscing by an apical lid; seeds minute.

Flowers and fruits: March–May.

Uses: culms used for making mats (Matthew, 1983).

Distribution: a weed of temperate and subtropical regions, now abundant in most swampy areas.

## Verbenaceae

### Lantana camara

*Lantana camara* L. Sp. Pl. 627.1753; var. *aculeata* (L.) Moldenke in Torreya 34: 9.1934; Bailey Man. Cult. Pl. 842. 1958; Hk.f. Fl. Brit. India 4:562. 1885; Gamble Fl. Madras 761. 1967(Rep.ed.2); Matthew Exot. Fl. Kod. 180. 1969; Rao & Razi Fl. Mysore 508. 1981.

Vern. names: Kan: *Lantavani gida*; Tel: *Pulikampe*; Tam; *Unnippoo*; Mal: *Arippu*.

Bushy shrubs; branches quadrangular, strigosely hairy; leaves aromatic, opposite, petiolate, ovate, acute, crenate–serrate, finely pubescent above and white villous below; flowers sessile, in axillary stalked spikes; peduncle long, slender, hairy; bracts hairy, ovate, acuminate; calyx small, truncate, very hairy; corolla variable in colour: white, pink, yellow, red, orange; ovary superior; 2-celled, style short; stigma subcapitate; fruit a drupe; seeds exalbuminous and tests reticulate.

Flowers and fruits: flowers year round.

Distribution: originated in tropical and subtropical America, now naturalized throughout the Western Ghats, growing along roadsides, railway lines, wastelands, forest margins. This species is considered as one of the ten worst weeds in the world. In the Western Ghats this species has eliminated a number of local species by physically encroaching over vast areas, and also due to its leaf and root leachates. The species can alter forest structure by replacing native understorey species, and this in turn can affect the distribution and behaviour patterns of animal populations.

Notes: *Lantana camara* was introduced to India in the 19th century (Krishnamurthy, 2003). The spread of this plant in the forests of South India is a special cause for concern as there is now apparently widespread change in forest structure. Pink-flowered varieties are not regarded as toxic but other varieties are said to be toxic. In forestry operations the weed is hugely problematic: the prickles cause skin and eye injuries and its extensive growth is an impediment to the free movement of animals.

## Conclusion

Forty-two invasive alien species from the Western Ghats region are described in this chapter and, among these, compositaceous weeds are found to be dominant. Although all these weeds pose threats to Indian agriculture – as well as to the native biodiversity – 12 species are included as being the worst alien weeds in the Western Ghats, and requiring serious efforts towards their eradication.

## Acknowledgements

The authors are grateful to the Council of Scientific and Industrial Research (CSIR), New Delhi, for financial support. The authors would also like to thank the Director and other authorities for the facilities offered and to Dr R.K. Kohli (Chandigarh) for his kind invitation to write this article. The senior author is thankful to the Indian National Science Academy, New Delhi for financial assistance through his position as INSA Honorary Scientist.

## References

Anonymous (1985–1995) *The Wealth of India – Raw materials*, 11 vols. CSIR, New Delhi, India.

Caceres, A., Lopez, B.R., Juarez, X., del Aguila, J. and Garcia, S. (1993) Plants used in Guatemala for treatment for dermatophytic infections. 2: Evaluation of antifungal activity of seven American plants. *Journal of Ethnopharmacology* 40, 207–213.

Chatterjee, D. (1940) Studies on the endemic flora of India and Burma. *Journal of the Royal Asiatic Society of Bengal* (*N.S.*) 5, 19–67.

Clarke, C. (1898) Sub-areas of British India illustrated by the detailed distribution of the Cyperaceae in that empire. *Journal of the Linnean Society* (*Bot.*) 25, 1–107.

Gamble, J.S. (1967) *The Flora of the Presidency of Madras*, 3 vols. Botanical Survey of India, Calcutta, India.

Heywood, V.H. (1989) Patterns, extents and modes of invasions by terrestrial plants. In: Drake, J.A., Mooney, H.A., di Castri, F., Groves, R.H., Kruger, F.J. and Williamson, M. (eds) *Biological Invasions: a Global Perspective*. John Wiley & Sons, Chichester, UK, pp. 31–60.

Hoshovsky, M. (1989) Element stewardship. Abstract For *Ulex europaeus*. ©THE NATURE CONSERVANCY. http//:www.oztion.com. http://www.gisinetwork.org/ India Invasive Plants/ index.html. http://www.medicineatyourfeet.com/ bidenspilosa.html

http://www.rain-tree.com/ageratum.htm

Karthigeyan, K., Sumathi, R., Jayanthi, J., Diwakar, P.G. and Lakra. G.S. (2004) *Limnocharis flava* (L.) Buchenau (Alismataceae) – a little known and troublesome weed in the Andaman Islands. *Current Science* 87, 25.

Kirtikar, K.R and Basu, B.D. (1981) *Indian Medicinal Plants* (revised issue), vol. III. Allahabad, India, pp. 2156–2161.

Krishnamurthy, K.V. (2003) *Textbook of Biodiversity.* Science Publishers Inc., Enfiels, New Hampshire.

Madhusoodanan, P.V. (1987) Taxonomy and distribution of the water fern *Salvinia* Segeuier in Asia. *Journal of Economic and Taxonomic Botany* 11, 497–504.

Matthew, K.M. (1983) *The Flora of Tamilnadu Carnatic.* Rapinat Herbarium, St. Joseph's College, Tiruchirapalli, India.

Mitchell, D.S. (1972) The kariba weed: *Salvinia molesta. British Fern Gazette* 10, 251–252.

Newsome, A.E. and Noble, I.R. (1986) Ecological and physiological characters of invading species. In: Groves, R.H. and Burdon, J. (eds) *Ecology of Biological Invasions.* Cambridge University Press, Cambridge, UK, pp. 10–20.

Ramaswamy, S.V., Rao, R.R. and Bhaskar, V. (1972–73) Contributions from the Herbarium, Mysore. VI. Notes on some adventive plants of Bangalore and Mysore Districts. *Journal of Mysore University* B 25, 40–44.

Rao, R.R. (1973) Studies on flowering plants of Mysore District. PhD thesis, Mysore University, India.

Rao, R.R. (1996) Compositae in the conservation of genetic diversity in wild plants in India. In: Caligari, P.D.S. and Hind, D.J.N. (eds). *Compositae: Biology & Utilization. Proceedings of the International Compositae Conference*, Kew, UK. Royal Botanic Gardens, Kew 2, 269–275.

Rao, R.R. and Razi, B.A. (1981) *A Synoptic Flora of Mysore District*. Today & Tomorrow Publishers, New Delhi, India.

Reddy, S.C., Reddy, B.G., Vatsavaya, K.N. and Raju, S. (undated) *Invasive Alien Flora of India*. An Internet publication for open review (http://www.gisinetwork.org/IndiaInvasivePlants/index.html).

Reinhard, F., Herle, M., Bastiansen, F. and Streit, B. (2003) *Economic Impact of the Spread of Alien Species in Germany*. Springer, Berlin, 229 pp.

Rotherham, I.D. (1990) Factors facilitating invasion by *Rhododendron ponticum*. In: *The Biology and Control of Invasive Plants*. British Ecological Society, London, pp. 86–94.

Singh, N.P. (1979) (publ. 1981) *Cassia sericea* Sw., a new record for India. *Bulletin of Botanical Survey of India* 21, 203–205.

Tadulingam, C., Venkatanarayana, G., Rajasekhara Mudaliar, C. and Sakharam Rao, J. (1955) *A Handbook of Some South Indian Weeds*. Government of Madras, India.

Vitousek, P.M., D'Antonio, C.M., Loope, L.L., Rejmanek, M. and Westbrooks, R. (1997) Introduced species: a significant component of human-caused global change. *New Zealand Journal of Ecology* 21, 1–16.

Yoganarasimhan, S.N. (2000) *Medicinal Plants of India*, vol II, Tamil Nadu. Interline Publishing Pvt Ltd, Bangalore, India, pp. 380–381.

# 13 Invasive Alien Plants in Tropical Forests of the South-eastern Ghats, India: Ecology and Management

**N. Parthasarathy, L. Arul Pragasan and C. Muthumperumal**

*Department of Ecology and Environmental Sciences, Pondicherry University, Puducherry, India*

## Introduction

Biological invasion is reported to be the second leading cause of species extinction after habitat loss, and is one of the major causes of biodiversity depletion (Jose *et al*., 2009). Invasion science has attracted attention from ecologists because of its significant ecological impacts and economic costs worldwide (Liu *et al*., 2005). Human activities have important influences on the dispersal of exotic plants (Mack and D'Antonio, 1998; Sax, 2002; Liu *et al*., 2005). Invasive species affect both biological and cultural systems. Study of the ecological and economic effects of invasive species has paralleled their progressively pervasive influence worldwide, yet their cultural impacts remain largely unexamined and therefore unrecognized. Invasive alien species pose a serious threat to the biodiversity of native species (Singh *et al*., 2006), particularly in the tropics, and the Indian subcontinent is no exception to this. Biological invasions have emerged as a major ecological and environmental policy issue, displacing native species in both terrestrial and marine habits at unprecedented rates (Mack *et al*., 2000; UNEP, 2001; Simberloff *et al*., 2005).

The objective of the research described in this chapter was to determine the diversity and ecology of invasive alien plants in the forests of the southern Eastern Ghats and to suggest control measures for invasive spread in order to save the native biodiversity.

## Study Area

The Eastern Ghats constitute an important biogeographic region as one of nine floristic zones in India, forming a broken chain of mountains covering the states of Orissa, Andhra Pradesh, Tamil Nadu and Karnataka, lying between the rivers Mahanadhi and Vaigai. The present study was carried out in the tropical forests of six major hill complexes, namely Bodamalais (BM), Chitteris (CH), Kalrayan (KA), Kolli hills (KO), Pachamalais (PM) and Shervarayans (SH) (10°93′–12°18′N), all falling within Tamil Nadu State (Fig. 13.1). The hills of the southern Eastern Ghats are composed mainly of charnockite, with gneissus and various metamorphic rocks. The climate data for Salem, the nearest station to the study sites, for the 20 years 1988 to 2007 obtained from the meteorological department, Government of India, show that the mean annual temperature is 28.3°C and mean annual rainfall is 1058 mm. The bulk of the rainfall is from August to October, with the mean annual number of days with rainfall for that period being 61. The vegetation of the southern Eastern Ghats has been affected by various human disturbances, including the introduction of invasive alien species. Photographs of the study sites and associated alien species are provided in Figs 13.2 and 13. 3.

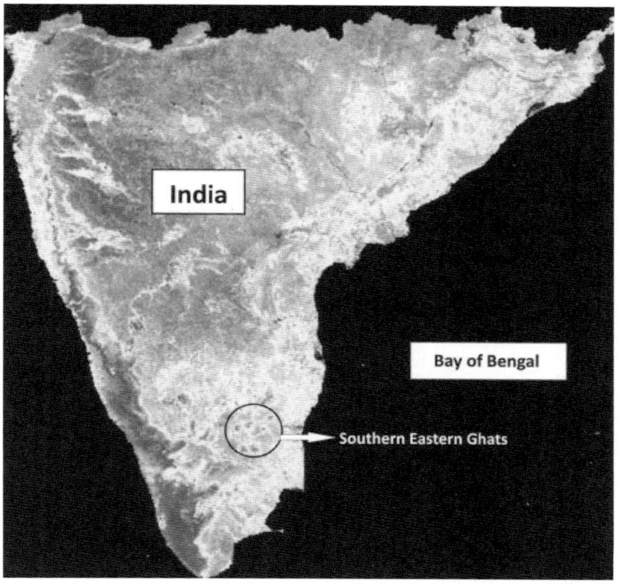


**Fig. 13.1.** Map showing the hill complexes of the southern Eastern Ghats, India.

**Fig. 13.2.** Views of intact forests (A, B) in the Eastern Ghats and others largely infested with invasive alien plants (C, D). A, evergreen forest in the Kolli Hills; B, semi-evergreen forest in the Chitteri Hills; C, invasion by *Lantana camara*, an aggressive alien species, in semi-evergreen forest in the Chitteri Hills; D, invasion by *Chromolaena odorata* in mixed deciduous forest in the Shervarayan Hills.

## Data Collection

The entire stretch of the southern Eastern Ghats was divided into grids of 6.25 × 6.25 km. A total of 120 grids was obtained from the six hill complexes and within each grid a transect of 0.5 ha (5 m × 1 km) was established. All native and alien plant species – trees ≥30 cm girth at breast height (gbh) and lianas ≥5 cm gbh were enumerated for each whole transect. Shrubs and herbs were enumerated from the beginning and end of each transect, in 5 × 5 m and 1 × 1 m quadrats, respectively. Site disturbance scores were obtained by assessing various disturbance attributes (on a scale of 1–5) for evaluation of conservation significance (Appendix 13.1).

## Results and Discussion

### Diversity and abundance of alien plants

Total plant diversity in the six sites was 648 species, comprising 601 natives and 47 aliens. By life-form, trees were most commonly recorded, with 272 species, followed by herbs (216), lianas (143) and shrubs (17). Of a total of 47 alien species, herb diversity was greatest (26 species) followed by 12 species of trees, 6 lianas and 3 shrubs (Appendix 13.2). Total and alien species diversity and abundance of woody species (trees and lianas) varied across the six study sites (Table 13.1). Spearman's rank correlation showed a positive correlation between total tree and liana diversity ($r^2$ =

**Fig. 13.3.** Selected invasive alien tree, liana, shrub and herbaceous species of the southern Eastern Ghats. A, *Manihot glaziovii* Muell. Arg. (*Euphorbiaceae*); B, *Coffea arabica* L. (*Rubiaceae*); C, *Lantana camara* L. (*Verbenaceae*); D, *Rubus niveus* Thunb. (*Rosaceae*); E, *Opuntia stricta* (Haw.) Haw var. *dillenii* (ker Gawler) L. Benson (*Cactaceae*); F, *Ageratum conyzoides* L. (*Asteraceae*).

**Table 13.1.** Site details of the six hill complexes studied.

| | BM | CH | KA | KO | PM | SH | Over 6 sites |
|---|---|---|---|---|---|---|---|
| Latitude (N) | 11°30.0′–11°39.0′ | 11°43.5′–12°06.0′ | 11°38.0′–12°04.0′ | 11°11.0′–11°28.0′ | 11°08.5′–11°28.5′ | 11°56.5′–11°42.5′ | 11°08.5′–12°06.0′ |
| Longitude (E) | 78°08.0′–78°21.0′ | 78°18.5′–78°40.5′ | 78°26.5′–78°53.5′ | 78°17.0′–78°29.0′ | 78°29.0′–78°48.5′ | 78°07.5′–78°22.5′ | 78°07.5′–78°53.5′ |
| Altitudinal range (m) | 300–1,155 | 400–1,211 | 220–1,250 | 200–1,352 | 200–1,021 | 354–1,649 | 200–1,649 |
| Area (km$^2$) | 156 | 742 | 1,055 | 508 | 586 | 430 | 3,477 |
| Hectares inventoried | 5 | 10 | 17 | 9 | 12 | 7 | 60 |
| Woody species diversity | | | | | | | |
| trees | 64 | 143 | 169 | 157 | 131 | 165 | 272 |
| lianas | 43 | 75 | 85 | 86 | 77 | 78 | 143 |
| Woody species abundance | | | | | | | |
| trees | 1,449 | 5,022 | 8,951 | 3,824 | 5,388 | 2,778 | 27,412 |
| lianas | 2,947 | 7,073 | 5,828 | 5,027 | 6,706 | 4,452 | 32,032 |
| Woody alien species diversity | | | | | | | |
| trees | 1 | 3 | 1 | 3 | 1 | 8 | 12 |
| lianas | 1 | 3 | 1 | 2 | 1 | 3 | 6 |
| Woody alien species abundance | | | | | | | |
| trees | 8 | 16 | 21 | 22 | 41 | 49 | 157 |
| lianas | 616 | 1,842 | 944 | 1,588 | 634 | 2,090 | 7,714 |
| Forest disturbance score[a] | 53 | 43 | 41 | 37 | 49 | 33 | 256 |

BM , Bodamalai; CH, Chitteri; KA, Kalrayan; KO, Kolli Hills; PM, Pachaimalai; SH, Shervarayan Hills.

[a] The criteria on which forest disturbance scores were obtained are detailed in Appendix 13.2.

0.958), between total tree and liana abundance ($r^2$ = 0.672) and between alien tree and alien liana diversity ($r^2$ = 0.812) across the six study sites, but not between alien tree and liana abundance ($r^2$ = 0.315). The most abundant alien woody tree species included *Acacia farnesiana* and *Tamarindus indica*, and *Lantana camara* and *Rubus niveus* among lianas (Table 13.2). Most invasive species spread rapidly and were fast growing, utilized available resources and possessed greater ecological plasticity than native species (Singh *et al.*, 2006). This trend was exemplified by the trees *Eriobotrya japonica* and *Annona squamosa*, the liana *Lantana camara*, the shrub *Opuntia stricta* and the herbs *Chromolaena odarata* and *Ageratum conyzoides*. The status of the major alien species and their characteristic features are summarized in Box 13.1.

## Dispersal modes of alien plants

In general, the diaspores of the alien invasive plants in the study area are dispersed by the autochorous mode (53%, 25 species), followed by zoochory (38%) and anemochory

**Table 13.2.** Abundance of invasive woody species from 120 transects in the six hill complexes.

| | | | | | | | Abundance | | |
|---|---|---|---|---|---|---|---|---|---|
| Invasive species | BM | CH | KA | KO | PM | SH | Total | (%) | Frequency[a] |
| Trees | | | | | | | | | |
| *Acacia farnesiana* (L.) Willd. | 0 | 0 | 0 | 0 | 41 | 5 | 46 | 29.30 | 7 |
| *Annona squamosa* L. | 0 | 0 | 0 | 0 | 0 | 2 | 2 | 1.27 | 2 |
| *Cassia siamea* Lam. | 0 | 1 | 0 | 0 | 0 | 0 | 1 | 0.64 | 1 |
| *Coffea arabica* L. | 0 | 0 | 0 | 0 | 0 | 2 | 2 | 1.27 | 2 |
| *Delonix regia* (Boj. ex Hook.) Rafin. | 0 | 2 | 0 | 0 | 0 | 0 | 2 | 1.27 | 1 |
| *Eriobotrya japonica* (Thunb.) Lindl. | 0 | 0 | 0 | 0 | 0 | 2 | 2 | 1.27 | 1 |
| *Grevillea robusta* A. Cunn. | 0 | 0 | 0 | 1 | 0 | 0 | 1 | 0.64 | 1 |
| *Maesopsis eminii* Engler | 0 | 0 | 0 | 0 | 0 | 1 | 1 | 0.64 | 1 |
| *Manihot glaziovii* Muell. Arg. | 0 | 0 | 0 | 0 | 0 | 2 | 2 | 1.27 | 1 |
| *Psidium guajava* L. | 0 | 0 | 0 | 1 | 0 | 0 | 1 | 0.64 | 1 |
| *Spathodea campanulata* Beauv. | 0 | 0 | 0 | 0 | 0 | 1 | 1 | 0.64 | 1 |
| *Tamarindus indica* L. | 8 | 13 | 21 | 20 | 0 | 34 | 96 | 61.15 | 29 |
| Total | 8 | 16 | 21 | 22 | 41 | 49 | 157 | 100 | |
| Lianas | | | | | | | | | |
| *Basella alba* L. | 0 | 1 | 0 | 0 | 0 | 0 | 1 | 0.01 | 1 |
| *Ipomoea asarifolia* (Desr.) Roemer & Shultes | 0 | 1 | 0 | 0 | 0 | 0 | 1 | 0.01 | 1 |
| *Rubus ellipticus* Smith | 0 | 0 | 0 | 1 | 0 | 0 | 1 | 0.01 | 1 |
| *Solanum seaforthianum* Andr. Bot. | 0 | 0 | 0 | 0 | 0 | 1 | 1 | 0.01 | 1 |
| *Rubus niveus* Thunb. | 0 | 0 | 0 | 0 | 0 | 7 | 7 | 0.09 | 4 |
| *Lantana camara* L. | 616 | 1840 | 944 | 1587 | 634 | 2082 | 7703 | 99.86 | 89 |
| Total | 616 | 1842 | 944 | 1588 | 634 | 2090 | 7714 | 100 | |

BM, Bodamalai; CH, Chitteri; KA, Kalrayan; KO, Kolli Hills; PM, Pachaimalai; SH, Shervarayan Hills.
[a] Number of transects where species recorded.

(9%). The woody species (trees and lianas) with largely fleshy fruits (berry or drupe) are largely dispersed by the zoochorous mode, followed by the shrubs. The autochorous mode of dispersal was prevalent in herbs (Fig. 13.4), which mostly produced capsules with innumerable seeds to increase their chances of dispersal.

According to Buckley *et al.* (2006) some of the most damaging invasive plants are dispersed by frugivores, and this is an area of emerging importance in weed management, which in our study sites was exemplified by *Lantana camara*, *Maesopsis eminii*, *Opuntia stricta*, *Physalis minima* and *Rubus niveus*. Fruit traits influence frugivore choice: fruit size, the presence of an inedible peel, defensive chemistry, crop size and phenology may all be useful traits for consideration in screening and eradication programmes (Buckley *et al.*, 2006). Diaspore dispersal by animals and wind is common in alien species; the seeds of *Lantana*, for example, are dispersed by frugivorous birds and several wild and domestic animals. Furthermore, many species have ecological amplitude to grow in a wide range of forest types, from tropical thorn to tropical wet evergreen, and *Lantana* is an aggressive colonizer, particularly at forest edges and

**Box. 13.1.** Status of major alien invasive plants of the southern Eastern Ghats and their key features.

*Lantana camara* L. (*Verbenaceae*)

*Lantana camara*, the thorny straggler, had the lion's share of 24% of total abundance of lianas in the forests of the southern Eastern Ghats. One of the world's 100 worst alien invasive species, it has been reported in many forest areas worldwide as the top-ranking and highest-impacting invasive species. Among the six major hill complexes in our study, invasion by *Lantana* was most severe in the Shervarayan Hills, followed by the Chitteri Hills.

*Rubus niveus* Thunb. (*Rosaceae*)

Another thorny straggler, this pecies has colonized forest edges in tropical evergreen forests of the Kolli and Shervarayan Hills > 1000 m elevation. Readily recognized by its characteristic leaves with white lower surface, the plant yields fleshy, edible achenes that are dispersed by vertebrates.

*Acacia farnesiana* (L.) Willd. (*Mimosaceae*)

The second most abundant alien thorny tree species in the southern Eastern Ghats, it is distributed among the thorn forests of the Pachamalai and Shervarayan hills at low elevation.

*Maesopsis eminii* Engler (*Rhamnaceae*)

Ecologically, this species was introduced as a shade tree in plantations, but its fruits are edible for vertebrate fauna and are thereby dispersed; occurs in high-elevation forests of the Shervarayan Hills.

*Ageratum conyzoides* L. (*Asteraceae*)

This common US weed, now distributed in all warm countries, prefers wet areas and has a wide distribution, from thorn forest to evergreen forest, particularly at the edges and forest trails of the southern Eastern Ghats.

*Chromolaena odarata* (L) King & Roanson (*Asteraceae*)

An aggressive colonizer in forest environments, it occupies large areas of disturbed forests and abandoned plantation sites. It mostly occurs in semi-evergreen to evergreen forests and in gap areas, with the appearance of snow cover when in full bloom at our study sites.

*Parthenium hysterophorus* L. (*Asteraceae*)

This North American alien species, now pantropical in distribution, was recorded in all six study sites. Notably, this species exists as ground cover in drier areas but grows to nearly 1 m under favourable ecological conditions.

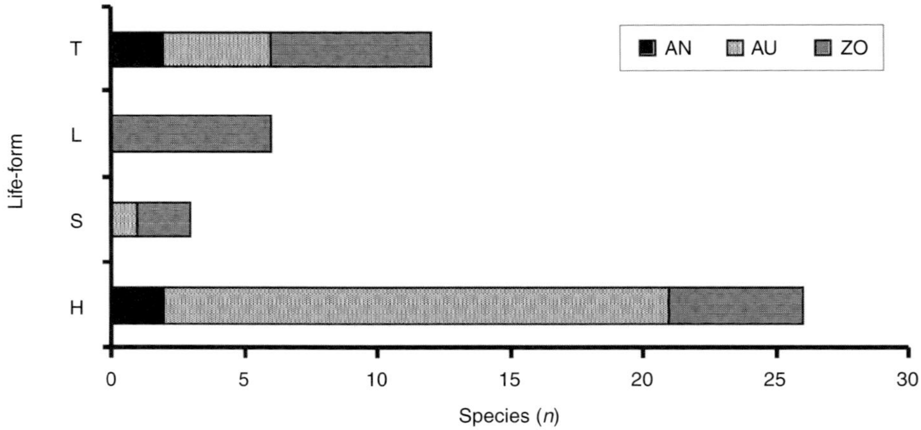


**Fig. 13.4.** Dispersal modes of the principal life-forms of invasive plants in the tropical forests of the southern Eastern Ghats, India. T, tree; L, liana; S, shrub; H, herb; AN, anemochorous; AU, autochorous; ZO, zoochorous.

disturbed and denuded areas. Combined modes of reproduction, both vegetative and sexual, promote successful invasion in alien environments, in addition to allelopathy (Singh *et al.*, 2006).

### Site disturbance and alien species

Among the six study sites, BM recorded the maximum disturbance score followed by PM, CH, KA, KO and SH (Appendix 13.1). The diversity of total and alien tree species was negatively correlated with site disturbance score across the six study sites (Fig. 13.5). This reveals that the diversity of tree species is mostly affected by anthropogenic disturbance. Also, total liana diversity and alien liana abundance were negatively correlated with site disturbance across the six sites (Fig. 13.5). The reason for the latter trend can be attributed to the fact that, irrespective of native or alien species, the level of disturbance by way of various anthropogenic impacts seems to be similar for both alien and native species, i.e. with no distinction between them resources are extracted, which ultimately leads to high disturbance scores.

### Economic importance and management of invasive plants

Of the 47 alien species, 39 (83%) are economically important and 15 (32%) are ecologically important (Appendix 13.2). Those alien species economically important can also be used with the aim of improving human welfare, well before any weed eradication programme is planned as part of management strategy; alternatively, they could be used as a resource on a sustainable basis when the species pose no problem in the locality concerned. For notoriously invasive species such as *Lantana*, this can be achieved by allowing local people to utilize the weed as a resource (e.g. stems for basket-weaving or as an alternative to rattan for furniture-making; as a source of fuel wood; or the aqueous leaf extract as a potential insect-repellent in agro-ecosystems, following field trials). Indigenous societies worldwide derive their cultural identity and physical well-being through intimate relationships with native biota (Posey, 1999; Grim, 2001; Steep *et al.*, 2002; Pfeiffer and Voeks, 2008).

Biological invasions are a significant economic issue, causing enormous losses in

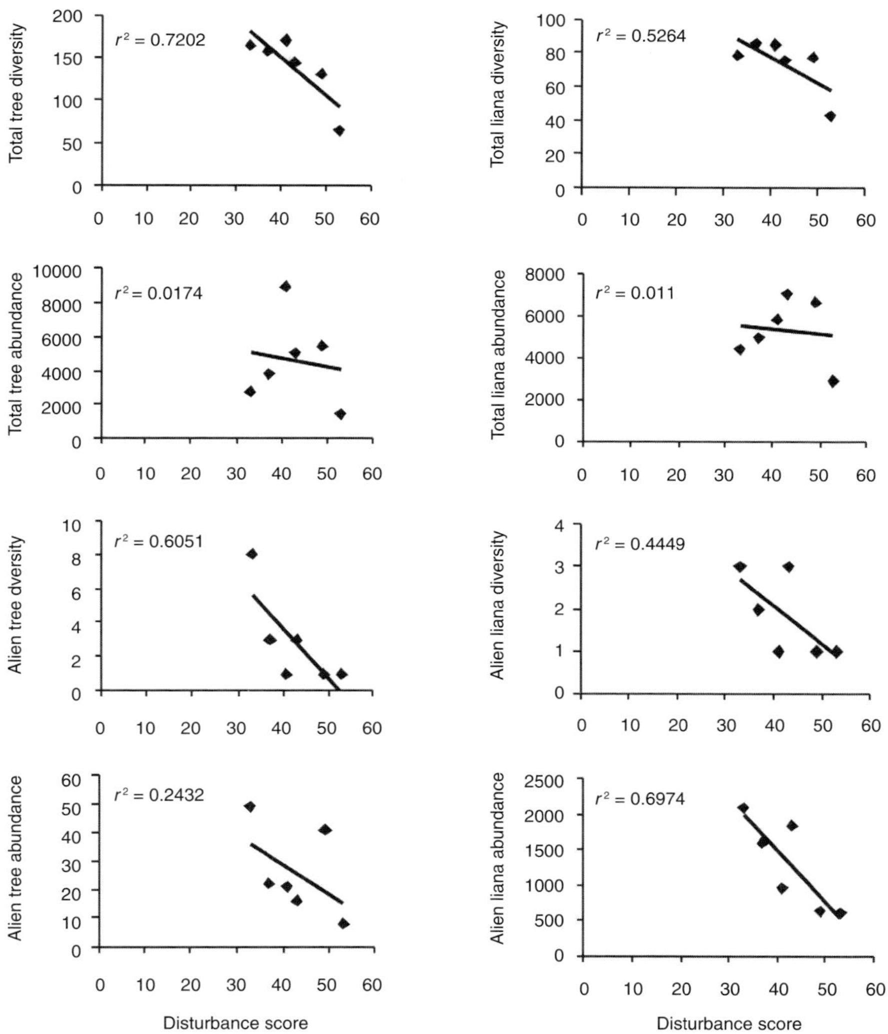


**Fig. 13.5.** Relationship between diversity and abundance of woody species, by disturbance score, across the six hill complexes of the southern Eastern Ghats.

the agricultural, conservation, fisheries, forestry, transportation and tourism sectors worldwide (Pimentel, 2002). Better-informed management of invasive plants is necessary to maintain forest ecosystems and biodiversity. Prevention and early eradication are often seen as cost-effective means of managing invasive species (Lodge *et al.*, 2006); eradication measures are most effective when populations are few and small (Groves and Panetta, 2002; Denslow, 2007). Furthermore, in most eradication programmes for invasive alien plants a management plan should be carried out well before the flowering and fruiting stage, which would save considerable effort in terms of time and resources, including manpower requirements, and spread would be controlled effectively before the propagules reached vast areas. As a general caution, all methods of weed eradication should be carried out without causing any major damage to native species and ecosystems; this is particularly crucial when the large-scale slash-and-burn practice of terrestrial weed eradication is attempted. The mechanical method of eradication of weedy invasive species is certainly a challenging task, especially when the extent of invasion is on a wide scale, calling for heavy investment in resource allocation, including manpower.

We also need to promote education and public awareness programmes to engage local communities and appropriate sector groups in the eradication of alien species, which obviously necessitates the provision of adequate funds and support, particularly for the control of major and aggressive alien species (Genovesi and Shine, 2003; Genovesi, 2005). As most forests of the Eastern Ghats are already badly affected by the invasion of *Lantana* sp., the need for its effective control must be emphasized. Hence, careful site-specific management strategies have to be developed, always bearing in mind site biodiversity and peoples' cultural diversity and not simply the scale of bioinvasion.

## References

Buckley, Y.M. *et al.* (2006) Management of plant invasions mediated by frugivore interactions. *Journal of Applied Ecology* 43, 848–857.

Denslow, J.S. (2007) Managing dominance of invasive plants in wildlands. *Current Science* 93, 1579–1586.

Genovesi, P. (2005) Eradications of invasive alien species in Europe: a review. *Biological Invasions* 7, 127–133.

Genovesi, P. and Shine, C. (2003) *European Strategy on Invasive Alien Species.* Council of Europe, Strasbourg, France, 50 pp.

Grim, J.A. (ed.) (2001) *Indigenous Traditions and Ecology: the Inter-being of Cosmology and Community.* Harvard University Press, Cambridge, Massachusetts.

Groves, R.H. and Panetta, F.D. (2002) Some general principles for weed eradication programs. *Proceedings of the XIII Australian Weed Conference*, pp. 307–310.

Jose, S., Kohli, R.K., Singh, H.P., Batish, D.R. and Pieterson, E.C. (2009) Invasive plants: a threat to the integrity and sustainability of forest ecosystems. In: Kohli, R.K., Jose, S., Singh, H.P. and Batish, D.R. (eds) *Invasive Plants and Forest Ecosystems*. CRC Press, Boca Raton, Florida, pp. 3–10.

Khare, C.P. (ed.) (2007) *Indian Medicinal Plants: an Illustrated Dictionary.* Springer, New York, 836 pp.

Liu, J., Liang, S., Liu, F., Wang, R. and Dong, M. (2005) Invasive alien plant species in China: regional distribution patterns. *Diversity and Distributions* 11, 341–347.

Lodge, D.J. *et al.* (2006) Biological invasions: recommendations for US policy and management. *Ecological Applications* 16, 2035–2054.

Mack, M.C. and D'Antonio, C.M. (1998) Impacts of biological invasions on disturbance regimes. *Trends in Ecology and Evolution* 13, 195–198.

Mack, R.N., Simberloff, D., Lonsdale, W.M., Evans, H., Clout, M. and Bazzaz, F.A. (2000) Biotic invasions: causes, epidemiology, global consequences, and control. *Ecological Applications* 10, 689–710.

Pfeiffer, J.M. and Voeks, R.A. (2008) Biological invasions and bio-cultural diversity: linking ecological and cultural system. *Environmental Conservation* 4, 281–293.

Pimentel, D. (ed.) (2002) *Biological Invasions: Economic and Environmental Costs of Alien Plant, Animal and Microbe Species.* CRC Press, Boca Raton, Florida.

Posey, D. (ed.) (1999) *Cultural and Spiritual Values of Biodiversity.* Zed Books, London.

Sax, D.F. (2002) Native and naturalized plant diversity are positively correlated in scrub communities of California and Chile. *Diversity and Distributions* 8, 193–210.

Simberloff, D., Parker, I.M. and Windle, P.N. (2005) Introduced species policy, management, and future research needs. *Frontiers in Ecology and the Environmental Conservation* 3, 12–20.

Singh, J.S., Singh, S.P. and Gupta, S.R. (2006) *Ecology Environment and Resource Conservation.* Anamaya Publishers, New Delhi, India, 668 pp.

Steep, J.R., Wyndham, F.S. and Zarger, R.K. (eds) (2002) *Ethnobiology and Biocultural Diversity.* International Society of Ethnobiology, Athens, Georgia.

UNEP (2001) Governments seek strategies for battling invasive alien species. Available at http://www.unep.org/Documents.multilingual/Default.asp?DocumentID=193&ArticleID=2787 (accessed 18 August 2009).

**Appendix 13.1.** Site disturbance scores (ranking: 1, rare; 2, low; 3, medium; 4, high; 5, very high) for the six hill complexes in the southern Eastern Ghats.

| | Site | | | | | |
|---|---|---|---|---|---|---|
| Disturbance | BM | CH | KA | KO | PM | SH |
| 1. Resource extraction | | | | | | |
| (i) firewood | 5 | 3 | 3 | 2 | 3 | 2 |
| (ii) timber | 4 | 4 | 3 | 3 | 4 | 3 |
| (iii) medicinal plants | 4 | 3 | 3 | 5 | 2 | 3 |
| (iv) edible fruits | 3 | 4 | 3 | 4 | 4 | 2 |
| (v) soil removal | 4 | 4 | 5 | 1 | 3 | 1 |
| (vi) herding cattle/goats for grazing | 5 | 3 | 4 | 2 | 4 | 2 |
| 2. Forest land use change | | | | | | |
| (i) hill cultivation | 3 | 4 | 4 | 3 | 5 | 3 |
| (ii) factories | 4 | 2 | 2 | 2 | 4 | 3 |
| (iii) quarrying and mining | 4 | 3 | 2 | 3 | 3 | 3 |
| (iv) roads and buildings | 4 | 3 | 2 | 3 | 3 | 4 |
| (v) enchroachment of forest land | 5 | 4 | 4 | 3 | 5 | 3 |
| 3. Tribal settlements & their dependence on forests | 3 | 3 | 3 | 4 | 4 | 3 |
| 4. Degraded forest area | 5 | 3 | 3 | 2 | 5 | 1 |
| Total score | 53 | 43 | 41 | 37 | 49 | 33 |

BM, Bodamalai; CH, Chitteri; KA, Kalrayan; KO, Kolli Hills; PM, Pachaimalai; SH, Shervarayan Hills.

**Appendix 13.2.** List of all invasive plant species recorded over a total of 120 transects, along with botanical details and indicators of major ecological, economic and medicinal[a] importance.

| Species (family) | Habit | Plant type | Forest type | Fruit type | Dispersal Mode | Ecological importance | Economic importance |
|---|---|---|---|---|---|---|---|
| *Annona squamosa* L. (*Annonaceae*) | T | E | MD, SE, E | B | ZO | Fruit | Edible fruit; seeds medicinal |
| *Asclepias curassavica* L. (*Asclepiadaceae*) | H | NA | E, SE, MD | F | AN | | Whole plant medicinal |
| *Acanthospermum hispidum* DC. (*Asteraceae*) | H | NA | SE, MD, D, T | A | AU | | |
| *Ageratum conyzoides* L. (*Asteraceae*) | H | NA | E, SE, MD, D, T | A | AU | | Medicinal |
| *Ageratum honstonianum* Mill. (*Asteraceae*) | H | NA | SE, MD, D, T | A | AU | | |
| *Bidens pilosa* L. var. *minor* (Blume) Sherff (*Asteraceae*) | H | NA | E, SE, MD, D, T | A | AU | | Medicinal |
| *Chromolaena odarata* (L) King & Roanson (*Asteraceae*) | H | NA | E, SE, MD, D, T | A | AU | | |
| *Parthenium hysterophorus* L. (*Asteraceae*) | H | NA | E, SE, MD, D, T | A | AU | | Medicinal |
| *Synedrella nodiflora* (L.) Gaertn. (*Asteraceae*) | H | NA | D, T | A | AU | | |
| *Tridax procumbens* L. (*Asteraceae*) | H | NA | D, T | A | AN | | Medicinal |
| *Basella alba* L. (*Basellaceae*) | L | E | MD | B | ZO | Fruit | Leafy vegetable |
| *Spathodea campanulata* Beauv. (*Bignoniaceae*) | T | E | E | C | AN | | Ornamental |
| *Opuntia stricta* (Haw.) Haw. var. *dillenii* (Ker Gawler) L. Benson (*Cactaceae*) | S | E | D, T | B | ZO | Fruit | Whole plant medicinal |
| *Cassia siamea* Lam. (*Caesalpiniaceae*) | T | E | SE, E | P | AU | | Fuel wood |
| *Delonix regia* (Boj. ex Hook.) Rafin. (*Caesalpiniaceae*) | T | BD | D, MD | P | AU | | Ornamental |
| *Tamarindus indica* L. (*Caesalpiniaceae*) | T | BD | D, DM | P | ZO | Fruit | Timber; medicinal; edible |
| *Cleome viscosa* L. (*Cleomaceae*) | H | NA | D, T | C | AU | | Medicinal |
| *Ipomoea asarifolia* (Desr.) Roemer & Shultes (*Convolvulaceae*) | L | BD | T | C | AU | | |
| *Manihot glaziovii* Muell.Arg. (*Euphorbiaceae*) | T | E | E | C | AU | | Rubber |
| *Croton bonplandianum* Baill. (*Euphorbiaceae*) | H | NA | D, T | C | ZO | | Medicinal |
| *Phyllanthus amarus* Schum. & Thonn. (*Euphorbiaceae*) | H | NA | D, T | C | AU | | Medicinal |
| *Jatropha gossypifolia* L. (*Euphorbiaceae*) | S | E | D, T | C | AU | | Whole plant, medicinal |
| *Pogostemon purpurascens* Dalzell (*Lamiaceae*) | H | NA | E, SE, MD | C | AU | | |
| *Salvia coccinea* Buc'hoz ex Etl. (*Lamiaceae*) | H | NA | E | C | AU | | Ornamental |
| *Abutilon crispum* (L.) Medik. (*Malvaceae*) | H | NA | SE, MD, D | M | AU | | |
| *Sida acuta* Burm. (*Malvaceae*) | H | NA | D, T | M | AU | | Medicinal |
| *Sida cordifolia* L. (*Malvaceae*) | H | NA | MD, D, T | M | AU | | Medicinal |
| *Acacia farnesiana* (L.) Willd. (*Mimosaceae*) | T | D | T | P | AU | | Fuel wood |
| *Mimosa pudica* L. (*Mimosaceae*) | H | NA | MD, D, T | P | AU | | Medicinal |

| | | | | | | | |
|---|---|---|---|---|---|---|---|
| *Psidium guajava* L. (*Myrtaceae*) | T | E | T, D, MD, SE | B | ZO | Fruit | Edible fruit |
| *Boerhaavia diffusa* L. (*Nyctaginaceae*) | H | NA | MD, D, T | A | ZO | | Medicinal |
| *Boerhaavia erecta* L. (*Nyctaginaceae*) | H | NA | MD, D, T | A | ZO | | Medicinal |
| *Crotalaria verrucosa* L. (*Papilionaceae*) | H | NA | MD, D, T | P | AU | | Medicinal |
| *Passiflora foetida* L. (*Passifloraceae*) | H | NA | D, T | B | ZO | Fruit | Medicinal |
| *Grevillea robusta* A.Cunn. (*Proteaceae*) | T | E | E | F | AN | | Timber; shade tree |
| *Maesopsis eminii* Engler (*Rhamnaceae*) | T | E | E | D | ZO | Fruit | Shade tree in plantations |
| *Eriobotrya japonica* (Thunb.) Lindl.(*Rosaceae*) | T | E | E | B | ZO | Fruit | Edible fruit |
| *Rubus ellipticus* Smith (*Rosaceae*) | L | E | EG | D | ZO | Fruit | Edible fruit |
| *Rubus niveus* Thunb. (*Rosaceae*) | L | E | EG, SE | D | ZO | Fruit | Edible fruit |
| *Coffea arabica* L. (*Rubiaceae*) | T | E | E | B | ZO | Fruit | Coffee |
| *Cardiospermum halicacabum* L. var. *luridum* (*Sapindaceae*) | H | NA | D, T | C | AU | | Medicinal |
| *Solanum seaforthianum* Andr. Bot. (*Solanaceae*) | L | E | SE | B | ZO | Fruit | |
| *Physalis minima* L. (*Solanaceae*) | H | NA | D, T | B | ZO | Fruit | Medicinal |
| *Solanum torvum* Sw. (*Solanaceae*) | S | E | D, T | B | ZO | Fruit | Medicinal |
| *Waltheria indica* L. (*Sterculiaceae*) | H | NA | D, T | C | AU | | Medicinal |
| *Lantana camara* L. (*Verbenaceae*) | L | BD | EG, SE, MD, D, T | D | ZO | Fruit | Fuel wood; furniture, Basket-making |
| *Tribulus terresteris* L. (*Zygophyllaceae*) | H | NA | D, T | S | AU | | Medicinal |

Habit: T, tree; L, liana; S, shrub; H, herb. Plant type: E, evergreen; NA, not applicable; BD, brewi-deciduous; D, deciduous. Forest type: MD, mixed deciduous; SE, semi-evergreen; E, evergreen; D, deciduous; T, thorn. Fruit type: B, berry; F, follicle; A, achene; C, capsule; P, pod; M, mericarp; A, anthocarp; D, drupe; S, schizocarp. Dispersal mode: ZO, zoochorous; AN, anemochorous; AU, autochorous.

[a] Data on medicinal uses obtained mainly from Khare (2007), but also from our own records.

# 14 Status of Alien Plant Invasions in the North-eastern Region of India

**Uma Shankar[1], A.S. Yadav[2], J.P.N. Rai[3] and R.S. Tripathi[4]**

*[1]Department of Botany, North-Eastern Hill University, Shillong, India; [2]Department of Botany, R.R. College, Alwar, Rajasthan, India; [3]Department of Environmental Sciences, G.B. Pant University of Agriculture and Technology, Pantnagar, Uttarakhand, India; [4]National Botanical Research Institute, Lucknow, Uttar Pradesh, India*

## Introduction

Invasion by alien plants in natural forests is growing rapidly and is an accompaniment to the depletion of biodiversity, one of the most pressing issues in biology today (Tripathi, 2009). Although the major cause of invasion may vary depending on the regional setting, almost all tropical forests are seriously threatened by anthropogenic activities leading to irreversible loss of biodiversity. This problem is particularly acute in the north-eastern region of India, which is known for its diverse and extensive lush forest cover and the fact that it comprises over 50% of the total number of vascular plants in India. These forests are highly exposed to clear-cutting, grazing, burning and harvesting of various timber and non-timber forest products (Uma Shankar et al., 1998, 2001; Uma Shankar, 2001a, b). In recent years habitat fragmentation has extended to subtropical and temperate forests of the region. The invasion of alien plants is expected to deplete native biodiversity at all levels of organization. Accompanied by forest clearing and habitat fragmentation, invasion may cause the extinction of unique populations or species. This chapter reviews the diversity of invasive alien species in the north-eastern region of India, their dominance patterns and the mechanisms facilitating that dominance.

## Vegetation in the North-eastern Region of India

The north-eastern region of India, encompassing eight states (Arunachal Pradesh, Assam, Manipur, Meghalaya, Mizoram, Nagaland, Sikkim and Tripura) and three northern districts (Darjeeling, Jalpaiguri and Coochbehar) in the State of West Bengal is bestowed with a rich diversity of fauna, flora and microbes (Fig. 14.1).

This region contributes a substantial geographical area to the bounds of the two most celebrated hot spots of biodiversity in the world, i.e. the Himalaya and the Indo-Burma (Conservation International, 2004). The Himalaya is estimated to harbour 10,000 plant species, of which 3160 are endemic. The largest family of the flowering plants in this hot spot is the *Orchidaceae*, with 750 species and five plant families endemic to the region (*Tetracentraceae*, *Hamamelidaceae*, *Circaesteraceae*, *Butomaceae* and *Stachyuraceae*). The eastern Himalaya is also a centre of diversity for several widely

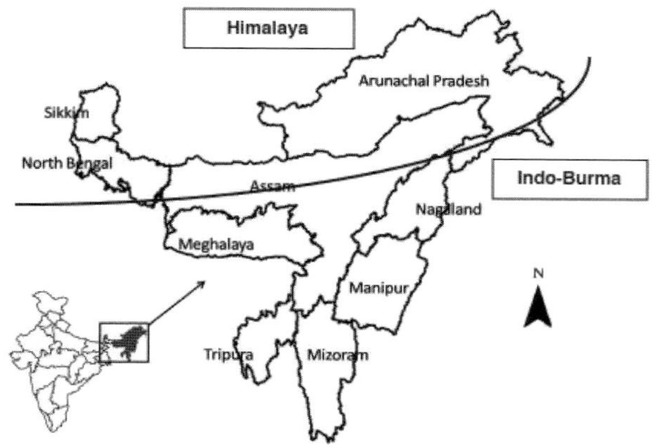


**Fig. 14.1.** Map of the north-eastern region of India, encompassing eight states and three districts of West Bengal. The arc shows the approximate delineation of the land mass contributing to two global hot spots of biodiversity, the Himalaya and Indo-Burma.

distributed plant taxa, such as *Rhododendron*, *Primula* and *Pedicularis*. The States of Arunachal Pradesh, Sikkim and Assam (part of) north of the Brahmaputra River contribute to this hot spot (Fig. 14.1).

The hot spot Indo-Burma boasts even more plant species (estimated at 13,500) and more endemics (7000). This hot spot has localized centres of endemism and is particularly rich in orchids and gingers. The States of Assam (south of the Brahmaputra River), Manipur, Meghalaya, Mizoram, Nagaland and Tripura contribute to this hot spot (Fig. 14.1). Indo-Burma is one of the first sites where mankind developed agriculture, and has a long history of using burning to clear land for agriculture and other needs. The requirement for agricultural products has only increased in recent years, with the expansion of both human and cattle populations and markets. This has contributed to widespread forest destruction, with tree plantations (teak, rubber and oil palm) replacing large areas of lowland forest and the introduction of cash crops (tea, coffee and vegetables) threatening montane and hill forests. Other threats to forests include logging, mining for mineral resources, firewood collection and charcoal production.

Forest cover statistics reveals that 66.1% of the total geographical area of all eight states of the north-eastern region is forest (Table 14.1). Of the total forest, however, only 10.4% is very dense, 46.4% is moderately dense and 43.2% is open. Arunachal Pradesh has nearly 39.1% but Sikkim only 1.88% of the total forest area. These statistics reveal that a large percentage of the geographical area (93.2%) of the forest is vulnerable to invasion by alien plant species, as the proportion of dense forest has reduced drastically due to disturbance of forest canopies.

## Native versus Alien Invasive Species

Most species have a home range of their own where they have evolved, bred and developed populations. As long as the species occur in their home range, they are designated as 'native'. Hence, 'native species' means, with respect to a particular ecosystem, a species that, other than as a result of an introduction, historically occurred or currently occurs in that ecosystem (Clinton, 1999). When species move out of their home range and establish in a new environment, they become 'alien' in that new habitat. In other words, 'alien species' means, with respect to a particular ecosystem, any species – including its seeds, eggs, spores or other biological material capable of propagating

**Table 14.1.** Forest cover in the north-eastern region of India as of 2005 (Anonymous, 2005).

| State | Geographical area (km$^2$) | Total forest cover (km$^2$) | Forest cover (% of geographical area) | Type of forest cover (km$^2$) | | |
|---|---|---|---|---|---|---|
| | | | | Very dense forest | Moderately dense forest | Open forest |
| Arunachal Pradesh | 83,743 | 67,777 | 80.93 | 14,411 | 37,977 | 15,389 |
| Assam | 78,438 | 27,645 | 35.24 | 1,444 | 11,387 | 14,814 |
| Manipur | 22,327 | 17,086 | 76.53 | 923 | 5,541 | 10,622 |
| Meghalaya | 22,429 | 16,988 | 75.74 | 338 | 6,808 | 9,842 |
| Mizoram | 21,081 | 18,684 | 88.63 | 133 | 6,173 | 12,378 |
| Nagaland | 16,579 | 13,719 | 82.75 | 236 | 5,602 | 7,881 |
| Sikkim | 7,096 | 3,262 | 45.97 | 498 | 1,912 | 852 |
| Tripura | 10,491 | 8,155 | 77.73 | 61 | 4,969 | 3,125 |
| Total | 262,184 | 173,316 | 66.10 | 18,044 | 80,369 | 74,903 |

that species – that is not native to that ecosystem (Clinton, 1999).

In the new environment, the alien species may quickly die if they fail to acclimatize, reproduce and produce fertile progenies. Alternatively, the alien species may find the new environment suitable for growth, reproduction and establishment of populations and may naturalize. Some alien species may even dominate the native flora and start replacing it. When this happens, the species is designated as 'invasive'. Hence, 'invasive species' means an alien species whose introduction is likely to cause economic or environmental damage or harm to human health (Clinton, 1999). The literature suggests the use of a number of terms to designate 'invasive' species with varied connotations. Some of the prevalent terms used include 'exotic', 'non-indigenous' and 'alien'. Notwithstanding, the invasive species are characterized by the following attributes: (i) new to the environment; (ii) capable of reproducing and establishing its own populations in the new environment; and (iii) capacity to replace native flora and cause harmful effects to the flora, fauna and humans in the new environment.

The invasiveness of alien plant species is determined by many life history traits such as (i) rapid growth rate (fast photosynthetic rate, early maturation); (ii) high reproductive effort (ability to reproduce at an early stage, ability to reproduce both sexually (through seeds) and asexually (through propagules), diverse dispersal mechanisms and high dispersal rate); (iii) better competitive ability (resource capture, allelopathy, pest exclusion, enemy release); (iv) stress tolerance (ability to survive in varied environmental conditions); and (v) ability to withstand human-induced disturbances such as grazing and burning (Bhowmik, 2005).

## Invasive Alien Plants in the North-eastern Region of India

Invasion of alien plants in the north-eastern region of India can be considered within the following ecosystems: (i) agricultural, (ii) forest and (iii) grasslands and marginal lands. The purely provisional classification adopted here is to facilitate the understanding of the distinctive features of diversity and distribution of alien weeds that invade these ecosystems. Presumably, species dominating agricultural systems may not be those dominating forest ecosystems and grasslands.

The predominant form of agriculture in the region is shifting cultivation (Borthakur, 1992; Ramakrishnan, 1992; Tawnenga *et al.*, 1996, 1997a, b). Over 100 tribes inhabiting the hilly tracts of this region have produced food by subsistence for millennia by maintaining long-duration rotation of cultivation cycles. Shifting cultivation, which is known in the region as *jhum*, was sustained until recently primarily due to low population pressure, abundance of forest cover and because grain formed only a meagre proportion of the human diet: the tribes consumed other food sources such as hunted meat, fish, tuber crops and a variety of other non-timber forest products (Uma Shankar, 2001a). Originally, *jhum* was a method of farming in tropical rainforests. Traditionally, both land and settlement were shifted, but now only land under cultivation is shifted (Uma Shankar and Sundriyal, 2007).

The other important method of cultivation in the region is terrace cultivation, which was introduced as an alternative to *jhum* primarily aiming at a reduction of nutrient losses through surface runoff. Yet another agricultural method popular among mountain farmers is the cultivation of paddy in the valleys. Valley cultivation is a monoculture of paddy, mostly during the rainy season. Both terrace and valley cultivation are forms of settled agriculture that depend mostly on rainfed water, or may be irrigated depending on the availability of resources.

All these forms of agriculture are increasingly exposed to invasion by alien weeds. Today, large landscapes anywhere in the region at any time of the year can be witnessed with luxuriant growth of invasive weeds. Precise statistics of how many hectares of land are affected by invasion remain to be derived. Remote sensing and GIS technologies can assist in estimating the area under different invasive species (Kandwal *et al.*, 2009). Many of the invasive weed species were not even recorded in the region until a few decades ago. However, they arrived following the intensification of agriculture in the 20th century to meet the increasing nutritional needs of a rising human population, became established and have now become invasive. The alien invasive species recorded most frequently in the region are listed in Table 14.2. It has been suggested that only a handful of these species occupy two-thirds or more of the agricultural landscape of the region across altitude gradients. The preliminary findings support the view that the USA is the primary geographical origin of invasive alien plant species in this region (Table 14.2), as is the case in China (Liu *et al.*, 2006).

Khonglam and Singh (1980) studied the distribution of three related species (*Ageratina adenophora*, *Ageratina riparia* and *Chromolaena odorata* (all previously *Eupatorium*)) between Guwahati and Shillong and found that these species showed a niche differentiation (spatial segregation) along an altitude gradient. *Chromolaena odorata* grows luxuriantly from 100 to 975 m (Burnihat to Barapani), *A. adenopora* from 550 to 1956 m (Nongpoh to Shillong peak) and *A. riparia* from 975 to 1706 m (Barapani to Upper Shillong). The three species are pioneer, early successional, invade quickly if light is not a limiting factor and use allelopathy as a competitive strategy (Tripathi *et al.*, 2006). Invasion by these three species is dealt with in detail in Chapters 4 and 6 of this volume.

A sympatric pair of annual weeds, *Galinsoga ciliata* (Raf.) Blake and *Galinsoga parviflora* Cav., characterized by a short life cycle, have been introduced to India from tropical America. Both these species have naturalized and occur abundantly in cultivated fields and disturbed ruderal habitats of north-eastern India, with *G. ciliata* being more aggressive than *G. parviflora* (Rai and Tripathi, 1984a). Both these species reproduce solely by seeds lacking dormancy. Seedling cohorts emerge at regular intervals, subject to favourable temperature and moisture conditions. Those emerging in spring, summer and the rainy seasons of the year follow Deevey's Type II and III survivorship curves (Rai and Tripathi, 1984a). Generally, late-emerging cohorts are less successful than early-emerging ones because seed germination, survivorship and growth may be reduced by competition from other plant species, as well as from previously established individuals of the same species (Rai and Tripathi, 1985a). Established cohorts of *G. ciliata* exercise a greater suppressive effect than those of *G. parviflora*. The emergence of seedling cohorts at different times reduces the degree of intra- and interspecific competition by avoiding crowding and ensures successful completion of the life cycle of a proportion of cohorts (Rai and Tripathi, 1984a). In an empirical study, seed germination and seedling growth of two species of *Galinsoga* were strongly

**Table 14.2.** Invasive alien plants most often recorded growing in different ecosystems in the north-eastern region of India.

| Species name (native range) | Common name/s | Impact/uses |
|---|---|---|
| *Asteraceae* | | |
| *Ageratina adenopora* (Spreng.) R. M. King and H. Rob. (North America) | Croftonweed, catweed, sticky-agrimony | Poisonous to mammals |
| *Ageratina riparia* (Regel) R. M. King and H. Rob. (North/South America) | Creeping croftonweed, mistflower, river eupatorium | Poisonous to mammals |
| *Ageratum conyzoides* L. (Pantropic) | Tropical whiteweed, goatweed | Poisonous to mammals |
| *Ageratum houstonianum* Mill. (Pantropic) | Flossflower, blue billy goatweed | Seed contaminant |
| *Artemisia nilagirica* (Clarke) Pamp. (Europe, Asia temperate) | Mugwort, green ginger, felon herb | Poisonous to mammals |
| *Bidens pilosa* L. (South America) | Hairy beggar-ticks, spanish needles, cobblers' pegs | Weed |
| *Chromolaena odorata* (L.) R. M. King and H. Rob. (North/South America) | Bitterbush, devilweed, Christmasbush | Seed contaminant |
| *Conyza bonariensis* (L.) Cronquist (Neotropics) | Argentine fleabane, wavy leaf fleabane | Weed |
| *Crassocephalum crepidiodes* (Benth.) S. Moore (Africa) | Redflower ragleaf, fireweed, thickhead | Weed |
| *Galinsoga ciliata* (Raf.) S. F. Blake (North America) | Hairy galinsoga, shaggy soldier | Seed contaminant |
| *Galinsoga parviflora* Cav. (North America) | Small flower galinsoga, gallant soldier | Seed contaminant |
| *Mikania micrantha* Kunth (North/South America) | Mile-a-minute, bittervine, climbing hempweed | Seed contaminant |
| *Parthenium hysterophorus* L. (North/South America) | Congress grass, carrot grass, false ragweed | Poisonous to mammals |
| *Synedrella nodiflora* (L.) Gaertn. (North/South America) | Synedrella | Host for crop pests |
| *Tithonia diversifolia* (Hemsl.) A. Gray (North/South America) | Mexican sunflower, shrub sunflower | Ornamental weed |
| *Caesalpinioideae* | | |
| *Senna alata* (L.) Roxb. (South America) | Christmas candle, ringworm shrub, candlebush | Ornamental weed |
| *Euphorbiaceae* | | |
| *Croton bonplandianus* Baill. (South America) | Croton | Weed |
| *Malvaceae* | | |
| *Urena lobata* L. (Pantropic) | Caesar weed, bur-mallow, paka | Seed contaminant |

| | | |
|---|---|---|
| *Mimosoideae* | | |
| *Acacia mearnsii* De Wild. (Africa, South America) | Black wattle | Fuel, soil stabilizer |
| *Mimosa invisa* Mart. ex Colla (South America) | Giant sensitive plant | Poisonous to mammals |
| *Oleaceae* | | |
| *Ligustrum robustum* (Roxb.) Blume (Asia tropical) | Privet | Intrudes in primary forest |
| *Oxalidaceae* | | |
| *Oxalis corniculata* L. (Worldwide) | Procumbent yellow sorrel, creeping lady's sorrel | Host for crop pests |
| *Papaveraceae* | | |
| *Argemone mexicana* L. (North/South America) | Mexican poppy, yellow thistle, prickly poppy | Medicinal, poisonous |
| *Poaceae* | | |
| *Imperata cylindrica* (L.) P. Beauv. (Pantropic) | Cotton wool grass, alang-alang, blady grass | Forage, soil stabilizer |
| *Saccharum spontaneum* L. (Africa, Asia) | Wild sugarcane, fodder cane | Host for crop pests |
| *Polygonaceae* | | |
| *Persicaria capitata* (Buch.-Ham. ex D. Don) H. Gross (Asia) | Pink knotweed | Ornamental weed |
| *Rumex nepalensis* Spreng. (Asia, Europe) | Knotweed | Seed contaminant |
| *Pontederiaceae* | | |
| *Eichhornia crassipes* (Mart.) Solms (South America) | Water hyacinth | Ornamental weed |
| *Rubiaceae* | | |
| *Borreria hispida* (L.) K. Schum. (Asia) | Shaggy button weed, Landrina, ligad-ligad | Seed contaminant |
| *Solanaceae* | | |
| *Solanum sisymbriifolium* Lam. (South America) | Wild tomato, sticky nightshade, red buffalo-bur | Poisonous to mammals |
| *Verbenaceae* | | |
| *Lantana camara* L. (North/South America) | Wild sage, lantana | Poisonous to mammals |

inhibited by the plant leachates of *E. riparia* (Rai and Tripathi, 1984b), indicating that the allelopathic influence of *E. riparia* plays an important role in regulating the population size of *Galinsoga* spp.

Soil moisture, texture and increasing population density may affect seed germination, seedling survival and growth of surviving individuals to varying degrees (Rai and Tripathi, 1983). In response to increasing population density, yield increased in *G. ciliata* but levelled off in *G. parviflora* at 16 plants per pot, revealing the greater stress-absorbing capacity of *G. ciliata*. However, a decrease in soil moisture due to an increasingly sandy texture reduced the yield of *G. ciliata* more than that of *G. parviflora*. Soil moisture stress caused such reduced seed production that seed output was even lower than the number of seeds sown, especially at high densities. The effect of various physical factors was exaggerated at high population densities, indicating the severity of intraspecific competition coupled with environmental constraints (Rai and Tripathi, 1983).

Competitive interactions affect reproductive allocation (Hickman, 1975; Holler and Abrahamson, 1977). *Galinsoga parviflora* exhibited a decrease in crude reproductive effort with increasing plant density (Rai and Tripathi, 1986a). On the other hand, the crude reproductive effort from *G. ciliata* increased with density in a low-nitrogen regime, suggesting an increase in reproductive allocation under the constraint of competition (Hickman, 1975). The crude reproductive effort of each species was also influenced by the presence of the other. Compared with *G. ciliata*, *G. parviflora* exhibited a lower crude reproductive effort in combination than in monoculture, suggesting that the former exercised a strong effect on the population of the latter. Such a strategy, coupled with competitive superiority and better reproductive growth in mixed stands, enables *G. ciliata* to dominate *G. parviflora* in dense populations under field conditions. However, the subtle differences in the plastic responses of the two species to the interacting influences of population density, soil nitrogen and competition allow the two weeds to coexist (Rai and Tripathi, 1986a).

A field study on plant-eating by the slug *Mariaella dussumieri* and by several other lepidopteran insects and their larvae revealed that these two species of *Galinsoga* produce a greater number of leaves in response to herbivory, but the leaf area per plant may decline (Rai and Tripathi, 1985b). Herbivory reduced survival, seed output, leaf area and dry matter yield and delayed flowering in *G. parviflora* as compared with *G. ciliata* in mixed stands, demonstrating that herbivores contribute considerably to the population regulation of *G. parviflora*, not only by consuming it but also by reducing its competitive ability (Rai and Tripathi, 1985b), as observed in other species (Van-Leewen, 1983).

The survival and growth of the two species were adversely affected by reduction in light intensity and increase in concentration of 2,4-D. Furthermore, *G. ciliata* was more susceptible to 2,4-D than *G. parviflora* was to a decrease in light (Rai and Tripathi, 1986b). Application of 2,4-D at the seedling stage under the high-light regime increased mortality and reduced growth. When applied at the flowering stage the formation of viable seeds was greatly impaired. This study demonstrated that the populations of the two species may be effectively regulated by herbicide application when environmental factors such as light and temperature are favourable for its efficacy and when the herbicide is applied at the vulnerable stage of plant growth.

## Mechanisms Facilitating Invasion of Alien Weeds in Agricultural Systems

What has gone wrong that has facilitated the current menace of alien invasive species in the north-eastern region of India? No one particular reason can be singled out; rather, a complex interaction of ecological, economic and social attributes has allowed the current dominance of invasive alien plants. The ecological reasons vary with the ecosystem, and within an ecosystem with the intensity and frequency of usage of bioresources. The economic reasons always drive farmers towards greater productivity with the least inputs of energy, time and money. Social development of societies is known to influence the resource use pattern, with increasingly greater preference for the introduction and use of non-native resources.

Innumerable workers have suggested that the shortening of the *jhum* cycle has exacerbated the problem of weeds under this system (Cutting *et al.*, 1959; Zinke *et al.*, 1978; Ramakrishnan, 1992; Tawnenga *et al.*, 1996, 1997a, b). Traditionally, a *jhum* cycle of 40–60 years was common in the region, meaning that any given generation of farmers cultivated a specific plot of forest land only once in their lifetime. This time span was adequate to allow the regrowth of the forest to its pre-slashed levels, and for restoration of almost all species of trees, shrubs, herbs and climbers. Hence, diversity and its distribution pattern remained unaltered. As *jhum* cycles began to shorten and were reduced to around 10–15 years, regrowth to attain pre-slashed levels of vegetation on land fallowed after cultivation was simply not possible. Consequently, communities of secondary successional species prevailed. Notwithstanding, fallowing of up to about 10 years in a *jhum* cycle kept weeds in check due to natural suppression (Saxena and Ramakrishnan, 1984). A further reduction of the *jhum* cycle to only 5 years or less exacerbated the succession of native herbs and shrubs and exposed the vulnerability of the ecosystem to invasion by alien weeds. Repeated short cycles of cultivation on the same patches of land presumably diminished the soil banks of propagules of native tree species, adversely affecting their regeneration. As a result, the weed community is maintained in a more or less permanent state of arrested succession (Kushwaha *et al.*, 1981a, b; Tawnenga *et al.*, 1996). Repeated short *jhum* cycle mean more frequent burning, which is known to stimulate populations of invasive species such as *E. adenophorum* and *E. riparia* (Tripathi and Yadav, 1987; Tripathi *et al.*, 2006).

Cultivation practices have also undergone a change with shortening of the *jhum* cycle. In the past, when cycles were longer, cultivators ensured that the propagules of most species in the standing vegetation were retained. Hence, they hoarded the individuals of rare species, put aside the knee-height stumps of the species that were known to re-sprout and thus spared the burning of many fruits and seeds that had fallen to the ground. These practices were extremely helpful in attaining pre-slashed levels of vegetation after crops had been cultivated. The presence of propagules of the native vegetation also checked the entry of alien species. The contemporary practices of *jhum* cultivation mostly ignore the indigenous knowledge gathered by previous generations and unhesitatingly assault the propagules of the native species, paving the way for alien weeds to invade. Hence, the extent of destruction of pre-farming vegetation and its propagules in the soil can determine the pattern of secondary succession and the rapidity with which a forest community develops. Toky and Ramakrishnan (1983) have distinguished four different patterns of weed dominance in relation to cultural practices and duration of the *jhum* cycle (Table 14.3).

The reproductive effort of a species plays an important role in imparting its dominance in an ecosystem. Successful reproduction means production of propagules (seeds or other vegetative bodies) capable of giving rise to newer individuals, effective dispersal of propagules and successful establishment of new individuals that can produce progenies. The most frequently recorded alien invasive plants listed in Table 14.2 are mostly species native to America and belonging to the family Asteraceae. The asteraceous species are characterized by a fruit type known as an 'achene'. The achene is light in weight and easily dispersed by wind and adherence to carriers, which makes these species very good colonizers (Hayashi and Numata, 1968; Raynal and Bazzaz, 1973). Alien invasive plants arriving as early successional species adopt an exploitative strategy to attain dominance in abandoned fields and *jhum* fallows that are temporarily enriched with nutrients and radiant energy. As succession proceeds, nutrient and light

**Table 14.3.** Four patterns of dominance shown by alien invasive weeds in shifting cultivation systems in the north-eastern region of India (based on data from Toky and Ramakrishnan, 1983).

| Type | *Jhum* cycle (years) | Weed dominance |
|---|---|---|
| Type I | 5 | *Chromolaena odorata* (*Eupatorium odoratum*) or *Imperata cylindrica*, or both, dominate the fallow. *C. odorata* dominates under a 5-year fallow cycle, particularly where the rhizomes of *I. cylindrica* and *Sachharum spontaneum* are removed by constant hoeing; otherwise *I. cylindrica* may exist alone or may coexist with *C. odorata* |
| Type II | 10 | *Mikania micrantha* dominates the fallow and, where abundant, forms a thick mat on the ground and may smother other herbaceous and shrub species |
| Type III | 5 | *Sachharum spontaneum* dominates the fallow, where cropping is usually for no more than one year and therefore its rhizomes have not been destroyed; this, however, is seen infrequently |
| Type IV | 30 | *Borreria hispida* dominates the fallow. This stoloniferous species has both prostrate and erect branches; it roots at the nodes and so forms a mat that inhibits the establishment of other species. It usually occurs where *I. cylindrica* and *C. odorata* have been eradicated by weeding |

availability reduce and other species get suppressed as the late-successional species take over. This generally happens between 10 and 20 years after the previous cropping has occurred. However, if cultivation is again commenced within 4–5 years, invasive weeds dominate on a continual basis and succession is arrested at the weed stage. One leading theory on the exceptional success of invasive plants is that they have escaped the natural enemies that hold them in check, freeing them to utilize their full competitive potential (Callaway and Aschehoug, 2000).

In shifting cultivation systems, invasion of weeds is stimulated if successive second-year cropping is carried out on the same plots. Nye and Greenland (1960) argued that the successive short-rotation cropping of rice–groundnuts caused a marked decline in the rate of regeneration of the woody fallow, because clearing and burning in the second year kills the suckers and seedlings and, in marginal areas of the forest, this practice leads to the establishment of grasses. The invasion of sites by grasses ('savannization') is encouraged by extended rotations, and grass invasion can readily be observed in sites slashed and cultivated consecutively for 2–3 years. In general, grasses fail to establish at a site unless farming operations intervene and delay the development of a closed tree canopy. An extra 1 or 2 years of clearing thereby suppresses coppice growth, and fallows treated in this way may lose not only their mature tree root masses, but also the mycorrhizal associations needed for nutrient uptake by trees (Janos, 1980). Tawnenga *et al.* (1996, 1997a, b) found that the harmful effects of successive second-year cropping were more intense on plots cropped after 20 years' fallow as compared with plots cropped after 6 years' fallow in Mizoram.

Under terraced systems, slashing of the weed biomass and crop residue arising from the previous cropping is burned, which results in the release of cations in one single flush (Patnaik and Ramakrishnan, 1990). However, under continuous cropping on the terraces cations are depleted through heavy leaching. Of all the cations, the loss of potassium is most pronounced (Mishra and Ramakrishnan, 1981; Toky and Ramakrishnan, 1981; Swamy and Ramakrishnan, 1988). Following cropping, the terraces are abandoned for several years and cropped subsequently. The cropped period determines the weed potential in a terrace system – twice the number of weeds in a 12-year cropped terrace than in a 4-year site (Patnaik and Ramakrishnan, 1990). Hence, continuous cropping on terraces results in increasing alien invasion of plants, in addition to adversely affecting soil fertility.

Valley agriculture is considered more tenable than terrace cultivation in the hills of the north-eastern region, due to a greater output of energy per unit of input of energy in the former than in the latter (Ramakrishnan, 1992). This sedentary form of agriculture experiences a fallow period for only for a few months in a year and, during this period, gives rise to more or less similar weed flora.

## Mechanisms Facilitating Invasion of Alien Plants in Forest Systems

Colonial foresters introduced *taungya* plantations in the hills of Darjeeling and Sikkim as an improvised version of *jhum* in the latter half of the 19th century, primarily in quest of cultivating good-quality timber (Uma Shankar *et al.*, 1998; Uma Shankar, 2001a). *Taungya* is a system of conversion of natural forest to timber plantations permitting the cultivation of crops for three consecutive years after slash-and-burn and planting of timber tree seeds or seedlings; thereafter, natural regeneration of the native vegetation is allowed, but is restrained by periodic thinning operations.

In the *taungya* system, clear-cut land is burned and planted with timber species, Agricultural crops are raised in the inter-planting space during the first two consecutive years, and natural regeneration is allowed in this space after crop abandonment. Ecosystem reconstruction is manifested through natural regeneration from three major pools: sprouting of roots/stumps of logged plants, seeds from the soil bank and seeds dispersed from other localities (Bormann and Likens, 1979). In tropical wet forests, a large portion of the woody cover that appears soon after clearing is commonly derived from coppicing of stems and rootstocks (Whitmore, 1985). If the cleared forest land is cultivated before it is allowed to recover, the seeds and vegetative propagules of many species may be damaged because cultivation on cleared forest lands is inevitably preceded by slash–burning.

A study of 35 years of recovery under *taungya* at the Mahananda Wildlife Sanctuary in Darjeeling concluded that *taungya*: (i) caused fragmentation of the forest landscape, posing a threat to mature forest species that are highly specialized; (ii) is evolutionarily conservative; and (iii) is sensitive to human-induced habitat change (Uma Shankar *et al.*, 1998). Only 28 of 87 species regenerated in the ≥10 cm cbh class of individuals, opening ecological niches for invasion by other species.

In Meghalaya, subtropical forests dominated by khasi pine (*Pinus kesiya*) are experiencing invasion by three shrub species, *Artemisia nilagirica*, *Lantana camara* and *Ligustrum robustum*. The preliminary observations indicate that these species are assuming an increasingly invasive role as they are replacing native flora, mostly at the forest fringes. However, *L. camara* is more aggressive and has already invaded sub-stantial ground cover under pine. The mechanisms stimulating the spread of *L. camara* and other species are well documented (Kohli *et al.*; 2006; Fig. 14.2).

In the Khasi Hills, *Artemisia nilagirica* is now spreading rapidly on landscapes vacated by the clear-cutting of *Pinus kesiya*. Recurrent burning of vegetation in spring, a common practice in this area, has added further to the threat posed by *A. nilagirica*. This species has not been seen as invasive to date, and an inventory of abundance, distribution and population dynamics of it is warranted. Since this species is a potential source of the drug artemisinin, its economic potential needs to be evaluated. Recently, an essential oil has been extracted from the leaves of *A. nilagirica* and its characterization is in progress (Aniruddha Sarma and Uma Shankar, 2009, unpublished; Fig. 14.2).

A native shrub, *Ligustrum robustum*, is fast approaching the status of invasive species in late-successional and primary forests (sacred groves) in Meghalaya. It is found in the lower canopy forming thickets around fringes, openings and disturbed patches. In a grid-based inventory of plant bioresources, we have recorded this species at high densities in at least 30% of sample plots (Uma Shankar, unpublished data). This species is included in the list of 100 of the

**Fig. 14.2.** Invasive weeds dominating forests, grasslands and marginal lands in the north-eastern region of India. A, *Lantana camara*; B, *Artemisia nilagirica*; C, *Ligustrum robustum*; D, *Tithonia diversifolia* (photographs courtesy of Uma Shankar).

world's worst invasive alien species, as it has now become invasive in the USA (www.issg.org/database).

## Mechanisms Facilitating Invasion of Alien Plants in Grasslands/Marginal Lands

The humid grasslands in north-eastern India, a result of savannization, are characterized by a mix of grasses and forbs that lack the dominance of leguminous species. The floristic composition of four grassland stands along an altitude gradient in Meghalaya showed a marked dominance of invasive species such as *Ageratum conyzoides*, *Chromolaena odorata*, *Imperata cylindrica*, *Mikania micrantha* (Uma Shankar *et al*., 1991). Zhang and Wen (2009) showed that *M. micrantha* is most likely to invade wet, well-lit sites, while *C. odorata* has the potential for expansion and invasion under strong light and moderate or even low soil water regimes. High structural and physiological plasticity in response to strong light and favourable soil water status may improve the growth advantages that benefit the invasion of these two invasive weeds (Fig. 14.3).

**Fig. 14.3.** Invasive weeds dominating agricultural systems in the north-eastern region of India. A, *Ageratum conyzoides*; B, *Ageratum houstonianum*; C, *Synedrella nodiflora*; D, *Bidens pilosa*; E, *Galinsoga parviflora*; F, *Chromolaena odorata*; G, *Mikania micrantha*; H, *Crassocephalum crepidiodes* (photographs courtesy of Uma Shankar).

Invasion of alien plants has also affected the natural wild parts of north-eastern India. The grasslands of Kaziranga National Park, a world heritage site, have been attacked by *Mimosa invisa* in recent decades (Uma Shankar, 2002). Some researchers have indicated the presence of another related species, *Mimosa rubicaulis* (Misra, 2005). *Mimosa* has established in Kaziranga following seed dispersal through floodwater from adjoining tea plantations (Uma Shankar. 2002). By virtue of its capacity to fix nitrogen, tea cultivators plant this leguminous species to boost growth of the tea bushes. However, this prickly *Mimosa* has become invasive and an obstacle to the movement of wild animals. Furthermore, the straggling *Mimosa* can cling to the top of the grasses that offer the best habitat in the world to the rhinoceros. *Mimosa* seeds can gain entry to the digestive tract of herbivorous animals, releasing the toxin mimosin. At present around 120 ha of prime grassland has been adversely affected, and scattered growth can be seen in many areas inside the park (Misra, 2005). The Wildlife Trust of India (WTI) has been helping the park in operations to eradicate *Mimosa* from grasslands.

Lately, the Mexican sunflower, *Tithonia diversifolia* has been recorded in the Khasi and Jaintia Hills at 800–1400 m altitude, most notably between Nongpoh and Mawlai. This autumn-flowering gives the landscape a spectacular appearance as it mostly occupies the marginal lands along roadsides, open wastelands and abandoned agricultural fields. The major cause of concern is its increasing intensity and the likelihood of replacing the local flora. Although its presence has been known in the hills for quite some time, no study has been carried out to determine its population distribution and regeneration dynamics.

## Conclusions

The invasion of alien species in the north-eastern region of India has emerged as a challenge for the 21st century, despite the fact that its extent has not been quantified and it has not been given subjected to scientific study. The point has now been reached, however, where management plans in regard to natural resources in different ecosystems must seriously address alien invasive species. Although a handful of species has been responsible for the majority of the damage to date, many other species cannot be overlooked. The most invasive and harmful species are from tropical American countries and belong to a single family of plants, the Asteraceae. Cultural changes in indigenous agricultural practices have primarily sheltered and stimulated the growth of alien invasive plants. The fear now is that even more species may become invasive in the region, as exemplified by *A. nilagirica*, *M. invisa* and *T. diversifolia*. There is an urgent need to investigate the mechanisms whereby plant species become invasive in local environments.

## Acknowledgements

The senior author (US) acknowledges the facilities afforded for this work at the Department of Botany, North-Eastern Hill University, Shillong, and partial funding through grant BT/PR7928/NDB/52/9/2006 from the Department of Biotechnology, Government of India, New Delhi. RST acknowledges support from INSA (Indian National Science Academy) in the form of an INSA Senior Scientist Award.

## References

Anonymous (2005) *State of Forest Report 2005*. Forest Survey of India. Dehradun, India.

Bhowmik, P.C. (2005) Characteristics, significance, and human dimension of global invasive weeds. In: Inderjit, L. (ed.) *Plant Invasions: Ecological and Agricultural Aspects*. Birkhäuser Verlag, Basel, Switzerland, pp. 251–268.

Bormann, F.M. and Likens, G.E. (1979) *Development of Vegetation after Clear-cutting: Species Strategies and Plant Community Dynamics. Patterns and Processes in a Forested Ecosystem*. Springer, New York, pp. 103–137.

Borthakur, D.N. (1992) *Agriculture of the North-eastern Region with Special Reference to Hill Agriculture*. BeeCee Prakashan, Guwahati, 264 pp.

Callaway, R.M. and Aschehoug, E.T. (2000) Invasive plants versus their new and old neighbors: a mechanism for exotic invasion. *Science* 290, 521–523.

Clinton, W.J. (1999) Executive Order 13112 of February 3, 1999 – Invasive species. The White House, Washington, DC (http://www.invasivespecies.gov/laws/exeorder.shtml).

Conservation International (CI) (2004) *Hot Spots Revisited*. Available online at http://www.biodiversityhot spots.org/XP/hot spots/himalaya

Cutting, C.V., Wood, R.A., Brown, P. and Ambrose, B.B. (1959) Assessment of fertility status and the maintenance of productivity of soils in Nyasaland. In: *3rd International African Soils Conference*, Dalaba, Guinea, pp. 817–824.

Hayashi, I. and Numata, M. (1968) Ecology of pioneer species of early stages in secondary succession. II. The seed production. *Botanical Magazine (Tokyo)* 81, 55–56.

Hickman, J.C. (1975) Environmental unpredictability and plastic energy allocation strategies in the annual *Polygonum cascadense*. *Journal of Ecology* 63, 689–701.

Holler, L.C. and Abrahamson, W.G. (1977) Seed and vegetative reproduction in relation to density in *Fragaria virginiana*. *American Journal of Botany* 64, 1003–1007.

Janos, D.P. (1980) Mycorrhizae influence tropical succession. *Biotropica, Special Issue 'Tropical Succession'* 12 (Suppl.), 56–64.

Kandwal, R., Jeganathan, C., Tolpekin, V. and Kushwaha, S.P.S. (2009) Discriminating the invasive species, 'Lantana' using vegetation indices. *Journal of the Indian Society of Remote Sensing* 37, 275–290.

Khonglam, A. and Singh, A. (1980) Altitudinal effect on chromosome associations in three species of *Eupatorium. Cell and Chromosome Newsletter* 3, 24–27.

Kohli, R.K., Daizy, R.B., Singh, H.P. and Dogra, K.S. (2006) Status, invasiveness and environmental threats of three tropical American invasive weeds (*Parthenium hysterophorus* L., *Ageratum conyzoides* L., *Lantana camara* L.) in India. *Biological Invasions* 8, 1501–1510.

Kushwaha, S.P.S., Ramakrishnan, P.S. and Tripathi, R.S. (1981a) Population dynamics of *Eupatorium odoratum* L. in successional environments following slash and burn agriculture (jhum). *Journal of Applied Ecology* 18, 529–536.

Kushwaha, S.P.S., Ramakrishnan, P.S. and Tripathi, R.S. (1981b) Population dynamics of *Imperata cylindrica* (L.) Beauv. var. *major* related to slash and burn agriculture (jhum) in north eastern India. *Proceedings of the Indian Academy of Sciences* 92, 313–321.

Liu, J., Dong, M., Miao, S.L., Li, Z.Y., Song, M.H. and Wang R.Q. (2006) Invasive alien plants in China: role of clonality and geographical origin. *Biological Invasions* 8, 1461–1470.

Mishra, B.K. and Ramakrishnan, P.S. (1981) The economic yield and energy efficiency of hill agroecosystems at high elevation of Meghalaya in north-eastern India. *Acta Oecologica (Oecologica Applicata)* 2, 369–389.

Misra, M.K. (2005) Improving protection and building capacity of staff at Kaziranga National Park. Technical Repot No. 4, UNESCO-IUCN-WII, Paris.

Nye, P.H. and Greenland, D.J. (1960) The soil under shifting cultivation. Commonwealth Bureau of Soil Science Technical Communication No. 51, Harpenden, UK, 156 pp.

Patnaik, S. and Ramakrishnan, P.S. (1990) Cation budget under terrace agroecosystem in Meghalaya in north-east India. *Proceedings of the Indian Academy of Sciences (Plant Sciences)* 100, 23–31.

Rai, J.P.N. and Tripathi, R.S. (1983) Population regulation of *Galinsoga ciliata* and *G. parviflora*: effect of sowing pattern, population density and soil moisture and texture. *Weed Research* 23, 151–163.

Rai, J.P.N. and Tripathi, R.S. (1984a) Population dynamics of different seedling cohorts of two coexisting annual weeds, *Galinsoga ciliata* and *G. parviflora* on two contacting sites. *Acta Oecologica* 5, 357–368.

Rai, J.P.N. and Tripathi, R.S. (1984b) Alleopathic effect of *Eupatorium riparium* on population regulation of two species of *Galinsoga* and soil microbes. *Plant and Soil* 23, 105–117.

Rai, J.P.N. and Tripathi, R.S. (1985a) Effect of established population of two annual weeds on emerged seedling cohorts. *Proceedings of the Indian Academy of Sciences* 95, 79–86.

Rai, J.P.N. and Tripathi, R.S. (1985b) Effect of herbivory by the slug, *Mariaella dussumieri* and certain insects on growth and competitive success of two sympatric annual weeds. *Agriculture Ecosystems and Environment* 13, 125–137.

Rai, J.P.N. and Tripathi, R.S. (1986a) Effect of density and soil nitrogen level on growth of *Galinsoga quadriradiata* and *G. parviflora* in pure and mixed stand. Canadian *Journal of Botany* 64, 1005–1017.

Rai, J.P.N. and Tripathi, R.S. (1986b) Population regulation of *Galinsoga ciliata* and *G. parviflora*: effect of 2,4-D application at different growth stage and light regimes. *Weed Research* 26, 59–67.

Ramakrishnan, P.S. (1992) *Shifting Agriculture and Sustainable Development – an Interdisciplinary Study from North-eastern India.* UNESCO and Oxford University Press, Oxford, UK, pp. 423.

Raynal, D.J. and Bazzaz, F.A. (1973) Establishment of early successional plant populations on forest and prairie soil. *Ecology* 54, 1335–1343.

Saxena, K.G. and Ramakrishnan, P.S. (1984) Patterns of herbaceous vegetation development and weed potential under slash and burn agriculture (jhum) in north-eastern India. *Weed Research* 24, 137–144.

Swamy, P.S. and Ramakrishnan, P.S. (1988) Nutrient budget under slash and burn agriculture (jhum) with different weeding regimes in north-east India. *Acta Oecologica (Oecologica Applicata)* 9, 85–102.

Tawnenga, Uma Shankar and Tripathi, R.S. (1996) Evaluating second year cropping on jhum fallows in Mizoram, north-eastern India: Phytomass dynamics and primary productivity. *Journal of Biosciences* 21, 563–575.

Tawnenga, Uma Shankar and Tripathi, R.S. (1997a) Evaluating second year cropping on jhum fallows in Mizoram, north-eastern India: Energy and economic efficiencies. *Journal of Biosciences* 23, 605–613.

Tawnenga, Uma Shankar and Tripathi, R.S. (1997b)

Evaluating second year cropping on jhum fallows in Mizoram, north-eastern India: Soil fertility. *Journal of Biosciences* 23, 615–625.

Toky, O.P. and Ramakrishnan, P.S. (1981) Cropping and yields in agricultural systems of the north-eastern hill region of India. *Agro-Ecosystems* 7, 11–25.

Toky, O.P. and Ramakrishnan, P.S. (1983) Cropping and yields in agricultural systems of the north-eastern hill region of India. *Agro-Ecosystems* 7, 11–25.

Tripathi, R.S. (2009) Alien plant invasion: a hot ecological issue. *Environews* 15, 7–10.

Tripathi, R.S. and Yadav, A.S. (1987) Population dynamics of *Eupatorium adenophorum* Spreng., and *E. riparium* Regel in relation to burning. *Weed Research* 27, 229–336.

Tripathi, R.S., Kushwaha, S.P.S. and Yadav, A.S. (2006) Ecology of three invasive species of *Eupatorium*: a review. *International Journal of Ecology and Environmental Sciences* 32, 301–326.

Uma Shankar (2001a) A case of high tree diversity in a 'sal' (*Shorea robusta*)-dominated lowland forest of eastern Himalaya: floristic composition, regeneration and conservation. *Current Science* 81, 776–786.

Uma Shankar (2001b) Increasing biomass scarcity in north-eastern India: options with the villagers. In: Arunachalam, A. and Khan, M.L. (eds) *Sustainable Management of Forests – India.* International Book Distributors, Dehra Dun, India, pp. 247–248.

Uma Shankar (2002) Wealthy wilds intimidated in Kaziranga. *Hima-Paryavaran* 14/15, 6–8.

Uma Shankar and Sundriyal, R.C. (2007) Reasons for marginalization and options for improving livelihoods of jhum farmers: a case study of Tangsas and Nishis in Arunachal Pradesh, India. In: Saxena, K.G., Luohui Liang and Kanok Rerksasem (eds) *Shifting Agriculture in Asia: Implications for Environmental Conservation and Sustainable Livelihoods.* Bishen Singh and Mahendra Pal Singh Publishers, Dehra Dun, India, pp. 65–74.

Uma Shankar, Tripathi, R.S. and Pandey, H.N. (1991) Structure and seasonal dynamics of humid tropical grasslands in Meghalaya, India. *Journal of Vegetation Science* 2, 711–714.

Uma Shankar, Lama, S.D. and Bawa, K.S. (1998) Ecosystem reconstruction through 'taungya' plantations following commercial logging of a dry, mixed deciduous forest in Darjeeling Himalaya. *Forest Ecology and Management* 102, 131–142.

Uma Shankar, Lama, S.D. and Bawa, K.S. (2001) Ecology and economics of domestication of non-timber forest products: an illustration of broomgrass in eastern Himalaya. *Journal of Tropical Forest Science* 13, 171–191.

Van-Leewen, B.H. (1983) The consequences of predation in the population of biology of the monocarpic species *Cirsium polustre* and *C. vulgare*. *Oecologia* 58, 178–187.

Whitmore, T.C. (1985) *Tropical Rain Forests of the Far East*, 2nd edn. Clarendon Press, Oxford, UK.

Zhang, L.-L. and Wen, D.-Z. (2009) Structural and physiological responses of two invasive weeds, *Mikania micrantha* and *Chromolaena odorata* to contrasting light and water conditions. *Journal of Plant Research* 122, 69–79.

Zinke, P.J., Sabhasri, S. and Kunstadter, P. (1978) Soil fertility aspects of the 'Lua' forest fallow system of shifting cultivation. In: Kunstadter, P., Chapman, E.C. and Sabhasri, S. (eds) *Farmer in the Forest.* East-West Center, Honolulu, Hawaii.

# 15 Invasive Alien Weeds in the Tropics: the Changing Pattern in the Herbaceous Flora of Meghalaya in North-east India

**R.R. Rao[1] and Kavitha Sagar[2]**

[1]*INSA Honorary Scientist, Bangalore, India;* [2]*Vidya Herbs, Bangalore, India*

## Introduction

Biological invasions worldwide by various notorious weedy species have become so widespread as to represent a major component of human-induced global environmental change. These species not only affect human health but also nations' economic scenarios. Invasive alien weeds in India are either intentional or unintentional introductions, mostly from the neo-tropical regions of the world, and have tremendous potential to establish, invade and outcompete native species. The threat to native biodiversity due to the spread of invasive alien weeds is considered second only to habitat destruction. Although extinction of local populations due to the spread of alien weeds was recognized as early as 1872 by Darwin and Wallace (Darwin, 1872; Wallace, 1902), it is only recently that the problem of biological invasions has received due attention. The Convention on Biological Diversity held in 1992 in Rio de Janeiro (Article 8 h) emphasizes the serious threat to native flora from exotic invasive weeds and urges the scientific community to take steps for their control and eradication.

India has shown keen interest in the study and control of invasive weeds since the 1900s (Muniyappan and Viraktamath, 1993), and various accounts of the introduction of exotic invasive weeds have been published over the years (Kashyap, 1922; Biswas, 1934; Maheshwari, 1962; Srivastava, 1964; Mathew, 1969; Ramaswamy *et al.*, 1972; Rao and Rao, 1976; Rao, 1978; Rao and Suryanarayana, 1979; Burhl, 1980; Rai and Gaur, 2006). Rao (1996) provides a list of major neo-tropical weeds that have greatly altered the floristic composition of the country. The ecology of biological invasions and threat to native habitats was discussed at length recently at a National Workshop at Banaras Hindu University, Varanasi (Rai and Gaur, 2006) and at Chandigarh in 2009. The ecological aspects of invasive aliens in north-east India have received considerable attention (Saxena and Ramakrishnan, 1984; Tripathi, 1985; Ramakrishnan, 1991). The present account deals with perceptible floristic changes, particularly in the herbaceous floras, that have occurred during recent years due to the introduction and spread of alien invasive weeds in Meghalaya and other states in north-east India.

## The Study Area

North-east India, including the East Himalaya (22–30°N, 89–97°E) forms the richest botanical zone of the country. East Himalaya is also considered as one of the global hot spots of biodiversity, where anthropogenic disturbances are at their maximum (Myers *et al.*, 2000). The mountains of East Himalaya and the sub-Himalaya have been subject to considerable development, posing a serious threat to a variety of fragile ecosystems whose existence is now in danger. A rapid increase in human population, particularly the tribal populations who exclusively practise shifting cultivation (*jhum*), has resulted in the loss of forest cover and has paved the way for the establishment of invasive weedy species in areas where several unique native species previously existed. The State of Meghalaya, comprising the Khasi, Garo and Jaintia Hills lying between 25°.00′ and 26°.10′ N and 89°.45′ and 92°.45′ E, is one of the richest and botanically interesting regions in the whole of the Indian subcontinent. This state is endowed with a heavy cover of natural forest, mainly falling under tropical and temperate types. A vast majority of the forests in Meghalaya are owned by tribal populations who practise *jhum* as part of their socio-religious customs. Hence, the Forest Department has no control over such forests, which has resulted in the dwindling of the original forest area. The resulting change in the floristic composition of various regions in Meghalaya State is conspicuous by the presence of invasive weeds such as: (i) *Mikania micrantha* H.B.K., all along the roadsides from Guwahati to Shillong, covering the ground as well as the tree canopies; (ii) *Chromolaena odorata* (L.) King & Robinson, as forest undergrowth in the Garo Hills; and (iii) *Ageratina adenophora* (Spreng.) King & Robinson, in pine forests at higher altitudes in Shillong, Upper Shillong, etc.

## Vegetation versus Exotic Weed Invasions

North-east India, predominantly within the humid tropics, exhibits maximum biological diversity. Around 17% of the country's total forest cover is found within this region, which support about 50% of the total number of vascular plants in India. The vegetation is quite varied, including cultivated plains, grasslands, meadows, marshes, swamps, scrub forests, mixed forests, deciduous forests and evergreen forests, each with its characteristic species composition. In this region we find tropical, subtropical, temperate and alpine vegetation (Rao and Murthy, 1990). There are also many sacred forests, representing relict vegetation. This rich diversity is largely attributable to the wide diversity of topography, climate and soil types – and to migration of floras from surrounding countries.

Exotic invasives of fairly recent introduction are gradually upsetting the original composition of floristic elements in this region, particularly the herbaceous flora, which is a serious concern for biologists. Although invasive weeds can out-compete and completely replace local herbaceous flora, these weeds can also affect the distribution of arborescent taxa by inhibiting seed germination and seedling establishment through allelopathic influences. Included among such aggressive invasive weeds in Meghalaya are *Ageratum conyzoides* L., *Chromolaena odorata* (L.) King & Robinson, *Ageratina adenophora* (Spreng.) King & Robinson, *Ageratina riparia* (Regel.) King and Robinson, *Erigeron karvinskianus* DC., *Galinsoga parviflora* L., *Galinsoga quadriradiata* Ruiz et Pavon, *Mikania micrantha* H.B.K. and *Parthenium hysterophorus* L., all belonging to the family *Asteraceae*. Other weeds such as *Lantana camara* L. (Verbenaceae), *Eichhornia crassipes* (L.) Solms. (*Pontederiaceae*), *Ipomoea fistulosa* Mart. ex Choisy (Convolvulaceae) and *Alternanthera paronychioides* A. St.-Hilaire (*Amaranthaceae*) are also dominant, all of which are instrumental in effecting drastic changes in floristic composition. *Galinsoga quadriradiata* Ruiz et Pavon, which was reported as a new record for Meghalaya (Rao and Rao, 1976) has today become one of the major invasive species. Steffen (1996), discussing global change and ecological complexity, emphasized the effects of biological invasions on ecological complexity and ecosystem functioning with the main objectives of determining the effects of biological invasions on biological diversity and determining the characteristics that make an ecosystem vulnerable to invasions and an alien weed a successful invader.

## Factors Favouring Biological Invasions

Biological invasion refers to exotic plants and animals that are introduced and become firmly established in the new soil. As such invasions are common in disturbed areas, it is quite often felt that invasions are primarily a consequence of disturbance. Biological invasions are thus directly linked with land use change, with humans the premier agents of disturbance. In the north-east region, the age-old practice of *jhum* has seriously disturbed the virgin forests resulting in denuded, often weedy, seral communities. The process of shifting cultivation involves clear-felling of a patch of forest, allowing it to dry and, before the onset of the monsoon season, the entire biomass is burnt and the land is cleared for cultivation for a limited period of 3–4 years; the area is then abandoned as a fallow and the tribal families move to some other forested area and repeat the process.

The forests are thus exposed to a two-pronged attack from this sort of shifting agriculture. First, it exterminates the primary vegetation and, secondly, it aids the colonization and spread of many weedy species. Once the forest-cleared area is left as fallow, the process of biological invasion sets in. Although such areas in the north-east are colonized by some of the local weedy species like *Dipsocus asper* L., *Leucas ciliata* Benth., *Justicia* spp., *Acalypha indica* L., *Emelia sonchifolia* (L.) DC., *Spilanthes calva* DC., *Brachiaria reptans* (L.) Gaud., *Cyrtococcum trigonum* A. Camus., *Eragrostis* spp. and many other grasses and sedges, these are soon pushed aside by the non-native invasives such as *Eupatorium* or *Mikania* (Tables 15.1 and 15.2).

Several native or naturalized grass species, that are otherwise eaten by cattle have today become comparatively rare, e.g. *Brachiaria mutica* Stapf., *B. reptans* (L.) Gard. et Hubb., *B. distachya* Stapf., *Paspalidium flavidum* A. Camus, *Cyrtococcum trigonum* A. Camus, *Panicum miliare* Lamk., *P. miliaceum* L., P. khasianum Munro ex Hk. f., Saccolepis *indica* A. Chase, *Digitaria* spp., *Arundinella setosa* Trin., *Perotis indica* (L.) Ktz. *Chloris barbata* Sw., *Microchloa indica* Beauv.,

**Table 15.1.** Some medicinally important species whose diversity has been adversely affected by invasive exotics in Meghalaya State.

| Species | Family |
|---|---|
| *Achyranthes aspera* L. | *Amaranthaceae* |
| *Acorus calamus* L. | *Zingiberaceae* |
| *Alpinia galanga* Sw. | *Zingiberaceae* |
| *Aristolochia tagala* Cham. | *Aristolochiaceae* |
| *Boerhaavia difusa* L. | *Nyctaginiaceae* |
| *Chenopodium ambrosioides* L. | *Chenopodiaceae* |
| *Cissampelos pareira* L. | *Menispermaceae* |
| *Clerodendron viscosum* L. | *Verbenaceae* |
| *Costus speciosus* L. | *Costaceae* |
| *Curcuma amada* Roxb. | *Zingiberaceae* |
| *Curcuma angustifolia* Roxb. | *Zingiberaceae* |
| *Kaempferia rotunda* L. | *Zingiberaceae* |
| *Paederia foetida* L. | *Rubiaceae* |
| *Plumbago zeylanica* L. | *Plumbaginaceae* |
| *Rubia cordifolia* L. | *Rubiaceae* |
| *Sida rhombifolia* L. | *Malvaceae* |
| *Smilax macrophylla* Roxb. | *Smilacaceae* |
| *Spilanthes calva* DC. | *Asteraceae* |
| *Valeriana hardwickii* Wall. | *Valerianaceae* |

**Table 15.2.** Some wild plant species of north-east India whose diversity has been adversely affected by the spread of invasive weeds.

| Species | Family |
|---|---|
| *Acalypha indica* L. | *Euphorbiaceae* |
| *Agrimonia eupatorium* L. | *Rosaceae* |
| *Alternanthera echinata* Sw. | *Amaranthaceae* |
| *Alysicarpus vaginalis* DC. | *Fabaceae* |
| *Amaranthus spinosus* L. | *Amaranthaceae* |
| *Andrographis paniculata* Nees | *Acanthaceae* |
| *Blumea balsamifera* DC. | *Asteraceae* |
| *Boerhaavia diffusa* L. | *Nyctaginaceae* |
| *Brunella vulgaris* L. | *Lamiaceae* |
| *Crotalaria alata* Ham. | *Fabaceae* |
| *C. mysorensis* Roth | *Fabaceae* |
| *C. juncea* L. | *Fabaceae* |
| *Cassia tora* L. | *Caesalpiniaceae* |
| *C. mimosoides* L. | *Caesalpiniaceae* |
| *C. occidentalis* L. | *Caesalpiniaceae* |
| *Chenopodium ambrosioides* L. | *Chenopodiaceae* |
| *Cleome ciliata* L. | *Capparidaceae* |
| *Clitoria ternatea* L. | *Fabaceae* |
| *Desmodium khasianum* Prain | *Fabaceae* |
| *D. sinuatum* Bl. | *Fabaceae* |
| *Dipsacus asper* Wall. | *Dipsacaceae* |
| *Emelia sonchifolia* DC. | *Asteraceae* |
| *Eriosema chinense* Vogel. | *Fabaceae* |
| *Euphorbia hirta* L. | *Euphorbiaceae* |
| *E. prostrata* Ait. | *Euphorbiaceae* |
| *Evolvulus alsinoides* L. | *Convolvulaceae* |
| *Flemingia ciliata* Roxb. | *Fabaceae* |
| *Fragaria indica* Andr. | *Rosaceae* |
| *Hedyotis rotundifolia* Roxb. | *Rubiaceae* |
| *H. aspera* Heyne | *Rubiaceae* |
| *Hydrocotyle javanica* Thumb. | *Apiaceae* |
| *Indigofera trita* L.f. | *Fabaceae* |
| *I. cordifolia* Heyne | *Fabaceae* |
| *Justicia* spp. | *Acanthaceae* |
| *Lantana indica* Roxb. | *Verbenaceae* |
| *Leucas ciliata* Benth. | *Lamiaceae* |
| *Mollugo pentaphylla* L. | *Aizoaceae* |
| *Osbeckia chinensis* L. | *Melastomaceae* |
| *Parochetus communis* Ham. | *Fabacrae* |
| *Passiflora foetida* L. | *Passifloraceae* |
| *P. nepalensis* Wall. | *Passifloraceae* |
| *Phlogacanthus curviflorus* Nees | *Acanthaceae* |
| *Polygala chinensis* L. | *Polygalaceae* |
| *Portulaca oleracea* L. | *Portulacaceae* |
| *Potentilla fulgens* Wall. | *Rosaceae* |
| *Priotropis cytisoides* W.&A. | *Fabaceae* |
| *Scoparia dulcis* L. | *Scrophulariaceae* |
| *Sigesbeckia orientalis* L. | *Asteraceae* |
| *Solanum nigrum* L. | *Solanaceae* |
| *Spilanthes acmella* L. | *Solanaceae* |
| *Tephrosia purpurea* | *Fabaceae* |
| *Teucrium quadrifarium* Buch-Ham. | *Lamiaceae* |
| *Tragia involucrata* L. | *Euphorbiaceae* |
| *Trianthema portulacostrum* L. | *Aizoaceae* |
| *Vigna vexillata* Benth. | *Fabaceae* |

*Eragrostis pilosa* Beauv., *E. gangetica* Steud., *E. unioloides* Nees ex Steud., *E. riparia* Nees, *E. tenella* R. & S. and *Bromus himalaicus* Stapf. ex Hk.f.

With the establishment of alien invasives in a new ecosystem, the natural succession process is hindered, resulting in changes in structure and composition. It is also known that, in the first three years of the fallow, predominantly invasive weeds such as *C. odorata* (L.) King & Robinson, *A. adenophora* (Spreng.) King & Robinson, *G. parviflora* Cav. and *G. quadriradiata* Ruiz et Pavon appear and, in many sites, even push out the native weeds (Rao, 1977). In addition, the invasive *Chromolaena* spp. interfere with the regeneration of timber trees in the forests and thus alter the floristic composition of the ecosystem.

Loss of a species from an area often attracts the attention of botanists, but loss of genetic variability within a species (due to population extinction) often goes unnoticed, as has happened with the majority of native flora. Species fragmentation, followed by decrease in population size, has a deleterious effect on their breeding structure, genetics and evolutionary dynamics, which forms the central focus of concern in conservation biology (Harper, 1977; Barrett and Kohn, 1991; Falk and Holsinger, 1991; Karron, 1991). In isolated populations, genetic drift operates and this results in loss of variation, which reduces the ability of populations to adapt to changing environments and increases their susceptibility to pests and diseases, thus leading to local extinction (Hamilton, 1982).

While exotic weeds have tremendous potential to multiply and invade new areas, human-induced factors in the north-east such as forest clearance for shifting agriculture, extension of townships, road construction, etc. are acting as catalysts for the rapid spread and colonization of invasive weeds in ever more new areas, altering floristic composition.

## Alien Invasives as Instruments of Floristic Change

The spread of non-native invasive weeds has brought about a perceptible decline in several populations of native herbaceous flora in the north-eastern region. A few such native or otherwise naturalized species that have been fragmented leading to loss of genetic diversity are listed in Tables 15.1 and 15.2. Although there are no publications dealing with the quantitative estimates of loss of native biodiversity from a region before and after the introduction of exotic invasives, it is quite clear from the massive spread of aliens in vast areas that the floristic composition of several ecosystems has greatly altered. Erosion of diversity (including genetic diversity) in a species gradually leads to fragmentation and finally extinction.

The impact of the spread of alien invasive weeds in north-east India can also be seen with regard to the floristic composition of weed flora in various agro-ecosystems, particularly in Meghalaya (Neogi and Rao, 1980). These workers have provided a complete list of weed flora of different crops in Meghalaya. Even a casual survey of harvested paddy fields reveals the disappearance of native weeds such as *Hydrolea zeylanica* (L.) Vahl., *Polygonum plebejum* (L.) R. Br., *Sphaeranthus indicus* L., *Rotala rotundifolia* Koehne and a few others, due to the invasion of exotic weeds such as *Mikania* and *Eupatorium*.

Haridasan *et al.* (1981) have discussed in detail the spread of forest weeds in Meghalaya, enumerating many important forest weeds belonging to diverse families in different forest ecosystems. In the pine forests at higher elevations of Meghalaya, it is believed that *A. adenophora* (Spreng.) King & Robinson, *C. odorata* (L.) King & Robinson and *A. riparia* (Regel.) King and Robinson have spread so abundantly that they now endanger many other forest weeds such as *Ambrosia artimisifolia* L., *Anaphalis cinnamomea* Cl., *Inula cappa* DC., *Lonicera macrantha* DC.,

*Drymaria cordata* (L.) Willd. ex R. & S., *Teucrium quadrifarium* Buch.-Ham., *Osbeckia nepalensis* Hk.f., *Rubia cordifolia* L., *Clerodendrum serratum* Spreng., etc. Even the grasslands around Shillong plateau, Riangdo, Ranikor, Mawphlang, Mawsynram and Cherrapunjee are now dominated by grasses such as *Saccharum spontaneum* L., *Neyraudia reynaudiana*, *Thysanolaena maxima* Kuntze., *Chrysopogon aciculatus* (Retz.) Trin., *Panicum khasianum* Munro., *Setaria palmifolia* (Koen.) Stapf., *Arundinella bengalensis* (Sprengel.) Druce, *Paspalum dsitichum* L. Many dicotyledonous species such as *Eriosema chinensis* Vogel, *Polygonum bistorta* Garcke., *Trifolium repens* L., *Centella asiatica* (l.) Urban, *Hemiphragma heterophyllum* Wall., *Impatiens chinensis* L. and some *Utricularia* spp., which are associated with grassland flora, are also affected by the spread of the invasive *Eupatorium* spp.; and *Aeginetia indica* Roxb., a curious member of the family *Orobanchaceae*, which normally appears in pine forests at higher elevations, has also become rare, perhaps again due to the spread of invasive *Eupatorium* species.

These weeds, while competing with the saplings of forest trees for space and nutrition, also block out much-needed sunlight for the latter, thereby obstructing the regeneration of forests. Similarly, in the tropical belt, invasives such as *M. micrantha* and *C. odorata* have taken a heavy toll of many other naturalized and native species such as *Crassocephalum crepidioides* (Benth.) S. Moore, *Spilanthes acmella* (L.) Murr., *Vernonia cinerea* (L.) Less., *Leea crispa* Willd., *Mimosa himalayana* Gamble, *Desmodium pulchellum* Benth., *Desmodium laburnifolium* DC., *Mussaenda frondosa* L., *Paedaria foetida* L. and *Pilea umbrosa* Wed., which were fairly abundant till recently. Haridasan and Rao (1985) have listed 45 species as being extremely rare, partly due to alterations in the natural habitat caused by invasive weeds.

## Examples of Serious Exotic Invasive Weeds in Meghalaya

Although the entire naturalized flora in India has taken its toll of native biodiversity, a few species, particularly of the family *Asteraceae*, have acquired invasive characteristics and these weeds have conspicuously altered species composition in many areas of north-east India. Some of the worst invasive weeds that have not only impacted human health but also affected the economic status of the tribal populations of the region are discussed below.

### *Mikania micrantha* H.B.K.

From the family *Asteraceae*, this invasive species has tremendous potential for multiplication, both vegetatively and sexually. The long, trailing branches root at nodes and establish as independent plants, which in turn produce long branches that cover the entire herbaceous flora in the locality. It is said that a single stock of *Mikania* can produce 20,000–40,000 achenes in just one season (Dutta, 1977), which are dispersed over long distances by wind. The weed is quite abundant all along the roadsides on the Guwahati–Shillong road, and can be seen climbing not only on the herbaceous flora, but also on most of the shrubby and small tree species; being a climbing plant, this species poses a severe threat to these native flora. At ground level, the branches of this species spread quite extensively over all other herbaceous vegetation and forest saplings, killing them completely.

### *Lantana camara* L.

Believed to have been introduced to India around 1809 as an ornamental plant by the then Royal Botanic Garden, Howrah, today this is one of the worst invasive weeds in India; it has spread almost from Kanyakumari to the Himalayan region. It is noted for its enormous flowers of varied colours that are seen throughout most of the year, and it produces huge numbers of seeds that are spread far and wide by birds. Having once established, it forms a dense, impenetrable thicket that completely smothers the native vegetation and thus alters the forest structure of the understorey species. The species also has a strong allelopathic character, which helps in the successful invasion of most areas.

### *Chromolaena odorata* (L.) King & Robinson

This species is also native to tropical America and Mexico, and was first established in India around 1900. It is very aggressive at both lower and higher elevations over the entire north-eastern region. The spread of this weed in Meghalaya State is largely favoured by the practice of shifting agriculture by the local people; after the *jhumed* fields have been abandoned, this species is one of the primary colonizers of the area. Having very high ecological amplitude, this species can even withstand burning. Like most other Asteraceae weeds, this species also produces enormous numbers of seeds (153,710 seeds per plant, according to Tripathi, 1985). Although not totally fire resistant, this weed grows luxuriously on burned sites. As the areas under slash-and-burn agriculture in the north east are constantly increasing the spread of this noxious weed is also accelerating, posing a serious threat to native species (Rao and Murugan, 2006).

### *Ageratina adenophora* (Spreng.) King & Robinson

A close relative of *C. odorata*, this has become an aggressive weed at comparatively high elevations throughout the tropical belt in the north-eastern states, including Meghalaya. Like many other species this weed is also a native of Mexico, now being extensively distributed in open forests and on hill slopes and plateaux in Meghalaya (Fig. 15.1a).

### *Eichhornia crassipes* (Mart.) Solms.

While the above invasive weeds have completely altered the herbaceous flora in their respective terrestrial ecosystems, *Eichhornia crassipes* has done the same in aquatic and wetland ecosystems. This Brazilian weed was introduced as an ornamental plant because of its fine blue flowers and shining foliage, but today it has spread to all parts of the country in the plains. This free-floating species has tremendous potential for vegetative multiplication and can completely cover an entire water body within a short span of time, affecting the growth of other aquatic plants such as *Monochoria hastata* (L.) Solms, *Nasturtum indicum* L., *Ludwigia octovalvis* (Jacq.) Raven, *Echinochloa colonum* (L.) Link, *Leersia hexandra* Sw., *Ottelia alismoides* (L.) Pers. and *Rotala rotundifolia* (Don) Koehne (Fig. 15.1f).

### *Galinsoga quadriradiata* Ruiz et Pavon

A South American weed, first reported from India by Babu (1969) and a new record for Meghalaya in 1976 by Rao and Rao, this plant is very closely allied to and associated with the invasive species *G. parviflora* Cav.

### *Parthenium hysterophorus* L.

A tropical American weed, *Parthenium* or 'congress grass' is a relatively recent immigrant into India (Rao, 1956); since the arrival of *P. hysterophorus* L. from the New World, it has spread throughout the Indian subcontinent (Aneja *et al.*, 1991), and is now considered to be the principal terrestrial weed in India (Dhawan *et al.*, 1993). The species is toxic to livestock and causes allergic responses, such as respiratory malfunction and dermatitis, in susceptible humans (Towers *et al.*, 1977). The extensive spread of this weed, particularly in the South, has altered herbaceous species composition in many areas. A number of local species have already given way to this noxious weed (Rao, 1996).

## Concluding Remarks

Alien invasive weeds in tropical regions represent a major threat to native biodiversity.

**Fig. 15.1.** A, Pure patch of *Ageratina adenophora* along the roadside in the hills; B, *Galinsoga parviflora* invasive in a hilly area; C, *Erigeron karvinskianus* occupying road cuttings in the hills; D, *Ageratina adenophora* covering the native *Nephrolepis* sp.; E, *Ageratum conyzoides* replacing the native vegetation; F, *Eichhornea crassipes* covering an entire wetland, pushing out the native weeds.

Meghalaya State, with its enormous biodiversity, has suffered most with the recent spread of invasive alien weeds such as *Mikania*, *Chromolaena*, *Ageratina* and *Parthenium*, which have greatly altered the floristic composition of the native herbaceous flora in many localities. This is very much evident all along the Guahati–Shillong road, where the extensive advance of *M. micrantha* has replaced literally every other weed, including the native species. Although the degree of loss of native species is not yet clearly established, the disappearance of a number of native species is clearly perceptible. Future flora investigators and floristic botanists must therefore concentrate on assessing the extent of decline in native species/populations by comparing past floristic accounts with current, to evaluate the rapidity of change in floral patterns due to the spread of invasive alien species over the years.

## Acknowledgements

The authors are grateful to the Head of HRDG, CSIR (Human Resource Development Group, Council of Scientific and Industrial Research), New Delhi for financial support and also to the Director, CIMAP, Lucknow for the facilities provided. The senior author is thankful to the Indian National Science Academy, New Delhi for financial assistance through his position as INSA Honorary Scientist.

## References

Aneja, K.R. Dhawan, S.R. and Sharma, A.B. (1991) Deadly weed, *Parthenium hysterophorus* L. – and its distribution. *Indian Journal of Weed Science* 23, 14–18.

Babu, C.R. (1969) *Galinsoga ciliata* (Rafn.) Blake, a new record for India. *Bulletin of Botanical Survey India* 11, 184–185.

Barrett, S.C.H and Kohn, J.R. (1991) Genetic and evolutionary consequences of small population size in plants: implications for conservation. In: Falk, D.A. and Holsinger, K.E. (eds) *Genetics and Conservation of Rare Plants*. Oxford University Press, New York and Oxford, pp. 3–30.

Biswas, K. (1934) Some foreign weeds and their distribution in India and Burma. *Indian Forestry* 60, 861–865.

Burhl, P. (1980) Recent plant immigrants. *Journal of the Asiatic Society of Bengal* II 4, 603–655.

Darwin, C. (1872) *On the Origin of Species*, 6th edn. John Murray, London.

Dhawan, S.R., Aneja, K.R. and Dhawan, P. (1993) *Parthenium hysterophorus* Linn. – the danger weed and its control. *Biome* 6, 117–122.

Dutta, A.C. (1977) Biology and control of *Mikania. Two Bud* 24, 17–20.

Falk, D.A. and Holsinger, K.E. (1991) *Genetics and Conservation of Rare Plants*. Oxford University Press, New York and Oxford.

Hamilton, W.D. (1982) Pathogens as causes of genetic diversity in their host population. In: Anderson, R.M. and May, R.M. (eds) *Population Biology of Infectious Diseases*. Springer-Verlag, New York, pp. 269–296.

Haridasan, K. and Rao, R.R. (1985) *Forest Flora of Meghalaya*, vol. 1. Bishen Singh & Mahendra Pal Singh, Dehra Dun, India.

Haridasan, K., Kumar, Y. and Rao, R.R. (1981) Enumeration of some important forest weeds of Meghalaya with special emphasis on their role and distribution. *Journal of Economic and Taxonomic Botany* 2, 161–171.

Harper, J.L. (1977) *Population Biology of Plants*. Academic Press, London.

Karron, D. (1991) Pattern of genetic variation and breeding systems in rare plant species. In: Falk, D.A. and Holsinger, K.E. (eds) *Genetics and Conservation of Rare Plants*. Oxford University Press, New York and Oxford, pp. 87–98.

Kashyap, S.R. (1922) Notes on some foreign plants which have recently established themselves about Lahore. *Journal of the Indian Botanical Society* 3, 68–71.

Maheshwari, J.K. (1962) Studies on naturalized flora of India. In: Maheshwari, P., Johri, B. and Vasil, M. (eds) *Proceedings of a Summer School of Botany*, Darjeeling, India, pp. 156–170.

Mathew, K.M. (1969) Exotic flora of Kodaikanal, Palni Hills. *Records Botanical Survey India* 20, 1–241.

Muniyappan, R. and Viraktamath, C.A. (1993) Invasive alien weeds in the Western Ghats. *Current Science* 64, 555–558.

Myers, N., Mittermeir, R.A., Mittermeir, C.G., Da Fonesca, G.A.B. and Kent, J. (2000) Biodiversity and hotspots for conservation priorities. *Nature* 403, 853–858.

Neogi, B. and Rao, R.R. (1980) Floristic composition of the weed flora, seasonal variation and phenology of some weeds of agriculture lands in Khasi hills, Meghalaya. *Proceedings of the Indian National Science Academy B* 46, 579–586.

Rai, L.C. and Gaur, J.P. (2006) (eds) *Invasive Alien Species and Biodiversity in India.* BHU, Varanasi, India.

Ramakrishnan, P.S. (1991) *Ecology of Biological Invasion in the Tropics*. International Scientific Publications, New Delhi, India, 195 pp.

Ramaswamy, S.V., Rao, R.R. and Bhaskar, V. (1972) Contributions from the Herbarium, Manasagangotri, Mysore. VI. Notes on some adventive plants of Bangalore and Mysore districts. *Journal of Mysore University B* 25, 40–44.

Rao, R.S. (1956) *Parthenium hysterophorus* Linn. – a new record for India. *Journal of the Bombay Natural History Society* 54, 218–220.

Rao, R.R. (1977) Changing patterns in Indian flora. *Bulletin of the Botanical Survey of India* 19, 156–166.

Rao, R.R. (1978) *Parthenium hysterophorus* Linn. (Asteraceae) from Guwahati, Assam. – a new record for north-east India. *Journal of the Indian Botanical Society* 58, 44–46.

Rao, R.R. (1996) Compositae in the conservation of genetic diversity in wild plants in India. In: Calgari, P.D.S. and Hind, D.J.N. (eds) *Asteraceae: Biology & Utilization. Proceedings of the International Asteraceae Conference*, Kew, 1994. Royal Botanic Gardens, Kew, 2, 269–275.

Rao, R.R. and Murthy, S.K. (1990) North-east India – a major center for plant diversity in India. *Indian Journal of Forestry* 13, 214–222.

Rao, R.R. and Murugan, R. (2006) Impact of exotic adventive weeds on native biodiversity in India: Implications for conservation. In: Rai, L.C. and Gaur, J.P. (eds) *Invasive Alien Species and Biodiversity in India.* BHU, Varanasi, India.

Rao, R.R. and Rao, M.K.V. (1976) *Galinsoga ciliata* (Rafn.) Blake in Shillong – a new record. *Geobios* 3, 71.

Rao, R.R. and Suryanarayana, K. (1979) Introduced weeds in the vegetation of Mysore district. *Journal of Bombay Natural History Society* 74, 688–697.

Saxena, K.G. and Ramakrishnan, P.S. (1984) Herbaceous vegetation development and weed potential in slash and burn agriculture (Jhum) in north-eastern India. *Weed Research* 24, 127–134.

Srivastava, J.C. (1964) Some tropical American and African weeds that have invaded the state of Bihar. *Journal of the Indian Botanical Society* 43, 102–112.

Steffen, W. (1996) GCTE's focus on global change and ecological complexity – an international research agenda. In: Ramakrishnan, P.S., Das, A.K. and Saxena, K.G. (eds) *Conserving Biodiversity for Sustainable Development.* Indian National Science Academy, New Delhi, India, pp. 131–139.

Towers, G.H.N., Mitchell, T.C., Rodriguez, E., Bennett, F.D. and Subba Rao, P.V. (1977) Biology and chemistry of *Parthenium hysterophorus* L., a problem weed in India. *Journal of Scientific and Industrial Research* 36, 672–684.

Tripathi, R.S. (1985) *Population Dynamics of a Few Exotic Weeds in North-east India.* Academic Press, Inc., London, pp. 157–170.

Wallace, A.R. (1902) *Island Life*, 3rd edn. Macmillan, New York.

# 16 Invasion by Alien Macrophytes in Freshwater Ecosystems of India

**Manzoor A. Shah and Zafar A. Reshi**

*Department of Botany, University of Kashmir, Srinagar, Jammu and Kashmir, India*

## Introduction

Freshwater ecosystems support almost 6% of all described species (Hawksworth and Kalin-Arroyo, 1995), despite comprising only a tiny fraction of the world's water (0.01%) and terrestrial (0.8%) surfaces (Gleick, 1996). Although freshwater biodiversity provides a wide variety of valuable economic goods and irreplaceable ecosystem services for humanity, it is increasingly threatened by overexploitation, pollution, water flow changes, habitat degradation and invasion by alien species (Dudgeon *et al.*, 2006). Of these threats, the spread of invasive species appears the most severe (Zedler and Kercher, 2004; Olden *et al.*, 2006) and causes considerable damage, with cascading effects on structural organization and functional integrity of freshwater ecosystems. There is a relatively greater degree of decline and extinction of species in freshwater ecosystems than in terrestrial or marine environments (Johnson *et al.*, 2008), mainly due to hydrologic alterations and biological invasions. It is for these reasons that freshwater biodiversity comprises a priority conservation concern during the United Nations' International Decade (2005–2015) for Action – '*Water for Life*' (Dudgeon *et al.*, 2006).

While the human species has greatly facilitated the invasion of aquatic alien species through intentional shipping and trade, aquarium releases and canal construction, etc. (Rahel, 2007), flowing river waters act as natural conduits for the transport of propagules of alien species (Lavoie *et al.*, 2003) across geographical and political boundaries. The relatively higher vulnerability of aquatic systems to plant invasions is attributed mainly to water being an efficient vector for dispersal and the relatively weaker restrictions on establishment and spread of plants. Aquatic habitats, such as lakes and wetlands, being mostly resource rich are especially more vulnerable to spread of invasive alien species (Zedler and Kercher, 2004). Besides, anthropogenic disturbance of established vegetation in these habitats further facilitates invasion meltdown, allowing aliens to compete with and displace useful native plant species, some of which are important sources of food and shelter for avifauna and other wildlife. Moreover, many aquatic plants are characteristically invasive due to their immense potential for rapidly increasing their spatial spread (Richardson *et al.*, 2000) and, once established, these invasive species impair ecological processes such as water table levels, fire regimes, soil quality and nutrient cycling, thereby influencing both the flow and chemistry of water, blocking drainage pipes, impeding navigation and hindering commercial and recreational fishing. Thus, invasive species affect aesthetics, wildlife and avifaunal habitability – and the overall ecology and socio-economics of infested waterbodies. For these reasons, governmental and non-governmental agencies have to spend millions of dollars annually to control the spread of invasive plants and reverse their impacts in affected habitats.

## Status of Freshwater Ecosystems in India

India, owing to its characteristic geological history, geographical setting and eco-climatic regimes, abounds in complex and diverse freshwater ecosystems that are rich repositories of native biodiversity. Aquatic habitats in India include 14 major, 44 medium and hundreds of minor river systems; many natural Himalayan and floodplain lakes and wetlands of diverse origin (Zutshi *et al.*, 1980); numerous thermal springs in the Himalaya, Western Ghats and other hills (Rao, 1975); a few salt lakes in the arid region of Rajasthan and cold desert of Ladakh; and several coastal lagoons. Moreover, the Indian landscape is dotted with over 4290 large (>15 m maximum depth or >1 million $m^3$ storage) and innumerable smaller man-made water-bodies (Sugunan 1995). In India, freshwater wetlands alone occupy 58.2 million ha (Directory of Indian Wetlands, 1993) and support 20% of the known biodiversity in the country (Deepa and Ramachandra, 1999).

The total number of aquatic plant species in Indian freshwaters exceeds 1200 (Gopal, 1995). These aquatic plants are characterized by a variety of life forms and functional groups, such as emergents, rooted- and floating-leaf types and submersed types that adapt to diverse habitat conditions, depending on the level, quality and availability of water during the growing season.

The freshwater ecosystems in India, together with their biodiversity, represent an evolutionary mosaic of interacting ecological, socio-economic, cultural and geopolitical factors. However, alteration in the natural disturbance regimes, mainly due to a shift from traditional, value-based to modern, market-driven social settings in the country, has provided opportunities for the introduction, establishment and spread of some alien species. India's fast-growing trade links in the globalized world have further accentuated the rate of plant invasion. Notwithstanding the well-known invasion-induced alteration of aquatic ecosystems, studies on aquatic plant invasions in India are inadequate, scattered and disorganized, leaving large information gaps on the taxonomic status of alien species and their current and potential distribution, and priority invasive species and their impact. Although several studies have been carried out on macrophytic diversity in different lentic and lotic freshwaterbodies of India (Mirashi, 1954; Chavan and Sabnis, 1961; Unni, 1971; Zutshi, 1975; Gopal and Sharma, 1979; Billore and Vyas, 1982; Dey and Kar, 1989; Gopal, 1990; Gopal and Chamanlal, 1991; Gopal and Zutshi, 1998; Kar and Arbhuiya, 2001), very little is known about the extent of invasion (Reshi *et al.*, 2008). A rich body of literature documents the widespread invasion of non-native species and their impacts in aquatic systems elsewhere, however (Stromberg *et al.*, 1997; Strayer, 1999; Gollasch, 2002; Genovesi and Shine, 2004; Gollasch and Nehring, 2006; Lodge *et al.*, 2006; Minchin, 2007a).

Lack of any systematic effort towards documenting alien plant species in the freshwater ecosystems of India has multifarious ecological and economic implications. It is in this context that an attempt, based on primary and secondary data sources, has been made to compile a conspectus of alien plant species harboured by different freshwater ecosystems. For the present study, alien species were considered as those that have been introduced intentionally or unintentionally in areas outside of their natural distributional range. Based on the available literature, herbarium material and field surveys, we tried to include all the alien plant species that have naturalized in India. Species occurring adjacent to aquatic environments with a dependency on water have also been included. Nativity of the species was established through all possible sources, both published and unpublished, and presented continent-wise as suggested by Pyšek *et al.* (2002). The important sources included Anonymous (1883–1970); Tutin *et al.* (1964–1980); Weber (2003); and Khuroo *et al.* (2007), in addition to consulting some relevant internet sources (www.efloras.org; Panov, 2004; Panov and Gollasch, 2004). For species of Asian nativity, Asia excludes the Indian subcontinent but includes the other Asian countries, including Central Asia, the Far East and South-east Asia. Nativity of those species traced to more than three continents was considered cosmopolitan. The documented plant species, in addition to a few trees, were classified into emergents, rooted floating-leaf types, submersed types and free-floating types. Despite being preliminary in nature, this would provide information on broad taxonomic and ecological patterns exhibited by alien aquatic plant species, and would also identify the potential gaps in knowledge requiring urgent attention in order to acquire a more complete assessment and characterization of alien species – which is pivotal to their management.

## Aquatic Alien Plant Species in India

The present study demonstrates that 223 alien aquatic plant species are present in India (Table 16.1), representing 116 genera from 60 families, mostly natives to Eurasia. *Cyperaceae* (30 spp.) was the largest family, followed by *Asteraceae* and *Polygonaceae* (13 spp. each) and *Potamogetonaceae* and *Poaceae* (11 and 10 spp., respectively). *Brassicaceae* and *Lemnaceae* (8 spp. each) were the next largest families, followed by *Amaranthaceae*, *Hydrocharitaceae* and *Lamiaceae* (7 spp. each). In addition, the list includes 24 monogeneric families. *Cyperus* and *Potamogeton* (15 and 11 spp., respectively) were the two largest genera. This inventory is based on an exhaustive survey of the literature (Duthie, 1903–29; Stewart, 1972; Babu, 1977; Zutshi *et al.*, 1980; Naqshi, 1981; Reshi, 1984; Kaul, 1986; Koul and Naqshi, 1988; Banerjee and Matai, 1990; Kak, 1990; Prasad *et al.*, 1996; Gupta, 2001; Khan *et al.*, 2004; Pandit *et al.*, 2005; Pradhan *et al.*, 2005; Adhikari and Babu, 2008) and field study surveys carried out in the Kashmir Himalaya, India.

The majority of the alien aquatic species in Indian waters are native to either Europe (37.16%) or Asia (25.28%), with those from either north or South America each ~10% (Table 16.2; Fig. 16.1) followed by Africa (8.42%) and Australia (6.51%). The highest proportion of European and North American native species among the Indian alien flora are found in the Kashmir Himalaya region, which may partly be attributed to the close climatic matching of these biogeographical regions; India's colonial past would have facilitated propagule transport from Europe to this region (Khuroo *et al.*, 2007) in the late-19th and early-20th century. Although neither mode nor purpose of introduction of the listed alien species was explored, most were probably introduced through frequent intercontinental voyages driven by economic interests. Besides increased trade and tourism activities, the construction of new canals and tributaries have also facilitated the exchange of global species. In addition, the possible role of migratory birds in alien introductions can not be ruled out, because thousands of such birds from different countries annually visit different Indian freshwater habitats; these birds can contribute significantly to influx by transporting seeds or other propagules of alien plant species.

Classification of alien plants in the present catalogue on the basis of their habit (Table 16.2; Fig. 16.2) demonstrated the predominance of perenniality (78.9%) in comparison with annuality (15.2%) and bienniality (4%). Plants were also classified into different life form types such as emergents, rooted floating-leaf types, free-floating types and submersed types. Emergents represent the dominant group (150 spp. (67.2%)), followed by rooted floating-leaf types, submersed types and free-floating types (27 (12.1%), 24 (10.7%) and 18 (8%) spp., respectively) (Fig. 16.3).

Whilst emergent alien macrophytic species fringe the littoral regions of lakes and wetlands, rooted floating-leaf types dominate

**Table 16.1.** Preliminary checklist of aquatic alien plant species found in the Indian freshwater ecosystems.

| Family/species | Origin | Growth form | Reference |
|---|---|---|---|
| *Acanthaceae* | | | |
| *Hygrophila polysperma* (Roxb.) T. Anders. | As | E | Pradhan *et al.* (2005) |
| *Rungia parviflora* (Retz.) Nees. | As | E | Pradhan *et al.* (2005) |
| *Alismataceae* | | | |
| *Alisma gramineum* Lej. | Eu | E | Kak (1990) |
| *Alisma lanceolatum* With. | Eu | E | Stewart (1972) |
| *Alisma plantago-aquatica* L. | Eu | E | Kak (1984); Khan *et al.* (2004) |
| *Saggitaria latifolia* Willd. | NAm | E | Kaul (1986) |
| *Saggitaria sagittifolia* L. | Eu | E | Reshi (1984) |
| *Amaranthaceae* | | | |
| *Achyranthes aspera* L. | As | E | Pradhan *et al.* (2005) |
| *Amaranthus lividus* L. | As, Af, SAm | E | Pandit *et al.* (2005) |
| *Alternanthera paronychioides* St. Hill. | SAm, Au | E | Adhikari and Babu (2008) |
| *Alternanthera peploides* (Hund & Bon) Ur | NAm | E | Khan *et al.* (2004) |
| *Alternanthera sessilis* Br. | SAm | E | Kaul (1986) |
| *Alternanthera caracasana* Kuwth. | NAm | E | Naqshi (1981) |
| *Alternanthera philoxeroides* (Mart.) Griseb. | SAm | E | Gupta (2001) |
| *Apiaceae* | | | |
| *Berula erecta* (Huds.) Coville | Eu | E | Stewart (1972) |
| *Hydrocotyle sibthorpioides* Lamk. | As, Af | E | Adhikari and Babu (2008) |
| *Aponogetonaceae* | | | |
| *Aponogeton crispus* Thunb. | Au | E | Gupta (2001) |
| *Araceae* | | | |
| *Crptocryne retrospiralis* (Roxb.) Fisher ex. Wydler | As | E | Gupta (2001) |
| *Colocasia esculenta* (L.) Schott var. | As | E | Gupta (2001) |
| *Lagenandra ovata* (L.) Thaw. | NAm | E | Gupta (2001) |
| *Pistia stratiotes* L. | SAm | FF | Gupta (2001) |
| *Asclepiadaceae* | | | |
| *Oxystelma secamone* (L.) Karst. | Au | E | Adhikari and Babu (2008) |
| *Eclipta alba* L. | SAm | E | Adhikari and Babu (2008) |
| *Asteraceae* | | | |
| *Ageratum conyzoides* L. | As | E | Adhikari and Babu (2008) |
| *Ageratum houstonianum* Mill | SAm | E | Adhikari and Babu (2008) |
| *Bidens cirnua* L. | NAm | E | Stewart (1972) |
| *Bidens tripartita* L. | Eu | E | Kaul (1986) |
| *Blumea laciniata* (Roxb.) DC. | As | E | Adhikari and Babu (2008) |
| *Bidens bitternata* (Lour.) Merr. & Sheriff | As | E | Adhikari and Babu (2008) |
| *Cirsium arvense* (L.) Scop. | Eu | E | Adhikari and Babu (2008) |
| *Centipeda minima* (L.) A. Br. & Asch. | As, Au | E | Adhikari and Babu (2008) |
| *Enydra fluctuans* Lour. | As | E | Adhikari and Babu (2008) |
| *Erigeron bonariensis* L. | SAm, Au | E | Adhikari and Babu (2008) |
| *Gnaphalium pensylvanicum* Willd. | As | E | Adhikari and Babu (2008) |
| *Gnaphalium polycaulon* Pers. | As | E | Adhikari and Babu (2008) |
| *Grangea maderaspatana* (L.) Poir. | As, Af | E | Adhikari and Babu (2008) |

**Table 16.1.** Continued.

| Family/species | Origin | Growth form | Reference |
|---|---|---|---|
| *Azollaceae* | | | |
| *Azolla pinnata* R. Brown | Af, As | FF | Pandit *et al.* (2005) |
| *Boraginaceae* | | | |
| *Myosotis caespitosa* Schultz | NAm | E | Kaul (1986) |
| *Myosotis scorpioides* L. Syn. *palustris* (L.) Nath | Eu | E | Kak (1990) |
| *Brassicaceae* | | | |
| *Barbarea intermedia* Boraue | Eu | E | Kaul (1986) |
| *Barbarea vulgaris* (L.) R. Br. | Eu | E | Kak (1990) |
| *Carmine flexuosa* | Eu | E | Reshi (1984) |
| *Cardamine hirsuta* L. | Eu | E | Kaul (1986) |
| *Nasturtium officinale* (Oeder) Barbus | Eu | E | Kaul (1986) |
| *Nasturtium microphyllum* Boenn. ex Rchb. | Eu | E | Gupta (2001) |
| *Rorripa islandica* Borbas | Eu | E | Naqshi and Javeid (1976) |
| *Rorippa nasturtium-aquaticum* (L.) Hayek | As | E | Adhikari and Babu (2008) |
| *Butomaceae* | | | |
| *Butomus umbellatus* L. | Eu | E | Gupta (2001) |
| *Callitrichaceae* | | | |
| *Callitriche stagnalis* Scop | Eu | E | Kak (1990) |
| *Caryophyllaceae* | | | |
| *Myosoton aquaticum* (L) Moench. | Eu | E | Kaul (1986) |
| *Sagina saginoides* (L) Karst. | Eu | E | Kaul (1986) |
| *Ceratophyllaceae* | | | |
| *Ceratophyllum demersum* L. | Eu | S | Kak (1990); Gupta (2001) |
| *Caesalpiniaceae* | | | |
| *Cassia tora* L. | Cosmop | E | Pradhan *et al.* (2005) |
| *Parkinsonia aculeate* L. | SAm | T | Pradhan *et al.* (2005) |
| *Tamarindus indica* L. | As, Af | T | Pradhan *et al.* (2005) |
| *Convolvulaceae* | | | |
| *Ipomoea aquatica* Forsk. | As | FF | Pradhan *et al.* (2005) |
| *Ipomoea carnea* Jacq. | SAm | E | Pradhan *et al.* (2005) |
| *Cyperaceae* | | | |
| *Carex diluta* Bieb. | Eu | E | Stewart (1972) |
| *Cladium jamaicense* Crantz. | Af | E | Gupta (2001) |
| *Cladium mariscus* (L.) Pohl. | Eu | E | Kak (1990) |
| *Cyperus alopecuroides* Rottb. | Af | E | Gupta (2001) |
| *Cyperus bulbosus* Vahl. | Af, As | E | Adhikari and Babu (2008) |
| *Cyperus compressus* L. | Cosmop | E | Adhikari and Babu (2008) |
| *Cyperus difformis* L. | Eu | E | Reshi (1984) |
| *Cyperus flabelliformis* Rottb | Af | E | Stewart (1972) |
| *Cyperus fuscus* L. | Eu | E | Kak (1990) |
| *Cyperus globosus* All. | Af, Eu | E | Reshi (1984) |
| *Cyperus glomeratus* L. | Eu | E | Kak (1990) |
| *Cyperus iria* L. | As, Af | E | Kaul (1986) |
| *Cyperus michelianus* (L.) Link | Eu | E | Kak (1990) |
| *Cyperus pumilus* L. | Au | E | Kak (1990) |
| *Cyperus pygmaeus* Rottb. | Cosmop | E | Adhikari and Babu (2008) |

*Continued*

**Table 16.1.** Continued.

| Family/species | Origin | Growth form | Reference |
|---|---|---|---|
| *Cyperus rotundus* L. | Eu | E | Kaul (1986) |
| *Cyperus sanguinolentus* L. | NAm,SAm | E | Kaul (1986) |
| *Cyperus triceps (Rottb.)* Endl. | As | E | Adhikari and Babu (2008) |
| *Eleocharis equisetoides* (Ell.) Torr. | SAm | E | Gupta (2001) |
| *Eleocharis atropurpurea* (Retz) Presl. | SAm | E | Reshi (1984) |
| *Eleocharis acicularis* (L.) Roem et Schlt. | NAm,SAm | E | Kak and Javeid (1976) |
| *Eleocharis palustris* R. Br. | Eu | E | Stewart (1972) |
| *Eleocharis parishii* Britton. | NAm | E | Kak (1990) |
| *Eleocharis pauciflora* Link | NAm | E | Kak (1990) |
| *Fimbristylis dichotoma* (L.) Vahl. | Af, As | E | Reshi (1984) |
| *Fimbristylis miliacea* (L.) Vahl | SAm | E | Gupta (2001) |
| *Scirpus juncoides* Roxb. | NAm | E | Kaul (1986) |
| *Scirpus martimus* L. | Eu | E | Paul (1977) |
| *Scirpus triqueter* L. | Eu | E | |
| *Scirpus tuberosus* Roxb. | Cosmop | E | Adhikari and Babu (2008) |
| *Elatinaceae* | | | |
| *Elatina triandra* Schkuht | Eu | S | |
| *Eriocaulaceae* | | | |
| *Eriocaulon siebealdianum* Sieb. | NAm | E | Kak (1990) |
| *Euphorbiaceae* | | | |
| *Trewia nudiflora* L. | As | T | Adhikari and Babu (2008) |
| *Fabaceae* | | | |
| *Aeschynomene aspera* L. | NAm | E | |
| *Acacia farnesiana* Willd. | As | T | Adhikari and Babu (2008) |
| *Haloragidaceae* | | | |
| *Myriophyllum verticillatum* L. | Eu | S | Pandit *et al.* (2005) |
| *Myriophyllum spicatum* L. | Eu | S | Kak (1990) |
| *Hippuridaceae* | | | |
| *Hippuris vulgaris* L. | Eu | E | Pandit *et al.* (2005) |
| *Hydrocharitaceae* | | | |
| *Hydrilla verticillata* Casp. | Eu | S | Gupta (2001) |
| *Hydrocharis dubia* Backer. | As | E | Kak (1990) |
| *Nechamandra alternifolia* (Roxb.)Thw. | Af | S | Gupta (2001) |
| *Ottelia alimoides* (L.) Per. | Au, Af | S | Gupta (2001) |
| *Vallisneria natans* (Lour) Hara | As | FF | Adhikari and Babu (2008) |
| *Vallisneria spiralis* L. | Eu | FF | Kak (1990) |
| *Vallisneria americana* Michx. | NAm | FF | Cook (1996) |
| *Juncaceae* | | | |
| *Juncus articulatus* L. | Eu | E | Kaul (1986) |
| *Juncus bufonius* L. | Eu | E | Reshi (1984) |
| *Juncus effusus* L. | Eu | E | Kak (1990) |
| *Juncus inflexus* L. | Eu | E | Stewart (1972) |
| *Juncaginaceae* | | | |
| *Triglochin palustris* L. | Eu | E | Kak (1990) |
| *Lamiaceae* | | | |
| *Lycopus europeus* L. | Eu | E | Kak (1990) |

**Table 16.1.** Continued.

| Family/species | Origin | Growth form | Reference |
|---|---|---|---|
| *Mentha aquatica* L. | Eu | E | Kak (1990) |
| *Mentha arvensis* L. | Eu | E | Kak (1990) |
| *Mentha longifolia* (L.) Huds. | Eu | E | Kak (1990) |
| *Mentha piperita* L. | Eu | E | Kak (1990) |
| *Mentha spicata* L. | NAm | E | Kak (1990) |
| *Salvia plebeia* R. Br. | As, Au | E | |
| *Lemnaceae* | | | |
| *Lemna gibba* L. | Eu | FF | Pandit *et al.* (2005) |
| *Lemna minor* L. | Eu | FF | Kaul (1986) |
| *Lemna perpusilla* Torrey | NAm | FF | Adhikari and Babu (2008) |
| *Lemna turionifera* Landolt | As, NAm | FF | Kak (1990) |
| *Lemna trisulca* L. | Af, As | FF | Kak (1990) |
| *Spirodela polyrhiza* Schleid. | Eu | FF | Reshi (1984) |
| *Wolffia arrhiza* Horkel ex Wimmer | Eu | FF | Kak (1990) |
| *Wolffia columbiana* Karsten | NAm | FF | Kak (1990) |
| *Lentibulariaceae* | | | |
| *Utricularia aurea* Lour. | As, Au | S | Pandit *et al.* (2005) |
| *Lytharaceae* | | | |
| *Ammania auriculata* Willd. | As, Au | E | Kaul (1986) |
| *Ammania baccifera* L. | As | E | Kaul (1986) |
| *Lytharum salicaria* L. | Eu | E | Kak (1990) |
| *Rotala densiflora* (Willd.) Koehne | As, Au | E | Kaul (1986) |
| *Marsiliaceae* | | | |
| *Marsilia minuta* L. | As, Eu | RF | Gupta (2001) |
| *Marsilia quadrifolia* L. | Eu | RF | Reshi (1984) |
| *Marsilia uncinata* A. Br. | Nam | RF | Venu (1998) |
| *Martyniaceae* | | | |
| *Martynia annua* L. | NAm | H | Adhikari and Babu (2008) |
| *Mimosaceae* | | | |
| *Pithecellobium dulce* (Roxb.) Benth. | SAm | E | Adhikari and Babu (2008) |
| *Menyanthaceae* | | | |
| *Menyanthes trifoliata* L. | As | E | Kaul (1986) |
| *Nymphoides peltata* Kuntze | Eu | RF | Kaul (1986) |
| *Najadaceae* | | | |
| *Najas gramineum* Del. | NAm, SAm | S | Kak (1990) |
| *Najas marina* L. | Eu | S | Pandit *et al.* (2005) |
| *Najas minor* All. | As | S | Adhikari and Babu (2008) |
| *Nelumbonaceae* | | | |
| *Nelumbo nucifera* Garetn. | As, Eu | RF | Zutshi *et al.* (1980) |
| *Nymphaeceae* | | | |
| *Cabomba aquatica* Aublet. | Af | S | Gupta (2001) |
| *Nymphaea alba* L. | Eu | RF | Koul and Naqshi (1988) |
| *Nymphaea nouchali* Burm.f. | As, Af | RF | Adhikari and Babu (2008) |
| *Nymphaea lotus* L. | Eu | RF | Koul and Naqshi (1988) |
| *Nymphaea mexicana* Zucc. | NAm | RF | Kak (1990) |
| *Nymphaea stellata* Willd. | Af | RF | Gupta (2001) |

*Continued*

**Table 16.1.** Continued.

| Family/species | Origin | Growth form | Reference |
|---|---|---|---|
| *Nymphaea tetragona* Georgi. | Eu | RF | Koul and Naqshi (1988) |
| *Nymphaea tuberosa* Paine. | NAm | RF | Pandit *et al.* (2005) |
| *Nymphoides cristata* (Roxb.) Kuntze | As | RF | Adhikari and Babu (2008) |
| *Nymphoides indica* (L.) Kuntze | As | RF | Adhikari and Babu (2008) |
| *Nymphoides peltatum* (Britt. & Bend.) | As | RF | |
| *Onagraceae* | | | |
| *Epilobium hirsutum* (L.) Gray | Eu | E | Kaul (1986) |
| *Epilobium palustre* L. | Eu | E | Cook (1996) |
| *Jussiea repens* L. | Sam | RF | Gupta (2001) |
| *Ludwigia adscendens* (L.) Hara | Sam | RF | Adhikari and Babu (2008) |
| *Ludwigia palustris* (L.) Elliot | Am | RF | Gupta (2001) |
| *Ludwigia parviflora* Roxb. | As | RF | Gupta (2001) |
| *Orchidaceae* | | | |
| *Spiranthes lancea* (Thunb) Bacher | Eu | E | Cook (1996) |
| *Poaceae* | | | |
| *Echinochloa colonum* Link. | As | E | Stewart (1972) |
| *Echinochloa cruss-galli* Beauv. | Eu | E | Kaul (1986) |
| *Hymanachne acutigluma* (Steud) Gil | SAm | E | Naskar (1990) |
| *Paspalum distichum* L. | Eu | E | Cook (1996) |
| *Paspalum paspaloides* (Michx.) Scrib | SAm | E | Stewart (1972) |
| *Phalaris arundianacea* L. | Eu | E | Stewart (1972) |
| *Phragmites communis* Trin. | Eu | E | Khan *et al.* (2004) |
| *Phragmites karka* (Retz.) Trin. ex Steud. | As, Af, Au | E | Adhikari and Babu (2008) |
| *Phragmites australis* Trin. | Eu | E | Khan *et al.* (2004) |
| *Saccharum bengalense* Retz. | As | E | Cook (1996) |
| *Polygonaceae* | | | |
| *Polygonum hydropiper* L. | Eu | RF | Kaul (1986) |
| *Polygonum amphibium* L. | NAm | E | Gupta (2001) |
| *Polygonum glabrum* Willd. | Af | FF | Gupta (2001) |
| *Polygonum plebeium* R. Br. | NAm | E | Gupta (2001) |
| *Polygonum barbatum* L. | As | E | Duthie (1903–1929) |
| *Polygonum nepalensis* (Meisn) Gross | As | E | Duthie (1903–1929) |
| *Polygonum orientalis* L. | As, Au | E | Naskar (1990) |
| *Polygonum tomentosum* Schrank | As, Eu | E | Gupta (2001) |
| *Rumex aquaticus* L. | As, Eu | E | Duthie (1903–1929) |
| *Rumex challepensis* Mill | As | E | Duthie (1903–1929) |
| *Rumex conglomeratus* Murry | Eu | E | Munshi (1975) |
| *Rumex dentatus* L. | Eu | E | Duthie (1903–1929) |
| *Rumex paulsenianus* Rech. | As | E | Naskar (1990) |
| *Pontederiaceae* | | | |
| *Eichhornia crassipes* (Mart.) Solms | SAm | FF | Adhikari and Babu (2008) |
| *Monochoria vaginalis* (Burm. f.) Presl. | As, Af | E | Adhikari and Babu (2008) |
| *Monochoria hastata* (L.) Solms | SAm | E | Gupta (2001) |
| *Potamogetonaceae* | | | |
| *Potamogeton crispus* L. | Eu | S | Adhikari and Babu (2008) |
| *Potamogeton filiformis* Pers. | Eu | RF | Kak and Javeid (1992) |

**Table 16.1.** Continued.

| Family/species | Origin | Growth form | Reference |
|---|---|---|---|
| *Potamogeton fluitans* Roth | Eu | RF | Duthie (1903–1929) |
| *Potamogeton natans* L. | Eu | S | Kak (1984) |
| *Potamogeton nodosus* Poir. | Eu | RF | Kak (1984) |
| *Potamogeton lucens* L. | Eu | S | Zutshi *et al.* (1980) |
| *Potamogeton pectinatus* L. | Eu | RF | Adhikari and Babu (2008) |
| *Potamogeton perfoliatus* L. | Eu | S | Pandit *et al.* (2005) |
| *Potamogeton pusillus* L. | NAm | S | Kundangar and Zutshi (1987) |
| *Potamogeton zosterifolius* Schumach. | Eu | S | Cook (1996) |
| *Potamogeton zizi* Mert. | Eu | S | Pandit *et al.* (2005) |
| *Portulacaceae* | | | |
| *Portula cleraceae* L. | NAm | RF | Kaul (1986) |
| *Primulaceae* | | | |
| *Primula inayati* Duthei | As | E | Kak (1984) |
| *Ranunculaceae* | | | |
| *Aconitum violaceum* Staph | As | E | Duthie (1903–1929) |
| *Caltha alba* K. Jacq | Eu | E | Khan *et al.* (2004) |
| *Rananculus aquatilis* L. | Eu | RF | Gupta (2001) |
| *Rananculus lingua* L. | Eu | S | Zutshi and Wanganeo (1979) |
| *Ranunculus muricatus* L. | Eu | E | Pradhan *et al.* (2005) |
| *Ranunculus sceleratus* L. | As | E | Adhikari and Babu (2008) |
| *Rosaceae* | | | |
| *Potentilla supina* L. | As | E | Pradhan *et al.* (2005) |
| *Rubiaceae* | | | |
| *Hedyotis corymbosa* L. | As | E | Pradhan *et al.* (2005) |
| *Salviniaceae* | | | |
| *Salvinia auriculata* Aubl. | SAm | FF | Banerjee and Matai (1990) |
| *Salvinia natans* All. | Af, Eu | FF | Zutshi and Wanganeo (1979); Pandit *et al.* (2005) |
| *Scrophulariaceae* | | | |
| *Bacopa monnieri* | As, Au | E | Pradhan *et al.* (2005) |
| *Limnophylla heterophylla* (Roxb.) Benth. | Au | S | Gupta (2001) |
| *Vandellia crustacea* (L.) Benth. | Eu | E | Adhikari and Babu (2008) |
| *Verbascum chinense* (L.) Santap. | Au | E | Pradhan *et al.* (2005) |
| *Veronica anagallis-aquatica* L. | As | E | Pradhan *et al.* (2005) |
| *Veronica beccabunga* L. | Eu | E | Kaul (1986) |
| *Solanaceae* | | | |
| *Nicotiana plumbaginifolia* Viv. | As | E | Pradhan *et al.* (2005) |
| *Solanum viarum* Dunal | SAm | E | Pradhan *et al.* (2005) |
| *Sparginaceae* | | | |
| *Sparganium erectum* Huds. | Eu | E | Kaul (1986) |
| *Tamaricaceae* | | | |
| *Tamarix dioica* Roxb. | As | S | Pradhan *et al.* (2005) |
| *Trapaceae* | | | |
| *Trapa natans* var *bispinosa* (Roxb.) Makino | Eu | RF | Kak (1990) |

*Continued*

**Table 16.1.** Continued.

| Family/species | Origin | Growth form | Reference |
|---|---|---|---|
| *Typhaceae* | | | |
| *Typha angustata* Bory & Chaub. | Eu, NAm | E | Kak (1990) |
| *Typha laximanii* Lepech | Eu | S | Khan *et al.* (2004) |
| *Typha elephantina* Roxb. | As, Eu | E | Banerjee and Matai (1990) |
| *Urticaceae* | | | |
| *Pouzolzia pentandra* (Roxb.) Benn. | As | E | Adhikari and Babu (2008) |
| *Pouzolzia zeylanica* (L.) Benn. | As, Au | E | Adhikari and Babu (2008) |
| *UtriculaWriaceae* | | | |
| *Utricularia vulgaris* L. | Eu | E | Adhikari and Babu (2008) |
| *Verbenaceae* | | | |
| *Phyla nodiflora* (L.) Greene | SAm | E | Pradhan *et al.* (2005) |
| *Zannichelliaceae* | | | |
| *Zannichellia palustris* L. | Eu | S | Stewart (1972) |

As, Asia; Eu, Europe; NAm, North America; Af, Africa; SAm, South America; Au, Australia; E, emergent; FF, free-floating type; S, submersed; T, tree; R, rooted floating-leaf type.

**Table 16.2.** Alien plant species belonging to different categories.

| Category | Sub-category | Species (%) |
|---|---|---|
| Origin | Asia | 25.28 |
| | Africa | 8.42 |
| | Australia | 6.51 |
| | Europe | 37.16 |
| | North America | 10.72 |
| | South America | 10.34 |
| Habit | Annual | 15.20 |
| | Biennial | 4.00 |
| | Perennial | 78.90 |
| | Annual/perennial | 1.70 |
| Growth form | Emergent | 67.26 |
| | Free-floating type | 8.07 |
| | Rooted floating-leaf type | 12.10 |
| | Submersed type | 10.76 |
| | Tree | 1.07 |

relatively shallower waters and submersed types are concentrated more in deeper waters. The distribution of the free-floating type of macrophytes is mainly determined by hydrological fluctuations and direction of water flow. These types of invasive species, however, characteristically form thick mats in eutrophic waters that are particularly rich in organic matter. The predominance of emergents amongst various life-form classes (Fig. 16.1) could be attributed to the decreasing depth of waterbodies throughout the country, due mainly to siltation and sedimentation (Khan *et al.*, 2004). The worst aquatic invasive species in India include *Azolla pinnata*, *Ceratophyllum demersum*, *Cyperus* spp., *Eichhornia crassipes*, *Hydrilla verticillata*, *Ipomea aquatica*, *Lemna minor*, *Monochoria vaginalis*, *Myriophyllum spicatum*, *Najas gramineum*, *Nymphoides peltatum*, *Phragmites australis*, *Pistia stratiotes*, *Potamogeton crispus*, *Potamogeton nodosus*, *Sagittaria sagittifolia*, *Salvinia molesta*, *Sparganium ramosum*, *Trapa bispinosa*, *Typha angustata* and *Vallisnaria americana*.

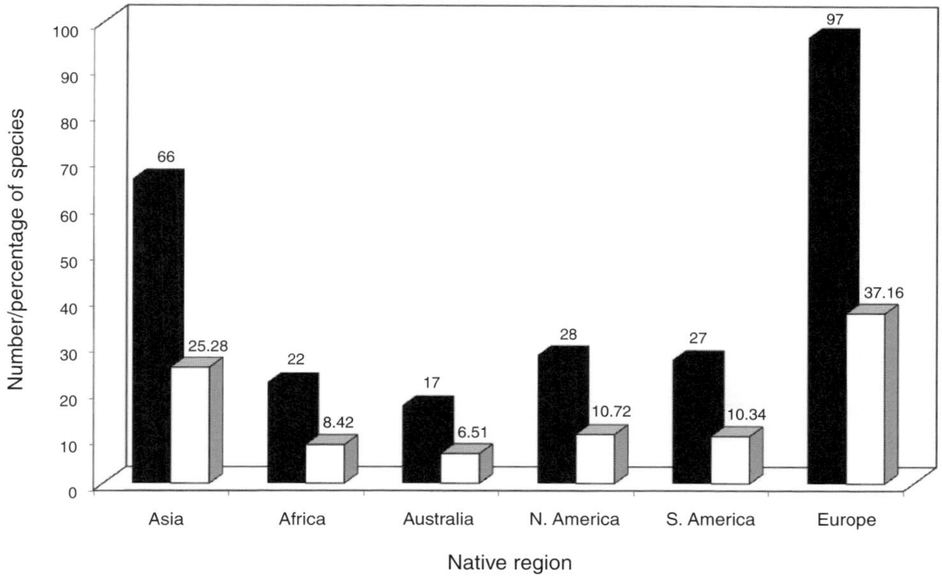


**Fig. 16.1.** Number (solid bars) and percentage (open bars) of species, by continent of origin.

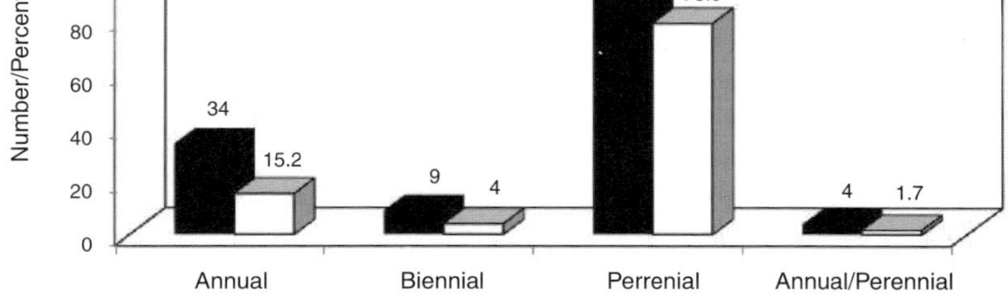


**Fig. 16.2.** Number (solid bars) and percentage (open bars) of species, by habit.

## Impacts of Plant Invasion on Freshwater Ecosystems

Invasive aquatic plants directly and indirectly impact the structural and functional attributes of freshwater ecosystems (Fig. 16.4), which in turn instigates a myriad of economic, social and ecological fall-outs (Zedler and Kercher, 2004). Economic impacts are manifested through reduced agricultural and livestock productivity, altered land use and land values, hampered

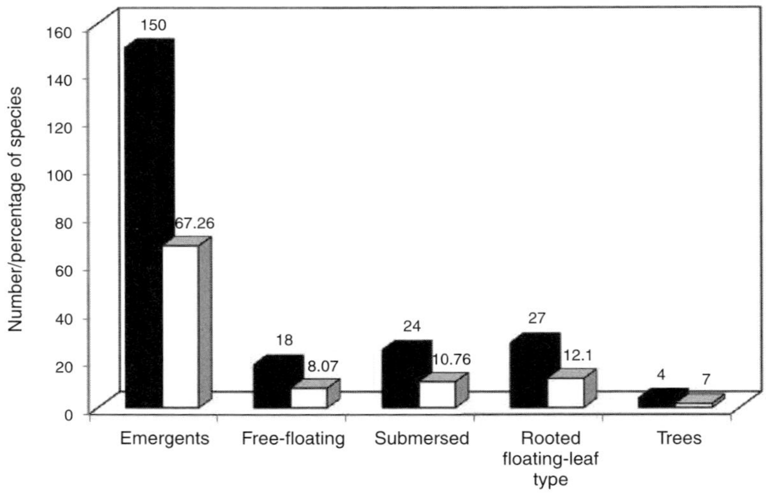


**Fig. 16.3.** Number (solid bars) and percentage (open bars) of species, by life form.

aquatic transportation and disrupted supply of economic goods and services rendered by fresh water ecosystems (Woo and Zedler, 2002). In the Indian context, with its agriculture-based economy, these changes directly and indirectly interfere with food security and economic sustainability. In addition, social impacts include decrease in value of public amenities and recreational activities such as tourism, swimming, boating, fishing and water-skiing, and deleterious effects on public health and safety. Moreover, invasive plants considerably influence ecosystem functions through impairment of water quality, alteration in nutrient cycling, change in habitat morphology, alteration in water flow and decrease in water retention. Aquatic invasive plants create dense mats of vegetation that shade and replace native aquatic plants, diminish habitat and food resource value for fish and birds and decrease water oxygen levels following their decay. Invasion-induced species extinctions change the structure of freshwater ecosystems, as reflected through reduced ecosystem stability and enhanced vulnerability to invasional meltdown.

## Strategies for Management of Aquatic Invasions

Strategies for the management of aquatic invasive species are often beset with difficulties, due to multiple interests and conflicting objectives of different stakeholders. However, efforts to manage aquatic invasive plant species in a culturally and geographically heterogenous country like India should be consistent with national goals or standards, while considering regional differences to the fullest possible extent. Exotic species currently present in Indian freshwater ecosystems need to be properly identified, mapped for their distribution and abundance and thoroughly researched with regard to their invasive status, mode and purpose of introduction, means of spread and ecological and socio-economic impacts. Better predictive tools and an operational preventive framework to guide invasion management policy need to be formulated. A unified conceptual management framework to reduce the associated economic, social and environmental costs of aquatic invasions is presented in Fig. 16.5; this framework needs to be integrated with

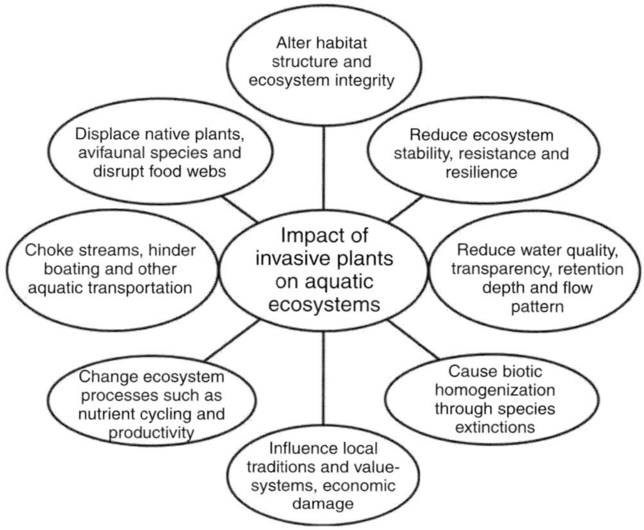


**Fig. 16.4.** Impact spectrum of invasive plants on aquatic ecosystems.

**Fig. 16.5.** Integrated framework for the management of aquatic plant invasions.

the steps, initiatives and plans for overall aquatic resource management within the country. Furthermore, models incorporating factors influencing community invasibility and species invasiveness, together with their complex feedback interactions, need to be developed for a better understanding of the invasion dynamics in order to overcome constraints to their management. Management strategies, based on such initiatives, will be most effective when a long-term, ecosystem-wide approach is employed rather than a tactical approach focused on battling individual invaders. Alternatively, failure to address the issue of plant invasions could result in wholesale loss of freshwater bioresources, disruption of ecological processes and the ecosystem services they provide, and the creation of homogeneous, impoverished landscape units rather than diverse communities. Moreover, at the national level, cooperation amongst regions and states in data acquisition on aquatic invasive species for effective management practices should be the priority consideration.

## Future Challenges

Since documentation and management of invasive species can present more complex challenges than anticipated for policy planners in India, there is an urgent need of a national panel of invasion experts for addressing the issue in a more effective and institutionalized manner. Giving the unprecedented rate of alien introductions in Indian freshwaters, early detection and monitoring the spatio-temporal spread of invasive plants, and understanding their extent, dynamics, and impacts is a challenging task for their management. Apart from the microcosm and mesocosm studies, extensive ground surveillance aided by the techniques such as remote sensing and GIS need to be undertaken on regional basis through a coordinated manner. Moreover, the sweeping waves of global environmental change, especially the climate change, present new challenges for individuals and organizations charged with invasive species management. This is because changing precipitation patterns, warmer water temperatures, receding glaciers, reduced ice cover, and altered flow regimes will influence the spread of invasive species often in unpredictable ways. Thus, in view of the possible range expansion of alien species in response to climate change, it is high time to develop a unified conceptual framework and integrated invasion management system to battle with invasive species in fresh water ecosystems of India, supported by a new set of decision-making tools and adequate infrastructural investment.

## Acknowledgement

We gratefully acknowledge the facilities provided by the Department of Botany, University of Kashmir, Srinagar, J&K, India to carry out this work.

## References

Adhikari, B.S. and Babu, M.M. (2008) Floral diversity of Baanganga Wetland, Uttarakhand, India. *Check List* 4, 279–290.

Anonymous (1883–1970) *Index Kewensis Plantarum Phanerogamarum*, vols I–II (1883–1885 and 15 Suppl.). Clarendon Press, Oxford, UK.

Babu, C.R. (1977) *Herbaceous Flora of Dehradun*. CSIR Publications, New Delhi, India.

Banerjee, A. and Matai, S. (1990) Composition of Indian aquatic plants in relation to utilization as animal forage. *Journal of Aquatic Plant Management* 28, 69–73.

Billore, D. and Vyas, L.N. (1982) Distribution and production of macrophytes in Pichhola lake, Udaipur, India. In: Gopal, B., Turner, R.E., Wetzel, R.G. and Whigham, A. (eds) *Wetlands: Ecology and Management*. International Scientific Publications and National Institute of Ecology, Jaipur, India, pp 45–54.

Chaudhari, H., Ramaprabhu, T. and Ramachandran, V. (1994) *Ipomea carnea* Jacq., a new aquatic weed problem in India. *Journal of Aquatic Plant Management* 32, 37–38.

Chavan, A.R. and Sabnis, S.D. (1961) A study on the hydrophytes of Baroda and environs.

*Journal of the Indian Botanical Society* 40, 121–130.

Colautti, R.I. and MacIsaac, H.I. (2004) A neutral terminology to define 'invasive' species. *Diversity Distribution* 10, 135–141.

Cook, C.D.K. (1996) *Aquatic and Wetland Plants of India: a Reference Book and Identification Manual for the Vascular Plants found in Permanent or Seasonal Fresh Water in the Subcontinent of India South of the Himalayas.* Oxford University Press, New York.

Deepa, R.S. and Ramachandra, T.V. (1999) Impact of urbanization in the interconnectivity of wetlands. In: *National Symposium on Remote Sensing Applications for Natural Resources: Retrospective and Perspective (XIX–XXI 1999),* Indian Society of Remote Sensing, Banglore, India.

Dey, S.C. and Kar, D. (1989) Aquatic macrophytes of Lake Sone in Assam. *Environment and Ecology* 7, 253–254.

Dudgeon, D. *et al.* (2006) Freshwater biodiversity: importance, threats, status and conservation challenges. *Biological Review* 81, 163–182.

Duthie, J.F. (1903–1929) *Flora of the Upper Gangetic Plain and of the Adjacent Siwalik and Sub Himalayan Tracts,* I–III. Superintendent of Government Printing, Calcutta, India, 1049 pp.

Dutta, T.R. and Gupta J.N. (1973) Some experiments on aquatic weed control in fisheries lakes and streams in U.P. Aquatic weeds in south-east Asia. *Proceedings of a Regional Seminar on Noxious Aquatic Vegetation in New Delhi,* pp. 249–252.

Genovesi, P. and Shine, C. (2004) European strategy on invasive alien species. In: *Nature and Environment* 137. Council of Europe, Strasbourg, France, 67 pp.

Gleick, P.H. (1996) Basic water requirements for human activities: Meeting basic needs. *Water International* 21, 83–92.

Gollasch, S. (2002) Databases on aquatic alien species: North and Mediterranean Seas and non-European Initiatives. In: Leppäkoski. E., Gollasch, S. and Olenin, S. (eds) *Invasive Aquatic Species of Europe. Distribution, Impacts and Management.* Kluwer Academic Publishers, Dordrecht, Netherlands, pp. 520–524.

Gollasch, S. and Nehring, S. (2006) National checklist for aquatic alien species in Germany. *Aquatic Invasion* 4, 245–269.

Gopal, B. (1990) Indian Subcontinent and the aquatic habitats. In: Gopal, B. (ed.) *Ecology and Management of Aquatic Vegetation in the Indian Subcontinent,* Kluwar Academic Publishers, Dordrecht, Netherlands.

Gopal, B. (1995) *Biodiversity in Freshwater Ecosystems Including Wetlands, Biodiversity and Conservation in India, A Status Report,* vol. 4. Zoological Survey of India, Calcutta.

Gopal, B. and Chamanlal, T. (1991) Distribution of aquatic macrophytes in polluted waterbodies and their bioindicator value. *Verhandlungen International Verein Limnology* 24, 2125–2129.

Gopal, B. and Sharma, K.P. (1979) Aquatic weed control versus utilisation. *Economic Botany* 33, 340–346.

Gopal, B. and Zutshi, D.P. (1998) Fifty years of hydrobiological research in India. *Hydrobiologia* 384, 267–290.

Gupta, O.P. (2001) *Weedy Aquatic Plants: Their Utility, Menace and Management.* Agrobios, Jodhpur, India, 273 pp.

Hawksworth, D.L. and Kalin-Arroyo, M.T. (1995) Magnitude and distribution of biodiversity. In: Heywood, V.H. (ed.) *Global Biodiversity Assessment.* Cambridge University Press, Cambridge, UK, pp. 107–191.

Hulme, P.E. (2006) Beyond control: wider implications for the management of biological invasions. *Journal of Applied Ecology* 43, 835–847.

Johnson, P.T.J., Olden, J.D. and Vander Zanden, M.J. (2008) Dam invaders: impoundments facilitate biological invasions into freshwaters. *Frontiers of Ecology and Environment* 6, 357–363.

Kak, A.M. (1984) Family Alismataceae in Northwestern Himalaya. *Journal of the Bombay Natural History Society* 81, 506–508.

Kak, A.M. (1990) Aquatic and wetland vegetation of Kashmir Himalaya. *Journal of Economic Taxonomic Botany* 14, 1–14.

Kak, A.M. and Javeid, G.N. (1976) *Galium palustre* L. and *Eleocharis acicularis* Roem et Schult.: two new plant records for India. *Journal of the Bombay Natural History Society* 74, 208–209.

Kar, D. and Arbhuiya, M.H.B. (2001) Ecology of aqautic macrophytes of Chatla Haor, a floodplain wetland in Cachar district of Assam. *Environment and Ecology* 19, 231–233.

Kaul, M.K. (1986) *Weed Flora of Kashmir Valley.* Scientific Publishers, Jodhpur, India.

Khan, M.A., Shah M.A., Mir, S.S. and Bashir, S. (2004) The environmental status of a Kashmir Himalayan wetland game reserve: aquatic plant communities and eco-restoration measures. *Lakes and Reservoirs: Research and Management* 9, 125–132.

Khuroo, A.A., Rashid, I., Reshi, Z., Dar, G.H. and Wafai, B.A. (2007) The alien flora of Kashmir Himalaya. *Biological Invasions* 9, 269–292.

Koul, K. and Naqshi, A.R. (1988) Nymphaeaceae of Jammu and Kashmir. *Journal of the Bombay Natural History Society* 85, 454–455.

Kumar, M. and Singh, J. (1987) Environmental impacts of aquatic weeds and their classification. In: *Proceedings of Workshop on Management of Aquatic Weeds*, Amritsar, Punjab, India.

Lancar, L. and Krake, K. (2002) Aquatic weeds and their management. *International Commission on Irrigation and Drainage* March, 77 pp.

Lavoie, C., Martin, J., Fanny, D. and Guy, L. (2003) Exotic plant species of the St Lawrence River wetlands: a spatial and historical analysis. *Journal of Biogeography* 30, 537–549.

Lodge, D.M. *et al.* (2006) Biological invasions: recommendations for U.S. policy and management (ESA Position Paper). Public Affairs Office, Ecological Society of America, Washington, DC.

Mack, R.N., Simberloff, D., Lonsdale, W.M., Evans, H., Clout, M. and Bazzaz, FA. (2000) Biotic invasions: causes, epidemiology, global consequences and control. *Ecological Applications* 10, 689–710.

Maheshwari, J.K. (1960) Studies on the naturalized flora of India. In: Mheshwari, P., Johri, B.M. and Vasil, I.K. (eds) *Proceedings of the Summer School of Botany*, pp. 156–170.

Middleton, B. (1999) *Wetland Restoration: Flood Pulsing and Disturbance Dynamics*. John Wiley & Sons, New York.

Minchin, D. (2006) The transport and spread of living aquatic species. In: Davenport, J. and Davenport, J.L. (eds) *The Ecology of Transportation: Managing Mobility for the Environment.* Springer, Dordrecht, Netherlands, pp. 77–97.

Minchin, D. (2007a) A checklist of alien and cryptogenic aquatic species in Ireland. *Aquatic Invasions* 2, 341–366.

Minchin, D. (2007b) Aquaculture and transport in a changing environment: overlap and links in the spread of alien biota. *Marine Pollution Bulletin* 55, 302–313.

Mirashi, M.V. (1954) Studies in the hydrophytes of Nagpur. *Journal of the Indian Botanical Society* 33, 299–308.

Naqshi, A.R. (1981) *Alternanthera caracasana* Kunth. (Amaranthaceae) from Indian subcontinent. *Journal of Economic Taxonomic Botany* 2, 249–250.

Naqshi, A.R. and Javeid, G.N. (1976) On identity of Indian *Rorippa islandica* Borbas. *Geobios* 3, 165–166.

Naskar, K.R. (1990) *Aquatic and Semi-aquatic Plants of the Lower Ganga Delta.* Daya Publishing House, Delhi, India.

Nayar, M.P. (1977) Changing patterns of the Indian flora. *Bulletin of Botanical Survey of India* 19, 145–154.

Olden, J.D., McCarthy, J.M., Maxted, J.T., Fetzer, W.W. and Vander Zanen, J.M. (2006) The rapid spread of rusty crayfish (*Orconectus rusticus*) with observations on native crayfish declines in Wisconsin (U.S.A.) over the past 130 years. *Biological Invasions* 8, 1621–1628.

Pandit, A.K., Rather, G.H., Wani, S.A. and Javeed, J.A. (2005) Current status of macrophytic vegetation in various freshwaterbodies of Kashmir Himalaya. *Journal of Research and Development* 5, 63–69.

Panov, V. (2004) Internet-based information resources on aquatic alien species relevant to the Ponto-Caspian Region. In: Dumont, H.J., Shiganova, T.A. and Niermann, U. (eds) *Aquatic Invasions in the Black, Caspian, and Mediterranean Seas.* Kluwer Academic Publishers, Dordrecht, Netherlands, pp. 257–269.

Panov, V. and Gollasch, S. (2004) Informational resources on aquatic alien species in Europe on the internet: present developments and future perspectives. In: Vanden, B.E., Brown, M., Costello, M.J., Heip, C., Levitus, S. and Pissierssens, P. (eds) *Proceedings of 'The Colour of Ocean Data' Symposium*, IOC Workshop Report 188, UNESCO, Paris, pp. 115–124.

Paul, S.R. (1977) Medicinal Plants of Netarhat, Bihar (India). *Pharmaceutical Biology* 15, 79–97.

Pradhan, P., Mishra, S.S., Chakraborty, S.K. and Bhakat, R.K. (2005) Diversity of freshwater macrophytic vegetation of six rivers of south West Bengal. *Tropical Ecology* 46, 193–202.

Prasad, V.P., Mason, D., Marburger, J.E. and Ajith Kumar, C.R. (1996) *Illustrated Flora of Keoladeo National Park, Bharatpur, Rajasthan.* Oxford University Press, Oxford, UK.

Pyšek, P., Sadlo, J. and Mandek, B. (2002) Catalogue of alien plants of the Czech Republic. *Preslia Praha* 74, 97–186.

Pyšek P., Richardson, D.M. and Jarošík, V. (2006) Who cites who in the invasion zoo: insights from an analysis of the most highly cited papers in invasion ecology. *Preslia* 78, 437–468.

Rahel, F.J. (2007) Biogeographic barriers, connectivity, and biotic homogenization: it's a small world after all. *Freshwater Biology* 2, 696–710.

Rao, K.L. (1975) *India's Water Wealth.* Orient Longman, New Delhi, India, 255 pp.

Rejmanek, M., Richardson, D.M., Higgins, S.I., Pitcairn, M.J. and Grotkopp, E. (2006) Ecology of invasive plants: state of the art. In: Mooney, H.A., Mack, R.M., McNeely, J.A., Neville, L., Schei, P. and Waage, J. (eds) *Invasive Alien*

*Species: A New Synthesis*. Island Press, Washington, DC, pp. 104–161.

Reshi, Z. (1984) Ecology and taxonomy of weeds of District Anantnag, Kashmir Himalaya. MPhil dissertation, Department of Botany, University of Kashmir, Srinagar, India.

Reshi, Z., Shah, M.A., Rashid, I. and Khuroo, A. (2008) Alien plant invasions: Threat to wetland health and public wealth. *Environmental Information System* (ENVIS) *Newsletter* 8, 2–5.

Richardson, D.M. and Pyšek, P. (2006) Plant invasions: merging the concepts of species invasiveness and community invasibility. *Progress in Physical Geography* 30, 409–431.

Richardson, D.M., Pyšek, P., Rejmanek, M., Barbour, M.G., Panetta, F.D. and West, C.J. (2000) Naturalization and invasion of alien plants: concepts and definitions. *Diversity and Distribution* 6, 93–107.

Soukup, A., Votrubova, O. and Cizkova, H. (2000) Internal segmentation of rhizomes of *Phragmites australis*: protection of the internal aeration system against being flooded. *New Phytologist*, 145, 71–75.

Stewart, R.R. (1972) *An Annotated Catalogue of the Vascular Plants of West Pakistan and Kashmir*. Fakhri Press, Karachi, Pakistan.

Strayer, D.L. (1999) Effects of alien species on freshwater mollusks in North America. *Journal of the North American Benthological Society* 18, 74–98.

Stromberg, J., Gengarelly, L. and Rogers, B. (1997) Exotic herbaceous species in Arizona's riparian ecosystems. In: Brock, J.H., Wade, M. and Pyšek, P. (eds) *Plant Invasions: Studies from North America and Europe.* GreenBackhuys Publishers, Leiden, Netherlands, pp. 45–57.

Sugunan, V.V. (1995) Reservoir fisheries of India. FAO Fisheries Technical Paper No. 345, Rome.

Tutin, T.G., Heywood, V.H., Burges, N.A., Valentine, D.H., Walters, S.M. and Webb, D.A. (eds) (1964–1980) *Flora Europaea*, vols 1–5. Cambridge University Press, Cambridge, UK.

Unni, K.S. (1971) An ecological study of the macrophytic vegetation of Doodhadhari lake, Raipur, MP, India I. Distribution and seasonal changes in aquatic plants. *Hydrobiologia* 37, 139–155.

Venu, P. (1998) A review of floristic diversity inventory and monitoring methodology in India. *Proceedings of the Indian National Science Academy (PINSA) B64* 5/6, 281–292.

Weber, E. (2000) Switzerland and the invasive plant species issue. *Botanica Helvetica* 110, 11–24.

Weber, E. (2003) *Invasive Plant Species of the World: a Reference Guide to Environmental Weeds*. CABI Publishing, Wallingford, UK.

Woo, I. and Zedler, J.B. (2002) Can nutrients alone shift a sedge meadow towards dominance by the invasive *Typha × glauca*? *Wetlands* 22, 509–521.

Zedler, J.B. and Kercher, S.M. (2004) Causes and consequences of invasive plants in wetlands: opportunities, opportunists, and outcomes. *Critical Reviews in Plant Sciences* 23, 431–452.

Zutshi, D.P (1975) Association of macrophytic vegetation in Kashmir lakes. *Vegetatio* 30, 61–65.

Zutshi, D.P., Subla, B.A., Khan, M.A. and Wanganeo, A. (1980) Comparative limnology of nine lakes of Jammu and Kashmir, Himalayas. *Hydrobiologia* 72, 101–112.

# 17 Plant Invasions in Jammu and Kashmir State, India

**Anzar A. Khuroo[1], Zafar A. Reshi[1], G.H. Dar[1] and Irshad A. Hamal[2]**

*[1]Department of Botany, University of Kashmir, Srinagar, Jammu and Kashmir, India; [2]Vice-chancellor, Baba Ghulam Shah Badshah University, Rajouri, Jammu and Kashmir, India*

## Introduction

In the contemporary era of global environmental change, biological invasion by alien species is implicated as the second major driver in the loss of native biodiversity worldwide (Wilcove *et al.*, 1998; Sax *et al.*, 2002). With an unprecedented surge in global trade, travel and transport over the last decade or so, the risk of invasions by alien plant species has escalated due to increasing introductions of species much beyond their natural distribution ranges (Meyerson and Mooney, 2007; Westphal *et al.*, 2008). Invasive alien plants not only pose a serious threat to native biota but have serious consequences for the economy, ecology, public health and bio-security of the countries involved (Meyerson and Reaser, 2003; Pimentel *et al.*, 2005; Pejchar and Mooney, 2009).

There is a global scientific consensus in plant invasions research that the starting point must be the systematic inventory of invasive alien flora, because this provides the crucial baseline data required for scientific understanding, future research needs and effective management of plant invasions (Hulme *et al.*, 2009; Rashid *et al.*, 2009). Broadly speaking, an inventory of invasive alien flora helps in: (i) unravelling the extent, patterns, pathways and drivers of plant invasions; (ii) understanding the underlying mechanisms that promote plant invasions; (iii) identifying high-impact invasive alien plant species; and (iv) developing effective strategies for the management of plant invasions. Notwithstanding such an indispensable role for inventories of invasive flora in plant invasion research, there are both geographical and taxonomic knowledge gaps in this regard, particularly in the least-studied continents of Asia and Africa (excepting South Africa), where standard inventories of invasive species are uncommon (Pyšek *et al.*, 2008).

India, one of the most rapidly globalizing Asian countries, faces a higher risk of plant invasions in being a mega-diverse nation, due to an exponential increase in trade and travel by land, air and water (Saxena, 1991; Sharma *et al.*, 2005). In addition, its colonial past has resulted in the introduction (intentional or unintentional) of alien plant species from different regions of the world (Chatterjee, 1947), and many of these introduced species have escaped into the wild and are presently spreading as invasives (Reddy, 2008). Over the last two decades, while a large body of scientific literature on plant invasions has been made available in the developed world (Clement and Foster, 1994; Preston *et al.*, 2002; Lambdon *et al.*, 2008), and even in our neighbouring Asian countries, such as China (Liu, *et al.*, 2005; Weber *et al.*, 2008), only a few such research studies have been undertaken in India, and these are limited on account of either being restricted to a particular region (Khuroo *et al.*, 2007, 2008) or not having followed standardized terminology and methodology (Negi and Hajra, 2007; Reddy, 2008). This renders research results non-reproducible and of little use in comparative analysis across spatio-temporal scales, thereby hampering nation-level policymaking efforts and management practices.

The Himalayas represent one of the global biodiversity hot spots. The Indian Himalayan region (IHR), being 250–300 km wide, stretches over 2500 km from Jammu and Kashmir in the west to Arunachal Pradesh in the east. In this region the climate, and consequently the biodiversity, vary considerably along altitudinal gradients. The region supports a rich biodiversity with an appreciably high degree of endemism. For example, of a total of 17,500 species of angiosperms recorded in India, the IHR contributes 8000 species of which 40% of are endemic to the region (Sharma *et al.*, 1997).

Amongst the 12 Indian states that fall within the IHR, Jammu and Kashmir State, with an area of ~222,236 km$^2$, contributes the largest area (41.65%) (Nandy *et al.*, 2006). The wide range of physiographical and phytoclimatic conditions met within the state supports a rich and diverse flora and fauna; for example, around 4500 species of vascular plant species have been recorded (Singh *et al.*, 2000). Over the recent past, in addition to other human drivers of biodiversity loss, plant invasions have emerged as one of the main threats to native biodiversity in the state (Reshi *et al.*, 2007; Khuroo *et al.*, 2009). The local drivers of plant invasions include land-use changes and habitat disturbances such as reckless deforestation, unsustainable livestock grazing, unplanned plantations, road construction in the high alpines, unregulated ornamental trade and other tourism–infrastructural activities.

It is against this backdrop that we present a synthesis of the current understanding of plant invasions in the Jammu and Kashmir. First, a conceptual framework that has been followed to study plant invasions within the state is outlined. Secondly, a critical review of the progress achieved in plant invasion research in three distinct regions (Kashmir, Jammu and Ladakh) is presented. Thirdly, current knowledge gaps and future challenges to a better scientific understanding and management of plant invasions in the state are underlined.

## General profile of the state

Jammu and Kashmir State lies between 32° 17′ and 37° 20′N and 73° 25′ and 80° 30′E (Fig. 17. 1). The state is mostly mountainous and shares borders with Pakistan to the west, China to the north and north-east and Punjab and Himachal Pradesh (Indian states) to the south. There are extreme variations in climate due to its location and topography. Phyto-geographically, the state is divided into the three distinct provinces of Jammu, Kashmir and Ladakh, differing markedly in their climate and vegetation. Jammu province has broad-leaved deciduous forests at lower elevations and dry coniferous forests at higher elevations; the Kashmir Valley contains coniferous forests and alpine meadows; and the natural vegetation in Ladakh is primarily Asiatic steppes along water courses and stunted scrub, although trees in the form of plantation stands also occur. The recorded forest area is 20,230 km$^2$, comprising 9.2% of the geographical area. The five forest types seen are subtropical evergreen, Himalayan moist temperate, Himalayan dry temperate, subalpine and alpine. While annual precipitation in Jammu is ~1700 mm, mainly due to the summer monsoon, that of the Kashmir Valley is ~1050 mm, mostly from winter snow; and because the high mountain ranges of Greater Himalaya prevent rain clouds reaching the interior valleys, Ladakh receives only ~100 mm annual precipitation. Geologically, rocks varying from the oldest period of the earth's history to the most recent river and lake deposits are seen. The state has a number of lakes, rivers, streams and glacial regions, the most important rivers being the Indus, Chenab and Jhelum (Rodgers and Panwar, 1990).

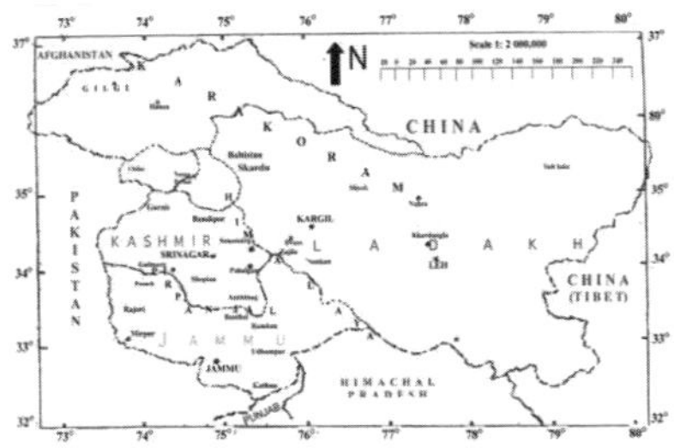


**Fig. 17.1.** Map showing the three regions of Jammu and Kashmir State: Jammu, Kashmir and Ladakh.

## Plant Invasions in Jammu and Kashmir State

### Conceptual framework adopted

In regard to characterization of invasive alien flora, alien species were first categorized into two groups – aliens under cultivation and those escaped into the wild. The latter group of alien plant species has been characterized under invasive alien flora because the actual process of invasion by an alien species in its non-native range starts when it escapes into the wild. As the extent and magnitude of invasiveness varies among alien plant species, species recognized as comprising invasive alien flora were categorized according to a stage-based invasion model (Fig. 17.2). In this classification scheme, modified after Colautti and MacIsaac (2004), stages 0 and I represented potential invaders (or their propagules) as residents in the native/donor regions and transport vectors, respectively. Although theoretically useful and relevant to management, these two stages were not recognized because of practical difficulties and hence the study recognized invasion stages only from II. The latter was represented by alien species that had recently escaped into the wild; this was followed by stage III species, comprising self-reproducing and established alien species. Subsequently, stage IV was subdivided into two stages – IVa, alien species that had become widespread but occasional, and IVb, those remaining localized but abundant. Stage V included those alien species that were both abundant and widespread throughout the region. Adoption of such a stage-based operational framework of plant invasions has the advantage of integrating alien plant species at a particular invasion stage with a relevant management strategy.

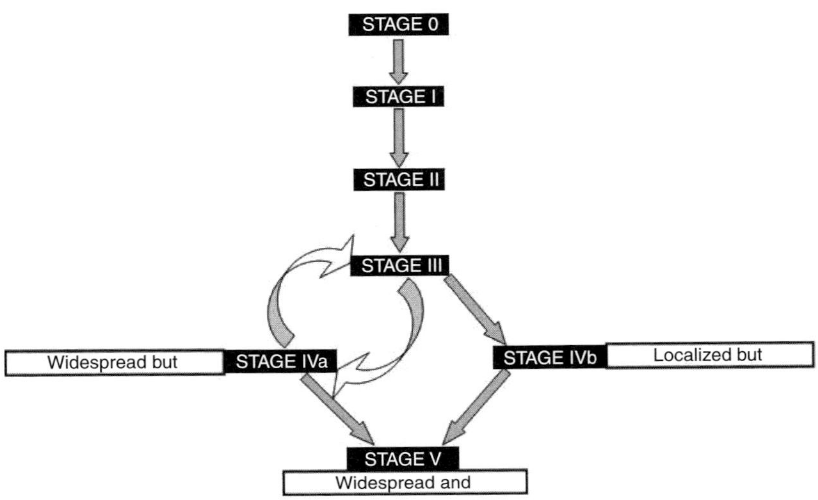


**Fig. 17.2.** Model showing the different stages recognized during the process of plant invasion (modified after Colautti and MacIsaac, 2004).

## Plant invasions in Kashmir region

In Kashmir region relatively significant progress has been achieved in plant invasion studies. An updated systematic inventory of invasive alien flora is available, wherein invasive alien plant species have been characterized, according to globally standardized terminology, reflecting the different stages of the invasion process. The inventory recognized 436 alien plant species at different stages of invasion (Table 17.1). The stages of invasion and number of species were as follows: II (119), III (107), IVa (56), IVb (77) and V (77) (Fig. 17.3). Such a stage-based approach has a clear advantage in precisely prioritizing the research and management efforts towards the high-impact invasive plant species, while at the same time keeping track of the lower-stage invasive plant species. For instance, managerial tools under the management strategy of prediction have little role in the management of the 77 species at stage V, whereas in the latter grouping the management strategy of prevention and

**Table 17.1.** Taxonomic overview of invasive alien flora of Kashmir region.

| Family | Number of species | Family | Number of species |
|---|---|---|---|
| *Agavaceae* | 1 | *Liliaceae* | 3 |
| *Alismataceae* | 5 | *Lythraceae* | 3 |
| *Amaranthaceae* | 15 | *Malvaceae* | 11 |
| *Amaryllidaceae* | 2 | *Marsileaceae* | 1 |
| *Apiaceae* | 10 | *Menyanthaceae* | 2 |
| *Apocyanaceae* | 1 | *Moraceae* | 4 |
| *Araliaceae* | 1 | *Najadaceae* | 1 |
| *Asteraceae* | 54 | *Nyctaginaceae* | 3 |
| *Balsaminaceae* | 1 | *Nymphaeaceae* | 5 |
| *Boraginaceae* | 5 | *Oleaceae* | 3 |
| *Brassicaceae* | 28 | *Onagraceae* | 6 |
| *Buddlejaceae* | 1 | *Oxalidaceae* | 1 |
| *Butomaceae* | 1 | *Papaveraceae* | 6 |
| *Buxaceae* | 1 | *Plantaginaceae* | 2 |
| *Campanulaceae* | 1 | *Plantanaceae* | 1 |
| *Cannabiaceae* | 1 | *Poaceae* | 56 |
| *Capparidaceae* | 1 | *Polemoniaceae* | 2 |
| *Caprifoliaceae* | 2 | *Polygonaceae* | 15 |
| *Caryophyllaceae* | 13 | *Portulaceae* | 1 |
| *Certophyllaceae* | 1 | *Potamogetonaceae* | 5 |
| *Chenopodiaceae* | 9 | *Primulaceae* | 1 |
| *Convolvulaceae* | 6 | *Ranunculaceae* | 13 |
| *Crassulaceae* | 1 | *Rosaceae* | 11 |
| *Cuscutaceae* | 2 | *Rubiaceae* | 8 |
| *Cyperaceae* | 15 | *Salicaceae* | 4 |
| *Elaeaganaceae* | 1 | *Salvinaceae* | 1 |
| *Euphorbiaceae* | 3 | *Scrophulariaceae* | 11 |
| *Fabaceae* | 19 | *Simaroubaceae* | 1 |
| *Fagaceae* | 1 | *Solanaceae* | 9 |
| *Geraniaceae* | 2 | *Sparganiaceae* | 1 |
| *Hippocastanaceae* | 1 | *Trapaceae* | 1 |
| *Hydrocharitaceae* | 2 | *Typhaceae* | 1 |
| *Hypericaceae* | 1 | *Urticaceae* | 1 |
| *Iridaceae* | 4 | *Valerianaceae* | 1 |
| *Juncaceae* | 3 | *Violaceae* | 2 |
| *Lamiaceae* | 19 | *Vitaceae* | 1 |
| *Lemnaceae* | 4 | Total | 436 |

**Table 17.2.** Management strategy and tools relevant at different stages of plant invasion (modified after Khuroo *et al.*, 2008).

| Management strategy | Management tools | Stages of invasion |
|---|---|---|
| Prediction | Quarantine regime<br>Evaluation of source flora<br>Climate matching<br>Precaution<br>Early detection | I, II, III |
| Prevention | Risk analysis<br>Early warning system<br>Rapid response and action<br>Eradication<br>Exclusion | III, IVb |
| Prescription | Containment<br>Control<br>Elimination<br>Mitigation<br>Adaptation<br>Restoration<br>Rehabilitation | IVa, IVb, V |
| Policy making | Monitoring<br>Public awareness<br>Policy initiatives and actions<br>Informed decision making<br>Legislative measures<br>Administrative measures | I, II, III, IVa, IVb, V |

prescription can be helpful (Table 17.2). In long-term invasive species monitoring programmes, temporal changes in the number of species moving through the stages of invasion can be used as a quantifiable indicator of management success (or failure).

In this region nearly half of the alien species have been introduced unintentionally – the majority may have been introduced by human hitchhikers, seed impurities, stowaways or other means. Among intentionally introduced species, the majority were for ornamental purposes (Fig. 17.4). While, as in case of the intentional introductions, risk assessment schemes can be applied for the a priori prediction of invasion threat posed by an alien plant species, it is difficult to do the same for unintentional examples. A decreasing trend is discernible in the distribution of invasive alien plant species along an elevational gradient: of the total of 436 species of invasive alien flora, 378 were distributed in the valley plain zone and just 14 species have reached the Montane alpine zone (Fig. 17.5).

Alien plant species with a herbaceous habit dominate, with annual, biennial and perennial herbs collectively constituting 65% of total alien flora (Fig. 17.6). In terms of source of alien flora in the region, Europe has the major share with 38%. This is probably on account of India's European colonial past and successful establishment due to climate matching (Fig. 17.7). In contrast, the major proportion (~55%) of the alien flora of India are reportedly of American origin. Interestingly, such divergent patterns in local-, regional- and national-level alien floras highlight the explicit mention of spatial scale when drawing conclusions in ecological studies. Although in terms of absolute number of species the three families *Poaceae*, *Asteraceae* and *Brassicaceae* together account for 25% of total alien flora, some families, such as *Amaranthaceae* and *Chenopodiaceae* are

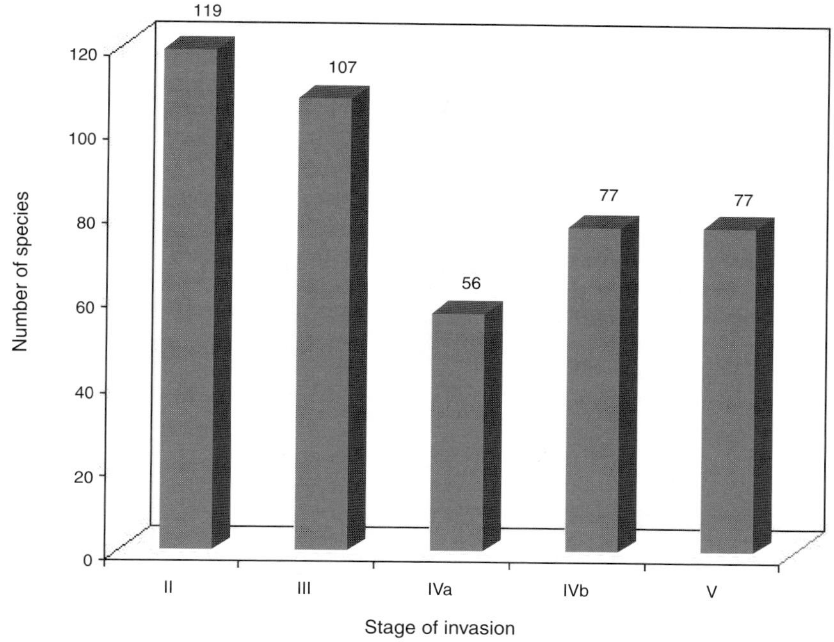


**Fig. 17.3.** Number of alien plant species at different stages of invasion in the Kashmir region.

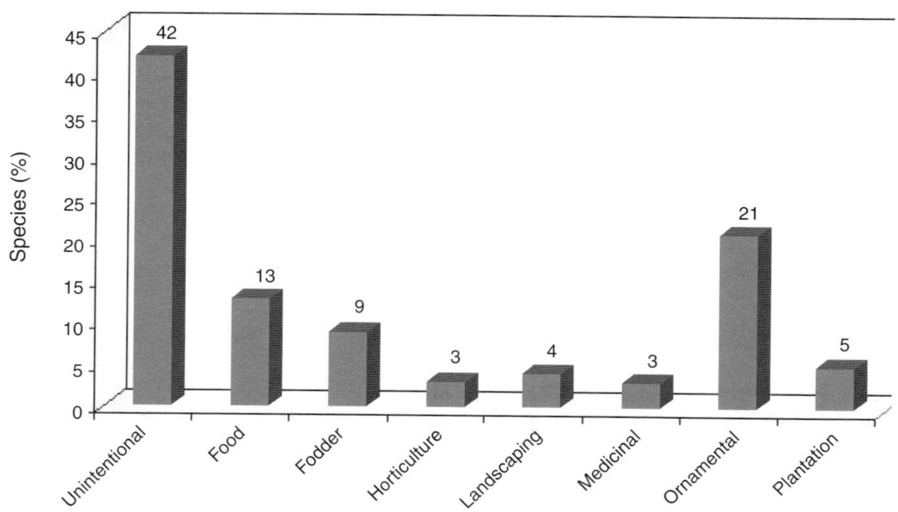


**Fig. 17.4.** Invasive alien flora in the Kashmir region by means of introduction.

disproportionately over-represented in terms of their relative proportion in the region (Khuroo *et al.*, 2007).

Recent research studies show that invasions by alien plants have altered the structure and functioning of the natural ecosystems in the region, such as grasslands (Reshi *et al.*, 2008a) and wetlands (Reshi *et al.*, 2008b). Detailed research studies have been carried out on a high-impact invasive

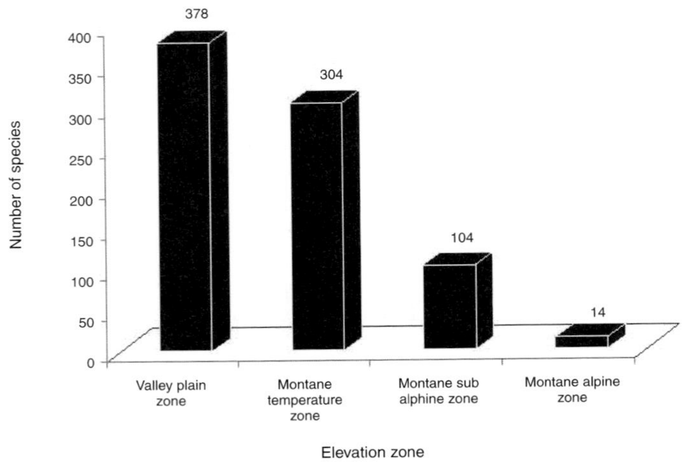


**Fig. 17.5.** Distribution of invasive alien plant species in the Kashmir region by elevation gradient.

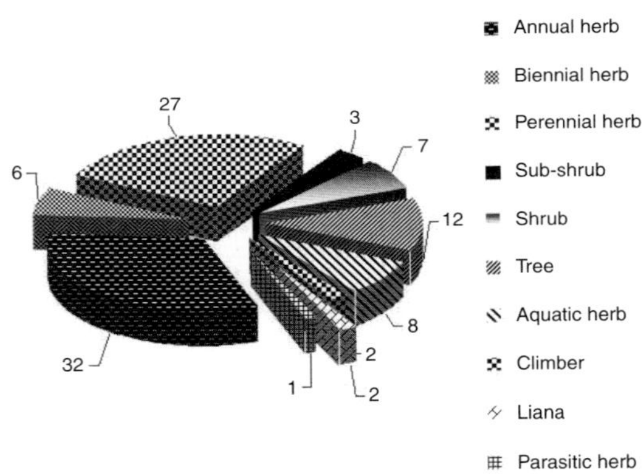


**Fig. 17.6.** Contribution to invasive alien flora of the Kashmir region, by different growth forms.

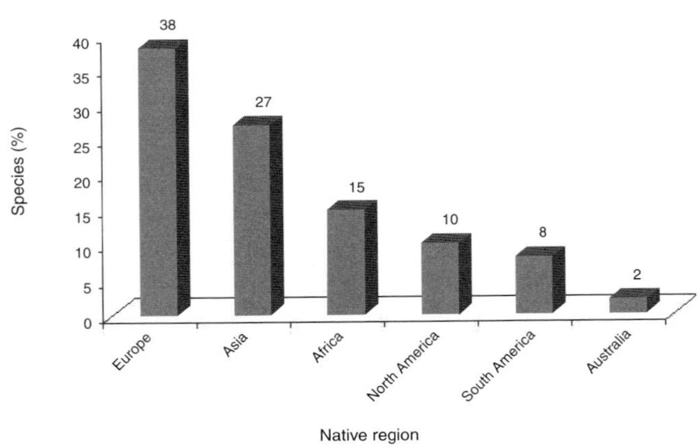


**Fig. 17.7.** Contribution to the invasive alien flora of the Kashmir region, by source.

alien plant species, *Anthemis cotula* L. (mayweed chamomile) native to Mediterranean Europe, which has invaded the entire Kashmir region. This annual herbaceous plant species, belonging to the family *Asteraceae*, not only forms very large populations but is also widespread in distribution, having now reached invasion stage V (Khuroo *et al.*, 2008). In its native range it is consumed by above-ground herbivorous species including aphids (Aphidae), spittlebugs (Cercopidae), bugs (Heteroptera), moths, slugs and snails (Erneberg, 1999). Based on extensive population surveys since 2000, an insect herbivore associated with this species triggered an over-compensatory response in its individuals (Rashid *et al.*, 2006). Invasion of ruderal habitats in the Kashmir region has been attributed to this species' protracted recruitment pattern, aided by habitat disturbance (Rashid *et al.*, 2009), high population size even after seedling mortality (Allaie *et al.*, 2005), allelopathic activity of its aqueous leaf leachate (Allaie *et al.*, 2006), AMF mutualism (Shah *et al.*, 2007, 2008a, b) and the profuse production of achenes and synchrony between their germination and favourable environmental conditions (Rashid *et al.*, 2007a, b). Similar detailed studies on other high-impact invasive plant species are warranted in the region, as these raise the possibility of identifying the syndrome of traits that predispose some alien species to becoming highly invasive (Pyšek and Richardson, 2007).

## Plant invasions in Jammu region

Unlike the Kashmir region, no systematic inventory of invasive alien flora is available for the Jammu region. Based on our preliminary field surveys, the following plant species are spreading as invasives (*sensu* stage V) in the region: *Ageratum conyzoides*, *Argimone mexicana*, *Eichhornia crassipes*, *Ipomea carnea*, *Lantana camara* and *Parthenium hysterophorus*. These troublesome invasive species oroginate from tropical America, and are presently rapidly invading both human-dominated habitats (e.g. roadsides, railway lines, agricultural fields, orchards) and natural areas (e.g. forests, grasslands, freshwater lakes) in the region. These species have caused adverse ecological, economic and health impacts.

Other alien plant species presently at the lower stages of invasion in the region include *Acanthospermum hispidum*, *Alternanthera philoxeroides*, *Antigonon leptopus*, *Blumea eriantha*, *Blumea obliqua*, *Calotropis gigantea*, *Cassia tora*, *Chamaesyce hirta*, *Chamaesyce indica*, *Chloris barbata*, *Conyza bipinnatifida*, *Crassocephalum crepidioides*, *Crotalaria pallida*, *Cuscuta reflexa*, *Datura innoxia*, *Datura metel*, *Echinops echinatus*, *Emilia sonchifolia*, *Euphorbia heterophylla*, *Galinsoga parviflora*, *Gnaphalium polycaulon*, *Gomphrena serrata*, *Imperata cylindrica*, *Ipomoea eriocarpa*, *Ipomoea pes-tigridis*, *Leucaena leucocephala*, *Malvastrum coromandelianum*, *Mimosa pudica*, *Mirabilis jalapa*, *Opuntia stricta*, *Passiflora foetida*, *Physalis angulata*, *Pistia stratiotes*, *Portulaca quadrifida*, *Potamogeton crispus*, *Prosopis juliflora*, *Salvinia molesta*, *Typha angustata* and *Xanthium strumarium*.

Broadly speaking, the taxonomic composition of the invasive plants in Jammu region closely matches those of other regions in the country where tropical and subtropical elements from tropical America are dominant plant invaders. The taxonomic composition of plant invaders is appreciably different in the Jammu region to that in Kashmir, and such divergent regional taxonomic patterns in plant invasions have wide biogeographical and management implications.

## Plant invasions in Ladakh region

In the Ladakh region, as in the Jammu region, a systematic inventory of invasive alien flora is unavailable. Although a number of alien plant species have been recorded in the region none of these are invasive (*sensu* stage V), unlike the other two regions of the state. Notable alien plant species include *Amaranthus caudatus*, *Amaranthus graecizans*, *Amaranthus powellii*, *Avena sativa*, *Bidens biternata*, *Callistephus chinensis*, *Chenopodium album*, *Chenopodium botrys*, *Convolvulus arvensis*, *Conyza canadensis*, *Datura stramonium*, *Daucus carota*, *Digitaria ischaemum*, *Galinsoga parviflora*, *Hordeum vulgare*, *Lactuca sativa*, *Malcolmia africana*, *Medicago luplina*, *Medicago sativa*, *Melilotus officinalis*, *Polypogon monospeliensis*, *Setaria viridis*, *Stellaria media* and *Trifolium repens*.

The majority of these alien species are restricted to agricultural fields, moist sites near human habitations and riparian habitats. It appears that, according to the process of plant invasion, a significant proportion of these species have crossed biological filters (reproductive capacity in wild) and achieved short-distance dispersal, but have failed to cross environmental filters (temperature, precipitation, soil) and achieve long-distance dispersal, due to the harsh climate of the region. In future, with predicted temperature increases due to climate change in the Himalayan highlands, these alien species may cross the environmental filter to spread within Ladakh region.

## Knowledge Gaps and Future Challenges

In regard to scientific understanding, informed policy making, systematic management and effective monitoring of plant invasions, the following questions need to be appropriately addressed in order to fill the ever-widening gaps in our understanding of alien plant invasions in this state: (i) What is the taxonomic composition of the invasive alien flora? (ii) What are the spatio-temporal distributional patterns of the invasive alien plant species? (iii) What are the current and potential ecological and economic impacts of these invasive species? and (iv) What policy, administrative and legal measures need to be formulated in managing plant invasions? In this regard, the following suggestions merit urgent attention and policy intervention.

- An integrated research, management and policy framework specific to plant invasions must be formulated for the state.
- There is an urgent need to generate a systematic inventory of invasive alien flora in Jammu and Ladakh regions, according to the stage-based conceptual framework as adopted for Kashmir region.
- An integrated research programme aimed at generating data on the spatial spread and population abundance of invasive alien plant species across different ecosystems/habitats in the three regions of the state needs to be launched.
- The pathways of introduction and drivers of spread of alien plant species need to identified and blocked.
- Research studies highlighting the ecological and socio-economic impacts of invasive alien plant species need to be publicized in order to invite the attention of policy makers and the general public.
- The extent and magnitude of plant invasions and their role in biotic homogenization need to be investigated.
- Predictive modelling tools and risk assessment frameworks now available need to be developed for prevention of future plant invasions. A stringent and effective regional quarantine regime needs to be institutionalized as soon as possible to prevent future plant invasions.
- Sustained public awareness and outreach activities and sensitization campaigns on the issue of plant invasions need to carried out to promote the proactive involvement of policy makers and stakeholders from the sectors of forestry, agriculture, horticulture, floriculture, tourism, livestock, wildlife, etc.

## Conclusions

In the current era of global change biology, the research, policy and management efforts required in meeting present and future challenges posed by plant invasions are going to manifoldly increase. As in recent decades, in the future too accelerated global trade and travel will enhance biotic exchange rates between nations, as well as within different climatic regions of the individual nation. India, with its varied climatic and physiographic regions, faces an increased risk of plant invasions. It could emerge as the 'hot spot' source (and recipient) of invasive alien species to (and from) the rest of the world.

Given the emerging threat of plant invasions in the Indian Himalayan State of Jammu and Kashmir, there is an imminent need for the compilation of a state-level inventory of invasive alien flora by adopting a single conceptual framework with standardized terminology and methodology. With escalating anthropogenic global change and increasing human-driven land-use changes brought about by unbridled urbanization, agricultural expansion, unsustainable livestock grazing, unregulated introduction of alien plants for plantation and ornamental purposes, unplanned road construction in the high alpines and other tourism–infrastructural activities, etc., the threat of plant invasions in this state is going to become stark reality if immediate management and policy measures are not undertaken at the earliest opportunity.

## Acknowledgements

We greatly acknowledge the critical suggestions and kind help given by our colleagues and supporting staff during the course of research studies over the past decade.

## References

Allaie, R.R., Reshi, Z. and Wafai, B.A. (2005) Demographic studies on alien invasive *Anthemis cotula* L. in Kashmir Himalaya. *Trends in Life Sciences* 20, 49–56.

Allaie, R.R., Reshi, Z., Rashid, I. and Wafai, B.A. (2006) Effects of aqueous leaf leachate of *Anthemis cotula* – alien invasive species on germination behaviour of some field crops. *Journal of Agronomy and Crop Science* 192, 186–191.

Chatterjee, D. (1947) Influence of East Mediterranean Region flora on that of India. *Science and Culture* 13, 9–11.

Clement, E.J. and Foster, M.C. (1994) *Alien Plants of the British Isles. A Provisional Catalogue of Vascular Plants (Excluding Grasses).* Botanical Society of the British Isles, London.

Colautti, R.I. and MacIsaac, H.I. (2004) A neutral terminology to define 'invasive' species. *Diversity and Distributions* 10, 135–141.

Erneberg, M. (1999) Effects of herbivory and competition on an introduced plant in decline. *Oecologia* 118, 203–209.

Hulme, P.E., Roy, D.B., Cunha, T. and Larsson, T.B. (2009) A pan-European inventory of alien species: rationale, implementation and implications for managing biological invasions. In: DAISIE (ed.) *The Handbook of Alien Species in Europe.* Springer, Berlin.

Khuroo, A.A., Rashid, I., Reshi, Z., Dar, G.H. and Wafai, B.A. (2007) The alien flora of Kashmir Himalaya. *Biological Invasions* 9, 269–292.

Khuroo, A.A., Reshi, Z., Rashid, I., Dar, G.H. and Khan, Z.S. (2008) Operational characterization of alien invasive flora and its management implications. *Biodiversity and Conservation* 17, 3181–3194.

Khuroo, A.A., Reshi, Z., Rashid, I., Dar, G.H. and Malik, A.H. (2009) Plant invasions in montane ecosystems. *Frontiers in Ecology and Environment* 7, 407–408.

Lambdon, P.W. *et al.* (2008) Alien flora of Europe: species diversity, temporal trends, geographical patterns and research needs. *Preslia* 80, 101–149.

Liu, J., Liang, S., Liu, F., Wang, R. and Dong, M. (2005) Invasive alien plant species in China: regional distribution patterns. *Diversity and Distributions* 11, 341–347.

Meyerson, L.A. and Mooney, H.A. (2007) Invasive alien species in an era of globalization. *Frontiers in Ecology and Environment* 5, 199–208.

Meyerson, L.A. and Reaser, J.K. (2003) Bioinvasions, bioterrorism, and biosecurity. *Frontiers in Ecology and Environment* 1, 307–314.

Nandy, S.N., Dhyani, P.P. and Samal, P.K. (2006) *Resource Information Database of the Indian Himalaya.* ENVIS Monograph 3, G.B. Pant Institute of Himalayan Environment and Development, Almora, India.

Negi, P.S. and Hajra, P.K. (2007) Alien flora of Doon Valley, Northwest Himalaya. *Current Science* 92, 968–978.

Pejchar, L. and Mooney, H.A. (2009) Invasive species, ecosystem services and human well-being. *Trends in Ecology and Evolution* 24, 497–504.

Pimentel, D., Zuniga, R. and Morrison, D. (2005) Update on the environmental and economic costs associated with alien-invasive species in the United States. *Ecological Economics* 52, 273–288.

Preston, C.D., Pearman, D.A. and Dines, T.D. (2002) *New Atlas of the British and Irish Flora*. Oxford University Press, Oxford, UK.

Pyšek, P. and Richardson, D.M. (2007) Traits associated with invasiveness in alien plants: where do we stand? In: Nentwig, W. (ed.) *Ecological Studies, Biological Invasions*. Springer, Berlin and Heidelberg, pp. 97–125.

Pyšek, P., Richardson, D.M., Pergl, J., Jarošík, V., Sixtová, Z. and Weber, E. (2008) Geographical and taxonomic biases in invasion ecology. *Trends in Ecology and Evolution* 23, 237–244.

Rashid, I., Reshi, Z., Shah, M.A. and Wafai, B.A. (2006) Does herbivory promote invasiveness of *Anthemis cotula* in Kashmir Himalaya, India? In: *International Symposium on Biology, Ecology and Management of the World's Worst Plant Invasive Species*, CEMDE, University of Delhi, India, pp. 83–84.

Rashid, I., Reshi, Z., Allaie, R.R. and Wafai, B.A. (2007a) Germination ecology of invasive alien *Anthemis cotula* helps it synchronise its successful recruitment with favourable habitat conditions. *Annals of Applied Biology* 150, 361–369.

Rashid, I., Reshi, Z., Shah, M.A. and Wafai, B.A. (2007b) What makes *Anthemis cotula* L. (Asteraceae) invasive in Kashmir Himalaya, India? In: *9th International Conference on the Ecology and Management of Alien Plant Invasions*, Perth, Western Australia, 205 pp.

Rashid, I., Sharma, G.P., Esler, K.J., Reshi, Z.A., Khuroo, A.A. and Simpson, A. (2009) A standardized response to biological invasions. *Science* 10, 146.

Reddy, C.S. (2008) Catalogue of invasive alien flora of India. *Life Science Journal* 5, 84–89.

Reshi, Z., Khuroo, A.A., Rashid I. and Dar, G.H. (2007) Alien plant invasions in the Kashmir Himalaya, India: an emerging threat to native biodiversity. In: *Proceedings of National Seminar on Himalayan Biodiversity*, 27–29 October, Rajouri, Jummai and Kashmir, India.

Reshi, Z., Rashid, I., Khuroo, A.A. and Wafai, B.A. (2008a) Effect of invasion by *Centaurea iberica* on community assembly of a mountain grassland of Kashmir Himalaya, India. *Tropical Ecology* 49, 147–156.

Reshi, Z., Shah, M.A., Rashid, I. and Khuroo, A.A. (2008b) Alien plant invasions: Threat to wetland health and public wealth. *Sarovar Saurabh* 4, 2–5.

Rodgers, W.A. and Panwar, H.S. (1990) *Biogeographical Classification of India*. Wildlife Institute of India, Dehradun, India.

Sax, D.F., Gaines, S.D. and Brown, J.H. (2002) Species invasions exceed extinctions on islands worldwide: a comparative study of plants and birds. *American Naturalist* 160, 766–783.

Saxena, K.G. (1991) Biological invasions in the Indian subcontinent: Review of invasion by plants. In: Ramakrishnan, P.S. (ed.) *Ecology of Biological Invasions in the Tropics*, pp. 53–73.

Shah, M.A. and Reshi, Z. (2007) Invasion by alien *Anthemis cotula* L. in a biodiversity hotspot: release from native foes or relief from alien friends. *Current Science* 92, 1–3.

Shah, M.A., Reshi, Z. and Rashid, I. (2008a) Mycorrhizal source and neighbour identity differently influence *Anthemis cotula* L. invasion in the Kashmir Himalaya, India. *Applied Soil Ecology* 40, 330–337.

Shah, M.A., Reshi, Z. and Rashid, I. (2008b) Mycorrhizosphere mediated Mayweed Chamomile invasion in the Kashmir Himalaya, India. *Plant and Soil* 312, 219–225.

Sharma, G.P., Singh, J.S. and Raghubanshi, A.S. (2005) Plant invasions: Emerging trends and future implications. *Current Science* 88, 726–734.

Sharma, J.R., Mudgal, V. and Hajra, P.K. (1997) Floristic diversity – review, scope and perspective. In: Mudgal, V. and Hajra, P.K. (eds) *Floristic Diversity and Conservation Strategies in India*, vol. 1. Botanical Survey of India, Calcutta, India, pp. 1–45.

Singh, D.K., Uniyal, B.P. and Mathur, R. (2000) Jammu and Kashmir. In: *Flora of India*. Botanical Survey of India, Calcutta, India.

Weber, E., Sun, S.G. and Li, B. (2008) Invasive alien plants in China: diversity and ecological insights. *Biological Invasions* 10, 1411–1429.

Westphal, M.I., Browne, M., MacKinnon, K. and Noble, I. (2008) The link between international trade and the global distribution of invasive alien species. *Biological Invasions* 10, 391–398.

Wilcove, D.S., Rothstein, D., Dubow, J., Phillips, A. and Losos, E. (1998) Quantifying threats to imperiled species in the United States. *Bioscience* 48, 607–615.

# 18 Risk Assessment for Management of Biological Invasions

**Zafar A. Reshi and Irfan Rashid**

*Department of Botany, University of Kashmir, Srinagar, Jammu and Kashmir, India*

## Introduction

Risk assessment schemes are now being developed and applied globally to predict not only the potential invasive alien species whose introduction could be prevented, but also to identify the high-risk species among those already introduced that would cause ecological and economic impact. Such screening protocols have assumed urgency in view of the enormous increase in the introduction and spread of invasive species due to intentional and unintentional transport of propagules of species beyond their natural biogeographical ranges (Wilson *et al.*, 2009), and to human-induced environmental changes (Thuiller *et al.*, 2008) that promote such biological invasions, with huge ecological and economic costs estimated at around US$1.4 trillion annually (Pimentel *et al.*, 2005). The need for such protocols also arises from the fact that importing of economically important species used as food, fodder, fibre, fuel, etc. cannot be altogether restricted or prevented; rather, species having the least likelihood of causing any harm need to be identified for use in agriculture, horticulture, forestry, etc. in non-native regions in order to promote human well-being and prosperity in a manner that does not jeopardize the structural and functional integrity of native ecosystems. In contrast to pests that are aggressively managed because of their adverse impacts and no potential benefits, species that are intentionally introduced for various purposes could become invasive, and risk assessment techniques are required to identify such species so that their introduction is not permitted.

Risk assessment schemes are also required to identify the potentially invasive species among those already introduced that are able to overcome the filters that interpose between various spatio-temporal stages of invasion – such as introduction, colonization, establishment and landscape spread – of the invasion process (Theoharides and Dukes, 2007). Early prediction and identification of such species would pave the way for rapid response and their control in a more cost-effective manner than is possible when these species increase in abundance and become widespread in occurrence.

While at present very few countries make use of risk assessment protocols, despite the unprecedented increase in the unintentional introduction of species into non-native areas, together with enhanced propagule pressure (Simberloff, 2009) due to enhanced trade, travel and connectivity between nations and societies and demonstrable economic benefits from the use of risk assessment (Keller *et al.*, 2007), other countries, particularly the rapidly developing countries such as India and China, need to focus attention on the development of prediction and risk assessment tools tailored to their needs in order to combat the menace of biological invasions.

## What is Risk Assessment?

Risk assessment was first defined in the report of the National Research Council of the USA in 1983, and the approach followed in this report was adopted by the US Environmental Protection Agency (EPA) in 1992 in defining risk assessment as a process that evaluates the likelihood of adverse ecological effects from exposure to a 'stressor'. Although the EPA risk assessment is the dominant framework in ecological risk assessment, it is specifically intended to deal with chemical and physical stressors; its applicability to biological stressors, such as invasive species, is uncertain. A risk assessment process allows for identification of the type of risk and analysis of its various dimensions and characteristics through the utilization of qualitative and quantitative data in a systematic manner. The ultimate objective of a risk assessment protocol is to produce quality risk assessments for specific invasive species, or to identify specific pathways that promote and contribute to the introduction and spread of alien species. A risk assessment scheme should also reveal the extent of uncertainty involved and, if possible, should recommend appropriate mitigation measures to reduce the risk.

Risk analysis is often divided into two components: risk assessment and risk management (NRC, 1983), the two components are interacting but functionally separate activities. However, traditional risk analysis schemes consider risk assessment and risk management separately (Leung *et al.*, 2002). Risk assessment measures the probability and severity of potential adverse effects of exposure to stressors. This measurement can be quantitative or qualitative, probabilistic or deterministic. Risk management is the process of identifying, evaluating, selecting and implementing necessary measures to reduce the risk. Risk assessment for invasive species is generally carried out to inform two classes of risk management decisions: (i) those regarding the introduction of potentially invasive species, their vectors or pathways prior to their establishment, allowing for decisions to permit or prohibit their introduction; and (ii) those regarding the allocation of scarce resources, not only for the control of established invasive species but also for the rapid-response measures necessary to control potentially invasive species (Anderson *et al.*, 2004).

## Risk Analysis Framework

A general risk analysis framework for biological invasions (Stohlgren and Schnase, 2006) involves risk analysis for both invasive species as well as invaded habitats, and its key elements are: (i) problem formulation; (ii) analysis of exposure and effects; (iii) risk characterization; and (iv) risk management (Table 18.1).

### Problem formulation

This stage of the process involves determination of the extent and severity of the problem. In regard to invasive species, it has been established beyond doubt that these represent a significant component of human-induced global environmental change and result in enormous economic losses and decreased biological diversity and ecosystem function. Invasive alien species also exacerbate poverty and threaten sustainable development through their impact on the agriculture, horticulture, forestry and fisheries sectors that support the livelihoods of the poorest of poor people in developing countries. The Millennium Ecosystem Assessment Board (2005) concluded that the impact of invasive alien species on biodiversity is evident in all the biomes, and is increasing alarmingly in certain biomes. Recent comprehensive data provided through the DAISIE (Delivering Alien Invasive Species Inventories for Europe) project (www.europe-aliens.org) reveal that Europe harbours more than 11,000 alien species, over half of which are terrestrial plants. Aquatic and terrestrial invertebrates account for >30% of species, whereas only ~5% are vertebrates. Economic impact in Europe due to these invasive species, according to conservative estimates,

**Table 18.1.** Elements of a risk analysis framework for invasive species (modified from Stohlgren and Schnase, 2006).

| Risk analysis step | | Specific information required |
|---|---|---|
| Problem formulation | | Scope and complexity of the problem |
| | | Defining assessment end points |
| Risk analysis | RISK ASSESSMENT | Identification of invasive species and their vectors |
| | | Surveys of current distribution and abundance of invasive species |
| | | Matching species traits to suitable habitats |
| | | Exposure assessment of native biodiversity |
| Risk characterization | | Understanding of data completeness |
| | | Estimates of the 'potential' distribution and abundance of invasive species |
| | | Estimates of the potential rate of spread of invasive species |
| | | Probable risks, impacts and costs |
| Risk management | | Prediction and prevention |
| | | Early detection and rapid response |
| | | Prescription for invasive species |
| | | Policy regulatory framework including legal provision and social considerations, together with costs and benefits |
| | | Information science and technology needs |

is close to €10 billion (~US$13 billion) annually. This figure is an underestimate, as potential economic and environmental impacts are unknown for almost 90% of the alien species found in Europe. In the USA alone, the yearly damage and control costs of invasive species are estimated to be more than US$138 billion (Pimentel *et al.*, 2005). This ecological and economic damage was the rationale for the selection of 'Trends in Invasive Alien Species' as one of 22 Headline Indicators to measure progress towards the Convention on Biological Diversity's target of reducing the rate of loss of biodiversity by 2010.

Although the occurrence of alien invasive species is ubiquitous across ecosystems, biomes, countries and continents, very limited information is available on the number, taxonomic affiliation, spatial spread and abundance of these species in the developing countries (Nuñez and Pauchard, 2009) in general, and in India in particular (Khuroo *et al.*, 2007). Pyšek *et al.* (2008) have demonstrated geographical and taxonomic biases in a study of biological invasions. Overall, the taxonomic bias in studies on invasive species is less pronounced than the geographical bias.

Another facet of problem formulation is defining an unambiguous assessment end point (USEPA, 1992), which is characteristic of an ecological component that can be affected by exposure and that is relevant to decisions about protecting the environment. However, introduced species pose problems in defining end points because there are often decades long lag times (Kowarik, 1995), during which a newly introduced species remains restricted in range and population size and has limited impact on the invaded ecosystem, after which it spreads and becomes a major environmental concern.

## Risk analysis

Risk analysis in regard to biological invasions requires detailed and reliable information about the invasive species, their traits that contribute to invasiveness, vulnerability of habitats to invasion, predictive modelling giving information on the potential distribution of species and the management costs of invasive species (Table 18.1). Biological traits contribute partly to invasion success (Vilá *et al.*, 2009); other important factors include residence time (Pyšek and Jarošík, 2005; Gassó *et al.*, 2009), chance events (Crawley, 1989), large native geographical range (Goodwin *et al.*, 1999) and propagule pressure, both spatially (by widespread dissemination or abundant plantings) and temporally (by long history of cultivation), which fundamentally influence the probability of invasions by alien species (Rouget and Richardson, 2003; Lockwood *et al.*, 2005; Catford *et al.*, 2009).

It is unlikely that a universal syndrome of traits will reliably predict invader success within and across ecosystems (Williamson, 1999; Thompson *et al.*, 2001, Davis *et al.*, 2005), because the interaction of spatially and temporally variable environmental conditions with invader qualities will vary (Sher and Hyatt, 1999; Alpert *et al.*, 2000; Austrheim and Eriksson, 2003). In fact, studies have revealed that community invasibility may be determined by temporal and spatial fluctuation of resource availability (Davis *et al.*, 2000); species richness and diversity (Levine and D'Antonio, 1999; Lonsdale, 1999); dominance (Wardle, 2001; Foster *et al.*, 2002; Smith *et al.*, 2004); degree of similarity of the new compared with the native climate (Cronk and Fuller, 1995); or species and functional group composition (Fargione *et al.*, 2003). A significant challenge in risk assessments of biological stressors is quantification, mapping and prediction of the interaction of species traits and habitat characteristics that promotes successful invasion (Lee, 2001).

For invasive species, exposure analysis involves estimating the likelihood of the introduction, establishment and/or spread of one or more potentially invasive species. It may also consider the quantity, timing, frequency, duration and routes of exposure as well as the numbers, species and other characteristics (e.g. susceptibility) of the populations exposed. In addition to these facets of risk, ecological portrayal of alien species, vis-à-vis their spatial distribution and population abundance in the non-native range, provides a framework for the characterization of alien species at different stages of the invasion process and holds promise for the formulation of effective management practices (Khuroo *et al.*, 2008).

## Risk characterization

Risk characterization in regard to invasive species envisages determination of the taxonomic, geographic and temporal completeness of data in a given region. Perusal of the relevant literature reveals that data in regard to taxonomic, geographical and temporal aspects of invasive species are far from complete and there is a distinct bias between the regions, countries, biomes and ecosystems (Pyšek *et al.*, 2008). For example, analysis of 2670 published papers dealing with 892 invasive species carried out by Pyšek *et al.* (2008) showed that most studies in regard to alien invasive species came from Europe and America (North, Central and South), while a limited number of studies were undertaken in Africa and Asia. Species that are more widespread and invasive have received greater attention in comparison with those at the establishment or naturalization stages of invasion. Thus, extensive and long-term studies at global, regional, country, state and provincial level are needed to offset the deficiency in data, which is pivotal to achieving a more complete picture of biological invasions. This would also improve our understanding of the mechanisms leading to successful invasion and provide broad, cross-taxon generalizations and principles.

Another critical facet of risk characterization is estimation of potential distribution, abundance and likely rate of spread of invasive species. Predictive habitat-based modelling of species' distribution now represents an important tool in invasive species management (Elith and Leathwick, 2009), and consequently a number of models have been developed that have their roots in quantifying species–environment relationships (Guisan and Thuiller, 2005).

After mapping the potential distribution of a species, spread of an invasive species needs to be estimated. Ideally, regular systematic field surveys are required to obtain vital information about the spread of alien species. Because of time, personnel and financial limitations and rapid rates of invasion, models have been developed to estimate potential distribution and spread of alien species after their introduction. Classic diffusion models (e.g. Skellam, 1951) assume that individuals disperse with equal probability in all directions by random, short movements across a homogeneous landscape. More recent stratified diffusion models (e.g. Hastings, 1996; Lewis, 1997; Shigesada and Kawasaki, 1997) include occasional long-distance 'jump' dispersal, believed to be important in the spread of many introduced species. Diffusion models have been used with some success to model the spread of established invaders (reviewed by Lockwood *et al.*, 2007). Gravity models (Fotheringham and O'Kelly, 1989) are also employed to explain and/or predict the spread of established introduced species. These consider not only characteristics of the source population (e.g. its size) but also those of potential colonization sites (e.g. their distance from source) and the spatial configuration of sites. Niche-based distribution models (e.g. Guisan and Thuiller, 2005; Peterson, 2006) are frequently used to predict the ultimate distribution of introduced species. Recent theoretical and empirical studies have revealed that the spatial spread of invasive species is a much more complex process than previously thought (Hastings *et al.*, 2005). Earlier models that considered only one non-evolving species in a homogenous habitat with random, short-range dispersal predicted a linear rate of spread (reviewed by Okubo and Levin, 2002), but such models are now being replaced by newer models that examine the role of long-range dispersal, spatial heterogeneity, temporal variability, other species and evolution in determining the rate of spread of invasive species (Hastings *et al.*, 2005).

### Risk management

Risk assessment and its characterization provide valuable information necessary for the management of invasive species. Not only is it possible to predict potentially invasive species through the use of various pre-introduction risk assessment schemes and prevent their introduction into non-native regions, but it is also possible to identify previously introduced species that, over the course of time, could become invasive and hence warrant rapid response.

Post-introduction risk assessment protocols aid in identifying priority invasive species that require prescription of appropriate management technologies for their control and containment. In addition to prevention, prediction and prescription, a policy framework encompassing legal provisions is required for effective management of invasive species. These management implications of risk assessment would be more useful if every country, region and state were to prepare an inventory of alien species of various taxa for development of a global database of alien invasive species, in view of its potential to yield robust generalizations about biological invasions and to help in taking informed policy decisions in regard to the management of invasive species. Besides, modelling the potential distribution of alien species would help in identifying habitats, ecosystems and areas susceptible to invasion by alien species and hence would pave the way for implementation of pro-active management policies in such habitats.

## Trait-based Risk Assessment Schemes

Trait-based risk assessment aims to discriminate invasive from non-invasive alien species (reviewed in Lodge, 1993; Mack, 1996; Kolar and Lodge, 2001; Rejmánek *et al.*, 2005; Pyšek and Richardson, 2007; Hayes and Barry, 2008). There are two types of trait-based risk assessment models (Table 18.2), namely pre-introduction (predictive) and post-introduction (prioritization) models (Randall *et al.*, 2008). Such models often use statistical discrimination analysis and classification and regression trees (CARTs) and are based on either rating systems (e.g. Pheloung *et al.*, 1999) or hierarchical decision trees (e.g. Reichard and Hamilton, 1997).

One of the most promising pre-introduction risk assessment schemes is the Australian weed risk assessment (WRA) (Pheloung *et al.*, 1999). This has been tested for the state of Hawaii and the Pacific Islands (Daehler and Carino, 2000; Daehler *et al.*, 2004), the Czech Republic (Krivánek and

**Table 18.2.** Pre- and post-introduction trait-based risk assessment systems.

| Name of system | Purpose of system | Scale of application |
|---|---|---|
| Predictive systems | | |
| Pheloung *et al.* (1999) (Australian weed risk assessment) | Accept or reject species for importation/ introduction | National |
| Reichard and Hamilton (1997) | Accept or reject species for importation/ introduction | National |
| Rejmánek and Richardson (1996) | Predict risk of woody plant invasions | National/ landscape |
| Williams *et al.* (2001) | Accept or reject species for importation/ introduction | National |
| Weber and Gut (2004) | Accept or reject species for importation/ introduction | National/ regional |
| Daehler *et al.* (2004) | Accept or reject species for importation/ introduction | Regional |
| Tucker and Richardson (1995) | Screening alien woody plants for their invasive potential | Regional |
| Prioritization systems | | |
| Champion and Clayton (2001) | Rank current and potential aquatic weeds by category | National |
| Fox *et al.* (2001) (Florida) | Categorize non-native plants | State/zones |
| Heffernan *et al.* (2001) (Virginia) | Prioritize state list of invasives | State |
| Hiebert and Stubbendieck (1993, 2007) | Rank by threat to site | State |
| Randall *et al.* (2008) (Invasive species assessment protocol) | Rank non-native plants by impacts to biodiversity; US list | National/ state/ regional |
| Orr *et al.* (1993); Orr (2003) | Rank by need for mitigation efforts | National/ regional |
| Robertson *et al.* (2003) (South Africa) | Determine priorities for national management and research | National |
| Thorp and Lynch (2000) | Determine priorities for national management | National |
| Virtue *et al.* (2001) (Australia) | Rank all weeds (agriculture, forestry, environment) for national significance | National |
| Weiss (1999) (Australia) | Categorize existing and emerging weeds by biological characteristics | State |

Pyšek, 2006), the Bonin (Ogasawara) Islands (Kato *et al.*, 2006), Japan (Nishida *et al.*, 2009), Florida (Gordon *et al.*, 2008), the USA (Randall *et al.*, 2008), Spain (Andreu and Vilá, 2009) and the Mediterranean region (Gassó *et al.*, 2009), and its consistent accuracy across six geographical regions has been reported by Gordon *et al.* (2008). This system consists of 49 questions encompassing: (i) biogeography; (ii) traits potentially contributing to invasiveness; and (iii) biology/ecology (Table 18.3). Responses to these questions are converted into scores ranging from –3 to +5. Based on total scores, a species is accepted for introduction (score <1), rejected (score >6) or recommended for further analysis (score 1–6). A minimum of ten answers (at least two from the biogeography section, two from the traits section and six from biology/ ecology section) are required for a species to be evaluated (Appendix 18.1).

A modified version of the above model was developed by Daehler and Carino (2000) and Daehler *et al.* (2004), better to suit the conditions of Hawaii and the Pacific Islands. Four of the 49 questions from the Australian WRA system (Pheloung *et al.*, 1999) were

**Table 18.3.** Performance of three assessment schemes in predicting invasive plant species in Kashmir Himalaya, India.

| Risk assessment scheme tested | Number | | | Accuracy (%) |
|---|---|---|---|---|
| | Invasive species rejected (Ir) | Invasive species accepted (Ia) | Total number of invasive species assessed (It) | |
| Pheloung *et al.* (1999) | 77 | 0 | 77 | 100 |
| Daehler *et al.* (2004) | 77 | 0 | 77 | 100 |
| Tucker and Richardson (1995) | 5 | 72 | 77 | 6.49 |

modified prior to use. Question 2.04, 'Native or naturalized in regions with extended dry periods', was modified to 'Native or naturalized in regions with tropical or subtropical climates'. Question 2.01, 'Australian climates', was replaced by 'Tropical or subtropical climates'. Question 4.10, 'Grows on infertile soils' was replaced by 'Tolerates limestone or a wide range of soils', to reflect conditions in Hawaii and the other Pacific Islands. Question 8.05, 'Effective natural enemies present in Australia', was replaced by 'Effective natural enemies present in Hawaii or the Pacific Islands'.

In comparison with the above rating systems, the South African fynbos screening system (Tucker and Richardson, 1995) is a hierarchical decision tree protocol that consists of a linear series of five modules, each of which contains multiple questions (Fig. 18.1). Depending on the answers to questions within each module, a species is classified as either 'low risk' or the procedure deferred to the questions in the following module. If a species advances through all modules without being classified as 'low risk', that species is considered 'high risk'.

After appropriate modifications, the above three schemes were assessed for prediction of invasive species in the Kashmir Himalayan region of India. In the Australian WRA scheme, 'Suitability of species to the Australian climate' was changed to 'Suitability to the Kashmir Himalayan climate' (question 2.01), and 'Origin or naturalization in regions with extended dry periods' was changed to 'Those with temperate climate' (question 2.04); 'Nitrogen-fixing woody plants' (question 5.03) was changed to 'Nitrogen-fixing plants' and 'The presence of effective natural enemies in Australia' was changed to 'The presence of effective natural enemies in the Kashmir Himalaya' (question 8.05). The risk assessment scheme of Daehler *et al.* (2004) was used, with minor modifications to three questions relating to climate and fire hazard, since the Kashmir Himalayan climate is temperate with occasional forest fires. Thus, the reply to question 4.08 ('Creates a fire hazard in natural ecosystems') was 'no' for all species and 'fire' was omitted from question 8.04 ('tolerates, or benefits from mutilation, cultivation or fire'). In the Tucker–Richardson scheme, fynbos defaults were changed to Kashmir Himalayan defaults and the reply to the question relating to fire was 'no' for all species.

The accuracy of each scheme in regard to the Kashmir Himalaya was calculated following the method of Smith *et al.* (1999). Accuracy (Ai) indicates the probability of correct classification, i.e. the proportion of known invasive species that were correctly assessed as invasive.

Where $I_r$ is the number of invaders that were rejected by the system and

$I_t$ is the total number of invaders assessed,

$$Ai = (I_r/I_t) \times 100$$

The Tucker–Richardson decision tree model developed for South African fynbos was least successful in the Kashmir Himalaya, with 6.49% accuracy (Table 18.3). This model was designed for application to very specific conditions – fire-prone, nutrient-poor shrublands – that do not occur in the Kashmir Himalaya, and this highlights the need for selection of a risk assessment system that is appropriate for a given region,

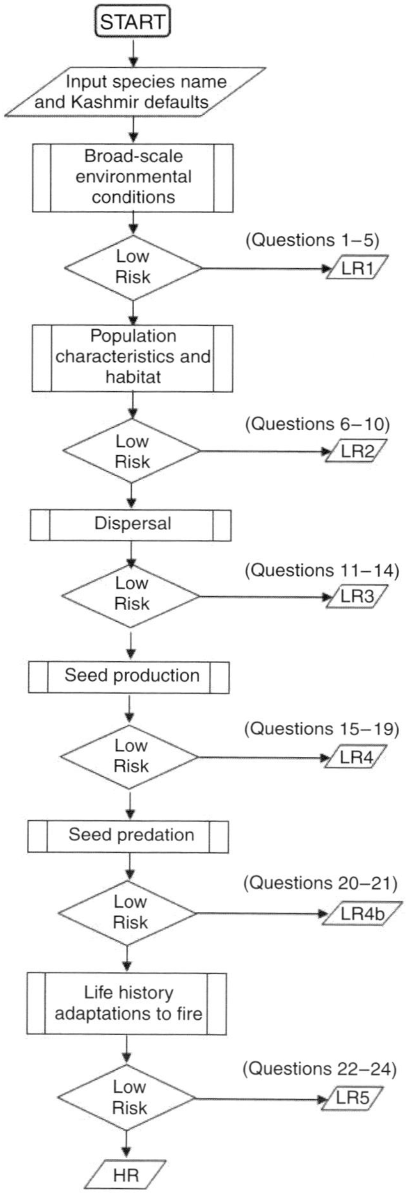


**Fig. 18.1.** A hierarchical decision tree protocol (Tucker and Richardson, 1995). LR, low risk (grades of); HR, high risk.

country or taxa. The average WRA score for Kashmir Himalaya was much greater than that found by some other studies (Table 18.4). Based on WRA scoring, individual species were classified as either 'very high risk' (≥18 points), 'high risk' (12–17 points) or 'intermediate risk' (6–11 points).

## Ranking Potential Invaders of the Kashmir Himalaya

Of the total of 77 invasive species investigated in the present study, only three (*Sambucus wightiana*, *Iris ensata* and *Juncus articulatus*) qualified as high-risk species, while the other species were very high-risk (Table 18.5). *Typha angustifolia* was given the highest WRA score (32), followed by *Anthemis cotula* and *Conyza canadensis* (each 31).

Leaving aside the pre-introduction risk assessment schemes discussed above, the most recent and comprehensive post-introduction risk assessment scheme is that of Randall *et al.* (2008), which categorizes alien species according to their impacts on biodiversity. This consists of 20 multiple-choice questions (Table 18.6) grouped into four sections. Based on total scores, an overall 'Invasive Species Impact Rank' or I-Rank is obtained and thereby species causing high (score 76–100), medium (score 51–75), low (score 26–50) or insignificant (score 0–25) negative impact are identified. This protocol is now used in the USA for prioritizing alien species requiring immediate attention.

## Habitat Suitability Models for Risk Assessment

A variety of techniques are available for modelling species distributions (Elith *et al.*, 2006), and the statistical prediction of species distribution is based on the assumption that the current distribution is reflective of a species' response to a set of environmental factors. Habitat suitability models relate environmental variables to the likelihood of occurrence of the species. The development of various modelling tools for predicting species distributions (Table 18.7) and new, fine-scale environmental data have significantly helped in predicting

**Table 18.4.** Weed risk assessment (WRA) scores (mean ± SE) for major invaders in different geographical regions for which risk assessment was tested.

| Geographical region | Major invader | Reference |
|---|---|---|
| Australia | 13.24 ± 0.51 | Pheloung *et al.* (1999) |
| Hawaii | 14.04 ± 1.02 | Daehler and Carino (2000) |
| Hawaii and Pacific Is. | 11.77 ± 1.30 | Daehler *et al.* (2004) |
| Czech Republic | 9.91 ± 0.75 | Krivánek and Pyšek (2006) |
| Bonin Is. | 14.78 ± 1.12 | Kato *et al.* (2006) |
| Florida | 14.90 ± 0.82 | Gordon *et al.* (2008) |
| Kashmir Himalaya | 25.14 ± 0.41 | Reshi and Rashid (this volume) |

**Table 18.5.** Weed risk assessment (WRA) scores for 77 invasive alien species of the Kashmir Himalaya.

| Species | WRA score | Species | WRA score |
|---|---|---|---|
| *Alisma plantago-aquatica* L. | 30 | *Thymus serpyllum* L. | 24 |
| *Sagittaria sagittifolia* L. | 26 | *Lemna minor* L. | 28 |
| *Amaranthus caudatus* L. | 28 | *Spirodela polyrhiza* Schleid. | 26 |
| *Amaranthus spinosus* L. | 29 | *Marsilea quadrifolia* L. | 29 |
| *Daucus carota* L. | 30 | *Nymphoides peltatum* Kuntze | 27 |
| *Scandix pectin-veneris* L. | 25 | *Morus alba* L. | 20 |
| *Anthemis cotula* L. | 31 | *Epilobium hirsutum* L. | 25 |
| *Artemisia absinthium* L. | 28 | *Oenothera rosea* Ait. | 24 |
| *Carduus edelbergii* Rech. f. | 30 | *Oxalis corniculata* L. PV | 24 |
| *Cirsium arvense* Scop. | 29 | *Plantago lanceolata* L. | 24 |
| *Conyza canadensis* Cronquist | 31 | *Plantago major* L. | 24 |
| *Crepis sancta* Babc. | 26 | *Agrostis stolonifera* L. | 24 |
| *Taraxacum officinale* Weber | 26 | *Bothriochloa ischaemum* Keng | 28 |
| *Arabidopsis thaliana* Heynh. | 24 | *Bromus inermis* Leyss. | 26 |
| *Lithospermum arvense* L. | 29 | *Dactylis glomerata* L. | 24 |
| *Capsella bursa-pastoris* Medic. | 24 | *Phragmites australis* Trin. | 27 |
| *Sisymbrium loeselii* L. | 24 | *Poa annua* L. | 25 |
| *Butomus umbellatus* L. | 21 | *Setaria viridis* P. Beauv. | 28 |
| *Cannabis sativa* L. | 29 | *Sorghum halepense* Pers. | 24 |
| *Sambucus wightiana* Wall. ex Wt. and Arn. | 12 | *Sorghum vulgare* Pers. | 24 |
| *Arenaria serpyllifolia* L. | 23 | *Vulpia myuros* Gmel. | 24 |
| *Stellaria media* Cyr. | 22 | *Polygonum hydropiper* L. | 23 |
| *Ceratophyllum demersum* L. | 28 | *Rumex hastatus* D. Don | 27 |
| *Chenopodium album* L. | 24 | *Potamogeton crispus* L. var. *serrulatus* Reichb. | 28 |
| *Chenopodium foliosum* Aschers. | 24 | *Anagalis arvensis* L. | 24 |
| *Convolvulus arvensis* L. | 26 | *Ranunculus arvensis* L. | 24 |
| *Carex notha* Kunth. | 22 | *Ranunculus laetus* Wall. ex Hk. f. and T. | 24 |
| *Cyperus difformis* L. | 22 | *Ranunculus muricatus* L. | 23 |
| *Cyperus globosus* All. | 22 | *Ranunculus sceleratus* L. A V | 24 |
| *Cyperus rotundus* L. | 22 | *Rubus ulmifolius* Schott. | 26 |
| *Euphorbia helioscopia* L. | 24 | *Salix alba* L. | 21 |
| *Medicago polymorpha* L. | 26 | *Salvinia natans* All. | 28 |
| *Robinia pseudoacacia* L. | 23 | *Veronica persica* Poir. | 24 |
| *Trifolium pratense* L. | 26 | *Ailanthus altissima* Sw. | 27 |
| *Trifolium repens* L. | 26 | *Sparganium ramosum* Huds. | 30 |
| *Aesculus indica* Hook.f. | 25 | *Trapa natans* L. | 28 |
| *Iris ensata* Thunb. | 12 | *Typha angustifolia* L. | 32 |
| *Juncus articulatus* L. | 15 | *Urtica dioica* L. | 28 |
| *Mentha longifolia* L. | 22 | | |

**Table 18.6.** Summary of questions protocol in post-introduction Invasive Species Assessment[a] (Randall *et al.*, 2008).

| |
|---|
| Section I. Ecological impact (5 questions, 50% of I-Rank score) |
| 1. Impact on ecosystem processes and system-wide parameters (33 points max.) |
| 2. Impact on ecological community structure (18 points max.) |
| 3. Impact on ecological community composition (18 points max.) |
| 4. Impact on individual native plant or animal species (9 points max.) |
| 5. Conservation significance of communities and native species threatened (24 points max.) |
| Section II. Current distribution and abundance (4 questions, 25% of I-Rank score) |
| 6. Current range size in region (15 points max.) |
| 7. Proportion of current range where it negatively impacts biodiversity (15 points max.) |
| 8. Proportion of region's biogeographic units invaded (3 points max.) |
| 9. Diversity of habitats or ecological systems invaded in region (3 points max.) |
| Section III. Trends in distribution and abundance (7 questions, 15% of I-Rank score) |
| 10. Current trend in total range within the region (18 points max.) |
| 11. Proportion of potential range currently occupied (3 points max.) |
| 12. Long-distance dispersal potential within region (9 points max.) |
| 13. Local-range expansion or change in abundance (18 points max.) |
| 14. Inherent ability to invade conservation areas and other habitats of native spp. (6 points max.) |
| 15. Similar habitats invaded elsewhere (9 points max.) |
| 16. Reproductive characteristics (9 points max.) |
| Section IV. Management difficulty (4 questions, 10% of I-Rank score) |
| 17. General management difficulty (18 points max.) |
| 18. Minimum time commitment (15 points max.) |
| 19. Impacts of management on native species (15 points max.) |
| 20. Accessibility of invaded areas (3 points max.) |

[a] There were five possible answers for each question: A–D and unknown.

the risk of invasive alien species – even at the global scale (Thuiller *et al.*, 2005). Ordinary multiple regression and its generalized form (GLM) are very popular for modelling species distributions.

Other methods include neural networks, ordination and classification methods, Bayesian models, locally weighted approaches (e.g. GAM), environmental envelopes or combinations of these models. New methods, including threshold-independent measures (e.g. receiver operating characteristic (ROC)-plots) and resampling techniques (e.g. bootstrap, cross-validation), have been introduced in ecology for testing the accuracy of predictive models (Guisan and Zimmermann, 2000). These predictive models require data in regard to the simple presence, presence/absence or abundance of species in locations where the species has been observed (Tsoar *et al.*, 2007). In addition, data on environmental variables, such as climate, elevation, land uses, etc., which have been measured or estimated across the region of interest, are also needed (Table 18.8). The spatial distributions of these variables (usually captured in a set of GIS data layers), along with a set of georeferenced sites of known species occurrence, are then evaluated by one of several possible modelling algorithms (Table 18.7). Each algorithm is designed to extract the relationship between environmental variation and species occurrence, although they differ in methodology and input formats (see Elith *et al.*, 2006, for a recent review and comparison of techniques). This relationship is then used to predict the *potential distribution* of species given the environmental conditions of the area and time period of interest.

Although the above-mentioned models have been widely used in predicting the risk of alien invasive species (Peterson, 2003; Thuiller *et al.*, 2005), in India they have yet to be adopted on any scale. Ganeshaiah *et al.* (2003), however, used two modelling

**Table 18.7.** List of habitat suitability models used for predicting species distribution (modified from Elith *et al.*, 2006).

| Method | Tool | Data | Software required | Reference |
|---|---|---|---|---|
| Climate envelope | BIOCLIM | p | DIVA-GIS | Busby (1991) |
| Bayesian approach | Bayes | p | R. package | Aspinall (1992) |
| Ecological niche factor analysis | BIOMAPPER | p | ENFA | Hirzel *et al.* (2002) |
| Regression (generalized linear model: GLM) | BIOMOD | pa | S-Plus, GRASP add-on | Thuiller (2003) |
| Regression (generalized additive model: GAM) | BIOMOD | pa | S-Plus, GRASP add-on | Thuiller (2003) |
| Multivariate distance | DOMAIN | p | DIVA-GIS | Carpenter *et al.* (1993) |
| Genetic algorithm for rule-set prediction | GARP (Desktop version) | pa | DektopGarp | Stockwell and Peters (1999) |
| Genetic algorithm for rule-set prediction (open modeller version) | OM-GARP | pa | GARP (new version) | Stockwell and Peters (1999) |
| Generalized dissimilarity modelling | GDM | pa (community) | Specialized program | Ferrier *et al.* (2002) |
| Maximum entropy | MAXENT | pa | Maxent | Philips *et al.* (2006) |
| Multivariate adaptive regression splines | MARS | pa | R. mda package | Thuiller (2003) |

p, presence; a, absence.

**Table 18.8.** Some commonly used environmental data sets.

| Data set | Type of data | Source |
|---|---|---|
| WORLDCLIM | Interpolated climate layers for global land areas | Hijmans *et al.* (2005) |
| SRTM 90m DEMs | 90 m resolution digital elevation data for global land areas | The Consultative Group for International Agriculture Research's Consortium for Spatial Information (CGIAR-CSI) |
| HYDRO1k | Global topographically derived data (e.g. streams, drainage basins, etc.) | United States Geological Service (USGS) |
| Several | Global current climate, environmental variables and future climate scenarios | Intergovernmental Panel on Climate Change (IPCC) |

approaches, namely GARP and DIVA-GIS, to predict the potential geographical distribution of sugar cane woolly aphid, which is a serious pest in Maharashtra and Karnataka, and demonstrated the usefulness of this approach in developing strategies for monitoring and managing this important agricultural pest. The time has now arrived when appropriate ecological data concerning the invasive species of India and environmental variables should be collected, in order to identify those species likely to become invasive and the regions/habitats likely to be invaded, so that proactive management policies can be put in place to minimize, if not eliminate, debilitating ecological and economic losses caused by invasive alien species.

## Conclusions

Risk assessment schemes are valuable tools to diminish the risk of invasion and to concentrate resources on tackling the pathways relevant to those species posing higher risk of invasion. In fact, such protocols represent the first step towards developing policies that predict potentially invasive species and their pathways of introduction, in order to restrict their introduction and spread. Besides, this should also facilitate the transport and introduction of economically important species devoid of any attendant risk of becoming invasive. The high cost:benefit ratio associated with invasive species risk assessments (i.e. the cost of allowing an invasion:benefit of allowing the introduction of a presumed non-invasive) suggests the development of robust risk assessment protocols tailored to local needs and requirements. Among the various risk assessment schemes, those developed for Australia (with some necessary modifications reflecting climatic and edaphic conditions) has been successfully tested for adoption in the Kashmir Himalaya, but its use at the national level requires further examination to demonstrate its efficacy in predicting invasive alien plant species for different areas of India. Besides, protocols for other taxa (e.g. invertebrates, mammals, etc.) also need to be developed in order to establish a more comprehensive pre-introduction risk assessment protocol for use in India. Similarly a post-introduction risk assessment scheme, such as that of Randall *et al.* (2008), needs to developed and adopted for identifying the worst invasive species requiring urgent attention and predicting alien species that have the potential to become invasive over a period of time, so that appropriate responses are initiated at a very early stage of the invasion process.

The use of prediction modelling is now top priority, as this is a vital tool for predicting the likely distribution of organisms based on their habitat and other environmental requirements. Since species-based risk assessment protocols assume that species and ecosystems are static over time, the application of species distribution modelling has become necessary to predict the likely distribution of invasive alien species in view of changing environmental conditions caused by natural and anthropogenic factors.

Notwithstanding the extent of economic and ecological damage caused by invasive alien species, little work has been done in India. A starting point should be the creation of an inventory of alien species of all taxa, followed by investigation of the factors that promote invasion and invasibility of habitats, as this would provide sound foundations for the development of an effective framework for risk assessment and management of biological invasions.

## Acknowledgements

Thanks are due to Prof. R.K. Kohli, Chairman, Department of Botany, Punjab University, Chandigarh, India for providing us with the opportunity to write this chapter. We are also grateful to a number of fellow scientists, both within India and abroad, whose help enabled us to compile the chapter in its present form.

## References

Alpert, P., Bone, E. and Holzapfel, C. (2000) Invasiveness, invasibility and the role of environmental stress in the spread of non-native plants. *Perspectives in Plant Ecology, Evolution and Systematics* 3, 52–66.

Anderson, M.C., Adams, H., Hope, B. and Powell, M. (2004) Risk assessment for invasive species. *Risk Analysis* 24, 787–793.

Andreu, J. and Vilá, M. (2009) Risk analysis of potential invasive plants in Spain. *Journal for Nature Conservation* 17, 129–141.

Aspinall, R. (1992) An inductive modelling procedure based on Bayes Theorem for analysis of pattern in spatial data. *International Journal of Geographic Information Systems* 6, 105–121.

Austrheim, G. and Eriksson, O. (2003) Recruitment and life history traits of sparse plant species in subalpine grasslands. *Canadian Journal of Botany* 81, 171–182.

Busby, J.R. (1991) BIOCLIM – a bioclimatic analysis

and prediction system. In: *Nature Conservation: Cost Effective Biological Surveys and Data Analysis.* CSIRO, Melbourne, Australia.

Carpenter, G., Gillison, A.N. and Winter, J. (1993) DOMAIN: a flexible modelling procedure for mapping potential distributions of plants and animals. *Biodiversity and Conservation* 2, 667–680.

Catford, J.A., Jansson, R. and Nilsson, C. (2009) Reducing redundancy in invasion ecology by integrating hypotheses into a single theoretical framework. *Diversity and Distributions* 15, 22–40.

Champion, P.D. and Clayton, J.S. (2001) A weed risk assessment model for aquatic weeds in New Zealand. In: Groves, R.H., Panetta, F.D. and Virtue, J.G. (eds) *Weed Risk Assessment.* CSIRO, Collingwood, Victoria, Australia, pp. 194–202.

Crawley, M.J. (1989) Chance and timing in biological invasions. In: Drake, J.A. *et al.* (eds) *Biological Invasions: a Global Perspective.* Wiley, Chichester, UK, pp. 407–424.

Cronk, Q.C.B. and Fuller, J.B. (1995) *Plant Invaders: the Threat to Natural Ecosystems.* Chapman and Hall, London, 241 pp.

Daehler, C. and Carino, D.A. (2000) Predicting invasive plants: prospects for a general screening system based on current regional models. *Biological Invasions* 2, 93–102.

Daehler, C.C., Denslow, J.S., Ansari, S. and Kuo, H.-C. (2004) A risk assessment system for screening out invasive pest plants from Hawaii and other Pacific Islands. *Conservation Biology* 18, 360–368.

Davis, M.A., Grime, J.P. and Thompson, K. (2000) Fluctuating resources in plant communities: a general theory of invasibility. *Journal of Ecology* 88, 528–534.

Davis, M.A., Thompson, K. and Grime, J.P. (2005) Invasibility: the local mechanism driving community assembly and species diversity. *Ecography* 28, 696–704.

Elith, J. and Leathwick, J. (2009) Species distribution models: Ecological explanation and prediction across space and time. *Annual Review of Ecology, Evolution and Systematics* 40, 677–697.

Elith, J., Graham, C.H., Anderson, R.P., Dudik, M. *et al.* (2006) Novel methods to improve prediction of species distributions from occurrence data. *Ecography* 29, 129–151.

Fargione, J., Brown, C.S. and Tilman, D. (2003) Community assembly and invasion: an experimental test of neutral versus niche processes. *Proceedings of the National Academy of Sciences of the United States of America* 100, 8916–8920.

Ferrier, S., Drielsma, M., Manion, G. and Watson, G. (2002) Extended statistical approaches to modelling spatial pattern in biodiversity in north-east New South Wales: II. Community-level modelling. *Biodiversity and Conservation* 11, 2309–2338.

Foster, B.L., Smith, V.H., Dickson, T.L. and Hildebrand, T. (2002) Invasibility and compositional stability in a grassland community: relationships to diversity and extrinsic factors. *Oikos* 99, 300–307.

Fotheringham, A.S. and O'Kelly, M.E. (1989) *Spatial Interaction Models: Formulations and Applications.* Kluwer Academic Publishers, Dordrecht, Netherlands/Boston, Massachusetts.

Fox, A.M., Gordon, D.R., Dusky, J.A., Tyson, L. and Stocker, R.K. (2001) IFAS assessment of non-native plants in Florida's natural areas. SS-AGR-79. Agronomy Department, Florida Cooperative Extension Service, Institute of Food and Agricultural Sciences, University of Florida, Gainesville, Florida, 25 pp.

Ganeshaiah, K.N., Barve, N., Nath, N., Chandrashekara, K., Swamy, M. and Uma Shanker, R. (2003) Predicting the potential geographical distribution of the sugarcane woolly aphid using GARP and *DIVA*-GIS. *Current Science* 85, 1526–1528.

Gassó, N., Basnou, C. and Vilà, M. (2009) Predicting plant invaders in the Mediterranean through a weed risk assessment system. *Biological Invasions* DOI 1.1007/s10530-009-9451–9452.

Goodwin, B.J., McAllister, A.J. and Fahrig, L. (1999) Predicting invasiveness of plant species based on biological information. *Conservation Biology* 13, 422–426.

Gordon, D., Daphne, R., Onderdonk, A., Fox, A.M. and Stocker, R.K. (2008) Consistent accuracy of the Australian weed risk assessment system across varied geographies. *Diversity and Distributions* 14, 234–242.

Guisan, A. and Thuiller, W. (2005) Predicting species distribution: offering more than simple habitat models. *Ecology Letters* 8, 993–1009.

Guisan, A. and Zimmermann, N.E. (2000) Predictive habitat distribution models in ecology. *Ecological Modelling* 135, 147–186.

Hastings, A. (1996) Models of spatial spread: A synthesis. *Biological Conservation* 78, 143–148.

Hastings, A., Cuddington, K., Davies, K.F., Dugaw, C.J., Elmendorf, S., Freestone, A. *et al.* (2005) The spatial spread of invasions: new developments in theory and evidence. *Ecology Letters* 8, 91–101.

Hayes, K.R. and Barry, S.C. (2008) Are there any consistent predictors of invasion success? *Biological Invasions* 10, 483–506.

Heffernan, K.E., Coulling, P.P., Townsend, J.F. and Hutto, C.J. (2001) Ranking invasive exotic plant species in Virginia. Natural Heritage Technical Report 01-13, Virginia Department of Conservation and Recreation, Division of Natural Heritage, Richmond, Virginia, 144 pp.

Hiebert, R.D. and Stubbendieck, J. (1993) Handbook for ranking exotic plants for management and control. U.S. National Park Service Natural Resources Report NPS/NRMWRO/NRR – 93/08, Washington, DC, 29 pp.

Hiebert, R.D. and Stubbendieck, J. (2007) *Alien Plants Ranking System*, Version 5.1 (http://www.npwrc.usgs.gov/resource/literatr/aprs/index.htm, accessed 25 October 2007).

Hijmans, R.J., Cameron, S., Parra, J.L., Jones, P.G. and Jarvis, A. (2005) Very high resolution interpolated climate surfaces for global land areas. *International Journal of Climatology* 25, 1965–1978.

Hirzel, A.H., Hausser, J., Chessel, D. and Perrin, N. (2002) Ecological-niche factor analysis: How to compute habitat-suitability maps without absence data? *Ecology* 83, 2027–2036.

Kato, H., Hata, K., Yamamoto, H. and Yoshioka, T. (2006) Effectiveness of the weed risk assessment system for the Bonin Islands. In: Koike, F., Clout, M.N., Kawamichi, M., De Poorter, M. and Iwatsuki, K. (eds) *Assessment and Control of Biological Invasion Risk.* Shoukadoh Book Sellers and IUCN, Kyoto, Japan and Gland, Switzerland, pp. 65–72.

Keller, R.P., Lodge, D.M. and Finnoff, D.C. (2007) Risk assessment for invasive species produces net bioeconomic benefits. *Proceedings of the National Academy of Sciences, U.S.A.* 104, 203–207.

Khuroo, A.A., Rashid, I., Reshi, Z., Dar, G.H. and Wafai, B.A. (2007) The alien flora of Kashmir Himalaya. *Biological Invasions* 9, 269–292.

Khuroo, A.A., Reshi, Z., Rashid, I., Dar, G.H. and Khan, Z.S. (2008) Operational characterization of alien invasive flora and its management implications. *Biodiversity and Conservation* 17, 3181–3194.

Kolar, C.S. and Lodge, D.M. (2001) Progress in invasion biology: predicting invaders. *Trends in Ecology and Evolution* 16, 199–204.

Kowarik, I. (1995) Time lags in biological invasion with regard to the success and failure of alien species. In: Pyšek, P., Prach, K., Rejmànek, M. and Wade, M. (eds) *Plant Invasions, General Aspects and Special Problems.* SPB Academic Publishers, The Hague, Netherlands, pp. 15–38.

Krivánek, M. and Pyšek, P. (2006) Predicting invasions by woody species in a temperate zone: a test of three risk assessment schemes in the Czech Republic (Central Europe). *Diversity and Distributions* 12, 319–327.

Lee, M. (2001) Non-native plant invasions in Rocky Mountain National Park: Linking species traits and habitat characteristics. MS thesis, Colorado State University, Fort Collins, Colorado.

Leung, B., Lodge, D.M., Finnoff, D., Shorgren, J.F., Lewis, M.A. and Lamberti, G. (2002) An ounce of prevention or a pound of cure: bioeconomic risk analysis of invasive species. *Proceedings of the Royal Society, London, Biological Science* 269, 2407–2413.

Levine, J.M. and D'Antonio, C.M. (1999) Elton revisited: a review of evidence linking diversity and invasibility. *Oikos* 87, 15–26.

Lewis, M.A. (1997) Variability, patchiness, and jump dispersal in the spread of an invading population. In: Tilman, D. and Karieva, P. (eds) *Spatial Ecology: the Role of Space in Population Dynamics and Interspecific Interactions.* Princeton University Press, Princeton, New Jersey, pp. 46–69.

Lockwood, J.L., Cassey, P. and Blackburn, T. (2005) The role of propagule pressure in explaining species invasions. *Trends in Ecology and Evolution* 20, 223–228.

Lockwood, J.L., Hoopes, M.F. and Marchetti, M.P. (2007) *Invasion Ecology.* Blackwell, Malden, Massachusetts.

Lodge, D.M. (1993) Biological invasions: lessons from ecology. *Trends in Ecology and Evolution* 8, 133–137.

Lonsdale, W.M. (1999) Global patterns of plant invasions and the concept of invasibility. *Ecology* 80, 1522–1536.

Mack, R.N. (1996) Predicting the identity and fate of plant invaders: emergent and emerging approaches. *Biological Conservation* 78, 107–121.

Millennium Ecosystem Assessment Board (2005) *Ecosystems and Human Well-being: Biodiversity Synthesis.* World Resources Institute, Washington, DC.

National Research Council of the United States (NRC) (1983) *Risk Assessment in the Federal Government: Managing the Process.* National Academic Press, Washington, DC.

Nishida, T., Yamashita, N., Asai, M., Kurokawa, S., Enomoto, T., Pheloung, P.C. *et al.* (2009) Developing a pre-entry weed risk assessment system for use in Japan. *Biological Invasions* 11, 1319–1333.

Nuñez, M.A. and Pauchard, A. (2009) Biological invasions in developing and developed countries: does one model fit? *Biological Invasions* DOI 10.1007/s10530-009-9517-1.

Okubo, A. and Levin, S.A. (2002) *Diffusion and Ecological Problems: Modern Perspectives*. Springer-Verlag, New York.

Orr, R. (2003) Generic nonindigenous aquatic organisms risk analysis review process. In: Ruiz, G.M. and Carlton, J.M. (eds) *Invasive Species: Vectors and Management Strategies*. Island Press, London, pp. 415–438.

Orr, R.L., Cohen, S.D. and Griffin, R.L. (1993) Generic nonindigenous pest risk assessment process (for estimating pest risk associated with the introduction of nonindigenous organisms). Internal Report, U.S. Department of Agriculture, Animal and Plant Health Inspection Service, Riverdale, Maryland.

Peterson, A.T. (2003) Predicting the geography of species' invasions via ecological niche modelling. *Quarterly Review of Biology* 78, 419–433.

Peterson, A.T. (2006) Uses and requirements of ecological niche models and related distributional models. *Biodiversity Informatics* 3, 59–72.

Pheloung, P.C., Williams, P.A. and Halloy, S.R. (1999) A weed risk assessment model for use as a biosecurity tool evaluating plant introductions. *Journal of Environmental Management* 57, 239–251.

Philips, S.J., Anderson, R.P. and Schapire, R.E. (2006) Maximum entropy models of species geographic distribution. *Ecological Modelling* 190, 231–259.

Pimentel, D., Zuniga, R. and Morrison, D. (2005) Update on the environmental and economic costs associated with alien-invasive species in the United States. *Ecological Economics* 52, 273–288.

Pyšek, P. and Jarošík, V. (2005) Residence time determines the distribution of alien plants. In: Inderjit, G. (ed.) *Invasive Plants: Ecological and Agricultural Aspects*. Birkhäuser Verlag-AG, Basel, Switzerland, pp. 77–96.

Pyšek, P. and Richardson, D.M. (2007) Traits associated with invasiveness in alien plants: Where do we stand? In: Nentwig, W. (ed.) *Biological Invasions*. Ecological Studies, vol. 193, Springer-Verlag, Berlin and Heidelberg, pp. 97–125.

Pyšek, P.D., Richardson, M., Pergil, J., Jarošík, V., Sixtova, Z. and Weber, E. (2008) Geographical and taxonomic biases in invasion ecology. *Trends in Ecology and Evolution* 23, 237–244.

Randall, J.M., Norse, L.E., Benton, N., Hiebert, R., Lu, S. and Killeffer, T. (2008) The invasive species assessment protocol: a tool for creating regional and national lists of invasive non-native plants that negatively impact biodiversity. *Invasive Plant Science and Management* 1, 36–49.

Reichard, S.H. and Hamilton, C.W. (1997) Predicting invasions of woody plants introduced into North America. *Conservation Biology* 11, 193–203.

Rejmánek, M. and Richardson, D.M. (1996) What attributes make some plant species more invasive? *Ecology* 77, 1655–1661.

Rejmánek, M., Richardson, D.M., Higgins, S.I., Pitcairn, M.J. and Grotkopp, E. (2005) Ecology of invasive plants: state of the art. In: Mooney, H.A., Mack, R.N., McNeely, J.A., Neville, L.E., Schei, P.J. and Waage, J.K. (eds) *Invasive Alien Species: a New Synthesis*. Island Press, Washington, DC, pp. 104–161.

Robertson, M.P., Villet, M.H., Fairbanks, D.H.K., Henderson, L., Higgins, S.I., Hoffman, J.H. *et al.* (2003) A proposed prioritization system for the management of invasive plants in South Africa. *South African Journal of Science* 99, 37–43.

Rouget, M. and Richardson, D.M. (2003) Inferring process from pattern in plant invasions: a semimechanistic model incorporating propagule pressure and environmental factors. *American Naturalist* 162, 713–724.

Sher, A.A. and Hyatt, L.A. (1999) The disturbed resource-flux invasion matrix: a new framework for patterns of plant invasion. *Biological Invasions* 1, 107–114.

Shigesada, N. and Kawasaki, K. (1997) *Biological Invasions: Theory and Practice*. Oxford University Press, Oxford, UK.

Simberloff, D. (2009) The role of propagule pressure in biological invasions. *Annual Review of Ecology, Evolution and Systematics* 40, 81–102.

Skellam, J.G. (1951) Random dispersal in theoretical populations. *Biometrika* 38, 196–218.

Smith, C.S., Lonsdale, W.M. and Fortune, J. (1999) When to ignore advice: invasion predictions and decision theory. *Biological Invasions* 1, 89–96.

Smith, M.D., Wilcox, J.C., Kelly, T. and Knapp, A.K. (2004) Dominance not richness determines invasibility of tallgrass prairie. *Oikos* 106, 253–262.

Stockwell, D. and Peters, D. (1999) The GARP modelling system: problems and solutions to automated spatial prediction. *International Journal of Geographic Information Sciences* 13, 143–158.

Stohlgren, T.J. and Schnase, J.L. (2006) Risk analysis for biological hazards: what we need to know about invasive species. *Risk Analysis* 26, 163–173.

Theoharides, K.A. and Duke, J.S. (2007) Plant

invasion across space and time: factors affecting non-indigenous species success during four stages of invasion. *New Phytolologist* 176, 1–18.

Thompson, K., Hodgson, J.G., Grime, J.P. and Burke, M.J.W. (2001) Plant traits and temporal scale: evidence from a 5-year invasion experiment using native species. *Journal of Ecology* 89, 1054–1060.

Thorp, J.R. and Lynch, R. (2000) The determination of weeds of national significance. http://www.weeds.org.au/docs/WONS/

Thuiller, W. (2003) BIOMOD – optimizing predictions of species distributions and projecting potential future shifts under global change. *Global Change Biology* 9, 1353–1362.

Thuiller, W., Richardson, D.M., Pyšek, P., Midgley, G., Hughes, G.O. and Rouget, M. (2005) Niche-based modelling as a tool for predicting the risk of alien plant invasions at a global scale. *Global Change Biology* 11, 2234–2250.

Thuiller, W.C., Albert, M.B., Araújo, P., Berry, M., Cabeza, M., Guisan, A. *et al.* (2008) Predicting global change impacts on plant species' distributions: Future challenges. *Perspectives in Plant Ecology, Evolution and Systematics* 9, 137–152.

Tsoar, A., Allouche, O., Steinitz, O., Rotem, D. and Kadmon, R. (2007) A comparative evaluation of presence only methods for modelling species distribution. *Diversity and Distributions* 13, 397–405.

Tucker, K.C. and Richardson, D.M. (1995) An expert system for screening potentially invasive alien plants in South African Fynbos. *Journal of Environmental Management* 44, 309–338.

United States Environmental Protection Agency (EPA) (1992) *Framework for Ecological Risk Assessment.* Environmental Protection Agency, Washington, DC.

Vilá, M., Basnou, C., Pyšek, P., Josefsson, M., Genovesi, P., Gollasch, S. *et al.* and DAISIE partners (2009) How well do we understand the impacts of alien species on ecosystem services? A pan-European, cross-taxa assessment. *Frontiers in Ecology and Environment* DOI:10.1890/080083.

Virtue, J.G., Groves, R.H. and Panetta, F.D. (2001) Towards a system to determine the national significance of weeds in Australia. In: Grove, R.H., Panetta, F.D. and Virtue, J.G. (eds) *Weed Risk Assessment.* CSIRO, Collingwood, Victoria, Australia, pp. 124–152.

Wardle, D.A. (2001) Experimental demonstration that plant diversity reduces invasibility – evidence of a biological mechanism or a consequence of sampling effect? *Oikos* 95, 161–170.

Weber, E. and Gut, D. (2004) Assessing the risk of potentially invasive plant species in central Europe. *Journal for Nature Conservation* 12, 171–179.

Weiss, J. (1999) *Assessment for Invasive Potential of Victoria's State. Prohibited, Priority, and Regional Priority Weeds.* Agriculture Victoria, Pest Plant Assessment Project, Frankston, Victoria, Australia, 24 pp.

Williams, P.A., Nicol, E. and Newfield, M. (2001) Assessing the risk to indigenous biota of plant taxa new to New Zealand. In: Grove, R.H., Panetta, F.D. and Virtue, J.G. (eds) *Weed Risk Assessment.* CSIRO, Collingwood, Victoria, Australia, pp. 100–116.

Williamson, M. (1999) Invasions. *Ecography* 22, 5–12.

Wilson, J.R., Eleanor, U., Dormontt, E., Prentis, P.J., Lowe, A.J. and Richardson, D.M. (2009) Something in the way you move: dispersal pathways affect invasion success. *Trends in Ecology and Evolution* 24, 136–144.

**Appendix 18.1.** Questions forming the basis of the Australian Weed Risk Assessment model.

| | | |
|---|---|---|
| *General* | | |
| Domestication/ cultivation | 1.01 | Is the species highly domesticated? If 'no' go to question 2.01 |
| | 1.02 | Has the species become naturalized where grown? |
| | 1.03 | Does the species have weedy races? |
| Climate and distribution | 2.01 | Species suited to Australian climates? (0, low; 1, intermediate; 2, high) |
| | 2.02 | Quality of climate match data? (0, low; 1, intermediate; 2, high) |
| | 2.03 | Broad climate suitability? (environmental versatility) |
| | 2.04 | Native or naturalized in regions with extended dry periods? |
| | 2.05 | History of repeated introductions outside its natural range? |
| Established elsewhere | 3.01 | Naturalized beyond native range? |
| | 3.02 | Garden/amenity/disturbance weed? |
| | 3.03 | Weed of agriculture/horticulture/forestry? |
| | 3.04 | Environmental weed? |
| | 3.05 | Congeneric weed? |
| *Biology/ecology* | | |
| Undesirable traits | 4.01 | Produces spines, thorns or burrs? |
| | 4.02 | Allelopathic? |
| | 4.03 | Parasitic? |
| | 4.04 | Unpalatable to grazing animals? |
| | 4.05 | Toxic to animals? |
| | 4.06 | Host for recognized pests and pathogens? |
| | 4.07 | Causes allergies or is otherwise toxic to humans? |
| | 4.08 | Creates a fire hazard in natural ecosystems? |
| | 4.09 | Is a shade-tolerant plant at some stage of its life cycle? |
| | 4.10 | Grows on infertile soils? |
| | 4.11 | Climbing or smothering growth habit? |
| | 4.12 | Forms dense thickets? |
| Plant type | 5.01 | Aquatic? |
| | 5.02 | Grass? |
| | 5.03 | Nitrogen-fixing woody plant? |
| | 5.04 | Geophyte? |
| Reproduction | 6.01 | Evidence of substantial reproductive failure in native habitat? |
| | 6.02 | Produces viable seed? |
| | 6.03 | Hybridizes naturally? |
| | 6.04 | Self-fertilization? |
| | 6.05 | Requires specialist pollinators? |
| | 6.06 | Reproduction by vegetative propagation? |
| | 6.07 | Minimum generative time (years)? |
| Dispersal mechanisms | 7.01 | Propagules likely to be dispersed unintentionally? |
| | 7.02 | Propagules dispersed intentionally by humans? |
| | 7.03 | Propagules likely to disperse as a produce contaminant? |
| | 7.04 | Propagules adapted to wind dispersal? |
| | 7.05 | Propagules buoyant? |
| | 7.06 | Propagules dispersed by birds? |
| | 7.07 | Propagules dispersed by other animals (externally)? |
| | 7.08 | Propagules dispersed by other animals (internally)? |
| Persistence attributes | 8.01 | Prolific seed production? |
| | 8.02 | Evidence that a persistent propagule bank is formed (>1 year)? |
| | 8.03 | Well controlled by herbicides? |
| | 8.04 | Tolerates or benefits from mutilation, cultivation or fire? |
| | 8.05 | Effective natural enemies present in Australia? |

# 19 Economics of Invasive Alien Species

**Vinish Kathuria[1] and S.P. Singh[2]**

[1]*Associate Professor, SJM School of Management, IIT Bombay, Mumbai, India;* [2]*Advisor, Planning Commission, Uttarakhand State Institute of Technology and Science, SGRR Education Mission, Dehradun, India*

## Introduction

Species – plants, animals and microbes – that spread to areas outside their native geographic locations are called non-indigenous species (NIS). Three fundamental drivers governing their spread are: (i) natural ecological process; (ii) trade in goods and services[1] and movement of people; and (iii) conversion of land from forest to other uses. Some of these NIS over time establish themselves as harmful invaders, termed invasive alien species (IAS)[2] which, according to the Convention on Biological Diversity (CBD), are 'alien species whose introduction and spread threaten ecosystems, habitats, or species with socio-cultural, economic and/or environmental harm and/or harm to human health'.[3] It has been estimated that of 256 vertebrate extinctions for which causes have been identified, 109 occurred due to biological invaders and 70 were caused by human exploitation (Cox, 1993, as cited in Olson, 2006: 178). It is estimated that in the USA, nearly 40 per cent of endangered or threatened species are under pressure from IAS (Wilcove *et al.*, 1998). A survey of 31 fish introductions in Europe, North America, Australia and New Zealand found that, in 77% of cases, native fish populations were reduced or eliminated following the introduction of non-native fish (Ross, 1991: 363).

Biological invasions are a major environmental concern, due to their negative impacts on biodiversity and economics: there is no science to predict which introduced species will become invasive and when (Agoramoorthy, 2007). Sufficient evidence is available to show that some species that were introduced to help humanity for agriculture purpose subsequently became invasive. For example, the golden apple snail, which was introduced to Asia from South America in 1980 for cultivation as a high-protein food source, has dispersed into the region's rice paddies where it feeds voraciously on rice seedlings, causing significant crop damage (Naylor, 1996: 443).

Estimates exist suggesting that nearly 50,000 alien species have been introduced to the USA (Pimentel *et al.*, 2005). The means or routes by which species are introduced into new ecosystems are termed 'pathways' or 'vectors'. The intentional release of species (such as brook trout introduced into cascade lakes to augment fishing opportunities) is used for livestock, pets, food crops, ornamental plants and ornamental fish.

Species that are unintentionally transported arrive via one of two pathways: (i) as

[1] For instance, in the USA, 85% of established non-native woody plant species were introduced through horticultural trade (Reichard and Hamilton, 1997), and 26% of non-indigenous freshwater fish were introduced through the aquarium trade (Fuller *et al.*, 1999, as cited in Krishnakumar *et al.*, 2009).

[2] A non-native species that is able to move aggressively into a habitat and monopolize resources such as light, nutrients, water and space to the detriment of other species is called an invasive species.

[3] Source: www.cbd.int/ (accessed January 2010).

hitchhikers on or within the bodies of other hosts/non-native species, such as livestock or fresh fruit and vegetable produce; and (ii) within packaging materials, cargo holds and ships' ballast (such as zebra mussel hitchhiking in the ballast water of ocean-going ships (Lackwood *et al.*, 2007; Cusack *et al.*, 2009). A ship must balance its cargo load for stability, enabling it to move smoothly through the water. The ballast in early cargo ships included rubble, gravel, stones and soil (Mack, 2004); this material used to be collected at the port of departure and offloaded casually at the port of arrival. Along with these materials were transported insects, plants, earthworms and other organisms (Carlton, 1987). According to an estimate by Carlton (1999), 10,000 or more species were transported each day within ballast water, also termed a biotic conveyer belt. That is why the percentage of non-native marine fish[4] unintentionally transported is very high (73%) compared with other groups of organisms, such as plants and birds, at generally <40% (Lockwood *et al.*, 2007).

The global transport of goods and people, however, is the primary vector for the introduction of IAS, ecological factors essentially explaining their propagation in new ecosystems. These include the lack of both controlling natural enemies and effective predators; environmental disturbances can also affect their spread.[5]

The economics of IAS are important, because these impose significant costs on policy makers and also affect consumer and social welfare. According to the CBD, invasive species are one of the leading causes of loss of natural biodiversity. Moreover, according to some (see, for example, Van Driesche and Van Driesche, 2001; Cusack *et al.*, 2009), the prevention, eradication and control of invasive species is an economic and policy issue and has less to do with biology and ecology.

This chapter discusses the economics of invasive species and is organized into five sections; the second section explains why the management or spread of IAS is an economic issue; the third section summarizes the economic impact of invasive species at the international scale; the fourth section focuses on the impact and efforts in India, while the next section explores the issues in regard to controlling IAS. The chapter's conclusion comprises the final section.

## Economic Activities and Vulnerability of an Area to IAS

The impact of invasive species is not uniform across the world's ecosystems. In terrestrial ecosystems, xeric (deserts, semi-deserts, tropical dry forest and woodlands) and northern Arctic systems are typically less affected. On the other hand, islands are more affected by IAS leading to the extinction of native species, the reason being, in part, because on average these have a greater percentage of merchandise imports as their share of GDP (43%) than continental countries (26.8%) (Dalmazzone, 2000). Similarly, in aquatic ecosystems, lake, river and near-shore marine systems are affected more than open oceans (pelagic) (Meyerson and Mooney, 2007).

The number of established non-native species is far lower in temperate compared with tropical zones (Sax, 2001). Generally Old World species are more successful as invaders than New World. Species movement between two regions connected commercially has thrown up strange results, one region receiving many more species than the other. Similarly, islands are more vulnerable than continents with which they are connected for commerce.

[4] In this context, it is important to note that aquarium fish are often not included in the debate on their role as IAS, but one third of the world's worst aquatic invasive species are ornamentals.

[5] Environmental disturbances such as weather events can either accelerate or slow the spread of invasive species. For example, a hurricane in 1938 blew the gypsy moth (*Lymantria dispar*) across a barrier zone that had been established along the Hudson River to slow its spread (Animal and Plant Health Inspection Service, 1985, cited in Olson, 2006). Similarly, hurricanes in the USA 2004 and 2005 resulted in the spread of citrus canker to Florida, which forced the US Department of Agriculture (USDA) to abandon an eradication programme that had begun in 1996 (Florida Department of Agriculture and Consumer Services, 2006).

Species movement largely depends on technological development of inter-continental commercial traffic. For example, early sailing ships had a limited route to follow as they depended on favourable winds for movement. Both the scale and direction of routes expanded with the arrival of steam-powered ships and advances in navigation systems. The rise in air transport dramatically expanded connecting pathways and species movements well beyond ports to the interior of continents (Mack, 2004). This is exemplified by the number of non-native species established within the San Francisco Bay estuary, which was reported to be one species every 55 weeks between 1851 and 1960 but had increased to one species every 14 weeks between 1961 and 1995 (Cohen and Carlton, 1998).

### IAS: an economics problem?

There are a number of reasons why the problem of IAS is an economic problem. First, it is economic forces – trade and transport of people – that are the primary vector of spread (Perrings *et al.*, 2000). Secondly, biological invasions are a classic example of a negative externality arising from economic decisions.[6] Thirdly, IAS management is a 'weakest link' public good (Cusack *et al.*, 2009). Management of IAS is non-exclusive (Summer, 2003), implying that one cannot exclude others. As a consequence, there exist incentives to 'free-ride' on the efforts of others and individuals bearing less than their fair share of control costs. Since collective gain or benefit from IAS management is significantly more than that received by a single individual or a region, this further reduces the incentive for individual action. Thus, IAS management is only as effective as the 'weakest link' in the chain. For instance, suppose all the seaports or airports of a country or region, bar one, have the best bio-security measures in place. The one port carrying out only bare minimum bio-security practices as mandated by law acts as the 'weakest link', resulting in invasive species introductions into the country/region despite the best efforts of other ports. Fourthly, the risks of biological invasions are endogenous in that they are affected by how countries value goods and services that can become vectors for invasive species, by how they protect themselves from IAS and by how they react to them after they occur (Jensen, 2002).

Once established, alien species exert huge economic costs worldwide. Although budgets for control of IAS are on the rise in most countries, the economics of the problem are still inadequately understood (Perrings *et al.*, 2000). The economics of invasive alien species (IAS) should consider all costs: direct costs from damages caused by IAS and those associated with the prevention, control and mitigation of invasions (*ibid.*).

As mentioned above, some species are intentionally introduced. However, the damage/economic impact caused by these species is less than that from those unintentionally introduced. This is largely because care is taken in the selection of intentionally introduced species that they are suited to the environment where they are to be introduced (Williamson, 1994; Smith *et al.*, 1999).

## Economic Impact of IAS: Estimations

The economics of control of pest and disease-inducing IAS have been well documented (see, for example, Williamson, 1998). Damage to industries in Europe and North America caused by zebra mussel has been well documented (Khalanski, 1997). Attempts made to aggregate the economic cost of invasions are few and their estimates vary widely, as we shall see later. One recent attempt to quantify the economic damages and control costs of invasive species in five countries – the USA, South Africa, the UK, Brazil and India – quoted a figure of US$336 billion per year (Pimentel *et al.*, 2000: 14).

[6] Negative externalities are the uncompensated costs to a third party arising from a particular decision or action. Examples include the human health effects of air pollution in the vicinity of a coal-burning factory.

Not only are the economic impacts of IAS high, but even the nature of impact can vary widely. To give an example, the coqui frog (*Eleutherodactylus coqui*) was introduced accidentally to Hawaii from Puerto Rico. Because of its piercing call (up to 70–100 decibels at a distance of 0.5 m is loud enough to cause hearing loss over long exposure), the property value of homes located near its habitats fell by 64% (Meyerson and Mooney, 2001).

Similarly, rats, which are commonly believed to be world's most widespread invasive mammals, have the greatest economic impact. They have been the cause of extinction of more birds, snakes and lizards on islands than any other predators (Mathews, 2000), besides causing widespread damage to agriculture. Pimentel *et al.* (2000) valued the economic damage caused by 250 million rats in the USA at US$19 billion per year, whereas Singleton (2003) quantified the extent of damage to rice production in Asia as 5–10% per annum. Rats also cause significant environmental impacts, and have contributed to the extinction of many species of wildlife. The current practice of controlling rats is through poison, which has a significant negative impact on both ecosystem services and humans.[7]

The economic impact of invasive species essentially means that the supply of some goods and services will diminish if invasive species have a detrimental impact on the ecosystems concerned. In this context, it is important to note that the total economic value of an ecosystem comprises both use and non-use values. Use values include the value of food production, recreation or climate amelioration, etc., whereas the existence value (i.e. the value of knowing that a natural ecosystem exists)[8] and bequest value (i.e. the value of leaving a natural and functional ecosystem to future generations) comprise the non-use values. A loss of total economic value is what one should measure when evaluating the impact of IAS. Table 19.1 shows the sources of invasives and their impacts, by ecosystem.

Table 19.2 summarizes some of the estimates of the impact of IAS in the context of developed countries. There are a few caveats before we look into these figures, howeever. First, most of these studies attempt to value existing invasions and disregard the value of preventing future occurrences, despite the fact that prevention of future invasions might be the most effective policy tool. Secondly, most measure the direct-use economic impacts, with only a handful taking into account the non-use impacts of IAS (Cusack *et al.*, 2009). Thirdly, it is not uncommon for aggregate values to be obtained by multiplying the constant marginal damage[9] per pest by an estimate of the total pest population/units; this may not be statistically reliable if the pest population is very large (Olson, 2006). Fourthly, damage can be non-linear, rather than the linear damage function assumed in many of these studies; this, however, requires estimation of a specific damage function. Lastly, economic valuation often neglects the loss of genetic information due to the invasion; invasions may damage ecosystem services, such as pollination and hydrological cycles (Daily, 1997). For example, in African *Fybbos* the hydrological cycle was disturbed substantially because of invasion by species of *Pinus*, *Acacia* and *Hakea* (Turpie and Heydenrych, 2000).

---

[7] By preying on other species or competing with them for food, rats have caused the decline of many small mammals, birds, reptiles and invertebrates. This effect has been particularly severe on islands; in the Seychelles, for example, rats have had more impact on endemic biodiversity than any other factor. In Mauritius, they are believed to have caused the extinction of a number of snakes and lizards, and contributed to the threatened status of many birds (Matthews and Brandt, 2004).

[8] The following example indicates how an invasive species can decrease the existence value of species-rich tropical rainforest by adversely affecting its endemic content and diversity. *Maesopsis eminii*, an invasive tree species, invaded the sub-mountain rainforests of the East Usambara Mountains of Tanzania. This invasion has drawn widespread attention, not because of any particular financial loss but because of the social loss of existence value (Lovett, 2000).

[9] Marginal damage is defined as the additional damage, in terms of money, when one more unit of the pest is introduced.

**Table 19.1.** Source of invasives and their impacts, by ecosystem.

| Ecosystem | Sources of invasives | Examples of invasive impacts |
|---|---|---|
| Agro-ecosystem | Intentional introductions for crops, livestock and horticulture; accidental transport of pests and diseases | Agricultural losses from bio-invasions worldwide range from US$55 to 248 billion per year (Bright, 1999: 51).<br>Asian gypsy moth and nun moth projected to cost US$35–58 billion in tree losses in the USA over 50 years (U.S. OTA, 1993:118).<br>46% of endangered plants and animals in USA negatively impacted by invasive species (Wilcove *et al.*, 1998). |
| Coastal | Through ballast water; intentional introductions for aquaculture, fish stocking and ornamental uses; species migrations via artificial canals | Invasion of Leidy's comb jelly, native to the western Atlantic, led to the collapse of the anchovy fishing industry in the Black Sea (Bright, 1998: 157). |
| Forest | Exotic tree species spread from forest plantations; horticultural introductions; transport of insect pests via traded forest products and other goods | S. Africa to spend ~US$900 million in next 20 years to control invasive trees and plants that consume ~3.3 billionn $m^3$ of water annually (WRI, 2000: 196–202).<br>In the eastern USA, native fir and hemlock forests are threatened by several species of adelgid, a tiny, aphid-like insect (Stein and Flack, 1996). |
| Freshwater | Fish, higher plants, invertebrates and microscopic plants accidentally introduced through shipping, commerce and aquaculture; and intentionally for fishing | Canada and the USA have spent US$8 million since 1991 to control sea lampreys, which have invaded the Great Lakes, plus US$12 million attempting to restore lake trout, a freshwater fish eaten by the sea lamprey (Fuller *et al.*, 1999: 19–21). Similarly, controlling the zebra mussel (a native of European rivers) in the Great Lakes cost US$300–400 million in 1989–1995 (Revenga *et al.*, 2000: 56).<br>Water hyacinth, native to the Amazon, has spread via rivers and lakes everywhere except Europe. It clogs waterways and infrastructure, reduces light and oxygen in freshwater systems and causes changes in water chemistry and species assemblages (Hill *et al.*, 1997). |
| Grassland | Intentional introductions for game hunting, improved rangeland productivity, and agriculture | In Australia *Parthenium*, a grass native to the Caribbean, invades pasture as well as cropland, causing allergic reactions in livestock and humans (Ricciari *et al.*, 2000: 239).<br>In Canada crested wheatgrass (*Agropyron cristatum*), native to Asia, displaces native species and, in turn, reduces soil nutrient content and organic matter (Christian and Wilson, 1999). |

Source: World Resources Institute (2001).

From Table 19.1 we can see that, even without accounting for non-use value and without anticipating future invasions, the economic impact of IAS is very high, which also implies significant social costs. In order to appreciate the significance of these values, Box 19.1 illustrates how different sectors have benefited from the control of invasion by the intentional introduction of European rabbits carrying rabbit calicivirus disease (RCD).

The case study of rabbit introduction to Australia and subsequent control provides invaluable knowledge about economics as well as wildlife management, which can be utilized in dealing with the problems of invasive mammals elsewhere.

**Table 19.2.** Economic impacts of selected invasive species and overall impact.

| Country | Species | Description of economic impact | Annual cost |
|---|---|---|---|
| USA | General estimate | Total direct and indirect use impacts | US$143 billion |
| | Weeds | Control costs, production losses | US$30.6 billion |
| | Invasive fish species | Depletion of natural stocks, other effects | US$6.03 billion |
| | Zebra mussel | Damage to infrastructure, control costs | US$1.12 billion |
| | Aquatic weeds | Losses, damages, control costs | US$122 million |
| | Purple loosestrife | Control costs, forage losses | US$50 million |
| | Rat | Consumption of: | |
| | | stored grains | US$19 billion |
| | | other materials | US$21.2 million |
| | | Loss of market due to quarantines | US$229 million |
| Germany | Plant, animal and microbial | Total | €168.2 million |
| | Dutch elm disease | | €5 million |
| Australia | Plant and animal | | AU$4.7 billion |
| | Plant | | AU$4.0 billion |
| | Ten vertebrates | Includes environmental costs | AU$704 million |
| Canada | Leafy spurge and knapweed | | CAN$38.21 million |
| | Three invertebrates | | CAN$101.3 million |
| | Emerald ash borer | | CAN$14–16 million |
| | Dutch elm disease | | CAN$1.5 million |
| | Potato wart fungus | | CAN$73.34 million |
| New Zealand | Plant | | NZ$100 million |
| | Vertebrates | | NZ$270 million |
| | Invertebrates | | NZ$2 billion |

Source: collated from different sources.

## Economic Impact of IAS and Efforts to Control Spread in India

Around 40% of Indian floral species are alien, of which, 25% are invasive (Gupta, 2004), according to certain estimates. A recent study by the Indian Council of Forestry Research and Education (ICFRE) has identified around 75 forest-invasive species (FIS) in different parts of the country, all posing a threat to the natural forest cover and, in addition, there are 36 other FIS localized in limited areas.[10] Of the 75 identified FIS, 61 are plant, including 12 fungi; the other 14 species are insects. These 75 FIS were found in various forest regions around the country. Interestingly, of these identified FIS, 28 are native to India but have taken invasive proportions in other bio-geographical regions of the country. Key IAS in India include *Lantana camara*, *Eupatorium* spp., *Parthenium* spp., *Ageratum conyzoides*, *Eichhornia crassipes*, *Salvinia molesta*, *Ipomoea carnea* and *Prosopis juliflora*.

Forest-invasive species can displace native plants, eliminate food and cover for wildlife and threaten rare plant and animal species. These species can change the functions of ecosystems and increase losses in forestry and natural resource management costs, while also posing a danger to biological diversity and other human enterprises – as well as to human health. Many invasive species have naturalized in India, although a few are now being used for various purposes such as medicines, furniture making and composting.

[10] Source: http://www.financialexpress.com/news/Invasive-species-are-threat-to-forest-cover-ICFRE/282361/0, Financial Express March 10 2008.

**Box 19.1.** Costs and benefits of controlling European rabbit (*Oryctolagus cuniculus*) by rabbit calicivirus disease (RCD) in Australia.

Introduced to Victoria in 1859 for game, the European rabbit spread at a rate of 70–100 km per year in the absence of competitors and predators. The economic damage – including that of control and loss of productivity – was estimated at AU$600 million per year (Wilson, 1995). The introduction of RCD reduced the rabbit population drastically and, because of the persistence of the epidemic effects of the virus, the rabbit population was maintained at 10–20% of its original level. This led to an increase in agricultural productivity and revival of biodiversity.

The potential benefit of RCD far exceeds its potential cost. The market benefits to agriculture and forestry were considerable, apart from several poorly quantified non-market values. There are some negative effects on wildlife, but these are far outweighed by the benefits. When rabbits are in abundance, red fox and dingoes (the largest mammalian predator in Australia) depend entirely on them for their diet, but with a reduced rabbit population dingoes increase their predation on the red kangaroo, which is a pest. Feral cats prefer young rabbits for their diet, and as many as 13 raptor species depend on rabbits for prey. In general, the populations of these predators would decrease, even though some would shift their habits to other prey, such as young lambs and small marsupials. The real threat, however, is that the effect of such an immuno-contraceptive may indirectly affect several other animal species. In that case, costs could be enormous.

*Public sector*

RCD persisted in the field, and thereby reduced subsequent rabbit control costs. The lifespan of dams increased because of the reduction in siltation (reduced rabbit population → increased plant cover → reduced erosion → reduced siltation rate of reservoirs) as the erosion rate decreased following recovery of plant cover.

*Agriculture*

A sharp decrease in crop losses due to rabbit grazing occurred, the use of poisoning to control rabbit populations reduced and the stocking capacity of the land was enhanced. Reduced losses of crops due to other herbivores, such as the red kangaroo, were noted as predators changed their hunting habits.

*Forestry*

Reduction in rabbit damage was seen.

*Non-market*

Increased regeneration of several plant species improved the flow of ecosystem services. In addition, there was reduced impact on several wild fauna because of the reduction in use of pesticides to control rabbits.

Source: Prian *et al.* (2000); White and Newton-Cross (2000).

In India, and for that matter in many developing countries, in general, human disturbance is of a chronic nature, resulting in forest degradation rather than clearance. Under such patchy and fragmented conditions large-scale invasion, excepting a few species, is relatively uncommon. Moreover, continuous human presence does not permit even IAS to expand without obstruction.

*Lantana camara*, one of the ten worst weeds worldwide and a native of tropical and subtropical America, was introduced in India as an ornamental shrub around

1809–1810. It is now found all over the Indian subcontinent, stretching from the submontane regions of the outer Himalaya to the southernmost part of India. The plant has spread rapidly due to human disturbances such as cultivation, road construction and forest fragmentation and degradation. *Lantana* has many adverse affects on forest ecosystems,[11] including: (i) loss of native biodiversity (Sharma *et al.*, 2005); (ii) replacement of native plant communities in forests ecosystems by the formation of dense, impenetrable thickets; (iii) contributon to the erosion of soil (Day *et al.*, 2003); (iv) adverse impacts on forest regeneration; (v) harbouring of vectors that carry infectious diseases (Syed and Gurin, 2004); and (vi) creation of a fire hazard (Hiremath and Sundaram, 2005; Love *et al.*, 2009).

*Ageratum conyzoides* is another weed native to South America that has invaded and now naturalized several parts of southern Asia, including India. This weed has become a problem in agroecosystems. Its invasive potential is attributed to its rapid growth, production of a large number of small, wind- and water-disseminated seeds and vegetative proliferation through stolons.

Insufficient information on Indian marine ecosystems exists in regard to the presence and impact of alien invasive species. Many freshwater species such as *E. crassipes*, *S. molesta*, *I. carnea*, etc. have become a nuisance in aquatic ecosystems. Of these, *E. crassipes*, introduced from Brazil in 1914–1916, is the most serious, as it causes hindrance to navigation, chokes irrigation systems and reduces the aesthetic value of water bodies. Many freshwater and marine algae, including species of *Microcystis*, *Caulerpa*, *Cladophora*, etc., cause extensive damage to ecosystems and adversely affect aquatic biodiversity.

We are not aware of any study that has investigated the economics of an IAS species in India. The quantification of such an impact would be somewhat difficult in India, because people are able to utilize any woody species to the extent that the expansion of any IAS might lead to a debate on whether it should be eradicated or accepted. *Prosopis juliflora* (mesquite) is a tropical example of such a species (Box 19.2).

Similarly, *Lantana camara* stems are used for furniture making in some areas; this practice helps restrict the uninterrupted spread of another invasive species.

## Issues in Controlling IAS

Exclusion and early detection are the most cost-effective methods of controlling and preventing IAS. Studies have indicated a very high benefit:cost ratio if IAS are detected, controlled and eradicated early. For instance, an OTA (1993) study in the USA context has found a benefit:cost ratio of 17:1 for early detection, control and eradication. This implies a potential saving of Rs. 17 for every Rs. 1 invested in early-detection programmes. Similarly, another study (ODA, 2000) has yielded a figure of 34:1 as a benefit:cost ratio.

The costs of controlling IAS rise rapidly as they gain a stronger foothold in ecosystems. Once species are established, eradication may no longer be a cost-effective possibility. Moreover, there is considerable uncertainty as to whether a particular species can really be eradicated once established. The cases of *Lymantria dispar* in Oregon and citrus canker in Florida are an interesting illustration of this uncertainty (Box 19.3). In that scenario, damage mitigation and control may be the only feasible policy responses.

When an economically useful species is introduced intentionally, its importers argue that its benefits would outweigh its likely harms; however, they do not pay when things go wrong. This is the classic externality problem, where social costs exceed the

[11] A recent study by ATREE (2010), however, did not find any negative effect of *Lantana* on the biodiversity of moist and dry deciduous forests in the Male Mahadeshwara Hills, Karnataka. Most parameters of biodiversity did not show the expected decrease with increased presence of *Lantana*. The results also suggest that *Lantana* did not affect the avian community as severely as it is believed to have affected other components of the ecosystem.

**Box 19.2.** Economic impact of *Prosposis julifora* (*ganda bawal*): an illustration.

*Prosopis juliflora* (mesquite), better known as *ganda bawal*, was planted to help the Gujarat State Forest Department fight salinity ingress and barrenness in the Banni region of Kutch. A ban was placed on the tree's harvesting in the 1980s, at which time it covered <10% of the Banni grasslands. However, it quickly became an invasive species, occupying >40% of the land by the late 1990s, thereby adversely impacting biodiversity and destroying the Banni grassland ecosystem. Early in 2004 the government lifted the ban, allowing the harvesting of *Prosopis* under Section 32 of the Indian Forest Act. The plan, on paper, was to make charcoal from the plant and thus help improve the economic conditions of the people of Banni; another aim was to contain the unchecked spread of the wild weed. The decision, however, backfired, as not only *Prosopis* being cut down, but also native trees, resulting in the crucial green cover of the region being reduced to <10% within 1 year.

On the other hand, the local people found charcoal production more profitable than their traditional profession of making cottage products: the species is now providing employment for nearly 30,000 people who are making charcoal. The poor use mesquite stems as fencing and construction poles, and its pods are used for both cattle and human food.

Mass cutting of *ganda bawal* trees and the pollution resulting from charcoal production have unexpectedly also reduced the wild honey bee population, with disastrous effects on honey production, crop pollination and crop yields in the Kutch region. The number of dwarf bee hives per 1 $km^2$ has reduced to only 20–25, from the earlier level of 60–70 in and around the Banni grasslands, following large-scale tree felling. Local honey hunters (the Koli community) who used to harvest ~300 t of wild honey annually from Kutch could collect only 50 t in 2008.

Thus, the economic benefits arising from *Prosopis juliflora* – the positive effect of reducing salinity ingress, provision of employment through charcoal production – must be set against its various negative effects – damage caused by its large thorns to domestic and wild animals, adverse impact on biodiversity, local extinction of economically useful species, effect on wild bee honey production, air pollution and depletion of groundwater.

Source: Maheswari (2008); Nair and Darji (2008); Pathan (1998).

private costs of introducing IAS. One of the ways to reduce such environmental externality is the issuing of assurance/environmental bonds, as suggested by Shogren *et al.* (1993). The two key difficulties associated with environmental bonds are: (i) regulator moral hazard, where the principal may impose liability without cause; and (ii) liquidity constraints that prevent the importer/agent from posting the requisite bond (Olson, 2006).

This policy change would shift the burden of proof of no harm to those responsible for introducing the species, which would also generate research necessary to prove that the species proposed for introduction is unlikely to be invasive (Perrings *et al.*, 2005).

Managing and controlling IAS becomes difficult when there are species involved such as fur-bearing animals, deer, feral pigs, etc. which, though causing damage to agricultural or environmental systems, are valuable from either the commercial or recreational aspect or for their contribution to biological diversity (Olson, 2006: 188).

The World Trade Organization (WTO) Agreement on the Application of Sanitary and Phytosanitary Measures (SPS Agreement) is in recognition of trade as a significant vector in the IAS problem. Article 2 of the SPS Agreement clearly states that it is important for individual countries to 'have the right to take sanitary and phytosanitary measures necessary for the protection of human, capital or plant life or health'. Similarly, the 156-nation International Maritime Organization (IMO) is trying to develop an international legal regime to control ballast water discharge, which is the source of many invasives in coastal estuaries. To help developing countries control the introduction of invasives through ballast water, the IMO, the Global Environmental Facility and the United Nations Development Programme committed US$10 million in

**Box 19.3.** Gypsy moth in Oregon and citrus canker in Florida: challenges in eradication.

The gypsy moth (*Lymantria dispar*) was originally introduced into the eastern USA in 1869 as part of a research project for the silk-producing industry. When reaching high population density this insect can cause extensive defoliation of trees and shrubs; millions of acres in that part of the country have now been defoliated by it. In 1983, three gypsy moths were discovered near the town of Lowell, Oregon (western USA). An extensive trapping programme in the area in 1984 resulted in a catch of over 1900 gypsy moths. The Oregon Department of Agriculture (ODA) implemented aerial spraying of a biological insecticide in 1985, to eradicate the moth. Nearly 100,000 acres were treated, and spraying continued into 1986. In 1987, extensive trapping revealed no gypsy moths in the area, and the pest was deemed eradicated. Since that time, the ODA has continued monitoring for the moth, deploying over 18,000 traps across the state. A few moths have been caught every year and limited spraying continues to be carried out: the species has therefore been effectively controlled.

Interestingly, in 2006, 66 moths were caught in the central part of the state. Investigations revealed that this new infestation had arrived from Connecticut in a car purchased through the Internet auction site eBay. This illustrates two things: (i) the difficult task faced by policy makers in trying to control invasive species, as other ways exist through which they may be transported; and (ii) control of an IAS does not end with its eradication – constant monitoring is required for years thereafter, with significant cost implications.

Similarly, citrus canker has been detected in Florida on three occasions and twice it has been declared eradicated. The first time it was deemed eradicated was in 1933 following detection in 1910. On the second occasion (1994) it was considered eradicated following detection in 1986. It was detected again near Miami in 1995, and a third eradication campaign ensued (Florida Department of Agriculture and Consumer Services, 2006). The USDA now believes that eradication is no longer feasible, due to the potential spread of the disease by hurricanes in 2004 and 2005, and a new citrus canker management plan is being formulated (Conner, 2006).

Source: Olson (2006); Cusack *et al.* (2009); www.oregon.gov/ODA/PLANT/plant_ann_rep06_ ippm_part2.shtml

2000 in technical assistance and pilot projects (IMO, 2001).[12]

Both control and prevention can involve a variety of policy instruments. The size of invasion can be reduced by harvesting, chemical or biological controls (as have been used in case of the gypsy moth in Oregon) or mechanical or manual removal (as in the case of *Lantana* in India). Whatever method is used for effective control, greater cooperation between private agencies and private stakeholders is important. In this context, it is important to indicate the linearity between eradication effort and the fact that historical attempts to eradicate invasive species indicates that it may cost as much to remove the final 1–10% of an invasion as it does to control the initial 90–99% (Myers *et al.*, 1998).

Recent times have seen biological control receiving increasing attention as a policy instrument for invasive species management. McConnachie *et al.* (2003) reviewed ten benefit:cost studies of successful biological control programmes (four insect pests, four terrestrial weeds and two aquatic weeds). For the terrestrial weeds, the benefit:cost ratio ranged from 1.9:1 to 24:1. Van Wilgen *et al.* (2004) estimated the costs and benefits of biocontrol of six invasive weed species in South Africa, where biocontrol has been practised since 1910. Their estimated benefit:cost ratios ranged from 8:1 for red sesbania to 709:1 for jointed cactus. Biological control programmes are not without risks, however. Control species may become invasive themselves and adversely impact non-target species. For example, feral cats introduced on many islands to control rats have proved so damaging to the island ecology that they are now subject to eradication programmes

12 As cited in World Resources Institute (2001).

(Nogales *et al.*, 2004). The cane toad (*Bufo marinus*) is another example of biological control gone awry. Introduced to Australia in 1935 as a biological control for the scarab beetle, a pest of sugar cane, they failed as a control and have subsequently become a significant ecological pest (McLeod, 2004).

Although there are a number of tools to control the spread of IAS, the control strategy depends on the nature of the species and the extent of the affected area. For instance, for control of weeds five different options exist – manual removal of weeds, herbicide application, biological control, control of animals that spread the weeds – such as wild pigs and, lastly, exclusion of tourists if a tourist area. Similarly, at very low weed and seed densities, the ideal policy would be to do nothing but, over a broad range of higher densities, the ideal policy involves a mix of manual removal, herbicides and biological control (Olson, 2006).

## Concluding Remarks

The spread of invasive alien species is second only to deforestation and other land use changes in damaging global biodiversity (Simberloff *et al.*, 2005). Increasing global connectedness and open economies have accelerated this global change. Generally, invasive alien species displace native species, impact their habitat, compete with them and can hybridize with and infect them. Some alien invasive species can drastically change even a biome: for example, *Phytophthora ramorum* killing the Californian oak (Rizzo and Garbelleto, 2003).

Since the loss of amenities caused by invasive species in natural systems is often incremental, the precise impact of existing invasive species on economic and recreational use is underestimated (Van Driesche and Van Driesche, 2000).[13] As a result, a greater level of coordination between local, state, national and international agencies is required. In this respect, a comprehensive bio-security framework is the only way to avoid, remedy and mitigate the economic and ecological risks posed by IAS.

Present economic conditions favour IAS, mainly on account of the opening up of nations' economies, particularly the agriculture, livestock and tourism sectors (Delmazzone *et al.*, 2000). A higher level of education of the public, government agencies, industry and non-governmental organizations is also needed to strengthen all links in the IAS management chain.

The problems associated with IAS are yet to appear on India's policy makers' radar. This is because there is no specific Act in India that deals with the management of IAS, unlike in the USA where the invasive species policy dates back to the Lacey Act of 1900. Of the nearly 30 federal US acts pertaining to invasive species, approximately half have been adopted since 1990 (National Agricultural Library, 2006).[14] However, no legislation exists for the import or introduction of aquatic organisms in India – there is no proper quarantine procedure in place even for imported fish. Exotic species introduced purposefully or accidentally into natural water bodies can adversely affect local fauna through genetic pollution, disease introduction and ecological impacts such as predation, competition and environmental modification. Although the National Committee for Exotic Species suggested the setting up of quarantine facilities for the import of aquarium fish near the international airports of Mumbai, Chennai and Kolkata (Sudhi, 2004) as early as 2004, there has been no concerted effort to implement such a policy to date (Krishnakumar *et al.*, 2009).

Managing and controlling IAS in India warrants improved biological knowledge, suitable policy changes and adequate budgets. In brief, the steps required for better control of IAS would be to: (i) prevent their entry; (ii) locate and eradicate them immediately; and (iii) manage them at a low level if eradication is not achievable.

Thus, to conclude, little is known of the economic impact of IAS in India. India's international trade has just begun to expand

[13] As cited in Cusack *et al.* (2009).

[14] As cited in Olson (2006).

rapidly, and therefore the impact of invasive species has yet to be seen. Unless their impact is understood, insufficient efforts will be made to manage and control them. For the effective management of IAS, knowledge of their ecology, morphology, phenology, reproductive biology, physiology and phytochemistry is essential (Raghubanshi *et al.*, 2005). Disturbance of ecosystems (e.g. habitat fragmentation), which is also related to economic activities, has played a key role in the spread of IAS elsewhere. There is also a need to improve our understanding of the interaction between human activities and IAS. For example, the spread of road network following the war in Uttarakhand, India led to the rapid spread of *Lantana* and *Eupatorium*.

## References


Agoramoorthy, G. (2007) Conservation awareness and research to protect rare plants. *Current Science* 92, 271–272.

Animal and Plant Health Inspection Service (APHIS) (1985) Gypsy moth: a pest of trees. Fact Sheet, APHIS, U.S. Department of Agriculture, Riverdale, Maryland.

ATREE (2010) *Consequences of Lantana Invasion.* http://www.atree.org/cons_lantana_invasion (accessed February 2010).

Bright, C. (1998) *Life Out of Bounds: Bioinvasion in a Borderless World.* W.W. Norton & Co., New York.

Bright, C. (1999) Invasive species: pathogens of globalization. *Foreign Policy* (Fall), 51–63.

Carlton, J.T. (1987) Patterns of trans-oceanic marine biological invasions in the Pacific Ocean. *Bulletin of Marine Science* 41, 452–465.

Christian, J.M. and Wilson, S.D. (1999) Long-term ecosystem impacts of an introduced grass in the Northern Great Plains. *Ecology* 80, 2397–2404.

Cohen, A.N. and Carlton, J.T. (1998) Accelerating invasion rate in a highly invaded estuary. *Science* 279, 555–557.

Cox, G.W. (1993) *Conservation Ecology.* William C. Brown Publishers, Dubuque, Iowa.

Day, M., Wiley, C.J., Playford, J. and Zalucki, M.P. (2003) *Lantana*: current management status and future prospects. *ACIAR Monographs* 102, 28.

Florida Department of Agriculture and Consumer Services, Division of Plant Industry (2006) Citrus Canker Eradiation Program history report. http://www.doacs.state.fl.us/pi/canker/pdf/cankerflorida.pdf (accessed January 2010).

Fuller, P.L., Nico, L.G. and Williams, J.D. (1999) *Non Indigenous Fishes Introduced into Inland Waters of the United States*, American Fisheries Society, Bethesda, Massachusetts.

Gupta, K. (2004) National Bureau of Plant Genetic Resources, New Delhi, India.

Hill, G., Waage, J. and Phiri, G. (1997) The water hyacinth problem in tropical Africa. In: Delfosse, E.S. and Spencer, N.R. (eds) *Proceedings of the International Water Hyacinth Consortium.* World Bank, Washington, DC.

Hiremath, A.J. and Sundaram, B. (2005) The fire–lantana cycle hypothesis in Indian forests. *Conservation Society* 3, 26–42.

International Maritime Organization (IMO) (2001) *Global Ballast Water Management Programme.* http://globallast.imo.org (accessed 21 June 2001).

Jensen, R. (2002) Economic policy for invasive species. www.nd.edu/~rjensen1/working papers/InvasiveSpecies.pdf (accessed January 2010).

Krishnakumar, K., Raghavan, R., Prasad, G., Bijukumar, A., Sekharan, M., Pereira, B. *et al.* (2009) When pets become pests – exotic aquarium fishes and biological invasions in Kerala, India. *Current Science* 97, 474–476.

Love, A., Suresh, B. and Babu, C.R. (2009) Management of *Lantana*, an invasive alien weed, in forest ecosystems of India. *Current Science* 97, 1421–1429.

Mack, R.N. (2004) Trade routes for commerce in plants or pathways for invasive species? The dualism of international commerce. In: Sindel, B.M. and Johnson, S.B. (eds) *Proceedings of the Fourteenth Australian Weeds Conference*, Weed Science Society of New South Wales, Sydney, pp. 72–77.

Maheswari, D.V. (2008) Felling of Gando Bawal banned in Kutch. http://www.expressindia.com/latest-news/felling-of-gando-bawal-banned-in-kutch/306495/ (accessed May 2007).

Matthews, S. and Brandt, K. (2004) Africa invaded: the growing danger of invasive alien species. In: *Global Invasive Species Programme.* http://www.gisp.org/casestudies/showcasestudy.asp?id=86 (accessed 14 February 2010).

Meyerson L.A. and Mooney H.A. (2007) Invasive alien species in an era of globalization. *Frontiers in Ecological Environment* 5, 199–208.

Nair, A. and Darji, H. (2008) Vanishing bees bring down crop yields in Kutch. *Times of India*, Ahmedabad, June 2006.

National Agricultural Library (2006) *Federal Laws and Regulations: Public Laws and Acts.*

Agricultural Research Service, U.S. Department of Agriculture, Washington, DC. Available online at http://www.invasivespeciesinfo.gov/laws/publiclaws.shtml (accessed January 2010).

Naylor, R. (1996) Invasions in agriculture: Assessing the cost of the golden apple snail in Asia. *Ambio* 25, 443–448.

Olson, L.J. (2006) The economics of terrestrial invasive species: a review of the literature. *Agricultural and Resource Economics Review* 35, 178–194.

Pathan, B. (1998) Govt wakes up to illegal charcoal manufacture. *Indian Express*, http://www.expressindia.com/news/ie/daily/19981026/29950904.html (accessed 26 October).

Perrings, C., Williamson, M. and Dalmazzone, S. (2000) *The Economics of Biological Invasions*, Edward Elgar, Cheltenham, UK; Northampton, M.A., USA. 249 pp.

Perrings, C.S., Dalamazzone, S. and Williamson, M. (2005) The economics of biological invasions. In: Mooney, H.A. *et al.* (eds) *Invasive Alien Species: a New Synthesis*. Island Press, Washington, DC.

Pimentel, D., Lach, L., Zuniga, R. and Morrison, D. (2000) Environmental and economic costs of non-indigenous species in the United States. *BioScience* 50, 53–65.

Pimentel, D., Zuniga, R. and Morrison, D. (2005) Update on the environmental and economic cost associated with alien invasive species in the United States. *Ecological Economics* 52, 273–288.

Raghubanshi, A.S., Rai, L.C., Gaur, J.P. and Singh, J.S. (2005) Invasive alien species and biodiversity in India. *Current Science* 88, 539–540.

Reichard, S.H. and Hamilton, C.W. (1997) Predicting invasions of woody plants introduced into North America. *Conservation Biology* 11, 193–203.

Revenga, C., Brunner, J., Henninger, N., Kassem, K. and Payne, R. (2000) *Pilot Analysis of Global Ecosystems: Freshwater Ecosystems*. World Resources Institute, Washington, DC.

Ricciari, A., Steiner, W.W.M., Mack, R.N. and Simberloff, D. (2000) Toward a global information system for invasive species. *BioScience* 50, 239–244.

Rizzo, D.M. and Garbellotto, M. (2003) Sudden oak death: endangering California and Oregon forest ecosystems. *Frontiers in Ecology and the Environment* 1, 197–204.

Ross, S.T. (1991) Mechanisms structuring stream fish assemblages: Are there lessons from introduced species? *Environmental Biology of Fishes* 30, 359–368.

Sax, D.F. (2001) Latitudinal gradients and geographic ranges of exotic species: implications for biogeography. *Journal of Biogeography* 28, 139–150.

Sharma, G.P., Raghubanshi, A.S. and Singh, J.S. (2005) *Lantana* invasion: an overview. *Weed Biology Management* 5, 157–165.

Simberloff, D. (2003) How much population biology is needed to manage introduced species? *Conservation Biology* 17, 83–92.

Simberloff, D., Parker, I.M. and Windle, P.N. (2005) Introduced species policy, management, and future needs. *Frontiers in Ecology and the Environment* 3, 12–20.

Stein, B.A. and Flack, S.R. (eds) (1996) *America's Least Wanted: Alien Species Invasions of U.S. Ecosystems.* Nature Conservancy, Arlington, Virginia, http://www.abi.org/publications/leastwanted/wooly.html (accessed 18 July 2001).

Sudhi, S. (2004) *The Hindu*, 19 November.

Syed, Z. and Guerin, P.M. (2004) Tsetse flies are attracted to the invasive plant *Lantana camara*. *Journal of Insect Physiology* 50, 43–50.

U.S. Congress, Office of Technology Assessment (OTA) (1993) *Harmful Non-Indigenous Species in the United States*, OTA-F-565, September. U.S. Government Printing Office, Washington, DC.

Van Driesche, J. and Van Driesche, R. (2000) *Nature out of Place: Biological Invasions in the Global Age*. Island Press, Washington, DC.

Van Driesche, J. and Van Driesche, R. (2001) Guilty until proven innocent. *Conservation Biology in Practice* 2 (http://www.hear.org/articles/cip_winter2001 v2n1_guilty.pdf).

Wilcove, D.S., Rothestein, D., Dubow, J., Phillips, A. and Losos, E. (1998) Quantifying threats to imperiled species in the United States. *BioScience* 48, 607–615.

Williamson, M. (1996) *Biological Invasions*. Chapman and Hall, London.

World Resources Institute (WRI), in collaboration with the United Nations Environment Programme, the United Nations Development Programme and the World Bank (2000) *World Resources 2000–01: People and Ecosystems: The Fraying Web of Life*. WRI, Washington, DC.

World Resources Institute (WRI) (2001) *Bioinvasions: Stemming the Tides of Exotic Species, in Earth Trends: Featured Topic, July*. WRI, Washington, DC, http://earthtrends.wri.org/pdf_library/feature/bio_fea_invasives.pdf (accessed January 2010).

# 20 Population Dynamics of Invasive Alien Species of *Eupatorium*

**R.S. Tripathi[1] and A.S. Yadav[2]**

[1]*National Botanical Research Institute, Lucknow, Uttar Pradesh, India;* [2]*Department of Botany, R. R. Autonomous College, Alwar, Rajasthan, India*

## Introduction

Several alien species have invaded the natural ecosystems and degraded habitats of India, including the north-eastern region, which is one of the hot spots of biodiversity. Some invasive alien plant species have become naturalized and have spread very rapidly in the region. Notable among these are *Eupatorium odoratum* L. (syn. *Chromolaena odorata* (L.) King & Robinson), *Eupatorium adenophorum* Spreng. (syn. *Ageratina adenophora* (Spreng.) King & Robinson), *Eupatorium riparium* Regel. (syn. *Ageratina riparia* (Regel.) King & Robinson), *Lantana camara* L., *Mikania micrantha*, *Imperata cylindrica*, *Eichhornia crassipes* and *Parthenium hysterophorus* (restricted to lower altitudes).

Excepting *E. crassipes* (common name water hyacinth; ACIAR, 2000), which has seriously infested the ponds, lakes and wetlands of the region, the other weeds mentioned above grow abundantly in deforested areas, abandoned fields, tree plantations, roadsides and wastelands. Because of their high reproductive potential and fast growth rate, these invasive aliens have reduced the productivity of natural ecosystems and suppressed the growth of indigenous plant species. It is, therefore, imperative to understand the population dynamics of these alien species, to check their spread to new areas and to manage the infested ecosystems. Although sufficient literature is available on the adverse effect of these weeds on indigenous plant species, and their control measures, only a few attempts have been made to investigate their population dynamics. However, a detailed study was carried out on the population biology of three invasive species of *Eupatorium* by the authors while working at the Department of Botany, North-Eastern Hill University, Shillong. The population dynamics of these invasive species reviewed in this chapter formed the major part of this study.

## Population Flux

Yadav and Tripathi (1981) studied the population dynamics of *E. odoratum* (Fig. 20.1) in a 5-year-old fallow at 100 m altitude at Burnihat in the state of Meghalaya in north-east India. The average population density of *E. odoratum* was 27 adult plants/$m^2$ at the beginning of study, but after a period of 2 years this figure had dropped to 23/$m^2$. Thus the rate of population growth per plant was 0.9 over the 2-year period (Table 20.1). Of the plants present at the beginning of the study, only ~57% had survived after 2 years; of 1577 plants recruited over the whole study period, only 8 remained at the end of the study, representing a total annual mortality of 98.6%. The expected turnover time for adult plants was found to be 4.6 years (Table 20.1). The adult populations, of course, showed

**Fig. 20.1.** *Eupatorium odoratum.*

**Table 20.1.** Population flux of *Eupatorium odoratum* over a 2-year period (24 June 1977–30 May 1979) in a 5-year-old fallow.

| Parameter | Population numbers |
|---|---|
| Number of plants/m$^2$ (start) | 27 |
| Number of plants/m$^2$ (finish) | 23 |
| Net change | –4 |
| Rate of increase (%) | –0.9 |
| Number of new plants established | 1577 |
| Number of plants lost | 1581 |
| Plants present from start to finish | 15 |
| Survival of plants (%) | 56.6 |
| Total number of plants recorded during study period | 1604 |
| Annual mortality for all individuals (%) | 98.6 |
| Expected time for complete turnover (years) | 4.6 |

greater survivorship as compared with the younger recruits, which experienced excessive mortality.

## Demographic Analysis of *E. adenophorum* and *E. riparium*

Tripathi and Yadav (1987) carried out a demographic analysis of *E. adenophorum* (Fig. 20.2) and *E. riparium* (Fig. 20.3) in an area (altitude 1500 m) near Shillong, the capital city of Meghalaya. In this study area vegetation had been slashed and burned 20 years previously for shifting cultivation by the local tribal people, and since that time the area had been left fallow, allowing regrowth of the forest. Unlike *E. odoratum*, both these species showed a high population density of adult plants: the adult population of *E. riparium* was 112 plants/m$^2$, showing a net gain of 44 plants over a 2-year period (Table 20.2). Thus, the population growth rate was 1.4 per plant, which may be attributed to its vegetative reproduction. The prostrate branches of *E. riparium* give rise to roots that become buried in the soil. The total number of plants recruited through vegetative reproduction was 102/m$^2$, of which 33% died over the 2-year period. The population of *E. adenophorum* was 50 plants/m$^2$ at the beginning of the study, showing a net loss of 24 plants over the 2-year period. The expected turnover time was 6.2 and 4.2 years for *E. riparium* and *E. adenophorum*, respectively.

## Survivorship of Seedling Populations

The seedling cohorts of *E. odoratum* from different years showed the same pattern of survivorship, although the cohorts from the 5-year-old fallow exhibited greater survivorship than those growing in the older *jhum* (shifting agriculture) fallows (Yadav and Tripathi, 1999). Thus, survivorship of the seedling population of *E. odoratum* showed rapid decline with increase in age of the secondary successional community (Fig. 20.4). These authors attributed this to increased exposure to shade that accompanies an increase in growth of vegetation. Kushwaha *et al.* (1981) made similar observations while evaluating seedling survivorship of *E. ododratum* in different secondary successional communities. They also found that the rate of

**Fig. 20.2.** *Eupatoriom adenophorum.*

**Fig. 20.3.** *Eupatorium riparium.*

**Table 20.2.** Population flux of adult plants of *Eupatorium riparium* and *E. adenophorum* in relation to burning over a 2-year period (April 1977–March 1979).

| Parameter | *E. riparium* | | *E. adenophorum* | |
|---|---|---|---|---|
| | Burned plot | Unburned plot | Burned plot | Unburned plot |
| Number of plants/m$^2$ (April 1977) | 66.7 ± 0.7 | 11.7 ± 14.5 | 18.7 ± 3.8 | 50.0 ± 6.1 |
| Number of plants/m$^2$ (March 1979) | 14.30 ± 16.0 | 156.0 ± 31.5 | 19.7 ± 6.7 | 26.0 ± 8.5 |
| Net change | 76.3 ± 14.4 | 44.3 ± 10.0 | 1.0 ± 3.0 | (-) 24.0 ± 7.5 |
| Rate of change (%) | 2.14 ± 0.6 | 1.4 ± 0.1 | 1.05 ± 0.1 | 0.52 ± 0.1 |
| Number of plants surviving from start to finish of study | 47.4 ± 7.4 | 75.0 ± 12.6 | 14.0 ± 1.2 | 26.0 ± 8.5 |
| Plant survival (%) | 71.5 ± 11.6 | 66.9 ± 3.9 | 75.0 ± 4.0 | 52.0 ± 4.7 |
| Expected time for complete turnover (years) | 7.6 ± 2.7 | 6.2 ± 0.8 | 8.0 ± 3.2 | 4.2 ± 1.0 |

seedling mortality was higher with increase in age of the successional community. Seedling mortality showed a degree of seasonality, with high mortality rates in the active growing season (July–October) and the dry winter season (December–February). This indicates that seedlings experience high mortality due to competition for resources during the active growing season, and due to low temperatures in the winter season (Yadav and Tripathi, 1981). Reduced light intensity under the vegetation canopy may be the chief factor in influencing seedling survivorship, which is in conformity with the findings of Bennett and Rao (1968) and Cruttwell (1968).

Seedlings of both *E. adenophorum* and *E. riparium* showed a similar survivorship pattern, although this was higher in the latter than in the former (Fig. 20.5). Seedling populations of both species experienced heavy mortality at the juvenile stage, but subsequently mortality became constant during the rest of the study period. Survivorship in both species was greater in the burned than in the unburned plot (Fig. 20.5). The heavy mortality recorded was probably due to the accumulation of a thick layer of pine litter (needles) and resource competition from the herbaceous vegetation. Tripathi and Yadav (1987) observed that seedlings derived from seeds germinating in

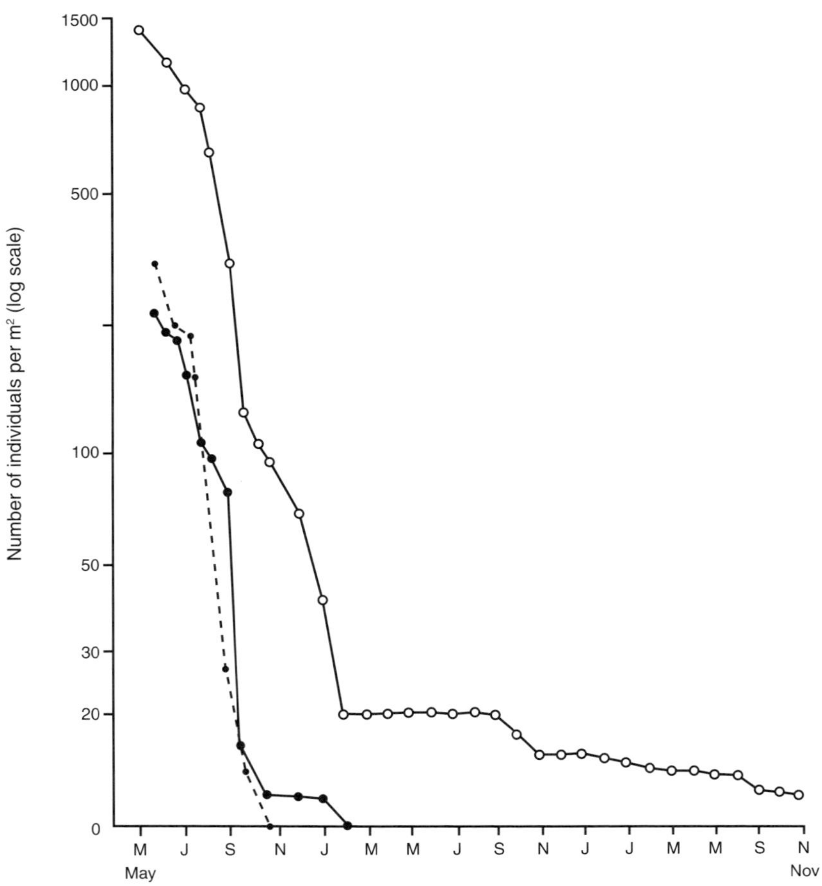


**Fig. 20.4.** Survival of *Eupatorium odoratum* seedling cohorts; seedling population in 1977 (open circles), 1978 (solid circles) and 1979 (broken line).

pine litter remained suspended over it and could not establish as their roots failed to come in contact with soil (Fig. 20.6). As stored food was quite limited due to the small size of the seeds, most seedlings died off as soon as the stored food had become exhausted. Similarly, seedlings germinating in dark cavities formed by the accumulation of pine needles on the horizontally growing shoots of *E. riparium* died off, presumably due to lack of sufficient light available for photosynthesis. The herbaceous vegetation formed a thick canopy, which also reduced to a great extent the amount of light that would have been available to the seedling population. Some seedlings were found to have been damaged by woodlice (*Oniscus* spp.), which occurred in great abundance at the time of seed germination. The possible role of other predators in seedling survival, however, cannot be excluded.

## Survivorship of Mature Plants

Survivorship of mature plants of *E. odoratum* was very high in comparison with seedling populations (Fig. 20.7). Adult mortality appears to be independent of prevailing soil moisture stress and temperature conditions during the winter months (Yadav and Tripathi, 1981). However, mature plants exhibited some mortality during the period of active vegetative growth, possibly triggered

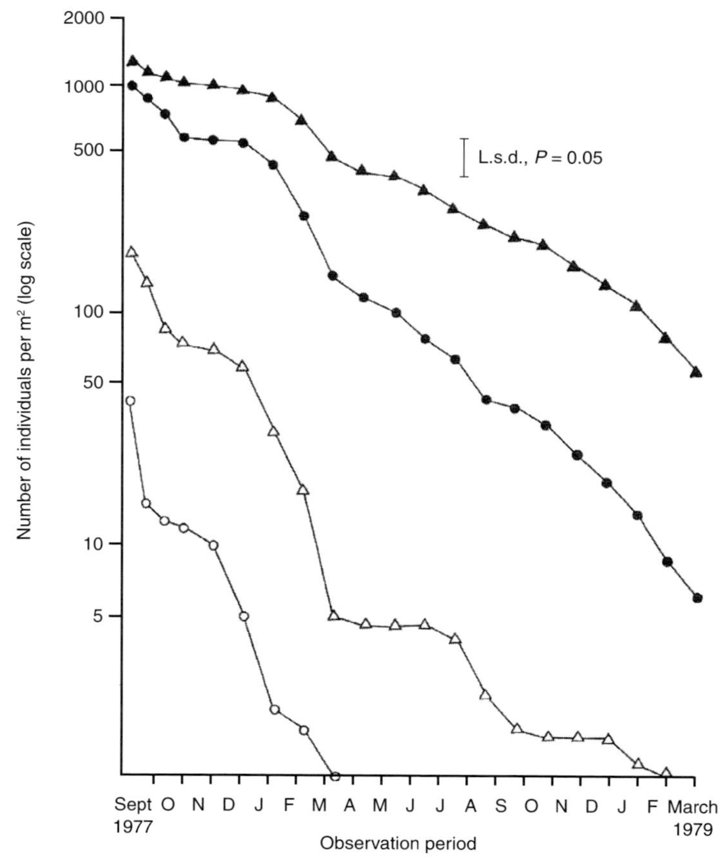


**Fig. 20.5.** Survival of *Eupatorium* spp. seedlings emerging shortly after burning: *E. adenophorum* in burned (closed circles) and unburned plots (open circles); *E. riparium* in burned (solid triangles) and unburned plots (open triangles).

**Fig. 20.6.** Layer of pine needles (a) with seedlings present (b) at the Shillong study site.

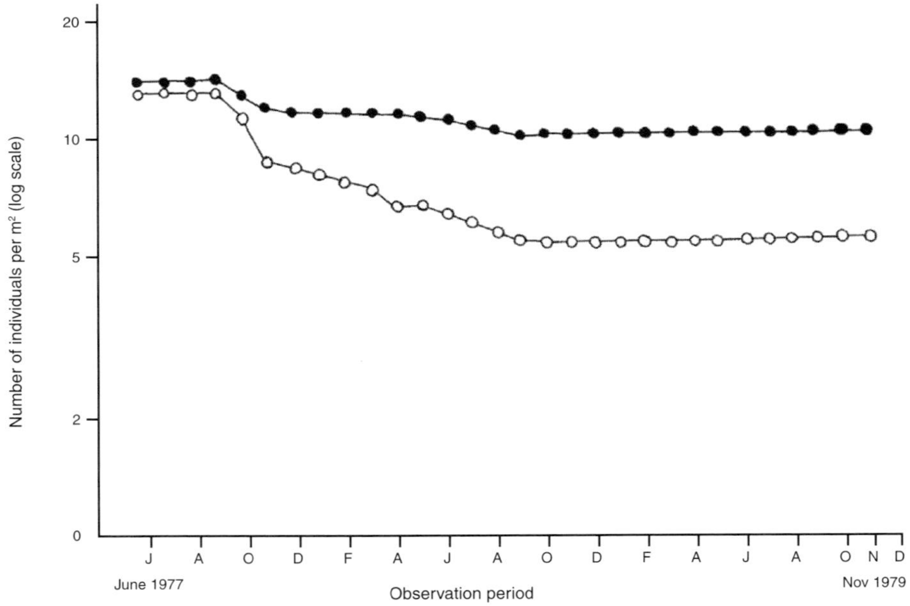


**Fig. 20.7.** Depletion curves for *Eupatorium odoratum*: 1-year-old population (open circles) and >1-year-old population (solid circles).

by more severe competition for resources. Yadav and Tripathi (1999) reported that mature *E. odoratum* plants died in the third year of their study due to senescence and termite attack. Kushwaha *et al.* (1981) evaluated survivorship of vegetative shoots of *E. odoratum* and suggested that the older shoots exhibited higher survivorship as compared with the younger; they reported high mortality in older shoots in a 5-year-old fallow but high survivorship in 10- and 20-year-old fallows. This is in conformity with the findings of Yadav and Tripathi (1999), who also suggested that survivorship of weed plants declined with age of the successional community.

In the case of *E. odoratum*, population size declined sharply in fallows older than 5 years, with such a marked reduction in seed production that those plants surviving in 10- and 20-year-old fallows had only sterile shoots (Kushwaha *et al.*, 1981). Saxena and Ramakrishnan (1984) also reported that allocation of biomass to reproduction in *E. odoratum* decreased with increasing age of the fallow. They further emphasized that the shortening of the *jhum* cycle to about 4–5 years was the main factor responsible for the spread and maintenance of *E. odoratum* populations at lower altitudes in the north-eastern region of India.

Survivorship of *E. adenophorum* and *E. riparium* mature plants was also high in the first year but decreased considerably during the second year of the study period, indicating age-dependent mortality (Fig. 20.8). Mature plants of both species experienced constant mortality throughout the year, with heaviest mortality in the active growing season (July–November, due to increased resource competition) and also during winter (December–March, when reproductive growth was taking place). The mortality pattern was similar in both species, but the rate was always higher in *E. adenophorum*. Tripathi and Yadav (1987) reported that high mortality in both the species could be attributed to their shallow and horizontally growing root system, and also to their sensitivity to soil moisture stress. Thus, mature plants of these species behave differently to those of *E. odoratum*.

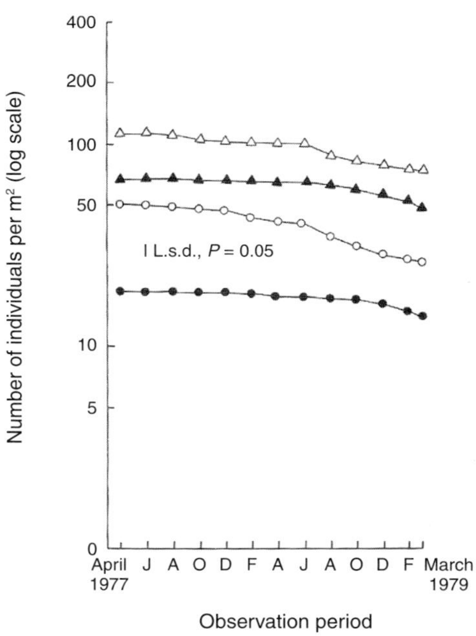


**Fig. 20.8.** Depletion curves for adult *Eupatorium* spp. in relation to burning: *E. adenophorum* in burned (closed circles) and unburned plots (open circles); *E. riparium* in burned (closed triangles) and unburned plots (open triangles).

The schematic summary of the population dynamics of *E. odoratum* over the 2-period indicates that its population showed a net loss of 13 individuals in spite of high seed production (Fig. 20.9). Obviously, there is constant reduction in the population of seeds and plants with the passage of time due to various environmental constraints operating at different stages of the life cycle. Yadav and Tripathi (1982) suggested the factors responsible for decline in population growth rate of these weeds: wind; rainfall; predatory organisms; non-availability of 'safe' microsites in sufficient number for germination of seeds and establishment of seedlings; competition for resources by growing vegetation; soil moisture stress during winter; low-temperature conditions prevailing in winter; certain pathogens causing stem rot diseases in seedlings; and 'natural senescence'.

## Seed Population Dynamics

Plant populations comprise actively growing individuals and seeds in the soil. The seed population dynamics of the three species of *Eupatorium* has been evaluated in Meghalaya (Yadav and Tripathi, 1982). The seeds of these weeds were mostly present in the surface layer of soil, while only a few were recovered from the deeper layers. A very small fraction of viable seeds was present in soil, although a good percentage of the freshly collected seeds remained viable when under laboratory conditions. The lower percentage of viable seeds in the soil seed bank may be due to the presence of non-viable seeds from the preceding year and the

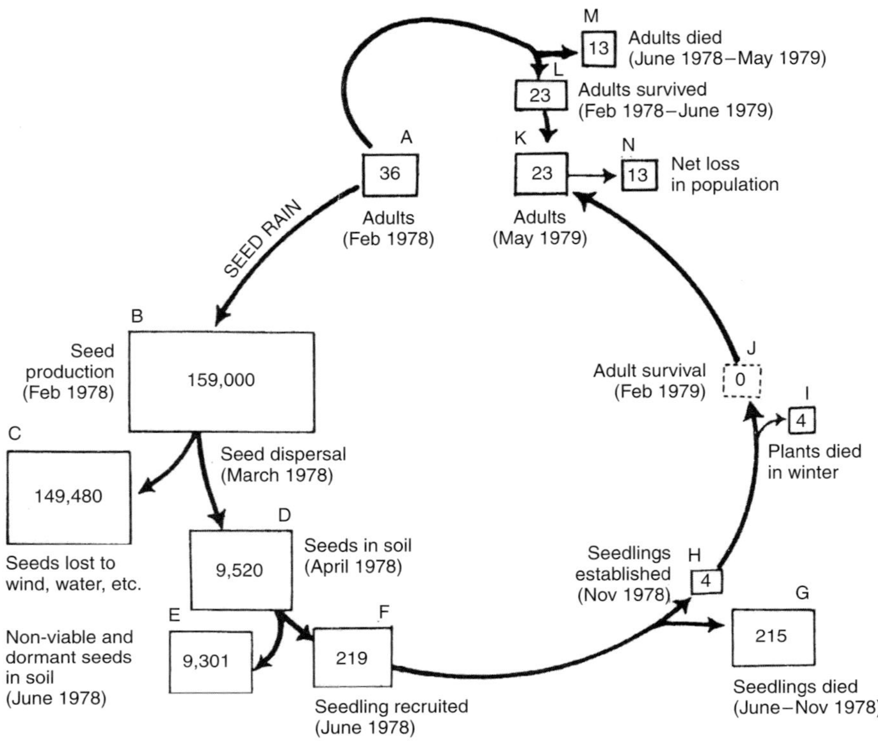


**Fig. 20.9.** Schematic summary of population flux for *Eupatorium odoratum* in the second year of study in a 5-year-old fallow at Burnihat, Meghalaya.

loss of some seeds through germination prior to collection of soil samples for determination of seed population.

Observations on the fate of buried seeds in soil indicate that survivorship of viable seeds in the three species of *Eupatorium* decreased exponentially throughout the study period. The mortality pattern of buried seeds was similar in all three species, with slight inter-species differences. Some seasonality was observed in viable seed mortality rates, as these were higher in the heat of summer and in the rainy season (April–September), when conditions were suitable for soil microbial activity, whereas low death rates were observed in the dry winter season (December–February), probably due to the lower activity of soil microorganisms under low-temperature and low-soil-moisture conditions. Although the proportion of viable buried seeds decreased considerably with time, recovery of seeds from all species was >50% even after 2 years. This suggests that the seeds of all three *Eupatorium* species when buried in soil are not highly susceptible to predators and decomposers.

A large proportion of viable seeds from the three species of *Eupatorium* entered enforced dormancy when buried in the soil, while a small fraction showed induced dormancy. Enforced dormancy occurs due to the non-availability of light below the soil surface. The proportion of seeds undergoing induced dormancy did not change throughout the first year, but increased considerably during the winter season of the second year after burial. These seeds again acquired enforced dormancy with the onset of summer in April, which might be due to their response to changes in temperature. The proportion of recovered non-viable seeds of *E. adenophorum* and *E. riparium* increased, indicating that with the passage of time loss of viability was faster than the decomposition process, while in *E. odoratum* the proportion of recovered dead seeds declined with time despite the considerable loss of viability of healthy seeds, suggesting that the decomposition of dead seeds is much faster at lower altitudes. The presence of a small cohort of seeds in the soil seed bank and rapid loss of seed viability when buried suggests that most seeds of all three *Eupatorium* species are wind-dispersed and may play a more important role – the colonization of new areas – than that of maintaining established populations.

## Factors Affecting Population Growth Rate

The chief factors influencing the population behaviour of *Eupatorium* spp. in the north-eastern region of India are the amount of seed rain levels, population density of established plants, soil moisture, soil fertility, light intensity, associated plant species and fire. The effects of associated plant species and controlled burning of vegetation have been discussed below, whereas the effects of other factors on the population growth of *Eupatorium* species are discussed elsewhere in the present volume.

Plant communities where *Eupatorium* spp. grow are dominated by mature plants of these species only. Hence, the effect of associated vegetation was evaluated by growing young transplants of *Eupatorium* spp. in the vicinity of adult plants and also with the associated vegetation (Yadav and Tripathi, 1983, 1984; Yadav, 1985). *Eupatorium odoratum*, along with *Imperata cylindrica*, dominated the vegetation in secondary successional communities at lower altitudes in a 5-year-old fallow at Burnihat, Meghalaya. The other key plant species comprising the vegetation, along with their respective density levels, are listed in Table 20.3. Less abundant species also growing as associates include *Setaria palmifolia*, *Paspalum conjugatum*, *Costus speciosus*, *Mimosa pudica*, *Sida rhombifolia*, *Wrightia tomentosa*, *Urena lobata*, *Dendrocalamus hamiltoni*, *Bauhinia purpurea* and *Cassia fistula*.

The vegetation in which *E. adenophorum* and *E. riparium* grow abundantly at higher altitudes is characterized by the sparsely growing *Pinus kesiya* in and around Shillong. The major associates of both these *Eupatorium* species are listed in Table 20.4. Other associated species having low population density values include *Cyperus rotundus*, *Cyperus sequiflora*, *Carex cruciata*, *Ophiopogon intermidius*, *Rubia cordifolia*, *Drymeria chordata*, *Lantana camara*, *Sida rhombifolia*, *Urena lobata*, *Galium rotundifolium* and *Plantago major*.

## Effect of Mature *Eupatorium* Plants

Transplants of *E. odoratum* grown with adult plants showed 80, 75 and 81% survivorship at a distance of 2, 10 and 20 cm, respectively; these values are not significantly different from survivorship of transplants grown in the absence of adult plants (Yadav, 1985). This indicates that survivorship of transplants is not influenced by the presence of

**Table 20.3.** Common plant species associated with *E. odoratum* in a 5-year-old fallow at Burnhat.

| Species | Density (number/m$^2$) |
|---|---|
| *Imperata cylindrica* | 203 ± 21.9 |
| *Rungia* sp. | 24 ± 2.3 |
| *Eupatorium odoratum* | 27 ± 4.5 |
| *Oftochloa nodusa* | 8 ± 3.6 |
| *Carex cubensis* | 3 ± 0.8 |

**Table 20.4.** Major associates of *E. adenophorum* and *E. riparium* at Shillong.

| Species | Density (number/m$^2$) |
|---|---|
| *E. adenophorum* | 37.0 ± 7.0 |
| *E. riparium* | 99.0 ± 22.0 |
| *Commelina benghalensis* | 53.0 ± 20.0 |
| *Arthraxon* sp. | 20.0 ± 11.0 |
| *Osbeckia crinita* | 6.0 ± 2.5 |
| *Pouzolzia hirta* | 6.0 ± 0.9 |

adult plants; however, the biomass per plant was drastically reduced when transplants were grown in the vicinity of adults (Table 20.5). The effect of adult plants was quite marked in all treatments, but the negative effect was most obvious at a distance of 2 cm. As transplant growth was reduced at the two greater distances, however, it could be assumed that *E. odoratum* adults may exercise their influence even beyond the 20 cm point. This may occur due to the extensively developed root system that might offer considerable competition to the root systems of transplants for below-ground resources, although this remains to be tested. Relative growth rates (expressed as percentage growth rate for a given treatment in relation to growth rate of controls) of transplants decreased with increased proximity between adults and transplants. Likewise, transplant reproductive growth was drastically suppressed as none of the transplants grown at varying distance from adult plants produced flowers (Yadav, 1985).

Yadav and Tripathi (1983) also studied the effect of adult plants of *E. adenophorum* and *E. riparium* on young transplants. The transplants responded to competition stress both through mortality and plasticity; transplants of both species showed greater mortality when grown with adults although *E. riparium* adults, however, exercised more adverse effects as compared with *E. adenophorum*. *Eupatorium riparium* transplants showed higher mortality than *E. adenophorum* in the presence of adults. Biomass accumulation by transplants of both species was suppressed when grown at a distance of 2 and 10 cm from the adult plants (Table 20.6). However, when grown at 20 cm distance, only *E. riparium* adults suppressed transplant growth while *E. adenophorum* had no affect. While the relative growth rate was also suppressed in the presence of adults of both species, however, the effect of *E. riparium* was more severe. Seed production by transplants was also suppressed by adults, but transplants of *E. riparium* grown at a distance of 10 and 20 cm did produce seeds (Table 20.7). This was also confirmed by greater reduction in growth of transplants of *E. adenophorum* when grown with adults of *E. riparium* compared with when *E. riparium* transplants were grown with adults of *E. adenophorum* (Tables 20.8 and 20.9). This greater regulatory influence of *E. riparium* might be due to its closed canopy appressed to the ground surface, in addition to its extensive shallow root system that is in sharp conflict with roots of young transplants. The greater suppression caused by *E. riparium* may also be attributed to its allelopathic effects on *E. adenophorum* (Rai and Tripathi, 1984). The relatively less marked effect of *E. adenophorum* adults on the transplants was, presumably, due to this species' erect habit and open canopy.

## Effect of Associated Vegetation on *Eupatorium* spp.

Yadav and Tripathi (1984) evaluated the effect of associated vegetation on young transplants of the three species of *Eupatorium*, and observed that survivorship

**Table 20.5.** Biomass accumulation and reproductive growth of *E. odoratum* transplants when grown at varying distance from adult plants for 17 months.

| Parameter | Adult plants present | | | Adult plants absent |
|---|---|---|---|---|
| | Distance from adults (cm) | | | |
| | 2 | 10 | 20 | |
| Biomass (g) | 1.63 | 2.46 | 2.66 | 99.6 |
| Fertile plants (%) | 0 | 0 | 0 | 69.1 |
| No. of capitula/plant | 0 | 0 | 0 | 465.0 |
| No.of seeds/capitulum | 0 | 0 | 0 | 30.7 |
| Seed output/plant | 0 | 0 | 0 | 1427.0 |

**Table 20.6.** Biomass accumulation (g) by *E. adenophorum* and *E. riparium* transplants when grown at varying distances from adult plants.

| Observation period (months from planting) | Transplant sp. | Control | Distance from adult *E. adenophorum* (cm) | | | Distance from adult *E. riparium* (cm) | | |
|---|---|---|---|---|---|---|---|---|
| | | | 2 | 10 | 20 | 2 | 10 | 20 |
| 7 | *E. adenophorum* | 0.94 | 0.04 | 0.23 | 0.86 | 0.02 | 0.04 | 0.25 |
| | *E. riparium* | 0.19 | 0.02 | 0.16 | 0.27 | 0.02 | 0.06 | 0.1 |
| 18 | *E. adenophorum* | 16.98 | 0.05 | 4.44 | 18.60 | 0.17 | 2.35 | 3.11 |
| | *E. riparium* | 7.07 | 0.30 | 3.02 | 7.66 | 0.05 | 2.60 | 5.47 |

**Table 20.7.** Reproductive growth of *E. adenophorum*[a] and *E. riparium* transplants when grown at varying distances from adult plants 7 months from planting.

| Parameter | Control | Distance between *E. riparium* transplants and *E. adenophorum* adults (cm) | | | Distance between *E. riparium* transplants and *E. riparium* adults (cm) | | |
|---|---|---|---|---|---|---|---|
| | | 2 | 10 | 20 | 2 | 10 | 20 |
| Fertile plants (%) | 68.0 ± 8.9 | 0 | 20.2 ± 10.2 | 28.2 ± 7.7 | 0 | 18.5 ± 1.8 | 20.0 ± 8.2 |
| No. of capitula/plant | 9.6 ± 1.4 | 0 | 3.6 ± 1.9 | 16.5 ± 3.1 | 0 | 2.8 ± 1.0 | 4.8 ± 1.7 |
| No. of seeds/ capitulum | 22.6 ± 1.4 | 0 | 15.5 ± 1.7 | 19.7 ± 1.6 | 0 | 17.4 ± 1.0 | 18.3 ± 1.2 |
| Seed output/plant | 216 ± 37 | 0 | 61 ± 23 | 33 ± 39 | 0 | 49 ± 17 | 89 ± 32 |

[a] No *E. adenophorum* transplants produced flowers when grown with adults of either species, while only 3.0% of transplants flowered in the control set.

**Table 20.8.** Reproductive growth of *E. adenophorum* transplants grown at varying distances from adult *E. adenophorum* and *E. riparium* 18 months from planting.

| Parameter | Control | Distance from adult *E. riparium* (cm) | | | Distance from adult *E. adenophorum* (cm) | | |
|---|---|---|---|---|---|---|---|
| | | 2 | 10 | 20 | 2 | 10 | 20 |
| Fertile plants (%) | 100 ± 0 | 0 | 25 ± 24.9 | 37.5 ± 7.2 | 0 | 50 ± 5.7 | 90.6 ± 5.5 |
| No. of capitula/ plant | 222 ± 101 | 0 | 17.6 ± 4.3 | 21 ± 6 | 0 | 47 ± 2.8 | 168 ± 71 |
| No. of seeds/ capitulum | 59.9 ± 2.2 | 0 | 38.7 ± 3.5 | 49.5 ± 1.4 | 0 | 38.2 ± 3.8 | 47.8 ± 2.6 |
| Seed output/ plant | 13,298 ± 6,025 | 0 | 681 ± 167 | 1,040 ± 271 | 0 | 1,795 ± 110 | 8,016 ± 3,362 |

of the transplants was severely reduced in the presence of associated species. In all three species the number of leaves per plant was reduced by growth of the associates (Table 20.10). However, the reduction in the number of leaves was more pronounced in *E. adenophorum* and *E. riparium*, especially with increased plant age. Associated vegetation also caused substantial reduction in leaf area, yield and reproductive growth of young transplants of all three *Eupatorium* species. Reproductive growth was so markedly reduced that no *Eupatorium* transplants produced seeds when grown with associates (Table 20.11). This resulted in the death of all *E. adenophorum* transplants after 10 months, and of *E. odoratum* after 19 months; however, some *E. riparium* transplants survived, indicating their ability to grow under low light intensity and high competition stress caused by the presence of associates.

**Table 20.9.** Reproductive growth of *E. riparium* transplants grown at varying distances from adult *E. adenophorum* and *E. riparium* 18 months from planting.

| | | Distance from adult *E. riparium* (cm) | | | Distance from adult *E. adenophorum* (cm) | | |
|---|---|---|---|---|---|---|---|
| Parameter | Control | 2 | 10 | 20 | 2 | 10 | 20 |
| Fertile plants (%) | 100 ± 0 | 0 | 90.4 ± 9.5 | 100 ± 0 | 100 ± 0 | 100 ± 0 | 100 ± 0 |
| No. of capitula/ plant | 304 ± 35 | 0 | 135 ± 115 | 268 ± 129 | 8.7 ± 2.2 | 123 ± 61 | 400 ± 183 |
| No. of seeds/ capitulum | 24.6 ± 1.0 | 0 | 21.8 ± 0.9 | 26.3 ± 0.3 | 23.2 ± 0.6 | 23.8 ± 0.9 | 24.7 ± 0.8 |
| Seed output/ plant | 7487 ± 930 | 0 | 2947 ± 250 | 7048 ± 3374 | 201 ± 31 | 2927 ± 1450 | 9880 ± 4517 |

**Table 20.10.** Number of leaves per plant of three species of *Eupatorium*, as affected by associated species.

| Time since planting (months) | *E. adenophorum* | | *E. riparium* | | *E. odoratum* | |
|---|---|---|---|---|---|---|
| | Associates present | Associates absent | Associates present | Associates absent | Associates present | Associates absent |
| 2 | 6.9 ± 0.5 | 4.8 ± 0.4 | 5.1 ± 0.2 | 6.6 ± 0.8 | – | – |
| 3 | – | – | – | – | – | – |
| 4 | 2.9 ± 0.2 | 7.8 ± 0.1 | 4.8 ± 0.4 | 8.7 ± 0.3 | – | – |
| 6 | 5.0 ± 0.4 | 10.3 ± 0.6 | 4.7 ± 0.4 | 34.6 ± 0.3 | 2.2 ± 0.3 | 11.3 ± 1.0 |
| 9 | – | – | – | – | 2.7 ± 0.7 | 16.5 ± 2.3 |
| 17 | 0 | 249 ± 67 | 8.0 ± 1.0 | 520 ± 83 | – | – |
| 19 | – | – | – | – | 0 | 50 ± 19 |

## Effect of Burning

Tripathi and Yadav (1987) studied the effect of burning on the population dynamics of *E. adenophorum* and *E. riparium* after burning a small plot of the study site in Shillong. The population of adult *E. riparium* was 112 plants/m$^2$ in the unburned area and 67 plants/m$^2$ in the burned area (Table 20.2). The population showed a net gain of 76 plants and 44 plants in the burned and unburned areas, respectively, over a 2-year period. The rate of population growth in the burned area was higher in comparison with that in the unburned area and the expected time of turnover for adults was longer in the burned area. The population of *E. adenophorum* was 50 plants/ m$^2$ in the unburned area and 19 plants/ m$^2$ in the burned area. The *E. adenophorum* population showed a loss of 24 plants in the unburned area and a net gain of one plant in the burned area. The rate of population growth in the burned area was almost double that of the unburned area. Survivorship of adult plants was high and their expected turnover period in the burned area was twice that of the unburned area. The expected turnover time increased in both species after burning, although there was a twofold increase in the case of *E. riparium*.

Seedling populations of *E. adenophorum* and *E. riparium* showed better survivorship and establishment in the burned area in comparison with the unburned area (Fig. 20.5). Greater seedling survivorship in the former may be attributed to the burning of the litter layer and herbaceous cover, as a consequence of which there was an increase in the number of 'safe sites' for seed

**Table 20.11.** Reproductive behaviour of three *Eupatorium* spp. as affected by associated species.

| Species | No. of capitula/plant | No. of seeds/capitulum | Seed output (no.)/plant |
|---|---|---|---|
| *E. adenophorum* | | | |
| Associates absent | 1,369 ± 549 | 64.6 ± 4.0 | 90,188 ± 36,221 |
| Associates present | 0 | 0 | 0 |
| *E. riparium* | | | |
| Associates absent | 2,797 ± 613 | 26.5 ± 0.8 | 74,115 ± 16,879 |
| Associates present | 0 | 0 | 0 |
| *E. odoratum* | | | |
| Associates absent | 769 ± 403 | 30.8 ± 0.8 | 22,441 ± 12,469 |
| Associates present | 0 | 0 | 0 |

germination and seedling establishment. Established seedlings in the burned area experienced constant mortality throughout the year. Although survivorship of adult plants of both weeds increased in the burned area, the population of *E. riparium* showed much greater survivorship as compared with *E. adenophorum* (Fig. 20.8).

## Conclusion

It may be concluded that *E. riparium* is better adapted to low light conditions, while *E. adenophorum* and *E. odoratum* prefer open habitats. A comparison of the extent of reduction in growth of *E. odoratum* and *E. adenophorum* due to associated vegetation revealed that the former species is more sensitive to competition and shade caused by associates, which shows that it requires greater light intensity in comparison with *E. adenophorum*. Survival of some of a proportion of *E. riparium* transplants while growing in competition with other plants indicates that this weed can withstand resource competition. This ecological attribute confers an additional advantage on this species in the sense that whenever associates are removed due to some disturbance to the vegetation, these suppressed transplants would grow and assist in the perpetuation of this weed. Our observations also suggest that these exotic species grow better in disturbed or open areas in north-east India and do not exhibit normal growth in established vegetation with a closed canopy. The frequent burning of vegetation, which is common in the north-eastern region of India due to the prevailing practice of shifting cultivation, may also contribute to maintaining the high population growth rates of these invasive *Eupatorium* species in the region.

## References


ACIAR (2000) Biological and integrated control of water hyacinth, *Eichhornia crassipes.* In: Julien, M.H., Hill, M.P, Center, T.D. and Jianqing, D. (eds) *Proceedings of the Second Meeting of the Global Working Group for the Biological and Integrated Control of Water Hyacinth*, Beijing, China, 9–12 October. *ACIAR Proceedings* 102, 96–104.

Bennett, F.D. and Rao, V.P. (1968) Distribution of an introduced weed *Eupatorium odoratum* L. in Asia and Africa and possibilities of its biological control. In: *Proceedings of the 9th Commonwealth Forestry Conference*, pp. 277–281.

Cruttwell, R.E. (1968) Preliminary survey of potential biological control agents of *Eupatorium odoratum* in Trinidad. *Proceedings of the 9th Britsh Weed Control Conference*, pp. 836–841.

Kushwaha, S.P.S., Ramakrishnan, P.S. and Tripathi, R.S. (1981) Population dynamics of *Eupatorium odoratum* in successional environments following slash and burn agriculture. *Journal of Applied Ecology* 18, 529–535.

Rai, J.P.N. and Tripathi, R.S. (1984) Allelopathic effect of *Eupatorium riparium* on population regulation of two species of *Galinsoga* and soil microbes. *Plant and Soil* 80, 105–117.

Saxena, K.G. and Ramakrishnan, P.S. (1984) Growth and patterns of resource allocation in *Eupatorium odoratum* L. in the secondary successional environments following slash and burn agriculture (jhum). *Weed Research* 24, 127–134.

Tripathi, R.S. and Yadav, A.S. (1987) Population

dynamics of *Eupatorium adenophorum* Spreng, and *E. riparium* Regel in relation to burning. *Weed Research* 27, 229–336.

Yadav, A.S. (1980) Studies on population dynamics of *Eupatorium odoratum* L., *E. adenophorum* Spreng. and *E. riparium* Regel. PhD thesis, North-Eastern Hill University, Shillong, India, 220 pp.

Yadav, A.S. (1985) Effect of the adult plants of *Eupatorium odoratum* L. on the survivorship and growth of its own transplants. *Geobios* 12, 20–23, 1985.

Yadav, A.S. and Tripathi, R.S. (1981) Population dynamics of the ruderal weed *Eupatorium odoratum* and its natural regulation. *Oikos* 35, 355–361.

Yadav, A.S. and Tripathi, R.S. (1982) A study on seed population dynamics of three weedy species of *Eupatorium*. *Weed Research* 22, 59–76.

Yadav, A.S. and Tripathi, R.S. (1983) The population of transplanted seedlings of *Eupatorium adenophorum* and *E. riparium* as regulated by their adult plants. *Tropical Ecology* 24, 201–215.

Yadav, A.S. and Tripathi, R.S. (1984) Effect of associated species on three *Eupatorium* species. *Indian Journal of Ecology* 11, 190–196.

Yadav, A.S. and Tripathi, R.S. (1999) Studies on the demography of *Eupatorium odoratum* in a secondary successional community. *Tropical Ecology* 40, 269–274.

# 21 Resource Utilization and Beneficial Aspects of Invasive Alien Weeds with Special Reference to the Western Ghats, India

**R.R. Rao[1], Kavitha Sagar[2] and N. Sathyanarayana[3]**

[1] *INSA Honorary Scientist, Bangalore, India;* [2] *Vidya Herbs, Bangalore, India;* [3] *Sir M. Visvesvaraya Institute of Technology, Yelahanka, Bangalore, India*

## Introduction

When one thinks of invasive weeds, naturally the control and eradication of such weeds come to mind and seldom the utility or beneficial aspects of these weeds for human welfare. Although invasive weeds are a menace to agriculture and forestry through their effects on the economic scenario of a nation, invasive weeds, like any other plant species, are not without beneficial properties. In fact, most invasive weeds do find uses in local herbal medicines, as wild vegetables, as raw materials for paper and pulp, for local handicrafts, as biofertilizers, as fodder for cattle, as agents of bio-remediation of polluted areas, as support for wildlife and some even as timber. In the present chapter the salient beneficial aspects of invasive alien weeds, particularly of the Western Ghats region, are discussed in the hope that some of these will be critically evaluated for their properties to facilitate large-scale collection and consumption and to aid in controlling their spread.

## What are Invasive Weeds?

Invasive species are those that, after introduction to a new environment, spread very rapidly and out-compete all native and other species in the area, some even by producing root leachates inhibiting the growth of all neighbouring flora. Invasive alien species have substantial negative impacts on the environment, ecosystems, native biodiversity, human health and even regional economies. Nearly 18% of Indian flora is constituted by invasive aliens of which 55% are from the tropical American region, 10% Asian, 20% Malaysian and 15% European and Central Asian (Rana, 2006).

## The Western Ghats Region

The Western Ghats biogeographic zone, which forms the Malabar province, is a narrow stretch of land running from the hills south of the Tapti River in the north to Kanyakumari in the south, along the west coast of India. This zone extends for approximately 1500 km, encompassing a wide variety of climatic conditions. The Ghats rise up abruptly in the west in a highly dissected plateau up to 2900 m in height and descend to the dry Deccan plains below (altitude 500 m) in the east. The extreme climatic and altitudinal gradient of the region has resulted in a variety of forest types, from evergreen to semi-evergreen, from moist deciduous to dry deciduous. At higher altitude stunted montane communities have also developed (Gamble, 1967). The tropical rain forests in the southern Western Ghats and the *sholas* in the Nilgiri and Anamalai regions are the most outstanding features of this region. The entire Western Ghats bio-geographic region is a major genetic estate with an enormous biodiversity of ancient lineage. Although the Western Ghats cover only 5% of India's total geographical area, the region contains more than 30% of the country's plant species, totalling ~12,000, from lower groups to flowering plants.

## Major invasive weeds in the Western Ghats

The authors, in a separate chapter (Chapter 12, this volume), have discussed in detail the invasive alien weeds of the Western Ghats from a taxonomic perspective. They have enumerated 42 invasive species, the vast majority belonging to the family *Asteraceae* and mostly drawn from tropical America and Mexico. Among the dominant invasive alien weeds in the region are *Parthenium hysterophorus* L, *Mikania micrantha* H.B.K., *Eichhornia crassipes* (Mart.) Solmns, *Lantana camara* L., *Ipomoea fistulosa* Mart. ex Choisy, *Synedrella vialis* (Less.) A. Gray., *Cassia uniflora* Mill., *Ricinus communis* L., *Chromolaena odorata* (L.) King & Robinson, *Eupatorium adenophorum* Spreng., *Pueraria phaseoloides* (Roxb.) Benth., *Mimosa invisa* Martius ex Colla, *Alternanthera philoxeroides* (Mart.) Griseb, *A. paronychioides* A. St.-Hilaire, *Hyptis suaveolens* (L.) Poit. and *Prosopis chilensis* (Molina) Stuntz. These invasives also show specific distribution patterns. While *M. invisa* Martius ex Colla, *P. phaseolioides* (Roxb.) Benth. and *M. micrantha* H.B.K. are fairly abundant in the Kerala region of the Western Ghats, *P. hysterophorus* L., *E. adenophorum* Spreng., *P. chilensis* (Molina) Stuntz, *C. odorata* (L.) King & Robinson, *H. suaveolens* (L.) Poit., *S. vialis* (Less.) A. Gray and *C. uniflora* Mill. are abundant in the Karnataka and Tamil Nadu regions.

## Resource utilization of invasive alien weeds in the Western Ghats

Although invasive aliens are a substantial threat and have several negative impacts, some of these weeds have also proved beneficial to human welfare. The Western Ghats region harbours a large number of ethnic tribes such as the Mannan, Palaiyan, Urali, Malayaryan, Malampandaran, Karuma, Kadar, Irulas, Kattunaikkas, Koyas, Yanadis, Paniyas, lambanis and several others, who largely depend on the surrounding flora for their day-to-day needs. They practise herbal medicine for most of their ailments, utilizing some of the alien weeds of the region; they also collect a number of wild herbs, including some invasive species, as green vegetables. The woody biomass of *Lantana* and *Prosopis*, for example, is also used in the manufacture of some handicrafts, baskets, handles for their instruments, etc. and for firewood. A few major uses of these invasive weeds with particular reference to the Western Ghats are discussed below.

### Invasive weeds as herbal remedies

There is no species that is not medicinal. In fact, all species are said to have one or more medicinal properties, some of which are well documented while others remain to be studied, and invasive weeds are no exception. Local villagers and the ethnic tribes throughout the Western Ghats region use many invasive species for various ailments. Some species such as *Ageratum conyzoides* L., *Xanthium indicum* Koenig., *Ricinus communis* L. (Fig. 21.1E) and *Eichhornia crassipes* (Mart.) Solmns. are so extensively used by local people as to raise doubts whether these weeds should be eradicated or cultivated. Similarly, the obnoxious weed *Parthenium hysterophorus* L. (Fig. 21.1G), which is considered to be among the worst weeds, has several medicinal and biological properties such as anti-amoebic (Sharma and Bhutani, 1988); anti-tumour (Mukherjee and Chatterjee, 1993); trypanocidal (Talakal *et al.*, 1995); and antimalarial (Anonymous, 2003). It also has herbicidal and antifungal

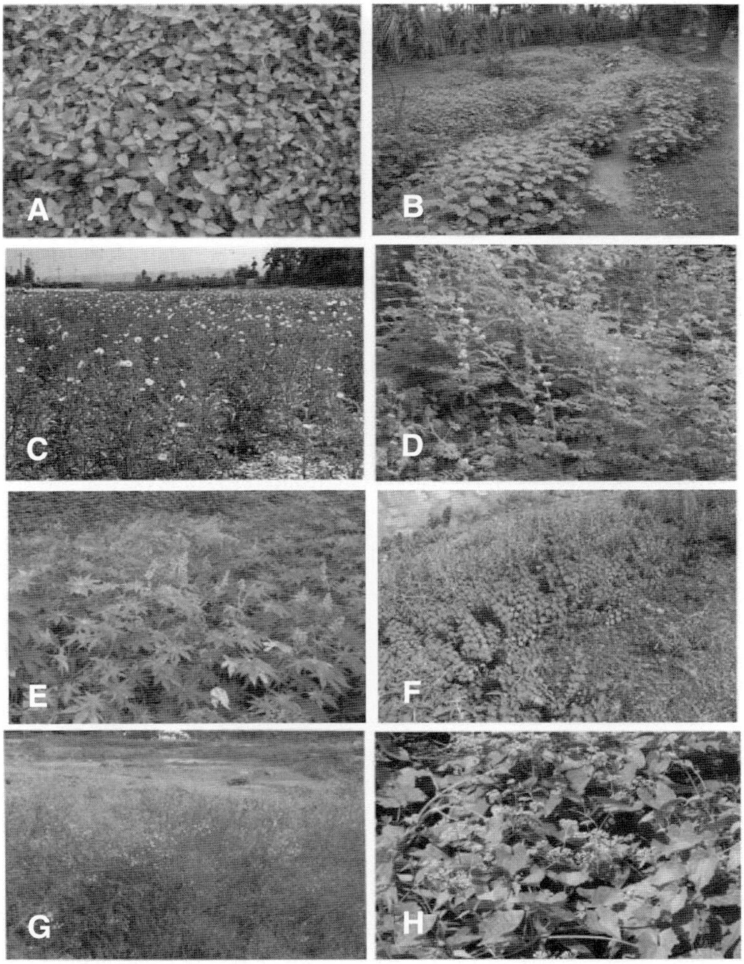


**Fig. 21.1.** Selected beneficial invasive alien weeds. A, *Synedrella vialis* (Less.) Gray.; B, *Cassia uniflora* Mill.; C, *Argemone mexicana* L.; D, *Mimosa invisa* Martius ex Colla; E, *Ricinus communis* L.; F, *Hyptis suaveolens* (L.) Poit.; G, *Parthenium hysterophorus* L.; H, *Mikania micrantha* H.B.K.

activities and produces parthenin, a toxin proved pharmacologically active against neuralgia and certain types of rheumatism (Sharma and Bhutani, 1988).

Another invasive species of the same family, *Xanthium indicum* Koenig, has also shown many medicinal properties. Its achenes are rich in vitamin C; it has cooling, laxative, fattening, anti-inflammatory, anthelmintic, tonic, alexiteric and antipyretic properties; and it improves the appetite and digestive functions, voice, complexion and memory. It cures leucoderma, urinogenital diseases, biliousness, poisonous insect bites, epilepsy, salivation and fever. The root is considered diuretic and the leaves are taken internally, while being applied externally for itch, abscesses and other cutaneous diseases (Kirtikar and Basu, 1981). It is also used in treating cancer and scrofula.

The euphorbiaceous species, *Ricinus communis* L. (Fig. 21.1E), has shown a wide range of excellent medicinal properties:

anodyne, antidote, aperient, bactericide, cathartic, cyanogenetic, discutient, emetic, emollient, expectorant, insecticide, lactagogue, larvicide, laxative, poison, purgative, tonic and vermifuge; castor or castor oil is a frequent ingredient in folk remedies for abscess, anasarca, arthritis, asthma, burns, cancer, carbuncles, catarrh, chancre, cholera, cold, colic, convulsions, corns, craw-craw, deafness, delirium, dermatitis, dog bites, dropsy, epilepsy, erysipelas, fever, flu, gout, guinea worm, headache, inflammation, moles, myalgia, nervous disorders, osteomyelitis, palsy, parturition, rash, rheumatism, scald, scrofula, seborrhea, skin sores, stomach ache, strabismus, swellings, toothache, tuberculosis, tumours, urethritis, uteritis, venereal disease, warts, whitlows and wounds (Duke and Wain, 1981).

*Lantana camara* L. which is one of the worst invasive alien weeds in India, has potential for exploitation as the species has wide-spectrum biocidal potential. Plant extracts revealed antibacterial (Sukul and Chaudhuri, 2001), nematicidal (Shaukat *et al.*, 2003), insecticidal (mosquito repellent) (Ogendo *et al.*, 2003), antifeedant, growth inhibitory (Sharma *et al.*, 2003) and molluscidal (Pal *et al.*, 2002) properties. Phytoextracts of *Lantana* completely inhibited sporulates of *Alternaria alternata*, the cause of potato early blight (Chaudhary *et al.*, 2001). *Lantana* biomass also helps in increasing the nutrient status of soil (Sharma *et al.*, 2003).

*Argemone mexicana* L. (Fig. 21.1C) commonly known as *satyanashi* (killer weed), is used as a medicine in traditional practice in some parts of India. The seed ash of *A. mexicana* L. is used externally to cure skin afflictions such as ringworm and dermatitis, and also hysteria, mental illness, insanity, asthma and conjunctivitis (Table 21.1). Even the Ayurveda and the Unani systems of medicine hold that *satyanashi* is diuretic, sedative, purgative, destroys worms and cures leprosy, skin diseases, inflammation, eye diseases and bilious fevers (http://www.gobartimes.org/20080715/weed_stories.asp).

The compositaceous weed, *Bidens biternata* (Lour.) Merr. *et* Sherff, has diverse ethnobotanical uses particularly in herbal treatments: astringent, anti-haemorrhagic, styptic, urogenital system tonic (diuretic, kidney and bladder tonic, anti-inflammatory), minor anti-microbial activity, neoplastic; it is also used for conjunctivitis, dysentery, haematuria, urethritis, cystitis, benign prostatic hypertrophy, renal urolithiasis, diarrhoea, peptic ulcer, ulcerative colitis, mucoid colitis, haemorrhoids, upper respiratory tract infections/irritations – hay fever, sinusitis, sinus headache due to allergies, infections and pollution, menorrhagia, and vaginal discharges (http://www.medicineatyourfeet.com/bidenspilosa.html).

Considering the diverse medicinal uses attributed to some of invasive species such as *Xanthium*, *Ricinus*, *Argemone*, *Chromolaena*, *Lantana* and *Mimosa*, there is a strong case for instigating cross-cultural ethnobotanical studies to short-list medicinal plants for further coordinated research programmes involving botanists, phytochemists, pharmacologists, pharmaceutical industries, biotechnologists, etc., for critical scientific scrutiny of the claims for safer, cheaper and acceptable drugs derived from alien species. If we are successful in this venture, not only will the spread of invasive weeds be controlled, but perhaps the commercial cultivation of these weeds may not be too far away.

## Invasive weeds as ornamental plants

Several invasive weeds have proved to be good ornamental plants, some having been introduced purely for ornamental purposes. *Eichhornia crassipes* (Mart.) Solmns, *L. camara* L., *I. fistulosa* Mart. ex Choisy and *T. diversifolia* (Hemsl.) A. Gray are a few such species. *Lantana camara* L., which has highly variably coloured flowers arranged in dense, compact spikes, is a suitable hedging and garden plant when cultivated under man-made conditions. Similarly, *E. crassipes* (Mart.) Solmns, with its thick, shining leaves and dark blue flowers, is quite ornamental in artificial ponds, aquaria, etc. *Erigeron karvinskianus* DC., with its white and pink florets and prostrate branches, if grown in man-made landscapes – particularly on

**Table 21.1.** Some important medicinal uses of selected invasive species.

| Species | Family | Medicinal uses |
|---|---|---|
| *Ageratum conyzoides* L. | *Asteraceae* | Pounded roots rubbed on the body for fever; extract for gonorrhoea, as an antidote for snake bite and other poisonous stings; leaf juice on cuts as haemostatic; decoction of whole plant for stomach complaints (Dagar and Singh, 1999); whole plant anthelmintic, antipyretic, styptic and used for burns, diarrhoea, headache and uterine problems (Jain and Sharma, 2000) |
| *Chromolaena odorata* (L.) King & Robinson | *Asteraceae* | Leaf juice applied on cuts and centipede bites (Dagar and Singh, 1999); crushed leaves applied for cuts and wounds as antiseptic, haemostatic, anthelmintic and for strains; as paste for fractures; taken for dysentery and cholera; in vitro efficacy of eupolin derived from plant tissues: repairs damage to cells caused by wounds, burns and skin infections |
| *Mikania micrantha* H.B.K. (Fig. 21.1H) | *Asteraceae* | Crushed plant used for stomach pains and dysentery; leaves for poulticing wounds and swellings; leaf juice on cuts; tender leaves eaten for constipation and indigestion |
| *Tithonia diversifolia* (Hemsl.) A. Gray | *Asteraceae* | Crushed capitula (inflorescence) used for wounds and bruises |
| *Xanthium indicum* Koenig. | *Asteraceae* | Powerful diaphoretic, diuretic and sedative; decoction for chronic malaria, urinary diseases, as a styptic and prophylactic against hydrophobia; leaf juice as ear drops, ringworm; bitter root juice is anti-neoplastic; crushed stem and leaves applied for chicken pox, insect bites; fruits are remedy for rheumatism, muscular paralysis, dropsy and malarial fever; seeds and leaves used in leprosy; fruit paste for eye inflammation; tincture from flowers for toothache; fruits slightly narcotic |
| *Ipomoea fistulosa* Mart. ex Choisy | *Convolvulaceae* | Leaves warmed with mustard oil applied on wounds; plants used for reducing blood pressure |
| *Ricinus communis* L. (Fig. 21.1E) | *Euphorbiaceae* | Medicinal uses of castor oil well established: purgative, ingredient of ayurvedic preparation 'boniol oil' for skincare; leaf poultice applied on boils and sores; root decoction for lumbago; crushed seeds used for deafness, headache, skin conditions, swellings; seed decoction for piles; leaves used for jaundice, smeared with groundnut oil, warmed and applied to abdomen to relieve menstrual pain |
| *Hyptis suaveolens* (L.) Poit. (Fig. 21.1F) | *Lamiaceae* | Considered to be stimulant, carminative, pseudorific and lactagogue; infusion for catarrhal conditions, uterine infections; leaf juice for colic, stomach ache and migraine; stem used as toothbrush in gingivitis and toothache |
| *Argemone mexicana* L. (Fig. 21.1C) | *Papavaraceae* | Leaf juice and roots used for jaundice and dropsy; leaves carminative and stimulant; seeds laxative, emetic, expectorant, antidote for snakebite; seed oil purgative and emetic; seed paste applied externally for skin diseases; bitter tea from roots for regulation of fertility; root paste applied in skin diseases; plant latex used in leprosy |
| *Lantana camara* L. | *Verbenaceae* | Leaves and fruits diaphoretic, carminative and antispasmodic; decoction for tetanus, rheumatism and malaria; decoction of fresh roots and fruits as gargle for toothache; leaf extract used for tuberculosis, asthma; also shows antifilarial, antimicrobial, fungicidal, insecticidal and nematicidal activities; essential oil used for skin itches, leprosy and scabies, and as antiseptic for wounds, swelling, eczema, tumours, bilious fevers, catarrhal infection and tetanus (Kirtikar and Basu, 1981) |

*Continued*

**Table 21.1.** Continued.

| Species | Family | Medicinal uses |
|---|---|---|
| *Acanthospermum hispidum* DC. | | Plant juice combined with bovine milk used for haematuria |
| *Eupatorium adenophorum* Spreng. | *Lamiaceae* | Crushed leaves applied to wounds |
| *Galinsoga parviflora* Cav. & *G. quadriradiata* Ruiz et Pavon. | *Asteraceae* | Crushed leaves used for cuts and wounds |
| *Parthenium hysterophorus* L.(Fig. 21.1 G) | *Asteraceae* | Leaf extract taken on empty stomach for piles; also useful as anti-amoebic (Sharma and Bhutani, 1988); anti-tumour (Mukherjee and Chatterjee, 1993); trypanocidal (Talakal *et al.*, 1995); antimalarial (Anonymous, 2003); has herbicidal, allelopathicin and antifungal activities; acts as remedy for hepatic amoebiases (Sharma and Bhutani, 1988); source of parthenin, a toxin proved pharmacologically active against neuralgia and certain types of rheumatism |
| *Cassia uniflora* Mill. (Fig. 21.1 B) | *Caesalpiniaceae* | Leaf, roots and seeds purgative; seeds febrifugal; roots diuretic, antiperiodic; seeds used for whooping cough, skin diseases, ringworm, fever, neuralgia, dropsy (Anjaria *et al.*, 2002) |
| *Cassia tora* L. | *Caesalpiniaceae* | Seed paste applied in skin diseases; seed powder to cattle as tonic; fresh leaves eaten; germicidal, mucilaginous and laxative; used as coffee substitute and has a naturant and anodyne action; useful in treating skin diseases such as ringworm, itch and psoriasis; has antibacterial, antiplatelet aggregation, anti-fungal and antibacterial activities (Caceres *et al.*, 1993) |
| *Luecaena latisiliqua* (L.) Gillis | *Mimosaceae* | Seed decoction is a good stimulant and tonic |
| *Synedrella vialis* (Less.) A. Gray (Fig. 21.1A) | *Caesalpiniaceae* | Leaves laxative and used for rheumatism |
| *Prosopis chilensis* (Molina) Stuntz | *Mimosaceae* | Fruit juice cathartic, cyanogenetic, discutient, emetic, poison, stomachic; used for colds, diarrhoea, dysentery, inflammation, sore throat (antibacterial) |
| *Mimosa pudica* L. | *Mimosaceae* | Root decoction as gargle to reduce toothache; diarrhoea (*athisaara*), amoebic dysentery (*raktaatisaara*), skin diseases, bleeding piles and urinary infections; leaf juice arrests bleeding and hastens wound healing process; also used in herbal preparations for gynaecological disorders; used in conditions such as bronchitis, general weakness and impotence, neurological problems; root is bitter, acrid, cooling, vulnerary, alexipharmic and used for treatment of biliousness, leprosy, dysentery, inflammation, burning sensation, fatigue, asthma, leucoderma, blood diseases, etc.; leaf decoction used for diabetes (www.motherherbs.com) |

garden embankments – can be pleasing to the eye. *Ipomoea fistulosa* Mart. ex Choisy and *T. diversifolia* (Hemsl.) A. Gray have conspicuous flowers and also form good hedging plants. The former has large, attractive pink–white flowers while the latter has large, attractive yellow capitula. Other invasive species such as *S. vialis* (Less.) A. Gray, (Fig. 21.1A), which forms compact green cushions dotted with small yellow capitula, can form attractive, eye-catching carpets along garden footpaths and in other man-made environments. *Ulex europaeus* L. forms an ornamental hedging shrub and has potential in land reclamation, having been used as a hedging plant and for binding soil on dry, sandy banks.

### Invasive weeds as edible plants

Many Adivasi tribes and other rural people consume a wide variety of wild leafy vegetables, including alien invasives. The tender leaves of *Alternanthera paronychioides* A. St.-Hilaire, *Alternanthera philoxerioides* (Mart.) Griseb., *M. micrantha* H.B.K., *A. conyzoides* L., *H. suaveolens* (L.) Poit., *S. vialis* (Less.) A.Gray and *Leucaena latisiliqua* (L.) Gillis are consumed as leafy vegetables. *Alternanthera philoxerioides* (Mart.) Griseb is known to contain essential nutrients such as Zn and Fe, and the leaves of *Mikania* are said to be rich in vitamins A and C. The shoot tips of *H. suaveolens* (L.) Poit are edible and sometimes used for flavouring. The young shoots and leaves of *X. indicum* Koenig. are cooked and eaten, this plant having sudorific and sedative properties. The seeds of *C. uniflora* Mill. and *L. latisiliqua* (L.) Gillis are used as a tea and coffee substitute. The fruits of *P. phaseoloides* (Roxb.) Benth. are edible and the leaves are chewed for toothache relief.

The mature fruits of *L. camara* L. and *P. chilensis* (Molina) Stuntz. are edible, the powdered seeds of the latter being used in the preparation of bread. Infusion of *H. suaveolens* (L.) Poit. leaves and the crushed pods of *P. chilensis* (Molina) Stuntz are occasionally used as beverages. The gum obtained from *P. chilensis* (Molina) Stuntz is also used in confectionery; the fruit pods are high in sugar and protein and are a rich food source for both man and beast. *Prosopis* honey is of the highest quality and exudate gum is comparable to gum arabic.

A recent study by scientists from the M.S. Swaminathan Research Foundation (MSSRF) has identified 101 edible greens in the forests of Wayanad, although not all are consumed regularly. The Paniyas consume around 83 species, the Kattunaikkas 43 and the Karumas 21 types of wild leafy species. Varieties of *maranths* are eaten quite frequently, as are other wild leaves such as *ponnamkanni* (*Alternanthera sessilis*). People of the Kattunaikka community are regular consumers of another wild plant, *maracheera* (*Embelia tsjeriam-cottam*); they also use corms of various *kattuchena* species (*Amorphophallus* spp.). Eaten raw, these corms are sure to cause a very unpleasant itching sensation in one's throat, but the Kattunaikka women wash the *kattuchena* corms several times in fresh water and then boil them in tamarind water. Likewise, the tender petioles of *karimthalu* (*Colocasia esculenta*) are harvested before the leaves unfold; they are then peeled, boiled in tamarind water and smeared with turmeric paste to remove their irritable properties. These are among several methods devised in the kitchens of the Wayanad tribes to make wild plants palatable (http://www.indiaenvironmentportal.org.in).

### Invasive weeds as fodder for livestock

India being predominantly an agriculture-based country, fodder requirements for livestock are very high. At times of scarcity or during famine some of invasive weeds such as *M. invisa* Martius ex Colla (Fig. 21.1D), *R. communis* L. (Fig. 21.1E), *P. phaseolioides* A. St.-Hilaire, *P. chilensis* (Molina) Stuntz, *E. crassipes* (Mart.) Solmns, *A. paronychioides* A. St.-Hilaire, *U. europaeus* L., *G. quadriradiata* Ruiz et Pavon and *C. odorata* (L.) King & Robinson, which grow abundantly everywhere, can supplement fodder requirements for cattle. Either these weeds can be fed directly or some sort of value-added products derived from them could be developed, simultaneously feeding livestock and limiting the spread of invasive weeds.

### Invasive species promoting phyto-technologies for bioremediation

Some invasive species are known to accumulate heavy metals from the soil, and are thus helpful in phyto-remediation of polluted environments. Prasad (2006) has discussed at length the importance of invasive plants in promoting phyto-technologies for removing heavy metals from the environment. He listed a number of alien invasive weeds including *A. philoxeroides* (Mart.) Griseb, *E. crassipes* (Mart.) Solmns., *S. molesta* Mitch., *S. rotundifolia*, which are all excellent absorbers of heavy metals. *Eichhornia crassipes* (Mart.) Solmns. absorbs various pollutants, including lead, mercury and strontium-90. Five types of uses are attributed to *Eichhornia*: medicinal, forage, vegetable, polluted water cleaner and ornamental. *Alternanthera philoxeroides* (Mart.) Griseb, which grows abundantly in contaminated water, has almost 2% iron content and can be utilized for removing lead and mercury from polluted waters (Wolverton and McDonald, 1975; Prasad, 2006). *Leucaena latisiliqua* (L.) Gillis can extract selenium from the soil and concentrates it in its seed.

## Invasive weeds as biofertilizers and in farming systems

Most invasive weeds in India act as biofertilizers. Species such as *L. camara* (L.), *C. odorata* (L.) King & Robinson, *E. adenophorum* Spreng., *C. uniflora* Mill., *C. tora* L., *E. crassipes* (Mart.) Solmns, *H. suaveolens* (L.) Poit, *Leucaena latisiliqua* (L.) Gillis and *P. hysterophorus* L., with their dense biomass, can serve as good organic fertilizers in croplands. *Ipomoea fistulosa* Mart. ex Choisy, *E. crassipes* (Mart.) Solmns, *M. micrantha* H.B.K., *C. occidentalis* L., *C. uniflora* Mill., *C. tora* L., *M. invisa* Martius ex Colla, *L. camara* L., *C. odorata* (L.) King & Robinson and *A. conyzoides* L. have shown their potential in for vermi composting (Rajkhowa *et al.*, 2005). Organic farmers, therefore, should not completely banish these species from their fields and farms; weeds can offer some benefits, particularly when occurring at low densities. Potential benefits of weeds (including alien weeds) in agroecosystems include the following (http://www.garden organic.org.uk):

- Help to conserve soil moisture and prevent erosion. A ground cover of weeds will reduce the amount of bare soil exposed, helping to conserve nutrients – particularly nitrogen, which could otherwise be leached away – especially on light soils.
- Food and shelter can be provided for natural enemies of pests and even alternative food sources for crop pests. The presence of weed cover may be a factor in increasing the effectiveness of biological control of pests and reducing pest damage.
- Weeds can also be valuable indicators of growing conditions in a field, e.g. water levels, compaction and pH.

In addition to the benefits conferred on farmlands, invasive weeds can also be an important source of food for wildlife, especially birds and insects (http://www.gardenorganic.org.uk).

## Invasive aliens as timber

Numbers of timber-yielding arborescent invasive species in India are sparse; only *Prosopis*, *Leucaena* and the thicker branches of *Lantana* are used. The stouter branches and trunks of *P. chilensis* (Molina) Stuntz yield a high-quality timber comparable to any other commercial hardwood in colour, finish and physical attributes; these are also used for posts and poles while the wood, known as 'wooden anthracite' in some areas, is an excellent fuel wood. The wood is also used for turnery, parquet floors and furniture.

## Miscellaneous uses of invasive weeds

Apart from the uses discussed above for invasive alien species, there are also some other interesting, but important, uses attributed to them (Table 21.2). For example, it is reported that capsules made from *C. odorata* (L.) King & Robinson help in reducing the desire for smoking (www.oztion.com). It is certainly worth investigating this property for product development, as smoking in public places has been banned in many countries. Similarly, the uses listed below could be investigated.

**Table 21.2.** Miscellaneous uses of selected invasive alien weeds.

| Species | Family | Uses |
|---|---|---|
| *Ricinus communis* L. | *Euphorbiaceae* | Dehydrated seed oil used in paper industry and in veterinary medicines; for coating fabrics and other protective coverings; lubricants, printing inks, textile dyeing, leather preservation; production of 'Rilson', a polyamide nylon-type fibre, paints and varnishes, waxes, polishes, carbon paper, candles and crayons; young leaves used for rearing Eri silk worms; castor oil pomace, the residue after crushing, is used as a high-nitrogen fertilizer; stems are made into paper and wallboard (Duke and Wain, 1981) |
| *Prosopis chilensis* (Molina) Stuntz | *Mimosaceae* | Charcoal made from wood; wood more valuable if processed, e.g. 30-fold by turning firewood into finished timber; bark and root rich in tannin (6.7%), which inhibits rhizobia |
| *Argemone mexicana* L. | *Papavaraceae* | Seed oil is a good lubricant |
| *Imperata cylindrica* (L.) P. Beauv. | *Poaceae* | Culms used for thatching |
| *Ageratum conyzoides* L. | *Asteraceae* | Leaves insect repellent due to terpenic compounds, mainly precocenes |
| *Leucaena latisiliqua* (L.) Gillis | *Mimosaceae* | Excellent protein source for cattle fodder; helpful in land reclamation, erosion control, soil improvement programmes; seed gum (25%) can be commercially exploited; seeds after softening are strung as beans into various jewellery items for tourists in some places |
| *Eupatorium adenophorum* Spreng. | *Asteraceae* | Leaves fodder for goats and sheep; insect repellent |
| *Ipomoea fistulosa* Mart. ex Choisy | *Convolvulaceae* | Forms good hedging as toxic to animals; stems used as toothbrush by some villagers |
| *Pueraria phaseoloides* (Roxb.) Benth. | *Fabaceae* | Grazing, hay and silage; suitable cover crop in rubber plantations |
| *Hyptis suaveolens* (L.) Piot. | *Lamiaceae* | Green manure; shoot tips for flavouring; infusion of leaves for beverage preparation; good insect repellent; seeds contain 13% fatty oils |
| *Parthenium hysterophorus* L. | *Asteraceae* | Ash can be used as an effective adsorbent material by its high Cr (VI) concentration in wastewater |
| *Ulex europaeus* L. | *Fabaceae* | Lectins extracted from seeds will bind selectively to certain glycoproteins and glycoplipids and are widely used in tissue typing |
| *Lantana camara* L. | *Verbenaceae* | $\alpha$-cellulose modified to carboxymethyl cellulose by typical carboxymethylation reaction; as a fuel for ethanol production; improves fertility of rocky, barren or hard laterite soils, retains humus in deforested areas, checks soil erosion; leaves used as green mulch; compounds from flowers act as strong repellents for *Aedes* mosquitoes; populations of *Lantana* greatly aid in apiary as 'butterfly gardening'; as a fuel for ethanol production |
| *Eichhornia crassipes* (Mart.) Solms. | *Pontederiaceae* | Excellent source of biomass for biogas production – better than cow dung since it decomposes faster and produces a greater volume of gas |
| *Chromolaena odorata* (L.) King & Robinson | *Asteraceae* | Luxuriant vegetative growth coupled with spreading root system extracts large quantities of nutrient elements from soil and acts as nutrient pump; capsules help in reducing desire for smoking (www.oztion.com); in China oil is extracted (Huai and Zhang, 2006) |
| *Xanthium indicum* Koenig. | *Asteraceae* | Leaves yield yellow dye |

## Conclusions

Without doubt, invasive alien species are a major threat to natural ecosystems and habitats and are responsible for the depletion of our indigenous biological resources. All efforts to check and control the spread of invasive aliens to date have proved unsatisfactory, and these weeds are now extending their domain to ever more new areas. Considering the multi-purpose uses of invasive alien species for human welfare, it would be worthwhile to initiate commercially exploitation of some of these major weeds, to reduce their population size. Pharmacologists, phytochemists, pharmaceutical industries, botanists, microbiologists and biotechnologists must cooperate in the challenging task of scientifically evaluating and authenticating some of the claims discussed above in regard to invasive aliens and establish marketable and value-added products from these alien weeds. *Lantana* stems for example, have been used in furniture making, woven hedging, hut roofs, basket weaving, etc. If exploited properly, the stress on *Calamus* spp. (canes) can be reduced and thus assist in the conservation of diminishing cane resources. These products can capture urban markets with an eco-friendly label, and could be utilized by corporate business, the general public and welfare organizations (ATREE, 2003; Sharma *et al.*, 2006). As *Lantana* grows throughout the country, several small-scale and even large-scale industries could be established in different regions, solely for the utilization of *Lantana* biomass for both socio-economic development and controlling the pest's spread. Perhaps their utilization could be one of the most efficient ways of controlling the problem of invasive weeds.

## Acknowledgements

The authors are grateful to the Council of Scientific and Industrial Research (CSIR), New Delhi, for financial support. The authors would also like to thank the Director and other authorities for the facilities offered and to Dr R.K. Kohli (Chandigarh) for his kind invitation to write this chapter. They are also thankful to the Director, Central Institute of Medical and Aromatic Plants, Lucknow for facilities and encouragement. The senior author is thankful to the Indian National Science Academy, New Delhi for financial assistance through his position as INSA Honorary Scientist.

## References

Anonymous (2003) *The Wealth of India (Raw Materials)*, vol. 4, NISCOM, New Delhi, India, pp. 282–284.

Anjaria, J., Parabia, M., Bhatt, G. and Kumar, R. (2002) *Nature Heals: a Glossory of Selected Indigenous Medicinal Plants of India*. Srishti Innovations, Ahmedabad, India.

ATREE (2003) Ashoka Trust for Research in Ecology and Environment. Retrieved from www.atree.org

Caceres, A., Lopez, B.R., Juarez, X., del Aguila, J. and Garcia, S. (1993) Plants used in Guatemala for treatment for dermatophytic infections. 2: Evaluation of antifungal activity of seven American plants. *Journal of Ethnopharmacology* 40, 207–213.

Chaudhary, R.F., Patel, R.L. and Chaudhary, S.M. (2001) In vitro evaluation of different plant extracts against *Alternaria alternata* causing early blight of potato. *Journal of the Indian Potato Association* 30, 141–142.

Duke, J.A. and Wain, K.K. (1981) *Medicinal Plants of the World* [computer index with more than 85,000 entries], 3 vols.

Gamble, J.S. (1967) *The Flora of the Presidency of Madras*, 2nd edn, 3 vols. Botanical Survey of India, Calcutta.

Huai, H.-Y. and Zhang, X.D. (2006) Ethnobotany of exotic weeds in China. *Ethnobotany* 18, 96–101.

Jain, S.K. and Sharma, P.P. (2000) A review of the parallel or unique indigenous uses of twenty five plants in India and some African countries. *Ethnobotany* 12, 51–55.

Kirtikar, K.R. and Basu, B.D. (1981) *Indian Medicinal Plants*, vols 1–11. Oriental Enterprises, Uttaranchal, India.

Mukherjee, B. and Chatterjee, M. (1993) Antitumor activity of *Parthenium hysterophorus* and its effect in the modulation of biotransforming enzymes in transplanted murine leukemia. *Planta Medica* 59, 513–516.

Ogendo, J.O., Belmain, S.R., Deng, A.L. and Walker, D.J. (2003) Comparison of toxic effects of *Lantana camara* L. with *Tephrosia vogelii* Hook. and synthetic pesticide against *Sitophilus zeamais* Motschulsky (Coleoptera: Curculionidae) in stored maize grain. *Insect Science Applications* 23, 127–135.

Pal, S., Koley, K.M. and Pathak, A.K. (2002) Molluscidal activity of *Azadirachta indica* leaves and *Lantana camara* leaves powder against *Indoplanorbis exustus*. *Indian Journal of Animal Science* 72, 561–562.

Prasad, M.N.V. (2006) Invasive plants of metalliferous substrates: biodiversity prospecting and metallomics to promote phytotechnologies for cleanup of metals in the environment. In: Rai, L.C. and Gaur, J.P. (eds) *Invasive Alien Species and Biodiversity in India.* Banaras Hindu University, Varanasi, India.

Rajkhowa, D.J., Gogoi, A.K. and Yaduraju, N.T. (2005) *Weed Utilization for Vermicomposting – Success Story*, NRC for *Weed Science*, Jabalpur, India.

Rana, R.S. (2006) Facing the threat of invasive alien species in India. In: Rai, L.C. and Gaur, J.P. (eds) *Invasive Alien Species and Biodiversity in India.* Banaras Hindu University, Varanasi, India, pp. 82–92.

Sharma, G.L. and Bhutani, K.K. (1988) Plant-based antiamoebic drugs. Part II. Amoebicidal activity of parthenin isolated from *Parthenium hysterophorus. Planta Medica* 54, 20–22.

Sharma, P.K., Ladha, J.K., Verma, T.S., Bhagat, R.M. and Padre, A.T. (2003) Rice wheat productivity and nutrient status in a *Lantana* (*Lantana* sp.) amended soil. *Biology of Fertile Soils* 37, 108–114.

Shaukat, S.S., Siddiqui, I.A., Ali, N.I. and Kan, G.H. (2003) Nematicidal and allelopathic responses of *Lantana camara* root extract. *Phytopathologia Mediterranea* 42, 71–78.

Sukul, S. and Chaudhuri, S. (2001) Antibacterial natural products from leaves of *Lantana camara* L. with activity comparable to some therapeutically used antibiotics. *Indian Journal of Pharmaceutical Sciences* 63, 20–23.

Talakal, T.S., Dwivedi, S.K. and Sharma, S.R. (1995) *In vitro* and *in vivo* therapeutic activity of *Parthenium hysterophorus* against *Trypanosoma evansi. Indian Journal of Experimental Biology* 33, 894–896.

Wolverton, B.C. and McDonald, R.C. (1975) Water hyacinths and alligator weeds for removal of silver, cobalt and strontium from polluted waters. NASA Technical Memo No. TM-X-71723, National Space Technology Lab, St. Louis, Missouri.

# 22 *Lantana* Mulching for Soil Fertility Improvement, Soil and Water Conservation and Crop Yield Enhancement in Rainfed Rice in the Kumaun Hills

**P. Kumar, M. Pant and G.C.S. Negi**

*G.B. Pant Institute of Himalayan Environment and Development, Kosi-Katarmal, Almora, India*

## Introduction

Rice is the predominant staple food crop of nearly half of the world's population, its cultivation covering 147 million ha in 2002 (IRRI, 2003). Worldwide, the area under rice cultivation is increasing rapidly due to ever-increasing human demand. In intensively cultivated rice fields the capacity of soil to supply plant-available nitrogen (N) may decline unless soils are managed properly (Kundu and Ladha, 1995). In order to enhance and/or maintain crop productivity rice cultivators have become increasingly dependent on chemical N fertilizers.

It has been reported that optimization of carbon (C) and N cycling through soil organic matter enhancement can improve soil fertility and crop yield while reducing negative environmental impacts (Blevins and Frye, 1993; Drinkwater *et al.*, 1998; Kushwaha and Singh, 2005). Sharma *et al.* (2005) reported that the application of organic matter residue provides better grain yield of rice in comparison with farmyard manure. Singh (1995) reported maximum effect from wheat residue inputs on microbial biomass and crop yield in dryland rice cultivation. Application of crop residue under reduced tillage has been frequently proposed as an alternative to soil fertility management to achieve higher crop yield, carbon sequestration and soil and water conservation (Tanaka *et al.*, 2006; Lal *et al.*, 2007; Maltas *et al.*, 2007).

In the Central Himalayan mountains in India, rainfed agriculture on tiny, sloping terraces is the mainstay of agriculture. Rainfed cropfields are characterized by low soil fertility, low soil moisture and low fertilizer input (Singh *et al.*, 1988; Kundu *et al.*, 2007), resulting in poor crop yield (0.1–1.3 t/ha/year: Mehta, 1999). Traditionally, huge quantities (~1.7 t/ha) of farmyard manure (FYM: oak/pine leaf litter used for animal bedding and composted with cattle urine and dung) are applied for replenishment of soil fertility at the time of seed bed preparation every season. The high lignin and low N and P levels of this type of litter (Table 22.1) slowly decompose and release nutrients in the temperate climate of the region (Upadhyay and Singh, 1989; Kandpal and Negi, 2003), leading to nutrient-poor soils (mean values N 0.16%, P (phosphorus) 0.06% and OC (organic carbon) 1.21%; Kumar, 2008). A high-quality organic mulch (*Lantana* leaves high in N and P and low in lignin) is therefore considered useful in enhancing the rate of decomposition and nutrient release when mixed with oak/

**Table 22.1.** Litter nutrient concentration for different mulching treatments applied in experimental plots FYM, farmyard manure; LSD, least significant difference.

| | Litter nutrient concentration (%) | | | | |
|---|---|---|---|---|---|
| Nutrient | Oak | Pine | *Lantana* | FYM | LSD |
| Carbon (C) | 48.50 ± 0.15 | 50.20 ± 0.38 | 40.25 ± 0.35 | 25.50 ± 0.30 | 1.004 |
| Nitrogen (N) | 1.10 ± 0.05 | 0.80 ± 0.04 | 1.30 ± 0.03 | 0.72 ± 0.02 | 1.121 |
| Phosphorus (P) | 0.060 ± 0.010 | 0.036 ± 0.009 | 0.257 ± 0.006 | 0.080 ± 0.002 | 0.024 |
| Potassium (K) | 0.220 ± 0.020 | 0.279 ± 0.002 | 1.23 ± 0.26 | 0.450 ± 0.021 | 0.064 |
| C:N ratio | 44.09 ± 2.04 | 60.48 ± 2.63 | 30.96 ± 0.81 | 35.40 ± 0.93 | 5.782 |
| Lignin | 17.54 ± 0.05 | 28.60 ± 0.11 | 6.30 ± 0.10 | 21.50[a] | 6.136 |
| Polyphenol:N ratio[a] | 7.46 | 3.76 | 4.73 | 0.42 | – |

[a] Source: Sinha *et al.* (2003).

pine leaf litter, as reported by Palm *et al.* (2001) and Singh *et al.* (2006).

Studies on the use of *Lantana* (a weed that grows profusely on croplands) as a mulch material have reported improvements in soil fertility and moisture retention (Acharya and Bhagat, 1984; Sharma and Acharya, 2000); enhanced root growth, nutrient uptake and grain yield in maize and wheat (Acharya and Sharma, 1994; Sharma *et al.*, 2003); and improved soil hydraulic properties (soil aeration and water infiltration) in wheat cultivation that are generally unfavourable after rice cropping (Bhardwaj and Bhagat, 2005; Bhushan and Sharma, 2005). Sharma and Parmar (1998) found that mulching with pine and *Lantana* leaves increased P use efficiency and yield in wheat. It is thus hypothesized that mulching with *Lantana* leaves may provide a readily accessible source of nutrients to the decomposer population, increase nutrient release and improve soil fertility and crop yield.

The objectives of the present work were to: (i) evaluate the effect of application of different combinations of three types of leaf litter (oak, pine and *Lantana*) on soil fertility and rice yield under rainfed conditions; and (ii) compare the traditional practice of crop cultivation with mulching to facilitate recommendation of suitable soil management practices for rainfed farming in the region.

## Materials and Methods

### Study site

The location of the experimental site was 29° 38′N and 79° 37′E, altitude 1155 m (G.B. Pant Institute of Himalayan Environment and Development, Kosi-Katarmal, Almora), in the lesser Himalaya, India. The region has a subtropical climate dominated by a typical monsoonal character. The year is divisible into a cold winter (November–February), a hot summer (April–June) and a wet and warm rainy season (July–September). During the two years of study (2006/2007) the mean annual rainfall was 723.7 mm in 2006 and 1039 mm in 2007, and the mean monthly minimum temperature ranged from a minimum of 0.35–20.18°C in January to a maximum of 12.0–31.31°C in June.

The soils of the study area are sandy loam to loamy sand with a neutral reaction. At the outset of the experiment the mean values of various physico-chemical characteristics of the soil were as follows: water-holding capacity 51.5%, bulk density 1.2 g/cm$^3$, sand 19.65%, silt 61.42%, clay 17.12%, pH 5.62, organic carbon (OC) 0.43%, total N, P and K 0.102, 0.062 and 0.560%, respectively; the C:N ratio was computed as 4.2; mean values were nitrate-N 1.0–1.02 μg/g, ammonium-N 1.54–1.55 μg/g, available P 3.23–3.26 μg/g and available K 0.92–0.95 μg/g. The rate of soil N mineralization was reported as ranging from 17.1 to 31.1 μg/g per month (Ghosh and Dhyani, 2005).

### Experimental design

Thirty-nine experimental plots (5 x 5 $m^2$, slope <5°) were prepared at the above site. Mulching material (oak and pine leaf litter and mature leaves of *Lantana*) was collected from the nearby forests and sun dried. Four mulch treatments ($L_1$–$L_4$) with nine replicate plots each were established in May 2006, in a randomized block design; the study and plots were under the same treatment regime the following year. The quantity of mulch materials applied to the experimental plots (surface mulching) just after sowing of the rice crop in late May 2006 was as follows: $L_1$ (25% oak, 1.22 kg; 25% pine, 1.63 kg; and 50% *Lantana*, 2.07 kg); $L_2$ (33% oak, 1.62 kg; 33% pine, 2.14 kg; and 33% *Lantana*, 1.37 kg); $L_3$ (12.5% oak, 0.612 kg; 12.5% pine, 0.812 kg; and 75% *Lantana*, 3.11 kg); and $L_4$ (100% *Lantana*, 4.116 kg). In addition, three control (conventional tillage, CT) plots were maintained as in the conventional practice of crop cultivation in the region involving FYM application (7.5 kg/plot). The amount of mulch material applied to each of the experimental plots was equated in terms of N input from FYM (1.7 t/ha), equivalent to 21.6 kg N/ha. The C:N ratio of the mulch materials was computed as 39.6, 42.9, 35.4 and 30.9 for $L_1$, $L_2$, $L_3$ and $L_4$, respectively, and seed input was 100–120 kg/ha, as followed by local agricultural practices.

### Soil analysis

The physico-chemical characteristics of the soil in the experimental plots were evaluated at the initiation of experiment (prior to sowing the rice crop in May 2006) and just after harvesting (October 2006). Soil pH was measured using an electronic pH meter (soil:water ratio of 1:2.5) and water-holding capacity by a perforated circular brass box system (Forster, 1995). Organic C content was analysed by the Walkley–Black method (Walkley, 1947). Soil was digested using a mixture of $H_2SO_4$, $H_2O_2$, selenium powder and lithium sulfate for analysis of total N, P and K (Anderson and Ingram, 1989). Total P was determined colorimetrically by the pH adjustment method (Okalebo *et al.*, 1993), while total K was determined using a flame photometer (Systronics Mediflame 128; Systronics, Dubai).

### Available N and N mineralization

Available nitrogen and N mineralization rates were estimated in May, July and September in both years of the study period. N mineralization was measured by the buried bag technique (Eno, 1960; Raghubanshi, 1992). Moist field soil samples (150 g) from each plot were enclosed in air-sealed polyethylene bags and incubated in respective soils at a depth of 10 cm. Coarse roots and large fragments of organic debris were removed to avoid any marked immobilization during incubation. $NO_3$-N and $NH_4$-N were determined immediately after soil collection and after 30 days of field incubation; increase in $NO_3$-N during incubation is referred to as nitrification and increase in the concentration of $NH_4$-N plus $NO_3$-N over the course of field incubation is defined as net N-mineralization. $NO_3$-N was measured by the phenol disulpfonic acid method after extraction by $CaSO_4$ (Jackson, 1958). $NH_4$-N was extracted in 2 M KCl and analysed by the phenate method (APHA, 1985). Alkaline sodium bicarbonate ($NaHCO_3$-Pi) was determined by the ammonium molybdate–stannous chloride method (Jackson, 1958).

### Plant biomass and grain yield

Crop yield was determined by the total harvest method. Crop biomass of each of the 39 plots was harvested manually at the time of crop maturity. Grain and straw from the two crops were separated manually, oven dried and weighed by component for each of the 39 plots and the yield was converted into kg/ha.

### Statistical analyses

Data were subjected to multifactorial ANOVA. Differences between treatments were analysed using Tukey's HSD as a range test. All analyses were performed using SPSS Statistical Software (SPSS/PC, 1986).

## Results and Discussion

### Total soil nutrients

Mean annual values for total soil nutrients of the experimental (control) and the mulch-treated plots ($L_1$–$L_4$) for the two study years are presented in Fig. 22.1. The 100% *Lantana*-mulched plots ($L_4$) recorded a significantly greater OC and total P, although total N was not found to be significantly different. Similarly, $L_3$ also recorded significantly greater soil OC. ANOVA analysis showed that mulching significantly affected pH ($F_{4,78}$, 3.103; *P*, 0.021), OC ($F_{4,78}$, 114.9; *P*, 0.000) and total P ($F_{4,78}$, 24.4; *P*, 0.000), but its effect on total N ($F_{4,78}$, 1.396; *P*, 0.245) was not significant. The year–mulch interaction was significant for pH ($F_{4,78}$, 3.928; *P*, 0.000) but not for OC ($F_{4,78}$, 1.448; *P*, 0.228), total N ($F_{4,78}$, 0.464; *P*, 0.762 ) and total P ($F_{4,78}$, 2.045; *P*, 0.098).

### Available soil nutrients

Decomposition of added residue releases nutrients in plant-available form. In this experiment the mean annual concentration (across sampling months and both study years) of $NO_3$-N, $NH_4$-N and N mineralization rate was found to be significantly high ($P$ <0.000) in the $L_3$ and $L_4$ treatment plots. The rate of N mineralization was highest in $L_4$ and lowest in $L_2$ (3.53 and 5.81 μg/g/month, respectively) (Fig. 22.2). ANOVA analysis showed a highly significant effect of residue quality on the available N pool ($NO_3$-N + $NH_4$-N) in soils and its supply rates (i.e. net N mineralization) (Table 22.2). The effect of month and year was significant for $NO_3$-N and $NH_4$-N, and the rate of N mineralization was generally higher during the second year of observation. ANOVA analysis indicated significant differences in $NO_3$-N and $NH_4$-N concentration due to both type of mulch and month. Similarly year–month, year–mulch and month–mulch interactions were significant for both $NO_3$-N and $NH_4$-N concentrations.

The positive effect of *Lantana* mulch on OC and total P in the soil of experimental plots as recorded in this experiment is in conformity with earlier reports (Larson *et al.*, 1972; Shi *et al.*, 1980; Koyama, 1981; Blevins *et al.*, 1983). Total soil N for $L_4$ and $L_3$ was also found to be higher (although not significant) than in control plots, while OC and N values were among the lowest for $L_2$ (high proportion of oak and pine litter). Evidently organic matter mineralization was lower in $L_2$; this treatment contains a greater proportion of pine litter, which is a poor-quality organic resource with low N and a high C:N ratio and decomposes slowly

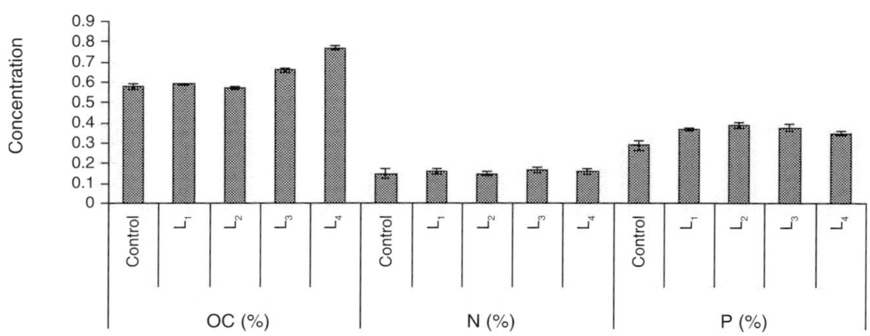


**Fig. 22.1.** Mean annual values for total soil nutrients of the experimental plots under different types of mu ching materials ($L_1$–$L_4$) and control. OC, organic carbon; N, nitrogen; P, phosphorus.

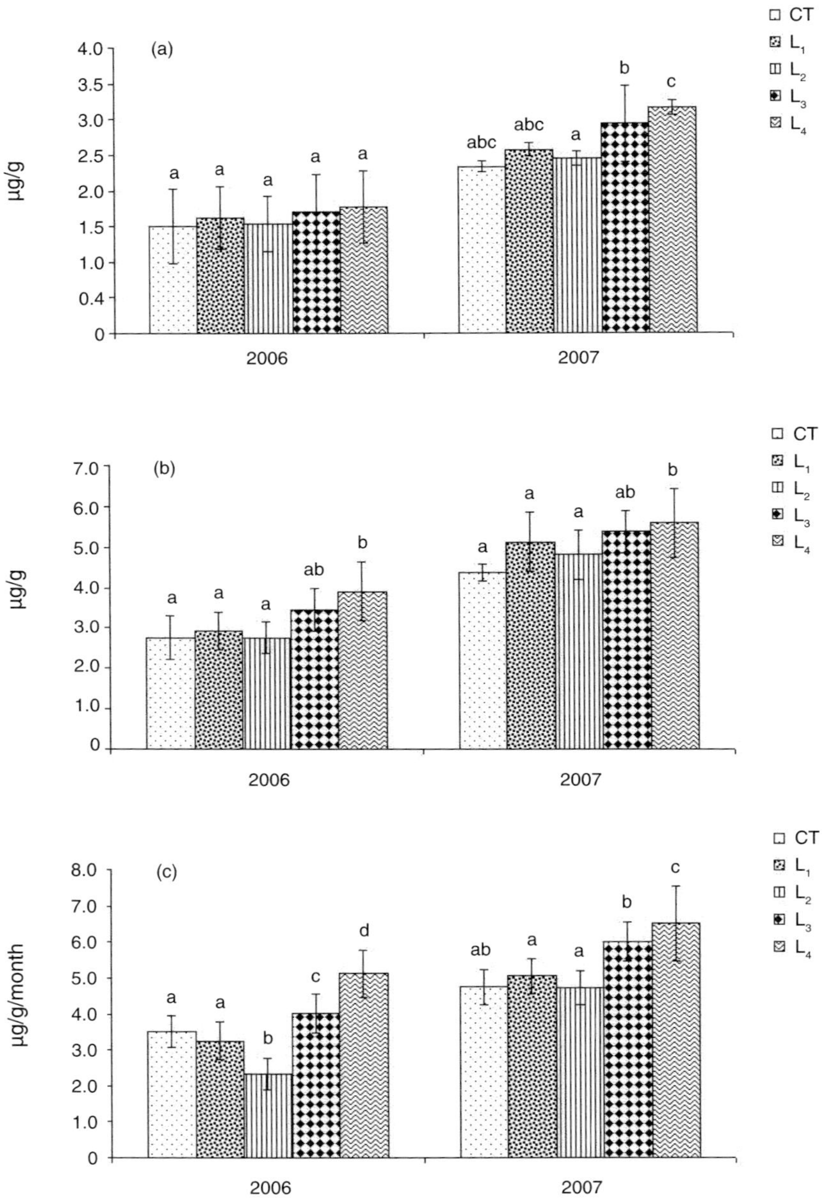


**Fig. 22.2.** Soil $NO_3$-N (a), $NH_4$-N (b) concentration and N mineralization rate (c) due to different mulch applications in experimental plots. Treatments differing significantly ($P$ <0.05) are indicated by different letters over the bars.

(Kandpal and Negi, 2003). Surface mulching of residues is widely known for improving the soil nutrient status, particularly in the top 10 cm (Bauer and Black, 1981; Saffigna *et al.*, 1989).

In dryland rice the application of wheat straw mulch enhanced the soil C level (Singh and Singh, 1995); however, a non-significant increase in soil organic C, total N and total P has also been reported (Singh and

**Table 22.2.** Summary of ANOVA analysis results for effect of mulching materials on available N and N mineralization in experimental plots.

| Source of variance | DF | Nitrate-N | Ammonium-N | N mineralization |
|---|---|---|---|---|
| Year | 1 | 495.8* | 1939.8* | 773.1* |
| Month | 2 | 143.0* | 566.8* | 328.8* |
| Mulch | 4 | 25.6* | 110.2* | 219.7* |
| Year–month | 2 | 133.5* | 297.5* | 19.150* |
| Year–mulch | 4 | 7.437* | 12.3* | 12.305* |
| Month–mulch | 8 | 3.351* | 10.6* | 7.926* |
| Year–month–mulch | 8 | 4.209* | 6.623* | 8.558* |
| Residual | 234 | | | |

DF, degree of freedom.
* Significant at 0.01.

Raghubanshi, 2003). Other studies have also reported soil N enrichment by residue incorporation in rice fields (Saito *et al.*, 1975; Ponnamperuma, 1984) dependent on the C:N ratio of the applied residue (Swift, 1987). In a tropical dryland ecosystem, Singh *et al.* (2007) reported that organic residue management increases soil N mineralization rate, N availability and microbial biomass in rice–barley rotation. In the Cerrado region of Brazil, Maltas *et al.* (2007) reported that under direct-seeding, mulch-based cropping soil N mineralization increased by ~2.0 kg N/ha/year. Increases in total soil N stock, thereby also increasing the pool of mineralizable N, have also been reported by Balesdent *et al.* (2000) and Sanchez *et al.* (2001); this supports our observation that N-rich plant residues are promising alternatives to mineral fertilizers for rice and other crops (Becker *et al.*, 1990). It should be pointed out that residues with a high C:N ratio, as investigated in our study ($L_1$ and $L_2$) may not be suitable for supplementing soil nutrients to support crop growth as they immobilize N during decomposition (Myers *et al.*, 1994). The present study indicates that application of residues, particularly those with a low C:N ratio, may eventually lead to an increase in soil nutrients (both total and available forms), and this practice should be encouraged over traditionally practised FYM application (control) to achieve higher soil fertility.

## Crop yield

The effect of improved soil fertility was evident in the yield of the rice crop in terms of both grain yield and residue yield (Fig. 22.3). The $L_4$ treatment plots recorded significantly higher grain yield (34% more compared with control) than other mulch treatments. Total yield (grain + straw) was also significantly greater for $L_3$ and $L_4$, where the proportion of *Lantana* was 75 and 100%, respectively. ANOVA analysis indicated significant differences in total crop biomass ($F_{4,78}$, 10.2; *P*, 0.000), grain yield ($F_{4,78}$, 11.1; *P*, 0.000) and residue yield ($F_{4,78}$, 5.4; *P*, 0.001) due to mulching treatment (Table 22.3). The effect of year was also significant on total crop biomass ($F_{1,78}$, 68.308; *P*, 0.000), residue yield ($F_{1,78}$, 15.677; *P*, 0.000) and grain yield ($F_{1,78}$, 222.8; *P*, 0.000), and these values were higher for the second year of study (2007).

The increase in rice yield probably resulted from both increased plant-available N during decomposition of the mulching materials and factors not related to N supply rather than from a net increase in total soil N (Buresh and de Datta, 1991). Total yield in the $L_4$ plots was recorded as being significantly higher because of high nutrient availability due to increased microbial biomass, a readily available source for nutrient mineralization. Furthermore, the higher crop yield recorded in 2007 was

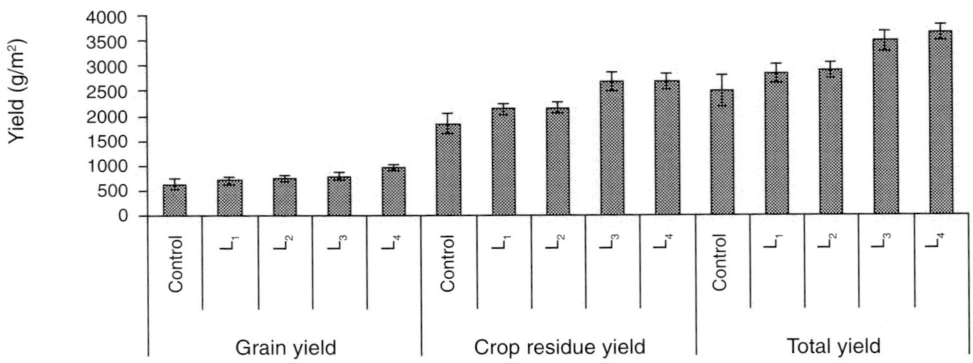


**Fig. 22.3.** Mean annual values for rice yield under different types of mulching materials ($L_1$–$L_4$) and control.

**Table 22.3.** Summary of ANOVA analysis on effects of mulching materials on crop biomass and grain yield of rice.

| Source of variance | DF | Grain weight | Residue weight | Total weight |
|---|---|---|---|---|
| Year | 1 | 222.8* | 15.6* | 64.3* |
| Mulch | 4 | 11.1* | 5.4* | 10.2* |
| Year–mulch | 4 | 1.8[a] | 0.619[a] | 1.2[a] |
| Residual | 68 | | | |

DF, degree of freedom.
[a] Not significant.
* Significant at 0.01.

presumably due to the higher levels of available forms of N compared with 2006. It should be pointed out that the addition of *Lantana* (a soft-leaf species) would have accelerated decomposition and nutrient release in the $L_3$ plots, resulting in higher crop yields. A strong positive correlation ($P < 0.001$) between grain yield and $NO_3$-N ($r$ = 0.583), $NH_4$-N ($r$ = 0.652) and N mineralization ($r$ = 0.614) was found. Similarly, a high grain yield for maize (3 t/ha) was achieved by applying *Crotolaria juncea* (4 t C/ha) relative to control on a sandy soil in Zimbabwe (Mtambanengwe and Mapfumo, 2006). Increase in rice yield by application of a *Lantana* mulch was also reported by Verma and Sharma (2000). The long-term effect of *Lantana* mulching studied by Sharma *et al.* (1995, 2003) and Bhushan and Sharma (2005) in north-west India was found to be improvement of soil water retention and transmission properties in a wheat–rice cropping pattern. Thus, the use of *Lantana* mulching can enable the enhancement of grain yield of rainfed rice and reduce dependency on inorganic fertilizers.

## Conclusions

Four combinations of leaf litter mulch (oak, pine and *Lantana*), when applied to a mountain rainfed rice cropland, positively affected soil fertility and crop yield over control (the traditional practice of crop cultivation using FYM). In the 100% *Lantana*-mulched plots significantly higher soil nutrient levels (OC, total P, available N and microbial biomass) and a significantly higher rate of N mineralization resulted in greater crop yield (34% more than control).

It may be concluded that the application of *Lantana* leaves (high N and low C:N ratio as compared with the traditionally used oak and pine leaves) at a rate of 1.7 t/ha/year (dry matter basis) presents a suitable alternative in soil fertility management in rainfed rice cropfields to achieve greater yield.

## Acknowledgements

This study was funded by DST, Government of India, New Delhi (Grant No. SP/SO/PS-30/2002). Thanks are due to Dr P.P. Dhyani, Group Head and Dr L.M.S. Palni, Director of the Institute for providing the necessary facilities to carry out this work.

## References

Acharya, C.L. and Bhagat, R.M. (1984) Infiltration behavior, root development and yield of rain fed maize under different soil management practices. *Proceedings of the Indian National Science Academy Part B* 50, 441–448.

Acharya, C.L. and Sharma, P.D. (1994) Tillage and mulch effects on soil physical environment, root growth, nutrient uptake and yield of maize and wheat on an Alfisol in north-west India. *Soil and Tillage Research* 32, 291–302.

Anderson, J.M. and Ingram, J.S. (eds) (1989) *Tropical Soil Biology and Fertility: A Hand Book of Methods.* CAB International, Wallingford, UK.

APHA (American Public Health Association) (1985) *Standard Methods for the Examination of Water and Waste Water.* American Public Health Association, Washington, DC.

Balesdent, J., Chenu, C. and Balabane, M. (2000) Relationship of soil organic matter dynamics to physical protection and tillage. *Soil and Tillage Research* 53, 215–230.

Bauer, A. and Black, A.L. (1981) Soil carbon, nitrogen and bulk density comparison in two cropland tillage systems after 25 years and in virgin grassland. *Soil Science Society of America Journal* 45, 1166–1177.

Becker, M., Ladha, J.K. and Ottow, J.C.G. (1990) Growth and nitrogen fixation of two stem-nodulating legumes and their effect as green manure on lowland rice. *Soil Biology and Biochemistry* 22, 1109–1119.

Bhardwaj, A.K. and Bhagat, R.M. (2005) Effect of long term *Lantana* (*Lantana camara* L.) amendment on water use and hydraulic properties of acid Alfisol. *Indian Journal of Soil Conservation* 33(3), 225–229.

Bhushan, L. and Sharma, P.K. (2005) Long-term effects of *Lantana* residue additions on water retention and transmission properties of a medium-textured soil under rice-wheat cropping in northwest India. *Soil Use and Management* 21, 32–37.

Blevins, R.L. and Frye, W.W. (1993) Conservation tillage: an ecological approach to soil management. *Advances in Agronomy* 41, 33–78.

Blevins, R.L., Smith, M.S., Thomas, G.W. and Frye, W. (1983) Influence of conservation tillage on soil properties. *Journal of Soil and Water Conservation* 38, 301–305.

Brookes, P.C., Powlson, D.S. and Jenkinson, D.S. (1982) Measurement of microbial biomass phosphorus in soil. *Soil Biology and Biochemistry* 14, 319–329.

Brookes, P.C., Landman, A., Pruden, G. and Jenkinson, D.S. (1985) Chloroform fumigation and release of soil nitrogen: a rapid direct extraction method to measure microbial biomass nitrogen in soil. *Soil Biology and Biochemistry* 17, 837–842.

Buresh, R.J and de Datta, S.K. (1991) Nitrogen dynamics and management in rice-legume cropping systems. *Advance Agronomy* 45, 1–59.

Drinkwater, L.E., Wagoner, P. and Sarrantonio, M. (1998) Legume-based cropping systems have reduced carbon and nitrogen losses. *Nature* 396, 262–265.

Eno, C.F. (1960) Nitrogen production in the field by incubating the soil in polyethylene bags. *Procedings of the Soil Science Society of America* 24, 277–279.

Ghosh, P. and Dhyani, P.P. (2005) Nitrogen mineralization, nitrification and nitrifier population in a protected grassland and rainfed agricultural soil. *Tropical Ecology* 46, 173–181.

IRRI (International Rice Research Institute) (2003) World rice statistics. Available at: http://www.irri.org/science/ricestat/index.asp

Jackson, M.L. (1958) *Soil Chemical Analysis.* Prentice-Hall, Englewood Cliffs, New Jersey.

Kandpal, K.D. and Negi, G.C.S. (2003) Studies on leaf litter decomposition rate for rainfed crop soil fertility management in the Western Himalaya. *Journal of Hill Research* 16, 35–38.

Koyama, T. (1981) The transformations and balance of nitrogen in Japanese paddy fields. *Fertility Research* 2, 261–278.

Kumar, P. (2008) Effect of reduced tillage and *Lantana* mulching on soil microbial processes, soil fertility and crop yield in the central Himalayan cropfields. PhD thesis, 207 pp.

Kundu, D.K. and Ladha, J.K. (1995) Enhancing soil nitrogen use and biological nitrogen fixation in wetland rice. *Experimental Agriculture* 31, 227–261.

Kundu, S., Bhattachraya, R., Prakash, V., Gupta, H.S., Pathatk, H. and Ladha, J.K. (2007) Long-term yield trend and sustainability of rainfed soybean-wheat system through farmyard manure application in sandy loam soil of the Indian Himalayas. *Biology and Fertility of Soils* 43, 271–280.

Kushwaha, C.P. and Singh, K.P. (2005) Crop productivity and soil fertility in a tropical dryland agro-ecosystem: Impact of residue and tillage management. *Experimental Agriculture* 41, 39–50.

Lal, R., Follett, R.F., Stewart, B.A. and Kimble, J.M. (2007) Soil carbon sequestration to mitigate climate change and advance food security. *Soil Science* 172(12), 943–956.

Larson, W.E., Clapp, C.E., Pierre, W.H. and Morachan, Y.B. (1972) Effects of increasing amounts of organic residues on continuous corn. II. Organic carbon, nitrogen, phosphorus and sulphur. *Agronomy Journal* 64, 204–208.

Maltas, A., Corbeels, M. and Scopel, E. (2007) Long-term effects on continuous direct seeding mulch-based cropping systems on soil nitrogen supply in the Cerrado region of Brazil. *Plant and Soil* 298, 161–173.

Mehta, G.S. (1999) Agriculture and rural development. In: *Development of Uttaranchal: Issues and Perspectives*. APH Publishing Corporation, New Delhi, India, pp. 25–35.

Mtambanengwe, F. and Mapfumo, P. (2006) Effects of organic resource quality on soil profile N dynamics and maize yields on sandy soils in Zimbabwe. *Plant and Soil* 281, 173–191.

Myers, R.J.K., Palm, C.A., Cueavas, E., Gunafillekey, I.U.A. and Brossard, M. (1994) The synchronization of nutrient mineralization and plant nutrient demand. In: Woomer, P.L. and Swift, M.J. (eds) *The Biological Management of Tropical Soil Fertility*. W. Wiley–Science Publications, Oxford, UK, pp. 81–116.

Ocio, J.A. and Brookes, P.C. (1990) An evaluation of methods for measuring the microbial biomass in soils following recent additions of rice straw and the characterization of the biomass that develops. *Soil Biology and Biochemistry* 22, 685–694.

Ocio, J.A., Brookes, P.C. and Jenkinson, D.S. (1991) Field incorporation of straw and its effects on soil microbial biomass and soil inorganic N. *Soil Biology and Biochemistry* 23, 171–176.

Okalebo, J.R., Gathua, K.W. and Woomer, P.L. (1993) *Laboratory Methods of Soil and Plant Analysis – A Working Manual*. Tropical Soil Biology and Fertility Programme (TSBF), UNESCO, Nairobi, Kenya.

Palm, C.A., Giller, K.E., Mafongoya, P.L. and Swift, M.J. (2001) Management of organic matter in the tropics: Transplanting theory into practice. *Nutrient Cycling in Agroecosystems* 61, 63–75.

Ponnomperuma, F.N. (1984) Straw as a source of nutrients for wetland rice. In: *Organic Matter and Rice*. International Rice Research Institute, Manila, Philippines, pp. 117–136.

Raghubanshi, A.S. (1992) Effect of topography on selected soil properties and nitrogen mineralization in a dry tropical forest. *Soil Biology and Biochemistry* 24, 145–150.

Saffigna, P.G., Powlson, D.S., Brookes, P.C. and Thomas, G.A. (1989) Influence of sorghum residues and tillage on soil organic matter and soil microbial biomass in an Australian vertisol. *Soil Biology and Biochemistry* 21, 759–765.

Saito, Y., Kousaka, I. and Kamada, K. (1975) Physico-chemical changes of paddy soils after long term incorporation of rice straw. *Technical Bulletin of the Aomori Agriculture Experimental Station* 20, 42–51.

Sanchez, J.E., Willson, T.C., Kizilkaya, K., Parker, E. and Harwood, R.R. (2001) Enhancing the mineralizable nitrogen pool through substrate diversity in long term cropping systems. *Soil Science Society of America Journal* 65, 1442–1447.

Sharma, D., Deka, J. and Talukdar, M.C. (2005) Effect of residual organics and tillage levels on wheat (*Triticum aestivum*) in a rice (*Oryza sativa*)–wheat sequence. *Indian Journal of Agricultural Science* 75, 673–675.

Sharma, P.K. and Acharya, C.L. (2000) Carry-over of residual soil moisture with mulching and conservation tillage practices for sowing of rainfed wheat (*Triticum aestivum* L.) in north-west India. *Soil and Tillage Research* 57, 43–52.

Sharma, P.K. and Parmar, D.K. (1998) The effect of phosphorus and mulching on the efficiency of phosphorus use and productivity of wheat grown on a mountain alfisol in the Western Himalayas. *Soil Use and Management* 14, 25–29.

Sharma, P.K., Verma, T.S. and Bhagat, R.M. (1995) Soil structural improvements with the addition of *Lantana camara* biomass in rice–wheat cropping. *Soil Use and Management* 11, 199–203.

Sharma, P.K., Ladha, J.K., Verma, T.S., Bhagat, R.M. and Padre, A.T. (2003) Rice wheat productivity and nutrient status in a *Lantana* (*Lantana* spp.)-amended soil. *Biology and Fertility of Soils* 37, 108–114.

Shi, S., Wen, Q. and Liao, H. (1980) The availability of nitrogen of green manures in relation to their chemical composition. *Acta Pedologica Sinica* 17, 240–246.

Singh, H. (1995) Nitrogen mineralization, microbial biomass and crop yield as affected by rice residue placement and fertilizer in a semi-arid tropical soil with minimum tillage. *Journal of Applied Ecology* 32, 588–595.

Singh, H. and Raghubanshi, A.S. (2003) Effect of variable quality residue on a tropical dryland rice soil. *Journal of Sustainable Agriculture* 22, 3–17.

Singh, H. and Singh, K.P. (1995) Effect of plant residue and fertilizer on grain yield of dry land rice under reduced tillage cultivation. *Soil and Tillage Research* 34, 115–125.

Singh, J.S., Raghubanshi, A.S., Reddy, V.S., Singh, S. and Kashyap, A.K. (1998) Methane flux from irrigated paddy and dryland rice fields, and from seasonally dry tropical forest and savanna soils of India. *Soil Biology and Biochemistry* 30, 135–139.

Singh, S., Mer, G.S. and Ralhan, P.K. (1988) Carbon balance for a Central Himalayan cropfield soil. *Pedobiologia* 32, 187–191.

Singh, S., Ghoshal, N. and Singh, K.P. (2006) Variation in soil microbial biomass and crop roots due to differing resource quality inputs in a tropical dryland agroecosystem. *Soil Biology and Biochemistry* 39, 76–86.

Singh, S., Ghoshal, N. and Singh, K.P. (2007) Synchronizing nitrogen availability through application of organic inputs of varying resource quality in a tropical dryland agroecosystem. *Applied Soil Ecology* 36, 164–175.

Sinha, B., Bhadauria, T., Ramakrishnan, P.S., Saxena, K.G. and Maikhuri, R.K. (2003) Impact of landscape modification on earthworm diversity and abundance in the Hariyali sacred landscape, Garhwal Himalaya. *Pedobiologia* 47, 357–370.

Sparling, G.P., Zhu, C. and Fillery, I.R.P. (1996) Microbial immobilization of $^{15}N$ from legume residues in soils of differing textures: measurement by persulphate oxidation and ammonia diffusion methods. *Soil Biology and Biochemistry* 28, 1707–1715.

SPSS/PC (1986) SPSS/PC for the IBM PC/XT/AT. SPSS Inc., Chicago, Illinois.

Swift, M.J. (1987) Tropical soil biology and fertility. In: *Interregional Research Planning Workshop*. Biology International, Special Issue No. 13.

Tanaka, H., Kyaw, K.M. and Motobayashi, T. (2006) Influence of application of rice straw, farmyard manure, and municipal biowastes on nitrogen fixation, soil microbial biomass N, and mineral N in a model paddy microcosm. *Biology Fertility of Soils* 42, 501–505.

Upadhyay, V.P. and Singh, J.S. (1989) Patterns of nutrient immobilization and release in decomposing forest litter in Central Himalaya. *Indian Journal of Ecology* 77, 147–161.

Vance, E.D., Brookes, P.C. and Jenkinson, D.S. (1987) An extraction method for measuring soil microbial biomass C. *Soil Biology and Biochemistry* 19, 703–707.

Verma, T.S. and Sharma, P.K. (2000) Effects of organic residue management on productivity of the rice-wheat cropping system. In: Abrol, I.P., Bronson, K.F., Duxbury, J.M. and Gupta, R.K. (eds) *Long-term Soil Fertility Experiments in Rice-Wheat Cropping Systems*. Rice-Wheat Consortium Paper Series 6, Rice-Wheat Consortium for the Indo-Gangetic Plains, New Delhi, India.

Walkley, A. (1947) A critical examination of a rapid method for determining organic carbon in soils – effects of variations in digestion conditions and inorganic soil constituents. *Soil Science* 63, 251–264.

Woomer, P.L. and Ingram, J.S. (1990) The biology and fertility of tropical soils. Report, TSBF Programme, Nairobi, Kenya.

# 23 Control of *Lantana* and Restoration of Biodiversity in Reserve Forests of Chandigarh: a Case Study

**Ishwar Singh**

*Conservator of Forests (Central), Northern Regional Office of the Ministry of Environment and Forests, Chandigarh, India*

## Introduction

*Lantana camara* (family *Verbenaceae*), a tropical shrub of forests, originated in tropical America and is regarded as a noxious weed in about 80 countries (Parsons and Cuthbertson, 2001). It is a pioneer in secondary succession, occurring in diverse habitats and agro-climatic conditions and on a variety of soil types. The plant generally grows best in open, unshaded situations, such as degraded land, pasture, edges of tropical and subtropical forests, warm temperate forests, beach fronts and forests recovering from fire or logging; it has also invaded plantations and riparian zones. It generally occurs in small clumps of about 1 m in diameter (Palmer and Pullen, 1995). In its naturalized range it often forms dense, mono-specific thickets, and at some sites it even climbs trees (Swarbrick *et al.*, 1998). Batianoff and Butler (2003) have rated *Lantana* among the highest-ranked invasive species in Queensland, Australia. Also, it is among the world's 100 worst invasive alien species, as recognized by the Invasive Species Specialist Group of IUCN.

In India, *Lantana* was first introduced in 1809 as an ornamental plant in Calcutta Botanical Gardens (Kohli *et al.*, 2006), from where it escaped to forests and is now growing rapidly throughout India (Thakur *et al.*, 1992; Fig. 23.1). In Australia the plant was first reported in 1841, but by 1897 it was recognized as being among the most troublesome weeds (Van Oosterhout *et al.*, 2004).

The Forest Department of U.T. Chandigarh, India, maintains and manages 26 km$^2$ of forest in the Shivalik Hills, in addition to another 8 km$^2$ on the periphery of Chandigarh City. The forested area in the Shivalik Hills, being a reserved area, has been declared the 'Sukhna Wildlife Sanctuary' and forms a major part of the hilly catchment area of Sukhna Lake. The other forest areas around the city are in the form of reserve forests, unclassed forests, botanical gardens and plantations.

The entire forest area of Shivalik hills and other forests around the city are now suffering form moderate to heavy infestation by *L. camara*, adversely affecting forest ecology and biodiversity. In 2001–2002 the Forest Department of U.T. Chandigarh rolled out a plan to completely and simultaneously eradicate the alien invader from the forests and restore biodiversity (Figs 23.2 and 23.3).

## Ecological and Environmental Impacts of *Lantana camara*

### *Lantana*: a weed

A plant species acquires the label 'weed' if it is unwanted and crowds out other 'more wanted' species. Weeds are aggressive colonizers that spread rapidly, out-competing

**Fig. 23.1.** Flowering *Lantana camara.*

other useful plant species, and they are usually not very palatable to either wild herbivores or domestic stock. In a wildlife sanctuary, a plant species is certainly unwanted if it is an exotic (not naturally belonging to the area). Even indigenous plants become weeds if they take over large areas and reduce natural species diversity in the process.

*Lantana camara* is one of most common and worst weeds, perhaps the most obnoxious in the Shivalik Hills and the forests around Chandigarh City, posing a serious threat to biodiversity. Following its introduction to India as a garden ornamental, *L. camara* adapted very well to local climatic and edaphic conditions and invaded large areas of forest, grasslands, grazing grounds and roadsides in all parts of the country, except for areas experiencing snowfall.

**Fig. 23.2.** Heavy infestation of *Lantana camara* in Sukhna Wildlife Sanctuary.

### Causes of weed infestation

The proliferation of weeds in a protected area is the result of disturbance such as overgrazing by domestic stock, extensive forestry (planting, thinning, clear-cutting) or abandoned cultivation. Any form of weed control is successful only if the cause of the weed infestation is brought under control and, even then, eradication is seldom practical or economically feasible over large areas. Weeds have very effective mechanisms for seed dispersal and often regrow vigorously from root cuttings.

### Impact of *Lantana* on ecology and environment

*Lantana* is a very hardy weed that grows rapidly in comparison with the indigenous plant species in wildlife sanctuaries and other forest areas. Natural growth of *Lantana* has now choked all natural regeneration in the forests of U.T. Chandigarh. Rapid proliferation of various species of *Lantana* has overrun the indigenous flora of Sukhna Wildlife Sanctuary, adversely impacting on the biodiversity as the invader kills all other undergrowth and damages trees, as clearly shown in photographs of vegetation in Sukhna Wildlife Sanctuary. Even trees such as *Acacia catechu* (*khair*), *Butea monosperma* (*Dhak*) and *Zizyphus* spp. were found to be completely covered by climbing *Lantana*, resulting in their slow death (Fig. 23.3).

The rapid spread of *Lantana* in Sukhna Wildlife Sanctuary enforced the extinction of a wide variety of endemic medicinal herbs, shrubs and trees and, due to the ensuing reduction in floral biodiversity, there was also a decline in faunal biodiversity.

*Lantana* catches fire very easily, thus posing yet another threat to Sukhna Wildlife Sanctuary and other forested areas (Fig. 23.4).

*Lantana* is not palatable for wild animals. It is poisonous, and many children in Africa are reported to have died after consuming its berries. It is strongly allelopathic, releasing certain toxic compounds into the environment that, in turn, adversely affect the growth of neighbouring trees and undergrowth – particularly *bhabbar* and other fodder grasses that are good soil binders.

**Fig. 23.3.** *Khair* (*Acacia catechu*) tree – a victim of climbing *Lantana*.

## Weed Management Strategies

### Removal

The entire forest area of U.T. Chandigarh, including Sukhna Wildlife Sanctuary, is heavily infested with *Lantana*, preventing regeneration of indigenous species on the forest floor. Although *Lantana* provides good soil cover and its flowers and fruit provide food for birds such as junglefowl and bulbuls, and honeybees and butterflies, it was deemed necessary to remove/eradicate *Lantana* to maintain the overall ecological and environmental balance.

With a view to the reintroduction the indigenous varieties of plants, conservation

**Fig. 23.4.** *Lantana* ablaze in a Nepli forest.

of biodiversity and development of an ideal natural habitat for wildlife in the sanctuary and other reserve forests, the Forest Department of Chandigarh Administration has taken a pioneering step by opting for the physical removal/suppression of *Lantana* weed in a phased manner.

## Control

A three-pronged strategy was adopted to control the proliferation of lantana in Sukhna Wildlife Sanctuary and other forest areas of U.T.

### *Removal of the causes of weed infestation*

Intensive forestry activities such as thinning, planting or clear-cutting were virtually prohibited, and no grazing by domestic stock was allowed. Weeds were removed before the fruiting and seeding stages so that seed dispersal could not take place.

### *Physical removal of* Lantana

The work of physical removal of the alien invader started in 2001–2002, with 131 ha of forest, including the Wildlife Sanctuary, declared free from *Lantana*. Many isolated patches of relatively smaller area (5–10 ha) infested with *Lantana* were chosen and the plants uprooted manually. During 2002–2003 *Lantana* was removed from 239 ha of forest, and this area was extended in subsequent years (Fig. 23.5).

It was decided to remove and eradicate *Lantana* in a phased manner to avoid partial loss of wildlife habitats.

**Fig. 23.5.** Forested area soon after removal of *Lantana*.

*Encouraging natural regeneration and afforestation (enrichment planting) of non-weed indigenous species*

During the manual removal of *Lantana*, care was taken to retain indigenous species of herbs, shrubs and trees. These plants acted as the seed source for regeneration, which speeded up the process of natural succession by diverse flora. As shown in Fig. 23.6, natural regeneration of indigenous species such as *Adhatoda vasica* (*Bansa*), *Muraya cognii* (*Kari patta*), *B. monosperma* (*Dhak*), *Tinospora cordifolia* (*Giloe*) and *Abrus precatorius* (*Ratti*) has now covered the land vacated by the removal of *Lantana*, which will prevent its reinfestation.

## Methodology Adopted for Removal of *Lantana* and Subsequent Regeneration

Isolated patches (5–10 ha) of forests infested with *Lantana* were selected. Manual removal – by digging out of the roots – was started from the edges, i.e. from either roads, firelines or *choe* banks, moving inwards to prevent the reseeding of patches cleared of weeds. Efforts were made to carry out the uprooting operation before the maturing of seeds, to avoid or minimize reseeding.

The first year's experience showed that *Lantana* had reappeared in around 25% of the area cleared by its removal, which necessitated further removal in the second year. In the third year, reappearance of the weed was at the level of <5%, as the forest floor was by then covered by natural regeneration and the planting of indigenous plant species. To prevent the reappearance of *Lantana* along *choe* banks, *Arundo-donax* (*Nada*) was planted to cover the cleared areas (Fig. 23.7), and has now very successfully out-competed Lantana. In addition to the regeneration of herbs, shrubs and tree species, several grasses have also established well following the removal of *Lantana*.

### Replanting following removal of *Lantana*

To supplement natural regeneration, fast-growing species such as *Dendrocalamus strictus* (bamboo), *Pythocelebium dulce* (jungle jalebi) and *Leucaena leucocephala* were planted to fill the areas cleared of

**Fig. 23.6.** Natural regeneration of non-weed species following removal of *Lantana*.

**Fig. 23.7.** *Arundo-donax* plantation.

*Lantana* so there was no time for the weed to re-establish itself. Palatable species planted have been protected against browsing, as shown in Fig. 23.8. Following planting of palatable grass species, several patches of land have now been developed as grassland for herbivores.

## Costs

Manual removal of *Lantana* has cost Rs.5000–6000/ha, as wage levels in U.T. Chandigarh are quite high. The success rate (i.e. non-recurrence of *Lantana*) was ~75% in the first year, with at least two further clearing operations being required.

## Results

Manual removal of *Lantana* by through uprooting and its biological suppression by planting indigenous fast-growing species have resulted in the following benefits.

### Restoration of biodiversity and improvement in wildlife habitats

Unchecked growth of *Lantana* had choked all natural regeneration in the forests, with a consequent adverse impact on biodiversity. Following complete eradication, prevention of its reappearance was ensured. The regeneration of indigenous species such as

**Fig. 23.8.** Successful restoration of palatable grasses and shrubs.

Bansa, Ratti, Karipatta, Giloe, Karaunda and other tree species is very encouraging; since 2005 good levels of flowering and fruiting has been noted in trees such as Ber, Dhak (Palas) and Karaunda. Due to the development of good grazing grounds for wildlife following the removal of *Lantana*, there has been a tremendous improvement in the faunal population and species in all forested areas, including Sukhna Wildlife Sanctuary. Honeybee and butterfly populations, in particular, have increased many-fold. The removal of *Lantana* has thus proved to be a great boon to the restoration of floral and faunal biodiversity of all the forests of Chandigarh.

Due to the regeneration of indigenous floral species supplemented by planting there has been a very marked improvement in wildlife habitats, not only in Sukhna Wildlife Sanctuary but also in all forested areas. Plentiful water holes, palatable grasses and shrubs and thick forests have made the forests of U.T. Chandigarh the permanent home of many varieties of fauna.

Due to excessive growth of *Lantana*, grasslands or grass patches had completely disappeared from the forests. Following the removal of *Lantana*, open patches of land were developed as grazing ground for wildlife by restoring palatable grasses, herbs and shrubs (Fig. 23.8). These patches of grassland have helped in the proliferation of wildlife in Sukhna Wildlife Sanctuary and all other forests; there is appreciable improvement in the general health of wildlife as the presence of *Lantana* was causing skin allergies in most types of fauna.

### Minimization of fire hazard

As mentioned above, *Lantana* is highly flammable; there was a major forest fire in Nepli in 2001, but thereafter serious fire incidences have reduced drastically.

## Conclusions

In considering the adverse ecological impact of *Lantana* on the biodiversity of flora and fauna, in 2001 the Forest Department of U.T. Chandigarh decided to eradicate this invasive weed completely by 2009–2010; this was a very challenging and tough task. The continuous motivation, dedication and hard work of Department officials have now made this a reality. Today the entire forests of Chandigarh, including Sukhna Wildlife Sanctuary, are completely free of *Lantana* and serve as a model for the successful restoration of biodiversity in India.

## References


Batianoff, G.N. and Butler, D.W. (2003) Impact assessment and analysis of sixty-six priority invasive weeds in southeast Queensland. *Plant Protection Quarantine* 18, 11–17.

Kohli, R.K., Batish, D.R., Singh, H.P. and Dogra, K.S. (2006) Status, invasiveness and environmental threats of three tropical American invasive weeds (*Parthenium hysterophorus* L., *Ageratum conyzoides* L., *Lantana camara* L.) in India. *Biological Invasions* 8, 1501–1510.

Palmer, W.A. and Pullen, K.R. (1995) The phytophagous arthropods associated with *Lantana camara*, *L. hirsuta*, *L. urticifolia* and *L. urticoides* (Verbenaceae) in North America. *Biological Control* 5, 54–72.

Parsons, W.T. and Cuthbertson, E.G. (2001) Common *Lantana*. In: *Noxious Weeds of Australia*. CSIRO, Melbourne, Australia, pp. 627–632.

Swarbrick, J.T., Willson, B.W. and Hannan-Jones, M.A. (1998) *Lantana camara* L. In: Panetta, F.D., Groves, R.H. and Shepherd, R.C.H. (eds) *The Biology of Australian Weeds*. R.G. and F.J. Richardson, Melbourne, Australia, pp. 119–136.

Thakur, M.L., Ahmad, M. and Thakur, R.K. (1992) Lantana weed (*Lantana camara* var. *aculeata* Linn) and its possible management through natural insect pests in India. *Indian Forester* 118, 466–488.

Van Oosterhout, E., Clark, A., Day, M.D. and Menzies, E. (2004) *Lantana Control Manual. Current Management and Control. Options for Lantana (Lantana camara) in the Australian State of Queensland*. Department of Natural Resources, Mines and Energy, Brisbane, Australia (http://www.nrm.qld.gov.au/pests/wons/Lantana, accessed 23 November 2004).

# 24 Woody, Alien and Invasive *Prosopis juliflora* (Swartz) D.C.: Management Dilemmas and Regulatory Issues in Gujarat

**C.N. Pandey[1], R. Pandey[2] and J.R. Bhatt[3]**

[1]*Additional Principal Chief Conservator of Forests, Aranya Bhawan, Gujarat, India;* [2]*Senior Research Fellow, Gujarat Ecological Education and Research (GEER) Foundation, Indroda Nature Park, Gujarat, India;* [3]*Director Environment, Ministry of Environment and Forests, New Delhi, India*

## Introduction

*Prosopis juliflora* (Swartz) D.C. is a woody tree (sometimes referred to as a woody shrub and the mesquite tree) that is an alien species in India. It was introduced to the Indian subcontinent at Sind province (now Pakistan) in 1877 (Muthan and Arora, 1983). The genus *Prosopis* comprises 44 recognized species; they have remarkable multiple adaptation capabilities in terms of a rapid rate of growth, salt tolerance and drought resistance, and can grow even in areas of scarce rainfall. Due to its rapid spread and adaptability to various environmental conditions, the species has been able to invade most parts of Gujarat, where arid to semi-arid conditions prevail.

*Prosopsis*, a native of Mexico, belongs to the family *Mimosaceae*, and its vernacular name in Gujarat is *Gandobaval*. It is a thorny shrub/small tree growing up to 10 m in height (usually 3–4 m), with a trunk of up to 1.2 m in diameter. Pods are generally curved or sickle-shaped, nutritious and are preferred by sheep, goats and cattle. It flowers twice a year, i.e. February–March and August–September. It coppices profusely when cut and, if cut or hacked repeatedly, it tends to become a woody shrub. It grows well under high salinity or alkalinity levels as high as pH 9.5. In the saline grasslands of Banni (*Kutch*), the density of *P. juliflora* is >2000 trees/ha in areas of soil surface pH 8.1–8.4.

## Management Dilemmas

### Ecological issues and significance

*Prosopis juliflora* is regarded as an invasive species in India, its seeds having been dispersed by cattle over large tracts in arid and semi-arid areas. This has both positive and negative impacts on the key species and habitat of the area, as illustrated by the following examples.

### Banni grasslands

The aggressive and invasive character of the species has resulted in changes in the floristic composition of the area. In many parts of western India higher salinity has favoured this species against other species that are less salinity tolerant; in fact, the rate of expansion of this species in the Banni grasslands of Gujarat is considered to be

 *Invasive Alien Plants: An Ecological Appraisal for the Indian Subcontinent* (eds J.R. Bhatt *et al.*)

alarming. It has been reported that this species has also spread to non-saline areas of Banni (Gavali, 2009), which has changed the ecological and biodiversity status of the area.

Estimations using remote sensing data technology (IRS LISS-III, 1997) have resulted in the land cover classification of Banni grassland (Table 24.1), showing that *P. juliflora* has extended to more than 16% of the total area.

Reduced grass production on account of the continued invasion of *P. juliflora* in Banni has drawn the attention of official bodies, and the removal of this species followed by grass regeneration have been prescribed in a Working Plan for the area. Consequently, the State Government decided in 2010 to initiate a programme of removal of this species through public participation, followed by grassland development. Gujarat State Forest Development Corporation will then convert the harvested *Prosopis* wood into charcoal for subsequent marketing.

### Jessore Wildlife Sanctuary

Jessore Wildlife Sanctuary, located in the north-eastern part of Gujarat, covers 18,060 ha and is dominated by *Prosopis juliflora* (220.32 individuals/ha, covering 35% of the forested area; GEER, 2009). The regeneration potential of the species is very high, with 77.51 seedlings/ha and 111.42 saplings/ha (total, 188.93 individuals/ha; GEER, 2009). The species now occupies a major part of the sanctuary and, as suggested by its regeneration potential, it is spreading and flourishing. The species, therefore, represents a threat to the indigenous species of the area and is likely to affect the overall structure of the forest.

### Velavadar National Park

Velavadar National Park is located in the Saurashtra region of Gujarat. Ecologically, the area is a combination of grassland ecosystem and scrub forest. Blackbuck, a small gazelle, is the flagship species here, for which open grasslands are important. The invasion of *P. juliflora* has been a matter of concern for the Park managers, who have adopted an eradication programme as part of the habitat management strategy; this has now contained the spread of this species in the area. *Prosopis juliflora* also has ecological relevance; it has been planted in saline areas and on desert borders to arrest the process of desertification, resulting in the creation of important habitats in these vegetation-deprived areas of Gujarat.

### Kachchh Great Indian Bustard Sanctuary

This small (2 $km^2$) sanctuary and its surrounding area represents grasslands where a healthy population of Great Indian Bustards (GIBs) has been established. It has been reported that one of the important habitats is the cover provided by *P. juliflora* in the area surrounding the grasslands. The extent of this cover has been estimated to be around 10% of the total area used by the birds (which extends beyond the small sanctuary area) (Pandey *et al.*, 2009). However, it has been reported that the bird's habitat preference is for this cover area (Pandey *et al.*, 2009), on account of the shade providing protection against the heat during the scorching summers. In winter, *Prosopis* cover provides protection against wind chill and during the monsoon, female birds have been found to use the edges of the cover area to conceal their young. Hence,

**Table 24.1.** Land cover classification of Banni grassland, Gujarat.

| Land cover | Area ($km^2$) | Proportion (%) |
|---|---|---|
| Grass | 702.60 | 27.06 |
| *Prosopis juliflora* | 434.11 | 16.72 |
| *Suaeda* and barren (saline patches) | 1300.82 | 50.09 |
| Water | 159.20 | 6.13 |
| Total | 2596.73 | 100.00 |

although *Prosopis* has been found to have a positive impact on this particular species, it is essential that its spread is contained to ensure that the floral structure of the habitat is not altered significantly.

### Little Rann Wild Ass Sanctuary

The Wild Ass Sanctuary (4953.71 km$^2$) in Little Rann (saline desert), Kachchh – also declared a Biosphere Reserve – largely comprises saline flatlands. The important habitats used by the flagship species – the Indian wild ass – are the saline flatlands, islands (elevated areas in the saline desert) and other areas. Remote sensing study has demonstrated that the cover provided by *Prosopis* is increasing at the rate of 106, 128 and 465 ha/year in the saline flats, islands and other areas, respectively. Thus the spread of this species is low in the saline areas and islands, which form the central area of the sanctuary, and higher in the areas constituting the perimeter and surrounding environment of the sanctuary (Singh *et al.*, 1998). Furthermore, movements of the animals have been found to be positively associated with *Prosopis*. It has been reported that during winter and the rainy season, sightings of the majority of the wild ass population are in the fringe areas dominated by *Prosopis* (Pandey, 2007). The movement corridors generally preferred by the animals for their seasonal movements are also through areas where *Prosopis* is well established, and around 20% of the wild ass population has been sighted in *Prosopis* scrubland (Pandey, 2007). Pods of *Prosopis*, and herbs associated with *Prosopis*, form a major dietary component during summer when most of the land becomes devoid of vegetation. The present distribution of *Prosopis* (largely concentrated in the fringe areas) is, therefore, not harmful to this flagship species of the Sanctuary, which is evident from its increasing population in the Sanctuary and its surrounding area (see Table 24.2).

**Table 24.2.** Population of Indian wild ass in the Wild Ass Sanctuary, 1983–2009.

| | 1983 | 1990 | 2009 |
|---|---|---|---|
| Population (*n*) | 1989 | 2075 | 4035 |

Source: Gujarat Forest Department census records.

## Economic Relevance

Apart from ecological impacts, *P. juliflora* is of significant socio-economic relevance. The species has created a green cover and bio-resource in the semi-arid and arid areas of Gujarat where the availability of bio-resources is very low. Its wood is of high calorific value (5000 kcal/kg) and it is therefore a preferred fuel in these areas – more than 70% of fuel demand is met by this species in these areas, and it is often the only source of firewood. Hence, it is a tree of considerable socio-economic significance for local people.

It is also the most preferred wood for charcoal preparation in Gujarat: the wood–charcoal conversion ratio is 25–27% using the traditional ground kiln method. In a typical village of the Kachchh district of Gujarat, it has been found that *Prosopis*-based charcoal contributes 79% of overall household income (Ayar *et al.*, 2009).

As its timber is hard, heavy, strong, close-grained and does not split easily, it is valuable in the manufacture of agricultural implements, household furniture and small utensils; however, its use in furniture making is limited in Gujarat primarily due to the non-availability of trees with long, straight trunks. The wood quality of *Prosopis* is, however, reported to be comparable to Indian rosewood (*Delbergia latifolia*) and teak (*Tectona grandis*) (Tewari *et.al.*, 2000).

*Prosopis* exudes gum from its sap, which is used for the sizing of cloth and paper. On average, ~40 g of gum is produced from one plant, although under drought conditions this may be greater. The gum contains D-galactose (45%), L-arabinose (24%), 6-37 L-rhamnose (13%) and glucuromic acid (13.7%). It possesses fairly good adhesive strength and can be used as an adhesive for brown paper and wallpaper. The gum has also been used in treatment of eye infections

(Vimal and Tyagi, 1986). In central Gujarat incomes are earned from the production of *P. juliflora* gum (Joshi and Oza, 2009).

*Prosopis* flowers profusely twice a year. From the vast areas established by *Prosopis* in Kachchh district, Gujarat State Forest Development Corporation (GSFDC) collects ~60 t of honey per year, generating ~100,000 man–days of employment. The honey is produced by the honeybee, *Apis florae*, a pollinator for the species. During purification of the honey, the wax is separated through filtration; ~3.0 t of wax is collected by GSFDC annually.

The ripe pods are consumed by livestock in large quantities and in drought conditions, serving as the main protein source for dairy cattle. Due to there being two flowering seasons (winter and summer), pods are collected twice a year. Maximum pod production is between March and June, and is estimated to be ~20 kg/tree/year (Shukla *et al.*, 1984). Vimal and Tyagi (1986) have reported that analysis of the pods yields protein (16.5%), fat (4.2%), carbohydrate (57.0%), fibre (16.8%), ash (5.4%), calcium (0.33%) and phosphorus (0.44%). The trace element composition of *Prosopis* pods has been estimated as 12.46–15.51 ppm copper, 22.11–22.30 ppm manganese, 18.30–28.01 ppm zinc and 203–638.8 ppm iron (Shukla *et al.*, 1984). For cattle and buffalo, the pods are not regarded as good fodder because of the high sugar content and indigestibility of the raw seeds (Shukla *et al.*, 1984). When fed in a dried and crushed state in the form of powder, the pods did not show any deleterious effect on cattle and, in fact, resulted in good performance indicators. During times of drought and scarcity, the pods are even used as food items by the poor in Kachchh.

## Policy and Regulatory Decisions in Gujarat Concerning *P. juliflora*

In view of the ecological and economic relevance of *P. juliflora*, it has been difficult to select management options. The species being alien and invasive, it has had adverse ecological impacts in many situations. On the other hand, the species has provided biotic resources in areas where such resources are minimal due to excessive salinity and aridity; in such situations, the species has also provided ecological benefits and created suitable habitats for other life forms. However, even in such situations there is a limit to the extent to which *P. juliflora* can be allowed to grow: the optimum level for *P. juliflora* is difficult to prescribe and, moreover, such levels are location-specific. Therefore, area- and situation-specific studies are necessary to determine how *Prosopis* is managed, and such studies available to date are far from adequate.

The ecological and economic issues mentioned above are echoed through various social processes in Gujarat; there is a social divide about the desirability or undesirability of species. Whereas one faction feels that the species has created green cover in barren, arid and saline areas of Gujarat with related ecological and economic benefits, the other is vocal about the fact that the aggressive invasion of the species has been at the expense of the precious grasslands on which animal husbandry – the traditional occupation of local communities in Gujarat – is directly dependent. Besides, there are opinions on the spread of the species in rural areas obstructing movements of people and cattle, due to its thorny nature. These issues are very contentious in relation to the Kachchh district, which contains one of the highest concentrations of *P. juliflora* in the state.

The species has spread to all types of terrain, irrespective of its ownership or use. It has spread to forests, government wastelands, community land and private land. As per the legal provisions made under the Indian Forest Act (1927), it is necessary to obtain transit permission from the Forest Department for transporting of *P. juliflora* timber or charcoal. In response to growing demands for eradication of *P. juliflora*, in 2004 the state government exempted *P. juliflora* from the requirement for the provision of transit passes, except from areas within 2 km of the coast or 1.6 km from Reserved Forests.

The issue of management of *P. juliflora* in

the Banni grasslands within Kachchh district has been addressed by a management plan (Working Plan) approved by the Government of India (Meena *et al.*, 2009), under which ecological as well as economic aspects of managing have been considered while prescribing management options. It has been prescribed that *P. juliflora* should be harvested according to a set plan, and village-level organizations would be allowed to convert the harvested wood to charcoal; subsequently GSFDC which would deal with the marketing of the charcoal. The densely covered *P. juliflora* areas would be managed for sustainable yields of wood on a 4-year rotation, while the more open and sparsely covered areas would be managed for improved grass production.

## References

Ayar, H., Patel, P. and Patel, R. (2009) Contribution of *Prosopis juliflora* to the economy of Mindhyari village in Western Kachchh, Gujarat. *Proceedings of the National Symposium on Prosopis: Ecological, Economic Significance and Management Challenges*. Gujarat Institute for Desert Ecology, Bhuj, India, pp. 54–59.

Gavali, D.J. (2009) Soil salinity versus *Prosopis juliflora* growth – the case of Banni Grassland. *Proceedings of the National Symposium on Prosopis: Ecological, Economic Significance and Management Challenges*. Gujarat Institute for Desert Ecology, Bhuj, India, pp. 33–36.

GEER (2009) *The Status of Biodiversity of Jassore Wildlife Sanctuary.* Gujarat Ecological Education and Research (GEER) Foundation, Gandhinagar, India.

Joshi, P. and Oza, B. (2009) Valuation of economic services provided by non-wood forest produce with special reference to *Prosopis juliflora* in Jaliyakuva village, Panchmahal district. *Proceedings of the National Symposium on Prosopis: Ecological, Economic Significance and Management Challenges*. Gujarat Institute of Desert Ecology, Bhuj, India, pp. 81–87.

Meena, R.L., Tikadar, S. and Vijaykumar, V. (2009) *Working Plan of Banni Protected Forests.* Forest Department, Gujarat State, Gandhinagar, India.

Muthan, K.D. and Arora, G.D. (1983) *P. juliflora* (Swartz) D.C. A fast-growing tree to bloom the desert. *CAZRI Monographs* 22, 19.

Pandey, C.N. and Jethva, B. (2008) *Movement Pattern of the Wild Ass Across the Proposed Kachchh Branch Canal on the Northern Fringe of the Wild Ass Sanctuary.* Gujarat Ecological Education and Research (GEER) Foundation, Gandhinagar, India.

Pandey, C.N., Jethva, B. and Munjpara, S. (2008) *Status, Distribution and Habitat Survey of Great Indian Bustard (Ardeotis nigriceps) in Gujarat.* Gujarat Ecological Education and Research (GEER) Foundation, Gandhinagar, India.

Sastry, K.L.N., Thakker, P.S. and Jadhav, R. (2003) Biodiversity threat through exotic species monitoring and management using remotely sensed data and GIS techniques. In: *6th Annual International Conference on MAP INDIA*, 28–31 January, New Delhi, India.

Singh, H., Soni, S., Shah, P.R. AND Neeta, V.C. (1998) *Ecologica; Study of the Wild Ass Sanctuary.* Gujarat Ecological Education and Research (GEER) Foundation, Gandhinagar, India.

Tewari, J.P., Harris, C., Harsh, L.N. and Cadoret, K.P. (2000) Managing *Prosopis juliflora* (*Viilayati babul*): a technical manual. CAZRI, Jodhpur, India and HDRA, UK.

# Index

10+50